S0-BHT-889

The Pre-1940 PhD's

HISTORY OF MATHEMATICS
VOLUME 34

The Pre-1940 PhD's

Judy Green
Jeanne LaDuke

Providence, Rhode Island

London, England

2000 *Mathematics Subject Classification.* Primary 01A60, 01A70, 01A80, 01A73, 01A55, 01A05, 01A99.

For additional information and updates on this book, visit
www.ams.org/bookpages/hmath-34

Library of Congress Cataloging-in-Publication Data

Green, Judy, 1943–
Pioneering women in American mathematics : the pre-1940 PhD's / Judy Green, Jeanne LaDuke.
p. cm. — (History of mathematics ; v. 34)
Includes bibliographical references and index.
ISBN 978-0-8218-4376-5 (alk. paper)
1. Women mathematicians—United States—History. 2. Women in higher education—United States—History. 3. Women pioneers—United States—History. I. LaDuke, Jeanne, 1938– II. Title.

QA28.G74 2008
510.82′0973—dc22 2008035318

∞ The paper used in this book is acid-free and falls within the guidelines
established to ensure permanence and durability.
The London Mathematical Society is incorporated under Royal Charter
and is registered with the Charity Commissioners.
Visit the AMS home page at http://www.ams.org/

10 9 8 7 6 5 4 3 2 1 14 13 12 11 10 09

Contents

List of Illustrations

List of Tables

List of Biographical Entries

Preface

During the joint mathematics meetings in January 1978 in Atlanta both of the authors attended a session on the history of American mathematics organized by Uta C. Merzbach, then curator of the Division of Mathematics, National Museum of History and Technology, Smithsonian Institution. Judy Green, then on the faculty of Rutgers University in Camden, New Jersey, was giving a talk on American women who had earned PhD's in mathematics before 1911; some of her material was based on interviews with a daughter of one of these women, Mary Winston Newson. In connection with her own interests in this area, Jeanne LaDuke, on the faculty of DePaul University in Chicago, had spoken with Newson's other daughter. During the meetings the authors had extensive conversations with Merzbach, who had already begun to compile a database on Americans active in mathematics before 1940. By the time the meetings were concluded it was decided that we would attempt to identify all American women who had earned PhD's in mathematics before 1940 and contribute the information relating to those women to the database. We eventually determined that 228 women fit the definition of those in the study, namely, those women who received PhD's from US institutions or US-born women who received foreign degrees. The first such degree was granted by Columbia in 1886. At the time of the January 1978 mathematics meetings, 118 had died, the first in 1907. Thus, techniques for the investigation would rely heavily on archival sources as well as communication, when possible, with the women who were still living. As of June 2008, three of the women are still living and are in their nineties.

Over the course of a number of years, we decided that we would attempt to reconstruct as complete a picture as possible of this community of women. It soon became the goal to identify each woman fitting the definition and attempt to trace each from birth to death by gathering biographical and bibliographical information. The biographical information includes material about family background, education, work history, special contributions, extracurricular interests, and professional recognition. We also tried to identify each woman's publications and professional presentations. We hoped to discover how each of these stories revealed insights into the larger mathematical, scientific, educational, and cultural communities, primarily in the United States, but also in a broader international setting when possible. Three decades later, although such a project has no natural end (there remain various elusive pieces of information), we now consolidate our understandings of this community of women.

We approached the project with no preconceived ideas about what we would find, but hoped to determine answers to questions, many of which are of importance today. Usually we cannot answer why a woman chose to do mathematics, but we can often tell something about her motivation from the choices that were made about the path of education and career. We can frequently surmise which factors appear to have been helpful and which were harmful in the course of study and work. Here we are dealing with those women who were persistent enough and fortunate enough to attain a doctorate; it is much harder to determine why various other individuals chose not to continue their studies or were unable to do so.

While we will describe our findings in some detail below, certain conclusions stand out. First, there were many more women who obtained PhD's in mathematics before 1940 than was generally thought. They came from a variety of backgrounds, not just families who were well-educated or well-to-do. Women's colleges played a prominent role in the undergraduate education of the women, most of whom taught at the high school or college level before receiving their doctorates. One school, the University of Chicago, stood out as a granter of PhD's in mathematics in the period in question. The women in the study typically were in strong graduate programs with respected advisors, but these degree-granting schools were generally reluctant to hire women. Thus, women's colleges and coeducational liberal arts colleges figured most prominently as employers. Most of the women had academic careers, but a few made important contributions while working for the government or in industry. A few women made significant research contributions to mathematics, but many did not, largely because they were in academic settings that emphasized teaching rather than research. The Great Depression created problems for obtaining jobs, apparently more for women than for men. Also, marriage presented its own set of difficulties, primarily because of anti-nepotism policies. On the other hand, several of the women had close female friendships. Finally, we find that the women lived much longer than would be expected from standard life tables.

This book contains a lengthy essay elaborating on many of our basic findings. In addition, there are concise biographical entries for each of the 228 women. Since we have gathered information that is much too detailed for a book of this sort, the American Mathematical Society is hosting a website that contains this additional material. The address of this website is on the copyright page and on the back cover of this book.

For each of the 228 women there is a much fuller biographical narrative than is possible here along with a list of publications by, and references to, the person. The individual bibliographies contain not only a complete description of the items authored but sources of reviews for the items when available and talks presented to professional meetings.[1] In addition, the expanded entries will contain detailed citations to sources. The material on the website is freely available and is searchable. Our hope is that the brief sketch we are able to provide in the book will prompt a visit to the website for a fuller picture. Furthermore, we will be delighted if the detail available on the website proves a source for further scholarship and insights into various aspects of this community of women.

In addition to the essay and the biographical entries in the book, we have included a bibliography and a list of archives and manuscript collections consulted. The bibliography contains primarily those items that have relevance for the essay. The list of archives and manuscript collections, however, is a list of all that were used in the preparation of the biographical entries. We have not included particular collections used at each of these depositories since this information will be available on the website.

[1] In a few cases the bibliographies are so wide-ranging that only selected publications are listed.

As we indicate in the introduction, a study of American women who held PhD's was undertaken by Helen Brewster Owens in the late 1930s. In 1937, and again in 1940, she sent questionnaires to American women mathematicians asking for basic biographical and bibliographical information. This material was invaluable in the early stages of our study. Where possible, the authors also used questionnaires to collect biographical and bibliographical information from the women themselves or from relatives and friends. Many other sources were necessary, however, in order to achieve the level of detail we wanted. The published sources we consulted include standard reference works, mathematical journals, and scholarly historical works. Particularly useful sources were *American Men of Science* and *American Men and Women of Science*, the *American Mathematical Monthly* and the *Bulletin of the American Mathematical Society*, and the reviewing journals *Jahrbuch über die Fortschritte der Mathematik*, *Revue Semestrielle des Publications Mathématiques*, *Zentralblatt für Mathematik*, and *Mathematical Reviews.*

Other sources include college and university archival and alumni records, mathematics departmental papers, various public records such as birth and death certificates and US census reports, and private materials. In the early years of the study, almost all of our research involved working in archives, reading through journals, going through college catalogs in the stacks of the Library of Congress and at the Center for Research Libraries, searching through volumes of the *National Union Catalog*, and scrolling through microfilms of census records at the National Archives in Washington, D.C., and in Chicago. In later years, many but not all of these resources have become available online. Archival records that have not been digitized still form the core of much of our work. However, library catalogs and online census records have lightened the load for checking facts. Similarly, the digital archives of JSTOR and Project Euclid have been used to verify most of the publication data listed in the individual bibliographies on the website. These sources have also provided some new information on the professional activities of the women in our study. Genealogical sites, especially Ancestry.com, have added to our ability to learn more biographical information about the women in the study and their families. We have had additional extensive correspondence and personal interviews with many of the women themselves, their family members, and other friends and associates when possible.

We have so many people to thank that we would exceed our page limitation were we to attempt to name them all. First, of course, we thank Uta C. Merzbach for the initial direction, ongoing support, and very concrete help throughout the years. We could only hope to approximate her high scholarly standards. Many of the women in the study and their friends and family have gone to great trouble to provide us with detailed information about their lives. Our results would have been much the poorer without the incredible help of dozens of public and college librarians and archivists. We cannot sufficiently express our gratitude for their knowledge and helpfulness. Many of our friends and relatives have gone off on fact-finding expeditions for us. We also wish to express our gratitude to the Smithsonian Institution for the utilization of numerous resources, especially when we were honorary research associates there.

The first author is grateful for a Rutgers University Research Council grant and particularly thanks Kyle Vaughan and Lynn McLaughlin at Marymount University for their extraordinary efforts on her behalf at retrieving items through interlibrary loan. We also thank DePaul University for assistance to the second author in the form of grants from the Faculty Research and Development Fund of the College of Liberal Arts and Sciences and from the University Research Council and for support for research leaves. We are grateful to the American Mathematical Society for taking on this somewhat unusual project, for publishing the book, and for providing the website for full biographical and bibliographical entries for the women in the study. We particularly thank the History of Mathematics Editorial Committee and Joseph W. Dauben, our contact person on the committee; Edward G. Dunne, our editor; Cristin Zannella, his editorial assistant; and the technical support staff, especially Barbara Beeton. Finally, we acknowledge the special, ongoing support of Paul Green and Carol Stukey. Paul provided excellent technical support, and Carol proofread, edited, and was our consultant for any and all writing questions. Furthermore, without these two, we might have starved.

CHAPTER 1

Introduction

In the 1950s about 5 percent of the US PhD's in mathematics were awarded to women. This proportion improved slightly to 6 percent in the 1960s, and, although the proportions and numbers continued to increase throughout the rest of the century, women in the profession were invisible to many. This invisibility apparently confirmed the notion that mathematics was essentially a man's field. Our findings refute this idea, for we have discovered that there was a large community of women mathematicians active throughout the last part of the nineteenth century and much of the twentieth. As noted in the preface, the first PhD in mathematics to a woman in the United States was awarded in 1886. Furthermore, more than 14 percent of the PhD's awarded in mathematics during the first four decades of the twentieth century went to women, a proportion not achieved again until the early 1980s.

In this book, we focus on the work and careers of the 228 American women we have identified as receiving PhD's in mathematics before 1940.[1] Included in this group are all women who were granted PhD's in mathematics by institutions in the United States, together with four women born in the United States who received PhD's from foreign institutions.[2] We do not include women who worked in the United States but were neither born nor received their doctorates there. While our definition does not include women who received PhD's in statistics, physics, or the history of mathematics, it does include women who wrote dissertations in these fields when the degree was obtained in a mathematics department whose requirements were the same for all who received PhD's. The definition also does not include women who earned PhD's in education even when the major was mathematics education, nor does it include the American women research mathematicians who were educated before World War II but did not possess a doctorate in mathematics. For example, of the twenty-two women who were members of the American Mathematical Society before 1900, only seven fit our definition.[3]

[1] In previous articles published by the authors, we have referred to 229 such women. One woman who was on our earlier list completed her course work and wrote a doctoral dissertation that was approved and sent to the Library of Congress. However, we have since learned that she withdrew from the program after failing to pass her final examination and, therefore, was not granted a PhD.

[2] We use the term "American" throughout to mean from the United States. While we have not made a study of women from other countries in the Americas who received PhD's in mathematics outside the United States, we know of three women mathematicians who earned PhD's from the University of Toronto before 1940: Cecelia Krieger in 1930, Alice W. Turner in 1932, and Mary J. Fisher in 1934.

[3] Of the others, some were astronomers, some had master's degrees in mathematics, and one had a doctorate in mathematics but was born and educated in England. For short descriptions of these twenty-two women see Whitman, "Women in the American Mathematical Society before 1900."

We have chosen 1939 to end our study because of the tremendous changes that occurred in the American mathematics community and larger society as a result of World War II. Furthermore, by considering this period, we have a sizeable collection of women who earned PhD's in mathematics, but it is also a collection small enough that we can document the personal and professional lives of each one of them.

In addition to identifying this large group of women, we describe the members of this community and place them within a broader historical context. One of the findings from this study is that these women come from a wide range of geographic, economic, and educational backgrounds. They are not racially diverse, however. No African-American women are represented, since the first African-American woman to earn a PhD in mathematics, Euphemia Lofton Haynes, was awarded the degree in 1943 by the Catholic University of America when she was in her early fifties. Of the 228 women, 212 were born in thirty-eight different states and the District of Columbia, while the remaining 16 were born in Canada, Europe, and Asia. The first was born in 1847, the last in 1915. They came from farms, cities, and small towns, and their parents had a broad mix of educational backgrounds, ranging from no formal schooling to professional and doctoral degrees.

These women received their undergraduate degrees from more than a hundred different institutions, mostly in the United States, about half of which were coeducational and the other half women's and coordinate colleges. They received their doctorates from all major schools granting such degrees in that period except Princeton University and the University of Virginia. Difficulties often arose when it came time to find employment, as many restrictions sharply curtailed their options. Even so, most had productive and apparently satisfying careers, frequently in women's colleges and other smaller coeducational colleges and rarely in schools with doctoral programs in mathematics. Several gave talks and published papers, despite working in schools with limited support for research. About two-thirds of the women in our study presented talks to the American Mathematical Society (AMS), and about half that number published at least two articles that had been presented at AMS meetings or were reviewed in one of the mathematics reviewing journals. Many were active in the larger mathematical community, especially in the Mathematical Association of America (MAA), where they frequently held leadership roles at the sectional and national levels.

There were many firsts among the women in our study. For example, Winifred Edgerton (Merrill) was the first woman to be awarded a PhD in any field at Columbia;[4] Mary Winston (Newson) was one of the first three women admitted as regular students to the university at Göttingen; Anna Pell Wheeler was the first woman to deliver an AMS Colloquium Lecture; Mina Rees was the first recipient of the MAA's Award for Distinguished Service to Mathematics; and Dorothy Bernstein was the first woman to be elected president of the MAA. Women were the first recipients, male or female, of PhD's in mathematics from Ohio State, Indiana, Texas, Brown, Duke, and Marquette; women and men shared the distinction of being the first recipients of PhD's in mathematics at Minnesota and Catholic University. Among other forms of recognition we find that both Olive C. Hazlett and Anna Pell Wheeler were starred in *American Men of Science* for mathematics, and

[4]More than forty years later Deborah Hickey (Maria) was the first woman to receive a PhD in any field at Rice.

Christine Ladd-Franklin was starred for psychology.[5] We also cannot fail to note that the USS *Hopper*, a guided missile destroyer, was named after Grace Murray Hopper.

Finally, the women in this study lived to a mean age of 80.5 and a median age of just over 84. The mean is, on average, more than eleven years longer than life tables predict, while the median is even more impressive. The youngest died at twenty-five, and the oldest, so far, lived to 104. Three of the women in our study are still living and are in their nineties as of this writing, so the mean age at death can only increase.

This is not the first study of American women with PhD's in mathematics. Such a study was undertaken by Helen Brewster Owens in the late 1930s. Owens was a mathematician and suffragist who had received a PhD from Cornell University in 1910. As the wife of F. W. Owens, the mathematics department chairman at Pennsylvania State College from the mid-1920s to the late 1940s, she was unable to hold a regular position on the faculty there until she was nearly sixty because of anti-nepotism policies. In 1937 and again in 1940, Owens sent questionnaires to American women mathematicians asking for basic biographical and bibliographical information.[6] In 1937 R. G. D. Richardson, secretary of the American Mathematical Society, was compiling a list of former and current members of the AMS for the 1938 semicentennial of the society. In connection with his project he offered Owens money for postage to help in her search for a complete list of American women with PhD's in mathematics. Although Richardson never published the biographical list, on September 8, 1937, a luncheon was held during the summer mathematics meetings in State College, Pennsylvania, honoring those American women who received PhD's in mathematics by 1900 as well as those who were professor emeritus of mathematics. The *Bulletin* of the AMS reported that sixty-seven attended and that the guests of honor at the luncheon were Mary Winston Newson who had received her doctorate from the university in Göttingen in 1897, and two professors emeriti, Clara E. Smith from Wellesley and Clara L. Bacon from Goucher; Smith had earned a PhD from Yale in 1904 and Bacon from Johns Hopkins in 1911.[7] The event was organized by the Committee for Study of Women's Work in Mathematics, which consisted of six women associated with the mathematics department at Penn State at the time.

We have used resources from two other investigations of women in mathematics, as well. Mary Elizabeth Williams collected information about nineteenth- and early

[5]The entry for Ladd-Franklin in *American Men of Science* appears under "Franklin, Mrs. Christine Ladd."

[6]The questionnaires that Helen Owens collected in 1937 and 1940 are in folders 134–38 of Series IV, Women in Mathematics and the Sciences, Helen Brewster Owens Papers, Schlesinger Library, Radcliffe Institute, Harvard University. Any such questionnaire will hereafter be cited as "Owens questionnaire."

[7]The meeting is described in T. R. Hollcroft, "The Summer Meeting in State College," *Bulletin of the American Mathematical Society* 43 (1937): 747. In addition to Newson, the original list of invited honored guests who had earned PhD's by 1900 included Winifred Edgerton Merrill (Columbia 1886), Annie MacKinnon Fitch (Cornell 1894), Elizabeth S. Dickerman (Yale 1896), Isabel Maddison (Bryn Mawr 1896), Leona May Peirce (Yale 1899), and Anne L. Bosworth Focke (Göttingen 1900); Ida M. Metcalf (Cornell 1893) was the only living woman who had received a PhD by 1900 who was missing from Owens's list. In addition to Bacon and Smith, the professors emeriti on the list of invited honored guests included Helen A. Merrill (Yale 1903) from Wellesley and Ruth G. Wood (Yale 1901) from Smith.

twentieth-century American women in mathematics.[8] Williams held a 1936 master's degree from the University of Kentucky and began her long career at Skidmore College in 1946. Her papers include the partial results of a 1970s study for which she had collected information about thirty-four American women mathematicians, most of whom did not have PhD's in mathematics. She was in the process of writing biographies of these women at the time of her death in 1976. In the early 1980s, Sister Miriam P. Cooney, of Saint Mary's College in Indiana, and Sister M. Stephanie Sloyan, of Georgian Court College in New Jersey, conducted a survey of "Religious Women in Mathematics" and shared some of their results with us. Both earned PhD's in mathematics, Cooney from the University of Chicago in 1969 and Sloyan from Catholic University in 1952.

The most recent historical study of American women in mathematics was conducted by Margaret A. M. Murray, who based her book, *Women Becoming Mathematicians*, on interviews with thirty-six of the approximately two hundred women who earned PhD's in the United States between 1940 and 1959. Another, more sociologically oriented, study using interviews with eleven women was conducted by Claudia Henrion and appeared as *Women in Mathematics: The Addition of Difference*. Many websites now include some information on women mathematicians.[9]

The current book is the result of a historical study of all American women who received PhD's in mathematics before 1940. Like Helen Owens, the authors helped organize a luncheon recognizing women mathematicians. That luncheon was part of a day-long meeting honoring American women who received PhD's in mathematics prior to World War II, which was sponsored by the Division of Mathematics of the Smithsonian Institution. The division's curator, Uta C. Merzbach, presided at the meeting, which took place on August 31, 1981; she was assisted by the authors. Fifteen women attended; all of the attendees had received their doctorates in mathematics in the period 1929 through 1940.[10] Unlike the 1937 luncheon, which was purely a way to honor pioneer women in mathematics, the 1981 meeting was an opportunity for the authors to collect first-hand information about the education and subsequent careers of the participants. Like Owens, the authors used questionnaires to collect biographical and bibliographical information from both the attendees and from as many as possible of those who were unable to attend the luncheon. Several of the personal stories that appear in this introductory essay and in the individual narratives come from those questionnaires or the transcription of the tapes made at the meeting.[11] Many of the other sources used in the study are indicated in the preface.

---

[8]The Mary Elizabeth Williams Papers are in the Schlesinger Library, Radcliffe Institute, Harvard University.

[9]Two websites that include biographies of women mathematicians are: the "MacTutor History of Mathematics Archive," http://www-history.mcs.st-and.ac.uk/history/, and "Biographies of Women Mathematicians," http://www.agnesscott.edu/lriddle/women/women.htm.

[10]All but two of the women honored at the meeting had earned PhD's by 1939, and so their biographies appear later in this volume. These thirteen women were: Beatrice Aitchison, Sister Leonarda Burke, Marion Greenbaum Epstein, Nola Anderson Haynes, Anna Stafford Henriques, Grace Murray Hopper, M. Gweneth Humphreys, Ruth Stauffer McKee, Irene Price, Grace Shover Quinn, Mabel Griffin Reavis, Sister Helen Sullivan, and Vera Ames Widder. In addition, Sister M. Elizabeth Frisch and Edith R. Schneckenburger, who received their PhD's in 1940 from Catholic University and the University of Michigan, respectively, also were present at the meeting.

[11]The questionnaires collected in connection with the 1981 Smithsonian meeting and the audiotapes made at the meeting are in the Early Women Doctorates Collection (number 2006.3037),

Figure 1.1: Meeting at the Smithsonian Institution, August 31, 1981. Photograph courtesy of the Smithsonian Institution, negative number 81-11284-15.

In what follows, we document the experiences of this large and diverse community of women mathematicians. We describe the members of this community, their lives, careers, and professional contributions, and place them within the larger setting of the formative years of American mathematics. We start with a brief discussion of the historical context for the study, including a short description of the state of American mathematics and of educational opportunities for women in the nineteenth century.

The first woman to earn a PhD in mathematics in the United States was Christine Ladd. She graduated from Vassar College in Poughkeepsie, New York, in 1869, and in 1882 she completed all the work for the PhD in mathematics at the Johns Hopkins University. However, the university trustees refused to grant the degree to Ladd solely because she was a woman. In 1926, at the fiftieth anniversary of the founding of Johns Hopkins, Ladd was offered an honorary degree as a result of her work in physiological optics. She insisted, instead, that it be the PhD she had earned forty-four years earlier or none. Hence, after most of a lifetime devoted to research and teaching in mathematics, symbolic logic, and physiological optics, and to organizing, writing, and speaking as an advocate of women's rights, especially the promotion of educational and professional opportunities for women, Ladd-Franklin was awarded her PhD in mathematics. She was born in Connecticut in 1847, a year before the Seneca Falls Convention that marks the inception of the women's

Mathematics Collections, Division of Information Technology and Communications, National Museum of American History, Smithsonian Institution. These questionnaires, which the authors continued to collect for several years after the meeting, will be cited as "Smithsonian questionnaires." The authors later collected additional questionnaires independently; these questionnaires, cited as "Authors' questionnaires," and the transcriptions of the audiotapes made by the authors, cited as "Smithsonian meeting tapes," are not part of the Early Women Doctorates Collection at the Smithsonian Institution.

rights movement in the United States and was, by a decade, the first born of these women who received PhD's in mathematics before 1940. She was the first to earn the degree in 1882; at age seventy-eight, she was the oldest to receive it. During the course of her life, Christine Ladd's activities intersected with many of the most significant people, institutions, and ideas and issues in late nineteenth- and early twentieth-century higher education, mathematics, and society, both in the United States and Europe. Her name will arise frequently as we continue our discussion of the early mathematical community and educational opportunities for women.

The last third of the nineteenth century was the period of both the maturing of American mathematics and the opening of higher education, both graduate and undergraduate, to women. By 1900 institutional mathematics in the United States was organized essentially as it is today, although on a much smaller scale, and professional journals and societies had been established.[12] Opportunities existed for first-class graduate work in mathematics for both men and women; moreover, there were women with the desire and background to take advantage of these opportunities.

In Europe, the nineteenth century was also, for the most part, the period during which women acquired access to higher education, but the course of development varied among different countries.[13] By the middle of the nineteenth century, British women had gained admission to schools calling themselves colleges. In London, Queen's College opened in 1848 as a branch of the Governesses' Benevolent Institution, initially to provide formal education and diplomas for governesses. The following year women were admitted to the College of Preceptors, and Bedford College in London was opened for the higher education of women. When Girton College, the first residential college for women in England, was established near Cambridge in 1869, most of the Cambridge University teaching faculty allowed women to listen to their lectures. Thus, these women students had opportunities to obtain a Cambridge education even though the university did not grant degrees to women until 1948. Beginning in 1878, however, women were able to obtain degrees based on examinations, as was the case with men, at the University of London.

Higher education in France and Switzerland was never officially closed to women. In the 1870s Switzerland began to regulate admission standards to universities but did not exclude women, many of whom were from Russia. By the early 1880s women were earning PhD's from the university at Zurich. In France the universities were public and some lectures at them were open to all. Other lectures were open only to those who had passed the *baccalauréat*. Since students prepare for the *baccalauréat* in *lycées* and it was not until 1881 that *lycées* for girls were established, the fact that the universities were open to women did not mean that French women were attending. In Germany, it was not until about 1890 that girls could attend the equivalent of a *Realgymnasium*, the school one needed to attend to gain admission

[12]See Parshall and Rowe, *The Emergence of the American Mathematical Research Community, 1876–1900*, for a discussion focusing on mathematical research in the last quarter of the nineteenth century.

[13]The discussion of women's higher education in Europe is based mainly on Bremner, *Education of Girls and Women in Great Britain*; Bridges, "Coeducation in Swiss Universities"; Johanson, *Women's Struggle for Higher Education in Russia*; Maddison, *Handbook of British, Continental and Canadian Universities* (1st and 2nd editions); and Zimmern, "Women in European Universities."

to a university. In Russia during the second half of the nineteenth century opportunities for women to get a higher education came and went. Women were first allowed to attend lectures in 1859 but were ordered expelled in 1863. In the early 1870s courses for women were established in parallel with those for men, but in 1886 admissions to most of these courses were closed. In about 1900 women finally gained lasting access to higher education. After the 1905 Russian Revolution, the number of women's higher education courses increased dramatically.

In the United States during the first half of the nineteenth century a number of institutions for girls or young women opened that were somewhat above secondary level but did not yet have full collegiate status. These were generally called seminaries in the Northeast, but often were called "colleges" in the South. Although these institutions tended to reflect the conservatism of the period regarding the education of women, many schools for girls provided more advanced preparation in science than comparable academies for boys. Furthermore, significant numbers of girls studied algebra, geometry, and trigonometry and became mathematics teachers in secondary schools.[14]

Some of these early women's seminaries in New England, most notably Mount Holyoke, eventually matured into genuine colleges. Christine Ladd-Franklin, writing in 1891 on the state of education of women in the South, enumerated a number of institutions there that she considered to have reached the status of semi-colleges. By this she meant that the requirements for beginning the junior year roughly paralleled the entrance requirements for regular colleges.[15]

It is generally considered that the first women to receive undergraduate degrees equivalent to those of men earned them in 1841 at the Oberlin Collegiate Institute, renamed Oberlin College in 1850. While Oberlin had opened and enrolled women in 1833, it did not admit women to the regular college course for another four years. The University of Iowa was the first public institution in the United States to admit men and women on an equal basis when it opened in 1855.[16] The Morrill Act of 1862 authorizing land-grant colleges paved the way for the opening of many state universities and provided funds for "the endowment, support, and maintenance of at least one college where the leading object shall be, without excluding other scientific and classical studies and including military tactics, to teach such branches of learning as are related to agriculture and the mechanic arts."[17] The Morrill Act included no mention of women or coeducation. Nonetheless, by 1870, eight state universities, all but one in the Midwest, accepted women in at least some programs.[18] A 1916

[14]For a discussion of mathematical education of girls see chapters three and four of Tolley, *The Science Education of American Girls*, 62–73, 78–92.

[15]Ladd-Franklin defined a regular college as one "which comes up to the standard set by the Association of Collegiate Alumnae for admission into its ranks." Franklin, "The Education of Woman in the Southern States," 94.

[16]In 1850 the University of Deseret, later to become the University of Utah, opened its doors and enrolled women during its second session; it closed in 1853 because of financial difficulties and did not reopen until after the Civil War.

[17]Act of July 2, 1862 (Morrill Act), Public Law 37-108; Enrolled Acts and Resolutions of Congress, 1789–1996; Record Group 11; General Records of the United States Government; National Archives.

[18]In the Midwest, Indiana, Iowa, Kansas, Minnesota, Michigan, Missouri, and Wisconsin enrolled women; in the West, California did. Utah had not yet reopened in 1870.

ruling by the secretary of the interior determined that "instruction in the industries for women is included in instruction in agriculture and mechanic arts."[19]

An important reason for the opening of higher education to women was the need to train women as teachers. Normal schools for training teachers sprang up in the middle of the nineteenth century. In several Midwestern states normal courses were also given on the campuses of state universities while the regular students were on vacation. This led first to a normal department within the university and eventually to full admission of women.

In 1855 Elmira College opened as the first women's college explicitly intended to have the same requirements as other institutions. It was chartered by New York State so that "no degree shall be conferred without a course of study equivalent to a full ordinary course of college study as pursued in the colleges of this state."[20] Other women's colleges with these standards followed and became important in the education of women. Among these are Vassar College, which opened in 1865, and Smith College and Wellesley College, both of which opened in 1875.

During the nineteenth century, some women's colleges also offered courses beyond their undergraduate curriculum.[21] However, since many of the universities that would not admit women as undergraduates did admit them to graduate programs, the women's colleges were not essential for the graduate education of women, even at the master's level, and, thus, played a relatively minor role.[22] In their article on women in the American mathematical research community, Fenster and Parshall list eighteen women whom they define as having been active during the period 1891 through 1906.[23] Of these eighteen women, eight had received master's degrees by 1906. However, only Elizabeth B. Cowley's 1902 master's degree from Vassar had been granted by a women's college. In the nineteenth century, Bryn Mawr College was the only women's college that conferred doctorates in mathematics; it granted two PhD's in the 1890s. Thus, most women who wanted doctorates in mathematics had to study at coeducational institutions or at men's schools that had opened their graduate programs to women. These men's schools made a distinction between allowing women to enroll as graduate students and permitting the enrollment of women as undergraduates, as can be seen from the following 1894 explanation. "The definiteness of aims, the increased earnestness and the more mature character which belong to the greater age of graduate students, may entirely or largely remove difficulties which are found in the way of men and women mingling in the undergraduate department. The Faculty of Yale University knows very well that to admit women to its graduate school is quite unlike opening the doors of Yale College to girls of the age of eighteen."[24]

---

[19]Henry S. Brunner, *Land-Grant Colleges and Universities 1862–1962*, Bulletin (US Office of Education) 1962, no. 13, 58.

[20]Taylor, *Before Vassar Opened*, 76. An early discussion of women's education in the United States can be found in Woody, *A History of Women's Education in the United States.*

[21]The astronomer Mary Watson Whitney received a master's degree in mathematics at Vassar College in 1872, and the Wellesley College mathematics department determined the requirements for a master's degree in the early 1880s.

[22]Yale officially opened its graduate courses to women in 1892, and two years later Harvard opened its graduate classes, but not its degrees, to women.

[23]Fenster and Parshall, "Women in the American Mathematical Research Community: 1891–1906," 241.

[24]Charles F. Thwing, *The College Woman* (New York: Baker and Taylor, 1894), 130–31.

During the late nineteenth century few of the leaders among those schools granting PhD's in mathematics were open to women. Before 1900 the Johns Hopkins University granted thirty PhD's in mathematics to men, by far the most of any school, and it was officially closed to women at both the undergraduate and graduate levels. Yale opened its graduate school to women in 1892 and was second in the production of mathematics PhD's with twenty-three doctorates before 1900, three of which were granted to women. Clark University and Harvard University were the next leading producers with eleven and ten respectively, and neither granted PhD's to women. The University of Pennsylvania followed with nine degrees, but its first PhD to a woman in mathematics was not granted until 1901. The next two schools were Columbia, with eight mathematics PhD's (one to a woman), and Cornell, with six (three to women). Lists of women in the study by date and by school of PhD appear as tables 1.1 and 4.3, respectively.

Between 1862 and 1899, American institutions conferred about 130 PhD's in mathematics; approximately 7 percent of these were awarded to women.[25] However, foreign-educated mathematicians, both American and foreign born, dominated American mathematics until well into the twentieth century. In the early 1880s Americans began to attend foreign universities with some frequency. Before 1900 about thirty US mathematicians, including one woman, earned doctorates at European universities.

In 1876 the British algebraist J. J. Sylvester was appointed as the first professor of mathematics at the newly established Johns Hopkins University, where he was to have a significant influence on the early development of American mathematics.[26] At that time few of the institutional characteristics we associate with a mathematical community existed in the United States, although many of the formal structures had been developed on the European continent and in Great Britain. Journals devoted solely to mathematical research became prominent in the second quarter of the nineteenth century: there were Crelle's *Journal für die reine und angewandte Mathematik* (1826) in Germany, Liouville's *Journal des Mathématiques Pures et Appliqués* (1836) in France, and the *Cambridge Mathematical Journal* (1839) in England followed by the *Quarterly Journal of Pure and Applied Mathematics* (1855), with Sylvester as an editor, in England. It was Sylvester at Johns Hopkins who established in 1878 the *American Journal of Mathematics*, the first US journal devoted to original mathematics investigations. In 1884 the *Annals of Mathematics* was started at the University of Virginia.

The first mathematical societies and their associated publications emerged in the third quarter of the nineteenth century. The first was the London Mathematical Society (1865) with its *Proceedings*, followed by La Société Mathématique de France and its *Bulletin* (1872). The organizational meeting of the New York Mathematical Society took place at Columbia University on Thanksgiving Day, 1888. Within three years of that meeting, the *Bulletin of the New York Mathematical Society* appeared, and the society increased its membership rolls to become national in character, if not yet in name. A major membership drive was undertaken to finance the publication

[25]Yale granted the first degree listed as a PhD in mathematics to John Hunter Worrall in 1862. Harvard granted its first such degree in 1873 to William E. Byerly. Also, in 1862 the first foreign mathematics graduate degree was awarded to an American when the university at Jena in Germany granted William Watson, later of Harvard, a PhD.

[26]For a discussion of Sylvester at Johns Hopkins see Parshall and Rowe, especially chapters 2 and 3.

of the *Bulletin*, and in 1891 six women joined the Society. Only one, Charlotte Scott of Bryn Mawr, of the six had a PhD in mathematics; the other five included two astronomers, two professors of mathematics, and one private school mathematics teacher. In 1894 the first issues of the *American Mathematical Monthly* appeared, and the New York Mathematical Society acknowledged its national character by changing its name to the American Mathematical Society.

The next forty-five years saw a tremendous growth in the mathematical community. Most of this growth took place after the turn of the century. The membership of the AMS grew from 337 at the end of 1899 to 2,283 at the end of 1939, an almost seven-fold increase. However, during this period the number of American women holding PhD's in mathematics grew even faster. At the end of 1899 there were 10 living American women with PhD's in mathematics, and forty years later there were 212. Even if we only consider 1920–39, the two full decades between World War I and World War II, when the membership in the AMS was growing at its fastest, we find that the membership somewhat more than tripled while the number of living women with PhD's in mathematics considerably more than quadrupled.[27]

In the following pages we will elaborate on this rapidly growing community of women mathematicians by providing a portrait of the pre-1940 women PhD's within the context of the larger educational and mathematical communities. Table 1.1 below documents the constantly increasing rate at which American women received PhD's during the period of our study.

Table 1.1: Doctorates awarded by year and school

Degree name with name used in the biographical entry, if different, in parentheses

1886: **Columbia:** Edgerton (W. E. Merrill)
1893: **Cornell:** Metcalf
1894: **Bryn Mawr:** Gentry; **Cornell:** MacKinnon (Fitch)
1895: **Cornell:** Baxter (A. B. Hill); **Yale:** Barnum
1896: **Bryn Mawr:** Maddison; **Yale:** Dickerman
1897: **Göttingen:** Winston (Newson)
1899: **Yale:** Peirce
1900: **Göttingen:** Bosworth (Focke)
1901: **Bryn Mawr:** Martin; **Columbia:** G. Andrews; **Pennsylvania:** Vivian; **Wisconsin:** Pengra (Crathorne); **Yale:** Wood
1903: **Yale:** H. A. Merrill
1904: **Yale:** Smith
1905: **Columbia:** Coddington (E. C. Williams); **Pennsylvania:** McKelden (Dimick)
1906: **Bryn Mawr:** Ragsdale
1907: **Wisconsin:** F. E. Allen
1908: **Chicago:** Sinclair; **Columbia:** Cowley; **Cornell:** Van Benschoten; **Yale:** Worthington
1909: **Ohio State:** Bareis; **Yale:** Walker (Hull)
1910: **Chicago:** Pell (Wheeler), White; **Cornell:** Owens; **Illinois:** Bennett (Grennan)
1911: **California:** Biddle (A. B. Andrews); **Johns Hopkins:** Bacon; **Yale:** Barney
1912: **Indiana:** Hennel
1913: **Chicago:** Sanderson; **Illinois:** Burns (Glasgow); **Johns Hopkins:** Lewis; **Pennsylvania:** Copeland
1914: **Bryn Mawr:** Cummings; **Cornell:** Tappan; **Johns Hopkins:** Miller, Young; **Michigan:** Benedict

[27]This period includes the 1920s, which was the only decade of the early twentieth century during which the growth in membership of the AMS more than doubled. The 1920s was also the only decade when the growth of the AMS membership exceeded the growth in the number of living women holding PhD's in mathematics.

1915: **Chicago:** Hazlett, Wells; **Columbia:** Seely; **Missouri:** Weeks (King)
1916: **Chicago:** Larew, Sperry; **Texas:** Horton (Porter)
1917: **Bryn Mawr:** Haseman; **Catholic:** Kelley; **Chicago:** LeStourgeon; **Columbia:** Cronin; **Cornell:** Howe; **Radcliffe:** Curtis (Graustein)
1918: **California:** Sznyter (Sagal); **Indiana:** McCain; **Johns Hopkins:** Cohen; **Syracuse:** Roe
1919: **Illinois:** Jacobs (Offermann)
1920: **Bryn Mawr:** B. M. Turner; **California:** McFarland (Buck); **Chicago:** Gibbens; **Michigan:** Rambo
1921: **California:** Alderton (Moore); **Chicago:** Kendall, Logsdon; **Illinois:** Armstrong; **Johns Hopkins:** Sutton; **Radcliffe:** Blodgett (Adams)
1922: **Bryn Mawr:** Buchanan (M. B. Cole); **Columbia:** Morenus; **Pennsylvania:** Mullikin; **Radcliffe:** Pairman (Brown)
1923: **Chicago:** Ballantine; **Johns Hopkins:** Whelan
1924: **Chicago:** Darkow, Hughes (Bushey), Hunt; **Cincinnati:** Eversull (B. E. Allen); **Cornell:** Colpitts, Dale, Torrey; **Minnesota:** Carlson
1925: **Bryn Mawr:** Lehr; **Chicago:** Pepper; **Yale:** Anderton
1926: **Bryn Mawr:** M. C. Gray; **Chicago:** Stark; **Cornell:** Farnum; **Johns Hopkins:** Barton, Ladd-Franklin (earned in 1882); **Kansas:** Babcock, Black
1927: **Bryn Mawr:** Guggenbuhl; **Chicago:** Griffiths; **Cornell:** Mears, Schoonmaker (Wilson); **Missouri:** Hightower; **Zürich:** Hofmann (Bechtolsheim)
1928: **Chicago:** Beenken, Jackson, Johnson (Yeaton); **Stanford:** Weiss
1929: **Brown:** Whelan (Sedgewick); **Catholic:** Mangold; **Iowa:** McCoy; **Johns Hopkins:** Dean; **Missouri:** Anderson (Haynes), Wyant; **Rice:** Hickey (Maria)
1930: **Bryn Mawr:** R. L. Anderson; **Chicago:** Hagen, Huke (Frink); **Columbia:** Kramer; **Cornell:** Moody; **Illinois:** Cooper, Harshbarger; **MIT:** Weeks; **Michigan:** Liu (Hsia); **Strasbourg:** Collier; **Wisconsin:** Stafford (Hirschfelder)
1931: **Catholic:** Burke, Gough, Morrison, Vaudreuil; **Chicago:** Chandler (Pixley), Rees; **Duke:** R. W. Stokes; **Illinois:** Taylor; **Johns Hopkins:** Karl; **Marquette:** Mayer; **Ohio State:** Schmeiser (Barnes), Shover (Quinn)
1932: **California:** McDonald; **Catholic:** Thuener; **Chicago:** Mason (Ballard), Olson; **Cornell:** Calkins, Carroll (Rusk); **Indiana:** Price; **Missouri:** Zeigel (Hedberg); **Radcliffe:** Cope, Russell, M. M. Sullivan
1933: **California:** Brady; **Catholic:** Arnoldy; **Chicago:** Bower, Newton, Stafford (Henriques); **Cornell:** Adams (Infeld), Johnson; **Duke:** Griffin (Reavis); **Illinois:** Chanler; **Johns Hopkins:** Aitchison; **Minnesota:** Wilder (Thornton); **Radcliffe:** Peters
1934: **Brown:** Gurney; **Bryn Mawr:** Hughes; **Catholic:** M. H. Sullivan; **Chicago:** Baker, Killen (Huston), Litzinger; **Illinois:** Ketchum, Terry (Nee); **Ohio State:** Leffler (Alden); **Pittsburgh:** Kanarik; **Radcliffe:** N. Cole; **Saint Louis:** Garvin; **Washington:** Haller; **Wisconsin:** Shea; **Yale:** Becker (Mazur), Hopper
1935: **Bryn Mawr:** Stauffer (McKee); **California:** Noble; **Catholic:** M. L. Hill; **Chicago:** Humphreys; **Cornell:** Blanch, Montague; **Michigan:** Schulte; **Pittsburgh:** Taylor (Speer); **Wisconsin:** L. Wolf, M. Wolf (Hopkins); **Yale:** Fox (Delevie)
1936: **Bryn Mawr:** Levin (Early); **Catholic:** Reilly; **Chicago:** M. R. Anderson, McFarland, Rasmusen, Wiggin (Casner); **Cincinnati:** Odoms (A. O. Gray); **Michigan:** Kloyda; **Pennsylvania:** Spencer
1937: **Bryn Mawr:** Grant, Tuller; **Chicago:** Rees (Bonner); **Illinois:** Modesitt (Reklis), Pence; **Ohio State:** Kohlmetz; **Stanford:** Manning (Little); **Wisconsin:** Lester
1938: **Bryn Mawr:** Ames (Widder), Greenebaum (Epstein); **Catholic:** Fowler, Galvin; **Chicago:** Boyce, Haberzetle (M. H. Turner), Mauch; **Michigan:** Simond; **Radcliffe:** Wishard (McMillan)

1939: **Brown:** Bernstein, O'Brien, Torrance; **Catholic:** Varnhorn;
Chicago: Landers, Rayl, E. C. Stokes; **Colorado:** Beaty, Johnson (Rosenbaum);
Cornell: Nelson; **MIT:** Plass (M. H. Williams); **Michigan:** Baxter;
New York: Griffin, Manel (Greenfield); **Purdue:** Fry

CHAPTER 2

Family Background and Precollege Education

The women in our study came from a broad range of geographic, economic, and educational backgrounds. Of the 228 women, 212 were born in the United States; of those remaining, 7 were born in Canada, 5 in Great Britain, and 1 each in China, Hungary, India, and Poland. Audrey Wishard McMillan, who was born in India, was the daughter of Americans living abroad and was educated in a school for children of American missionaries. Gertrude Blanch was born in Poland and immigrated sufficiently young to have been educated mainly in the United States. Rosella Kanarik was born in Hungary and came to the United States at about age three. We have no information about the precollege education of Shu Ting Liu Hsia, although we are fairly certain she was educated in China before doing her undergraduate and graduate work at the University of Michigan.

About one-fifth of the women who were born in the United States had at least one parent who was foreign born. Looked at slightly differently, about 15 percent of the parents of US-born women were foreign born, just slightly over the percent reported in relevant census data, and apparently not significant. However, among women who were born into Jewish families, almost all were either foreign born or had at least one parent who was foreign born.

Those women who were born in the United States were born in thirty-eight different states and the District of Columbia, with most of the states not represented being ones with low population and few opportunities for undergraduate education for women.[1] Among the states that are the birthplace of at least ten women in our study, New York, Illinois, Massachusetts, and Maryland produced considerably larger numbers of women than was to be expected from the population distribution in the country at the time.[2] Most strikingly, Maryland produced roughly three and a half times as many as would be expected, which is almost entirely attributable to the role of Goucher College in Baltimore as an undergraduate school and the neighboring Johns Hopkins as a graduate school. More generally, the Northeast produced considerably more future women doctorates than would be predicted by the population distribution and the South produced considerably fewer.[3] This distribution correlates with the relative accessibility for women of undergraduate education in those regions.

---

[1]Of the forty-eight states in 1915, the latest year of birth for women in the study, those not represented are Alabama, Arizona, Florida, Montana, Nevada, New Mexico, North Dakota, Utah, Vermont, and Wyoming. Dorothy McCoy was born in Oklahoma Territory.

[2]The leading states in terms of birth were New York (33), Illinois (20), Pennsylvania and Massachusetts (17), Maryland (12), and Ohio (10).

[3]We are using the regional designations of the US Bureau of the Census.

In general, the women in the study did not come from homes where the parents were highly educated or well off. We have confirmed information about the educational background of one or both of the parents for less than a third of the women in our study. However, we can infer a good deal more about their education from their occupations. Using data from census reports and other sources, we have been able to determine occupations for all but a half dozen of the fathers and for most of the mothers, for whom "housekeeping" or "none" was usually listed on census reports.

There is a broad mix of backgrounds ranging from no formal schooling to professional and doctoral level degrees. As would be expected, the fathers tended to have more formal schooling than the mothers, given the limited opportunities for women to obtain an education, especially in the nineteenth century. Even so, a number of the mothers appear to have been influential in directing their daughters' education. In the case of Mary Winston Newson, for example, we know that her father was a country doctor and that her mother, who had taught French, art, and mathematics at various schools in the East and in Illinois, prepared the children for college. Her mother learned geology through a correspondence course with the Field Museum in Chicago and learned Greek on her own in order to teach it to her children.

Even though relatively few of the women in our study came from homes where parents had earned advanced academic degrees, some had fathers, and, in a few instances, mothers, who earned professional degrees in education, medicine, or law. Olive B. Hazlett, the mother of Olive Clio Hazlett, graduated from the Laura Memorial Woman's Medical College in Cincinnati in 1898 and was then licensed to practice medicine in Ohio and Massachusetts. Olive C. Hazlett moved with her mother to the Boston area in 1899, where her mother became a physician at the Reformatory Prison for Women in Sherborn, Massachusetts, while her father, a postal clerk, remained in Ohio.

Six women in the study had fathers who earned PhD's, including two, W. A. Manning and R. P. Baker, with doctorates in mathematics. W. A. Manning, the father of Dorothy Manning Smiley Little, earned his doctorate in mathematics from Stanford University, where he spent his career and supervised his daughter's master's thesis and doctoral dissertation. Among his other students were two more of his four daughters, both at the master's level. R. P. Baker, the father of Frances Ellen Baker, joined the mathematics department at the University of Iowa when his daughter was two and completed his PhD at the University of Chicago when she was seven. The wives of these two mathematicians were also highly educated. Dorothy Manning's mother, Esther Crandall Manning, earned both a bachelor's and master's degree in Greek from Stanford and was a doctoral student in psychology there at the time of her marriage. Frances Baker's mother, Katherine Riedelbauch Baker, earned a diploma in music. She taught privately and, before her marriage, served briefly as the head of the music department of a small college.

William Henry Zeigel, the father of Marguerite Zeigel Hedberg, taught mathematics and headed the mathematics department at a state teachers college in Missouri. His PhD, earned at George Peabody College for Teachers when his daughter was in her teens, was in education rather than mathematics. He then became dean and head of the education department at Delta State Teachers College in Mississippi. In 1982 Marguerite Hedberg described her mother as a housewife.[4]

Margaret Pillsbury Baxter and Frances Thorndike Cope also came from academic families; their fathers both held doctorates in psychology earned in the late nineteenth century, before their marriages. Walter Bowers Pillsbury (PhD Cornell 1896) was in the philosophy and psychology department at the University of Michigan at the time of his daughter's birth and became chair of the psychology department when the disciplines separated in 1920. Edward L. Thorndike (PhD Columbia 1898) was one of the most influential educational psychologists of the first four decades of the twentieth century and was a member of a distinguished academic family. He spent his career at Columbia as did his two brothers, one in English and one in history. Both mothers were college educated; Margaret Pillsbury's mother had earned a bachelor's degree from Michigan in 1905, while Frances Thorndike's had completed the four-year classical course at Boston University before her marriage in 1900.

Finally, Beatrice Aitchison's father, Clyde Bruce Aitchison, earned a doctorate in economics from American University a year before his daughter received hers in mathematics from Johns Hopkins. He was an attorney and at the time that he earned his doctorate he was fifteen years into a period of service on the Interstate Commerce Commission that extended from the administration of Woodrow Wilson into that of Harry Truman.

The families we have described above exhaust the instances in which a parent held a doctorate. However, there were other instances of parents who taught at the college level, including some who taught mathematics. Virginia Modesitt Reklis's father taught mathematics at Eastern Illinois State Teachers College, and Helen Schlauch Infeld's father taught high school mathematics for most of his career, before he joined the faculty of New York University's School of Commerce, Accounts and Finance when he was in his mid-fifties. Although it was unusual for a mother to have taught at the college level, both of the parents of Mary Emily Sinclair taught at what is now Worcester Polytechnic Institute. Her father was professor of mathematics, and her mother was an instructor of English and modern languages for three years, 1869–72.

Professional and academic families make up about one quarter of those for whom we know the occupation of at least one parent. The remainder are roughly equally divided among three categories: farmers; managers, officials, and proprietors; and various workers, including clerical workers, craftsmen, salesmen, and service workers.[5]

Among the women in our study who were born into urban working-class families is the only pair of sisters, Louise A. Wolf and Margarete Wolf Hopkins. They were born thirteen years apart in Milwaukee, Wisconsin; their mother was born in Germany; both parents had completed their formal education through elementary school; and their father was first a streetcar conductor and then a truck gardener. Margarete Hopkins reported in a 1997 conversation with one of the authors that their interest in mathematics derived from their mother. The elder sister, Louise Wolf, left college after a year, held a variety of positions for the next dozen years,

[4]Smithsonian questionnaire.

[5]These categories are based on the occupational classifications introduced by the US Bureau of the Census in 1950.

and resumed her education when her younger sister began college at the University of Wisconsin. In her senior year, the mathematics department chair wrote to the dean to request an assistantship for Louise Wolf. He noted that Miss Wolf was a senior mathematics major with a straight A record in math. He ordinarily would not appoint a senior to such a position but "Miss Wolf is an exceedingly capable woman, over 30 years of age, who has had experience in teaching and whose university education has been delayed by the fact that she is helping educate a younger sister."[6] Despite the difference in their ages, they received their bachelor's degrees a year apart, and both earned their doctorates in 1933.

Other children of working-class families were able to complete their educations without delay because of the existence of public higher education, most notably Hunter College, which was free to residents of New York City. An exception was Gertrude Blanch, who lived in New York City but whose father's death at the time of her high school graduation obliged her to work to support herself and her mother. Only after her mother's death fifteen years later did Blanch resume her education. She worked as an office manager for a hat dealer in Manhattan during the day and went to New York University at night, with her tuition being paid by her employer.[7]

Blanch is not the only woman in our study whose family was fatherless for a significant period. In at least eighteen instances, the mother in the family was divorced, separated, or widowed early in the marriage. The effect of the father's death or absence on the daughter's educational prospects was not always as drastic as in the case of Blanch. In some cases the mother's commitment to education overrode the disadvantage of not having two parents present. For example, after Dorothy McCoy's father died when she was three, her mother moved to another town because the schools were better there. Dorothy McCoy's brother, Neal, younger by two and a half years, also became a mathematician and earned his undergraduate and graduate degrees at the same time and at the same schools as his older sister. Their BA's were from Baylor University in 1925, and their PhD's were from the University of Iowa in 1929.

Vivian Spencer, who received her PhD from the University of Pennsylvania in 1936, also had a younger sibling who became a mathematician. Domina Spencer was twelve years younger than Vivian and received her PhD from the Massachusetts Institute of Technology in 1942. Domina Spencer later reported that their "father was a traveling salesman, and as such saw his wife and children only at irregular intervals.... [The sisters] and their mother moved from place to place as Vivian ... attended college ... and graduate school.... Mr. Spencer would join the family in these various locations when work and finances permitted."[8]

In many cases, even without access to free public higher education or the advantages of a professional background, a parental commitment to education or the existence of a state university was key to the academic achievement of the next generation. A dramatic example is provided by the family of Mary Gertrude Haseman,

[6]Warren Weaver to Dean Sellery, October 1, 1930, folder 1929–31, box 1, Budget Files, Department of Mathematics, College of Letters and Science, University of Wisconsin-Madison Archives.

[7]David Alan Grier describes Blanch's background, education, and how she came to work on the Mathematical Tables Project in *When Computers Were Human* (Princeton, NJ: Princeton University Press, 2005), 206–10.

[8]Murray, *Women Becoming Mathematicians*, 49. Domina Spencer is one of the thirty-six women interviewed by Murray.

who was born into a farm family in central Indiana. She was the seventh of nine children, all of whom earned bachelor's degrees and all but one of whom earned master's degrees; almost all of these degrees were granted by Indiana University. Four of Haseman's six older brothers eventually earned doctorates in the sciences, one of them in mathematics from the university in Göttingen.

Another woman in our study, Vera Ames Widder, who grew up on a farm, spoke of her high school experience and her parents' support in 1981. She reported:

> My parents were pioneer farmers in Saskatchewan, and we went to a school that was not a country school but was very much like a country school. It just had the three years of high school, so that I had to go away for the fourth year. I thought I'd teach, perhaps in a country school, so I was going to take a teaching training program after I finished high school. But farming was prosperous in the late twenties and when my parents asked if I wanted to go to a university, I thought that was a nice idea, so I did.[9]

Starting at age twelve, Christine Ladd (Ladd-Franklin) kept a journal in which she writes of her desire for education. By age fifteen, after three years of public education in Portsmouth, New Hampshire, she was sent to a coeducational preparatory school, the Wesleyan Academy, in Wilbraham, Massachusetts. In a journal entry from 1863, she writes:

> In two weeks I am to go to Wilbraham. I am glad to go. I do not really suppose that it is a very superior school but experience is the best test and I shall not have long to wait to form an opinion. It may be very fine and I may learn something, at any rate it will be better than my life here for I can get no chance to study more than Latin lessons. It is an open question whether my education is to be finished there or not. I know nothing of the plans of my father, nothing of the state of his finances except that he has lost heavily in stocks this summer and that he has not gotten me a piano although he gave me the promise of one. It is my great desire to have a college education & I shall use every means to bring my plans to a consummation.[10]

Earlier that year she wrote to her aunt about her partiality to geometry, saying that she liked it "ever so much, better than any thing else," and expressed her belief that she got her "love of mathematics" from her father, who had been a merchant in New York City.[11]

Our information on the precollege schooling of the women in our study is incomplete, although we are certain of the type of secondary school for nearly three-quarters of them. Of the ones we know, at least 70 percent attended public secondary school. At least a dozen went to parochial school and another twenty or so attended other private schools or academies. Several attended both public and private schools and a few had private tutors.

---

[9]Smithsonian meeting tapes.

[10]Entry in 1860–66 Diary, August 4, 1863 (p. 257), Christine Ladd-Franklin Diaries, Archives and Special Collections Library, Vassar College Libraries.

[11]Christine Ladd to her aunt, February 13, 1863, box 2, Christine Ladd Franklin and Fabian Franklin Papers, Rare Book and Manuscript Library, Columbia University.

While it is difficult to quantify the significance of the high school experience for most of the women, some comments from our 1981 Smithsonian meeting may shed some light. The following are some responses given when the participants were asked why they went into mathematics. M. Gweneth Humphreys said, "I can't remember a time when I wasn't interested in arithmetic or mathematics. I tried to trace it but I never could quite succeed. I did have excellent training in high school, which was in Western Canada." Anna Stafford Henriques was one of the women who did not grow up in a household with two parents; both her parents died when she was about fourteen. Her response was, "Well, I guess it was my mother's fault.... She didn't mind that I couldn't do arithmetic, my mother didn't. But ... when I got only a 92 in algebra she was heartbroken. When I got to geometry I knew that was it. So I always told people I was going to be an architect because that was acceptable. I really intended to be an astronomer but you know that wasn't very much of a field. So I stayed in mathematics." Finally, Ruth Stauffer McKee, who had originally wanted to be a doctor, responded, "I went to a girls' high school; we had excellent math teachers. They were women who had gone to places like Wellesley and they were an inspiration. Geometry was thrilling, it was exciting and that was back there in my head, but the MD was a goal."

There were many for whom the path from secondary school through college was not straightforward. About 15 percent worked for one or more years before beginning their program at the school where they earned their bachelor's degree, Gertrude Blanch and Louise Wolf being among the most dramatic cases. Unlike Blanch and Wolf, most of the women who did not go directly to college spent the intervening time as primary or secondary teachers. This was particularly common for the Catholic sisters, almost all of whom taught before beginning their college work.

CHAPTER 3

Undergraduate Education

Well over one hundred different institutions, most in the United States, granted undergraduate degrees to the women in our study.[1] About two-thirds of these American schools were private and one-third public. Similarly, about two-thirds were coeducational, and the rest were women's colleges (24%), coordinate colleges (6%), or schools predominately for men (3%).[2] On the other hand, while about two-thirds of the women attended private colleges and one-third attended public schools, just over half of the women (52%) received their undergraduate degrees at coeducational institutions, while the remaining attended women's colleges (37%), coordinate colleges (11%), and colleges predominately for men (2%). The role of the women's colleges declined over time as more schools opened their doors to women.

Substantial opportunities for women to obtain a solid undergraduate education in the United States existed by about 1870. Private, often church-related, coeducational colleges proliferated after the mid-nineteenth century; state universities and land-grant colleges, especially in the Midwest, were open to women; and women's colleges opened in the East beginning with Elmira College in 1855 and Vassar College in 1865. In the South and Northeast, seminaries for women were maturing into institutions of higher education.

By the 1870s and 1880s the curriculum offered at many women's colleges was similar to that offered by the men's colleges. The traditional classical curriculum, which stressed Greek, Latin, philosophy, and religion, was being replaced in the men's colleges by the elective system at about the same time that the colleges for women were being established. In this new curriculum fewer courses were prescribed, and science achieved a new prominence. For example, when Vassar opened, four of its eight original professors were in the sciences, and the school prescribed four semesters of mathematics courses for graduation. While until the end of the nineteenth century Vassar required more mathematics than other women's colleges, it was not the only women's college to require first-year students to take mathematics both semesters. These mathematics courses were being prescribed at a time when many in the United States still believed that institutions of higher education should not be open to women. Mabel Newcomer discusses and refutes three arguments against higher education for women that were common in the mid- to late-nineteenth century and persisted to some extent into the twentieth century.

[1]See table 3.2 for a list of these schools and the women who obtained their undergraduate degrees there.

[2]A coordinate college is one that is officially separate from but closely associated with a men's college. Some coordinate colleges shared faculty with the associated men's college and some had separate faculty. Starting in 1908 a number of Catholic colleges for men allowed women to enroll in some programs. See Poulson and Higgins, "Gender, Coeducation, and the Transformation of Catholic Identity," 491–92, for a discussion of coeducation in Catholic colleges.

These arguments were: "women were mentally inferior to men and would be quite unable to meet the standards set for the men's higher education;" women "could not stand the physical strain of higher learning;" and "education would reduce the number of marriages and the size of families."[3]

In view of their emphasis on science and mathematics, it is not surprising that during the nineteenth century private women's colleges were so important for the undergraduate education of the women in this study. This was the case even though coeducational institutions enrolled about 70 percent of the total number of women pursuing postsecondary education in 1900. Furthermore, most of the women enrolled at coeducational institutions were not enrolled in liberal arts programs; 70 percent were enrolled in teacher training courses and another 3 percent were studying home economics.[4]

Of the thirty-two bachelor's degrees granted to the women in our study before 1900, fourteen were awarded by Eastern women's colleges and one by a Southern women's institution. The rest were distributed nearly evenly among state universities primarily in the Midwest, private coeducational colleges in the Midwest and East, and foreign institutions in Canada and Great Britain.

The first woman in our group to obtain a bachelor's degree was Christine Ladd (Ladd-Franklin), who earned her degree from Vassar College in 1869. Although she graduated from preparatory school in 1865, a lack of funds delayed Ladd's matriculation at Vassar College until 1866. A similar lack of funds caused her to take a year out of her college studies to teach. While at Vassar, Ladd was influenced by Maria Mitchell, professor of astronomy and notable nineteenth-century American scientist, and also by Charles S. Farrar, with whom she studied physics, the subject of greatest interest to her at that time.

The second undergraduate degree, also from Vassar, was awarded to Charlotte Barnum in 1881, while the third went to Winifred Edgerton (Merrill), who graduated from Wellesley College in 1883. As we have seen above, women's and coordinate colleges together conferred almost half of all undergraduate degrees to women in our group. As can be seen in table 3.1 below, women's colleges also lead among those granting the most undergraduate degrees to women in our group, with the six leading schools all being women's colleges in the East.

Wellesley College granted by far the largest number of baccalaureates in our group with seventeen. Wellesley was remarkable for the consistency with which it produced those graduates, granting between two and five degrees during each of five decades starting with the 1880s. Henry F. Durant, the founder, had emphasized mathematics and science from its beginning in 1875. This emphasis was supported by Eben N. Horsford, a chemist who provided money for the library, scientific apparatus, and travel funds for sabbatical leaves and pensions for the president and most department heads, provided that "the offices ... [were] held by ladies."[5]

---

[3]Newcomer, *A Century of Higher Education*, 25–31.

[4]Ibid., 91.

[5]Eben Norton Horsford, *Scheme Matured and Adopted by the Trustees in 1886, on the Basis of a Bequest Made to Wellesley College in 1878: Extract from a Letter to the Trustees, Dated January 1, 1886* (Cambridge: J. Wilson and Son, 1886), 9.

Table 3.1: Leading undergraduate institutions by decade, 1869–1936

	60s	70s	80s	90s	00s	10s	20s	30s	Total
Wellesley College			2	5	2	4	3	1	17
Goucher College					1	7	2		10
Hunter College						1	5	3	9
Mount Holyoke College					2	2	3	1	8
Vassar College	1		1		2		3		7
Smith College			1	3	3				7
University of Wisconsin			1	1	1			3	6
University of Kansas			1		1	2	2		6
Brown University (Women's College)						2	4		6
Bryn Mawr College				1		1	3		5
Leading institutions	1		6	10	12	19	25	8	81
Other US schools			1	9	21	25	62	17	135
Five Canadian schools				3			2	2	7
Three European schools*				2		1	2		5
Totals	1		7	24	33	45	91	27	228

*In three instances European degrees were not bachelor's degrees. Two of these are degrees from the University of Edinburgh, where the first degree given at that time was a master's degree. The third is an *Abitur* from the Realgymnasium der Schillerschule in Frankfort am Main. This is not an undergraduate degree, but represents work equivalent to high school plus about two years of college work in the United States.

It should be noted that these faculty women were expected to be unmarried. If they did marry, they were expected to resign.[6]

Wellesley enrolled more women than any other school in the country in the late 1880s and early 1890s. By the early 1910s, Wellesley was still the third largest educator of women in the United States, surpassed only by the University of Chicago and Smith College. The mathematics department, although initially small, grew rapidly in the early years and had at least six members from the late 1890s until the Depression years of the 1930s.[7] With only two exceptions for part-time faculty, the department was staffed entirely by women before World War II.[8] In addition, many Wellesley faculty members participated actively in the affairs of the larger mathematical community and, because of their contacts, were able to initiate opportunities for their most promising students to attend graduate school. For example, Helen Abbot Merrill and Clara Eliza Smith, members of the Wellesley mathematics

[6]A discussion of the issue of the marital status of women on the faculty of women's colleges through the early 1900s appears in Rossiter, *Women Scientists in America: Struggles and Strategies to 1940*, 15–16.

[7]Much of the information about the Wellesley department comes from a scrapbook compiled by Helen A. Merrill: "Scrapbook of the History of the Department of Mathematics," Mathematics Department, Wellesley College Archives (hereafter cited as Merrill Scrapbook).

[8]James M. Peirce of Harvard lectured on Quaternions 1887–88. A. Harry Wheeler, a high school teacher in Worcester, Massachusetts, gave courses on constructing models of surfaces for two years, 1926–28.

department since 1893 and 1908, respectively, were active at the national level and both became vice presidents of the MAA. In March 1922, Evelyn Wiggin (Casner), a 1921 Wellesley graduate, received a letter from R. G. D. Richardson, professor and head of the department at Brown and secretary of the AMS. Richardson, at the suggestion of "[his] good friends, Misses Merrill and Smith at Wellesley," invited Wiggin to apply for a position as a graduate assistant in the department at Brown. He wrote in April that he is glad she is coming and that she can stay with them [the Richardsons] until she finds lodging.[9]

Although Wellesley was the leading granter of baccalaureate degrees to our women, the roots for its influence emerged much earlier. Oberlin College, founded in 1833 as Oberlin Collegiate Institute, is generally considered to have been the first coeducational college to confer the baccalaureate degree on women. While two female graduates of Oberlin, Josephine Alberta Robinson (Roe) (1894) and Mary Emily Sinclair (1900) earned PhD's in mathematics early in the twentieth century, Oberlin's principal impact in the nineteenth century on the mathematical education of women was through two of its graduates, Helen A. Shafer (1863) and Ellen A. Hayes (1878). Although Oberlin awarded Shafer an honorary MA, neither she nor Hayes ever earned a degree beyond the baccalaureate. Nonetheless, both became mathematics faculty members at Wellesley and were particularly influential in building the extremely successful mathematics program there.

Figure 3.1: Helen Shafer's senior mathematics class at Wellesley College in 1886. Seated on Shafer's right is Helen A. Merrill and on her left is Ellen Fitz Pendleton. (Shafer was president of Wellesley 1887–94 and Pendleton was president 1911–36.) Photograph courtesy of Wellesley College Archives, photo by Seaver.

[9]R. G. D. Richardson to Evelyn Wiggin, March 28, 1922, Correspondence 1921–1925, R. G. D. Richardson Papers, Brown University Archives. Hereafter letters in this collection will be cited as "Richardson Correspondence."

Shafer and another member of the mathematics faculty, Ellen Fitz Pendleton (Wellesley BA 1886), served as Wellesley's third and sixth presidents. Together they presided over Wellesley for half of the college's first six decades. Shafer had been head of the department from 1877 until 1888 and was followed in that position by Ellen Hayes. During her tenure as head of the department, Hayes significantly increased the number of applied subjects taught by the mathematics department and often used instructors in the physics department to teach them. Hayes remained head of the department until 1897 when she became head and sole member of the applied mathematics department. The pure mathematics department at Wellesley, headed by Ellen Burrell (Wellesley BA 1880), was large, with between five and eight teaching faculty from the time of the split in 1897 until the departments were reunited upon Hayes's retirement in 1916. At that time, Helen Merrill became department chairman and remained so for the next sixteen years. The department remained large during most of her chairmanship, but by the late 1930s, presumably because of the Depression, its size had been cut in half.

Just over half of the seventeen Wellesley graduates who later earned PhD's in mathematics came from New England, including five from Massachusetts. The rest were from Illinois, New Jersey, New York, and the District of Columbia. The majority had attended public high schools and most fit the stereotype of students of the eastern women's colleges, described in 1923 as "a preponderance of the fine middle-class type, daughters of professional and business men, with a few representing working class families."[10] However, while most fit this stereotype, three of the seventeen Wellesley graduates were daughters of farmers. The mean age at graduation was about twenty-two, with only three being slightly older. After graduating from Wellesley, most did not go directly to graduate school; ten of them taught, most frequently in high schools, before beginning their graduate work. Indeed, the median time of eleven years between the bachelor's degree and the PhD is significantly higher for the Wellesley graduates than for the entire group of PhD's, for which the median is eight years.

The first of the Wellesley graduates, Winifred Edgerton (Merrill), class of 1883, was the first woman in this country to be granted a PhD in mathematics—in 1886 from Columbia. Helen Merrill reports that the Wellesley department was "greatly impressed and pleased to learn of the results of an effort on the part of some Columbia men to make things hard for this undesired student. They asked their professor to use the hardest possible text in their course in Celestial Mechanics. The book chosen was Watson's, which Miss Hayes' class, including Winifred Edgerton, had used at Wellesley."[11] Edgerton was twenty-three, among the youngest of all the women in our study, when she received her doctorate. Even so, before entering Columbia she had done independent work in mathematical astronomy, having calculated the orbit of a comet using data from the Harvard College Observatory.

A dozen of the Wellesley graduates earned master's degrees before their doctorates, and the seventeen doctorates were awarded from nine different institutions, with Yale and Radcliffe each granting three and the others one or two each. Wellesley also became a major employer of its graduates, with a dozen teaching there at some point, six of these in their first postdoctoral positions and four spending the

---

[10] John Palmer Gavit, *What Are Women's Colleges Doing?* (New York: New York Evening Post, 1923), 64.

[11] Merrill Scrapbook, 15.

bulk of their careers at Wellesley. Eleven of the Wellesley graduates eventually had continuous or nearly continuous careers; nearly half married.

Woman's College of Baltimore (Goucher College since 1910) graduated its first class of five students in 1892. Among these five students were two mathematics majors.[12] Although the college was roughly a third the size of Wellesley and rarely had a mathematics faculty of more than two, Goucher awarded ten undergraduate degrees to women in our study, more than any other institution except Wellesley. The first professor of mathematics at Woman's College of Baltimore, W. C. L. Gorton, had received his PhD in mathematics from Johns Hopkins, as did his successor, William H. Maltbie.

Two women with Johns Hopkins PhD's were particularly influential at Goucher: Clara L. Bacon was on the faculty from 1897 until 1934; Florence P. Lewis from 1908 until 1947. Bacon, originally from Illinois, had graduated from Wellesley in 1890 and taught at four schools in the Midwest before moving to Woman's College of Baltimore. She had studied at the University of Chicago for six summers and acquired a master's degree in 1904. Lewis had received her bachelor's and master's degrees from the University of Texas, where she had studied philosophy and mathematics. After receiving the master's degree in philosophy, Lewis studied philosophy and mathematics at a number of schools in the United States and Europe. She earned a second master's degree, this one in mathematics, from Radcliffe in 1906.

Figure 3.2: Florence P. Lewis and Clara L. Bacon. Photographs courtesy of Goucher College Archives.

In 1907 the trustees of the Johns Hopkins University voted to open graduate courses to women. In September of that year, Bacon, who had been teaching in Baltimore for ten years, and Lewis, who was at the time a tutor at the University of Texas, applied for admission to the graduate program in mathematics at Johns Hopkins. After Lewis had completed one year of full-time study at Johns Hopkins, Bacon hired her as an instructor at the Woman's College. In 1911 Bacon became the first woman to be granted a PhD in mathematics from Johns Hopkins. Lewis received her PhD, the second granted to a woman in mathematics by Johns Hopkins, in 1913. Both were students of the English algebraic geometer Frank Morley.

[12]For a history of the mathematics department at Goucher see Cockey, *Mathematics at Goucher: 1888–1979*.

Of the dozen women in our study who were born in Maryland, nine, all from Baltimore, attended Goucher. Except for Bessie Irving Miller (BA 1907), the first of the Goucher undergraduates, and Beatrice Aitchison (BA 1928), the last, the remaining Goucher students were taught primarily by Bacon and Lewis, who, with rare exceptions, comprised the entire mathematics department faculty from 1909 until the mid-1920s. Both Bacon and Lewis were revered by their students. Lewis wrote Bacon's obituary for the *Goucher Alumnae Quarterly* and in it quoted several students. One said, "It was her support that made graduate study possible for me. Her patience and understanding as a teacher opened up the beauty of mathematics. For many years her faith in all of us made life seem good."[13] A similarly touching passage, written by Helen Dodson Prince, a former student who became a well-known astronomer, appears in Lewis's obituary. "Dr. Florence P. Lewis came to the Goucher faculty in 1908. Thirty-nine years later, in 1947, she became Professor Emeritus of Mathematics and Astronomy. During these many years the mathematics majors rejoiced in knowing Dr. Lewis as a teacher. In 1953, though few in number, this group contributed most of the funds for the Lewis telescope, and promised more if needed. Love and admiration for Dr. Lewis did not diminish with passing time."[14]

Generally the Goucher graduates fit the previously quoted 1923 description of women who attended eastern women's colleges; their fathers were mainly merchants but also included a physician, a lawyer, and a leather hider. Six of the Goucher graduates earned master's degrees before their doctorates, and the ten doctorates were awarded by just four different institutions. Seven of the ten women in our study who graduated from Goucher received their PhD's at Johns Hopkins; five, all but the first and the last, wrote their dissertations under the direction of Frank Morley, who had been the dissertation advisor for both Bacon and Lewis. Although nine of the ten women had continuous or nearly continuous teaching careers, none taught at Goucher. These nine all remained single.

Hunter College of the City of New York was the first tuition-free and publicly-funded college for women in the United States and had only commuters as students. It was the third largest undergraduate producer of women in our study. Hunter began offering classes in 1870 as Female Normal and High School, within months became the Normal College of the City of New York, later expanded its program to four years, and offered its first bachelor's degree in 1892. It changed its name to Hunter College in 1914. By 1916 it was the largest women's college in the country and the third largest educator of women, exceeded only by the University of Chicago and the University of California in Berkeley. Furthermore, the mathematics department was large, with over twenty members starting in the late 1920s, and mathematics was reported to be one of the most popular majors.[15] Hunter became coeducational in 1951.

It was in 1916 that Elizabeth Karl (Sister Mary Cordia Karl) became the first woman in our study to graduate from Hunter. When Karl graduated, there were seven regular members of the mathematics department; all were women and none had a doctorate at that time. It was not until 1923, after three more women in our

---

[13]Florence P. Lewis, "Clara Latimer Bacon," *Goucher Alumnae Quarterly* (Spring 1948), 20.

[14]Helen Dodson Prince, "Florence P. Lewis," *Goucher Alumnae Quarterly* (Spring 1964), 18.

[15]Mina Rees, class of 1923, reported the popularity of the major during her undergraduate years in an interview with Rosamond Dana and Peter J. Hilton. "Mina Rees" in *Mathematical People* (Boston: Birkhäuser, 1985), eds. Donald J. Albers and G. L. Alexanderson, 258.

study had graduated, that Hunter hired a faculty member with a doctorate, Tomlinson Fort (Harvard 1912). At about the same time, Lao G. Simons, an assistant professor, received her PhD in mathematics education from Columbia University's Teachers College.[16] Fort was there for only three years, but the Hunter department continued to grow after Simons became department chair. By 1935, when the last woman in our study to get a Hunter bachelor's degree graduated, there were sixteen regular faculty, half of whom were women with doctorates.

The demographics of the Hunter undergraduates differed significantly from those who attended the other women's colleges in the East. Six of the nine were daughters of immigrants, mainly from Eastern Europe, and a seventh had a mother born in England. Hunter did not charge tuition and had no dormitories; about half of the women in our study who graduated from Hunter were daughters of working-class parents. The nine Hunter graduates obtained their doctorates from six different institutions, with Bryn Mawr College granting three, New York University two, and four other schools one each. Six of the Hunter graduates married, one entered religious life, and two remained single.

Mount Holyoke College, Smith College, and Vassar College were the next leading producers of undergraduates in our group. Mount Holyoke opened in 1837 as Mount Holyoke Female Seminary but did not confer its first bachelor's degree until 1889. The change of name to Mount Holyoke College in 1893 also signaled a change of emphasis, offering a more comprehensive curriculum rather than a modified seminary curriculum.[17] Between 1902 and 1931 it granted eight degrees to women in our study, four of whom earned PhD's from the University of Chicago. One of these, Mildred Leonora Sanderson, earned a national reputation for her dissertation, but she died the year after receiving her doctorate. Two other women who received PhD's from Chicago were among the most recognized women mathematicians of the early twentieth century and served sequentially on the Mount Holyoke faculty. Anna Pell (Wheeler) went to Mount Holyoke in 1911 and left in 1918 for Bryn Mawr. She was replaced by Olive C. Hazlett, who had taught at Bryn Mawr until 1918 and remained at Mount Holyoke until she accepted a position at the University of Illinois in 1925.

Clara Eliza Smith was an early graduate of Mount Holyoke Seminary. While there she took plane and solid geometry and plane and spherical trigonometry as requirements for her graduation in 1885, before bachelor's degrees were offered. She also studied higher algebra, analytic geometry, and calculus on her own before entering the Yale graduate school in 1901 to study mathematics. The following year Smith was awarded a bachelor's degree from Mount Holyoke College after passing an exam on one year's additional work in French and on the presentation of a certificate from Yale for courses in mathematics; in 1904 she received her doctorate from Yale. Smith later became one of the core faculty members at Wellesley. Three other Mount Holyoke graduates also had continuous full-time employment after receiving their PhD's. All were at women's colleges: Mary Evelyn Wells and Abba V. Newton were on the faculty at Vassar, 1915–48 and 1944–73, respectively; R. Lucile Anderson joined the faculty at Hunter in 1930 and remained until she retired about forty years later.

---

[16]Since her degree was in mathematics education, Simons is not in our study.

[17]For a discussion of the curriculum in the early women's colleges see Robinson, *The Curriculum of the Woman's College*.

Smith College offered its first instruction in 1875. Although initially somewhat smaller than Wellesley, Smith became, by the turn of the century, the largest women's college in the country and remained so until the mid-teens, when it was surpassed in size by Hunter. However, Smith, unlike the other leading women's colleges, produced all seven of its undergraduates in our study in its early history with none graduating after 1908.[18] During the school's first quarter century, the mathematics department consisted of one and occasionally two faculty members. Eleanor P. Cushing, a Smith alumna from the class of 1879, was usually the sole teacher of mathematics from 1881 until the early 1890s. From the mid-1890s until 1909, Cushing and Harriet R. Cobb, Smith class of 1889, were the only senior members of the department. Until the early 1930s the mathematics department consisted almost entirely of Smith alumnae. Ruth Goulding Wood, class of 1898, joined the department in 1902, the year after receiving her PhD from Yale, but remained an instructor until 1909. Suzan Rose Benedict, class of 1895, joined Cushing, Cobb, and Wood in 1906 as an assistant in mathematics, was shortly promoted to instructor, and then received her PhD in 1914 from the University of Michigan. Susan M. Rambo, class of 1905, became the fourth long-term member of the faculty in 1908 and received her doctorate in 1920, also from Michigan. For brief periods Pauline Sperry (BA 1906) and Ida Barney (BA 1908) also taught at Smith. Thus, five of the seven Smith graduates in our study later taught at Smith, with three of them, Wood, Benedict, and Rambo, spending their careers there. In a history of the college, written in part by the long-time president William Allan Neilson, it was noted, "When Miss Cobb became Emeritus in 1931 the department of Mathematics had been for fifty years one of the most conservative in the college. It had been dominated, even after she was no longer there, by the spirit of Eleanor Cushing."[19] Four of the Smith alumnae obtained their doctorates at Yale, two at Michigan, and one at Chicago. All in this group remained single.

Vassar College in Poughkeepsie, New York, also produced seven graduates who later earned PhD's. These bachelor's degrees were granted over the lengthy period 1869 through 1928 and include the first and second undergraduate degrees given to women in our study, to Christine Ladd (Ladd-Franklin) and Charlotte Barnum. All but one of the Vassar graduates were from New York State. Columbia, Radcliffe, and Yale granted two doctorates each to Vassar graduates, and Johns Hopkins finally granted the remaining one to Ladd-Franklin. Three of the Vassar graduates married and four remained single. Although Vassar did not grant PhD's in mathematics, it did grant at least sixteen master's degrees between 1896 and 1928, all but two to women who did not receive doctorates. The two who did, Elizabeth Cowley and Eugenie Morenus, both received BA's from Vassar and PhD's from Columbia. During much of this period Vassar's faculty included Henry Seely White, who was professor 1905–36, and who served as president of the AMS 1907–08.

---

[18]Euphemia Lofton (Haynes) graduated from Smith in 1914 and became, in 1943, the first African-American woman to receive a PhD in mathematics. It was nearly thirty years before Smith next graduated a woman who later received a PhD in mathematics.

[19]Harry Norman Gardiner and William Allan Neilson, *Smith College: The First Seventy Years*, manuscript, ca. 1946, Neilson personal papers, Smith College Archives. Also available at http://clio.fivecolleges.edu/smith/pres-neilson.

Bryn Mawr was the only women's college that awarded doctorates in mathematics.[20] Although it was the third largest producer of doctorates in mathematics to women before 1940, awarding nineteen between 1894 and 1938, it granted only five bachelor's degrees to women in our study. One of these undergraduate degrees was awarded in the nineteenth century; the other four were awarded between 1915 and 1928. Of the five undergraduates only the first, Emilie Martin (1894), received her doctorate from Bryn Mawr.

Together the seven women's colleges described above granted about 28 percent of the undergraduate degrees earned by the women in our study. Of these seven, five are among the colleges dubbed the "Seven Sisters."[21] The remaining two of the "Seven Sisters," Radcliffe and Barnard, each graduated only three women who earned PhD's in mathematics before 1940. These two colleges, unlike the other five, were founded as coordinate colleges, linked with Harvard and Columbia, respectively. It is particularly surprising that Radcliffe does not figure more prominently as an undergraduate school, since a report published in 1918 noted that "Radcliffe offers more courses in mathematics than in any other subject of its curriculum" even though "mathematics is not required for a degree."[22] Perhaps it is noteworthy that at both Barnard and Radcliffe the undergraduate classes were taught by the male faculty of the associated men's universities.

Eighteen women in the study received bachelor's degrees from thirteen different Catholic institutions. Of these, eight received degrees from seven different Catholic colleges for women. Fifteen of the eighteen Catholic sisters received their bachelor's degrees from Catholic institutions, five of them from women's colleges and six from institutions that were primarily for men.[23] These Catholic women's colleges were small and were distributed over a wide geographic area including California, Texas, Minnesota, and Illinois, as well as eastern states. A common pattern was the following. In the mid-nineteenth century, a community of sisters was called to the United States from Europe by a local bishop to provide educational and other social services to the diocese. Academies were established, which gradually expanded their offerings to include a two-year college curriculum, frequently with an emphasis on teacher training. These institutions generally operated as both academies and colleges until, in the first decades of the twentieth century, they became full colleges, officially recognized with a state charter and accreditation by the Catholic Education Association or one of the secular regional accrediting associations.[24]

It is generally agreed that the first true Catholic college for women in the United States is the College of Notre Dame of Maryland, in Baltimore. The School Sisters of Notre Dame came to this country from Bavaria in 1847 and opened an academy for girls in Baltimore that year. The academy's curriculum expanded until it was chartered as a college in 1896 and graduated its first class of six in 1899,

[20]Although Radcliffe also awarded doctorates, the women who received these degrees took classes at Harvard and had Harvard faculty as advisors.

[21]The "Seven Sisters" (Barnard, Bryn Mawr, Mount Holyoke, Radcliffe, Smith, Vassar, and Wellesley) organized in 1927 to promote private women's colleges.

[22]Robinson, *The Curriculum of the Woman's College*, 101.

[23]Five of these six women religious studied at Jesuit institutions, while one was at a Benedictine college.

[24]Broader issues of accreditation for colleges and universities are discussed at various points in Gleason, *Contending with Modernity: Catholic Higher Education in the Twentieth Century*.

about thirty years after the earliest secular women's colleges in the country.[25] For more than forty years the principal mathematics professor at the College of Notre Dame of Maryland was Sister Mary Cordia Karl (PhD Johns Hopkins 1931), an early graduate of Hunter College. A student of Sister Mary Cordia was Mary C. Varnhorn, a Baltimore native and 1936 graduate of the College of Notre Dame, who was the only lay woman in our study to earn a doctorate from Catholic University. Varnhorn spent her career as a faculty member at Trinity College in Washington, D.C.

Trinity College was the first of the few Catholic women's colleges that originated as a four-year college without a preexisting academy. Its origins were not unrelated to the history of the Catholic University of America. When, in 1895, Catholic University announced that it would admit lay persons, there was no reference to gender or race. A number of women submitted applications, and the local press even announced that women would be admitted to Catholic University as special students. At the inauguration of the new schools for the laity, however, the rector of the university announced that three "colored men" had been admitted but that it had not been possible to admit women.[26] With pressure mounting on the university to admit women, the administrative leadership urged the local Sisters of Notre Dame de Namur to open a college instead of an academy as they had planned. With the support of some of the more liberal bishops arguing for an expanded role for women, Trinity College began with seventeen students in 1900 and with a faculty of members of the local order supplemented by some faculty from nearby Catholic University.

Despite the existence of a number of Catholic colleges for women, the need to educate Catholic sisters for teaching responsibilities at the elementary, secondary, and collegiate level prompted Catholic University to offer a summer school for teaching sisters and local lay women teachers in 1911. In this first summer, 284 students registered, of whom 255 were sisters from thirty-one states and Quebec. Twenty-nine of these sisters, from eleven different communities, stayed over for the beginning of the Sisters College that October. Since many of the women religious entered with previous college work, eighteen of them received bachelor's degrees in June 1912. Almost immediately the College started granting graduate degrees, with twenty-five master's degrees in 1913 and two PhD's in 1914.[27] Although the Sisters College was physically separate from the University, the college was staffed by professors from Catholic University. Three of the eighteen women religious in our study earned their undergraduate degrees at the Catholic Sisters College. The college was dissolved in 1964.

Catholic Sisters College was one of the colleges for women that were closely associated with a men's institution. These coordinate colleges also played a significant role in the undergraduate education of women in our study. In addition to Catholic Sisters College, those we have identified are Barnard College, Columbia;

[25]Among sources consulted for the discussion on Catholic women's undergraduate education are Bowler, "A History of Catholic Colleges for Women"; Gleason, *Contending with Modernity*; and Oates, "The Development of Catholic Colleges for Women."

[26]Cohen, "Early Efforts to Admit Sisters and Lay Women to The Catholic University of America," 3–5.

[27]These figures are reported in C. Joseph Nuesse, *The Catholic University of America* (Washington, DC: Catholic University of America Press, 1990), 172–73. Slightly different numbers are reported in Cohen, "Early Efforts to Admit Sisters," 7–8.

Woman's College, Duke University; H. Sophie Newcomb College, Tulane University; Radcliffe College, Harvard University; Shepardson College, Denison University; and Women's College, Brown University.[28] These seven coordinate colleges or their associated universities graduated eighteen women, with Brown University the only one that granted more than three bachelor's degrees to women in our study.

Brown University (Women's College) graduated six women who received PhDs in mathematics before 1940. These six all graduated before the Women's College was named Pembroke College in 1928. They attended in pairs, with two graduating in 1916, two in 1923, and the final two in 1925 and 1926. Each of the six remained at Brown for one year to continue her studies towards a master's degree. More than seventy years after receiving her bachelor's and master's degrees, Elizabeth Hirschfelder remembered the encouragement of two particular members of Brown's faculty. When asked how likely it was for women to pursue a career in college teaching in a field like mathematics, she replied: "I think in a good many places it wasn't as good as it was at Brown on account of those two Canadian professors [R. G. D. Richardson and Raymond C. Archibald] who felt very strongly that women should be given a chance."[29] With very few exceptions, undergraduate women and men at Brown did not take classes together. Nevertheless, as one can see from the photograph below, the Mathematics Club was coeducational. Furthermore, a history of the Women's College noted that "there had grown up in the mathematics department (headed by Professor Archibald) a particularly close kind of association between students and Faculty."[30]

About half of the women in our study received their undergraduate degrees from nearly seventy different coeducational institutions in the United States and Canada, most of which graduated only one or two students in our group. The exceptions are the University of Kansas and the University of Wisconsin with six each; the University of California, Indiana University, the University of Missouri, and Oberlin College with four each; and the University of Chicago, the University of Michigan, New York University, and the University of Washington with three each. While seven of the ten coeducational schools that produced the largest number of graduates in our study were state universities in the Midwest and West, more than half of the American coeducational schools on our list of undergraduate institutions were private, and, of those, most graduated only one student in our group. Two of the public coeducational institutions, Delta State Teachers College and Western Kentucky State Normal School and Teachers College, enrolled mainly women; they, too, graduated only one student each in our group.

Private coeducational schools graduated just over 20 percent of the women in the study. The first woman in our study to graduate from such a school was Ida Metcalf, who received her PhB from Boston University in 1886. Metcalf was twenty-nine at the time and had taught in small schools in New Hampshire before being admitted to Boston University as a special student in 1883. Having originally been established as a Methodist seminary, Boston University offered its first instruction at the postsecondary level and awarded its first degree in 1871. After adopting the

[28]Some of these coordinate colleges did not grant degrees. In these cases the women received their degrees from the associated universities.

[29]Elizabeth Hirschfelder, interview by Barry Teicher, 16 October 1995, interview 465, Oral History Project, University of Wisconsin-Madison.

[30]Grace E. Hawk, *Pembroke College in Brown University* (Providence, RI: Brown University Press, 1967), 79.

Figure 3.3: Brown University Mathematics Club, 1917–18. Raymond C. Archibald is seated third from the left and R. G. D. Richardson is seated fourth from the right. C. Raymond Adams, later to marry Rachel Blodgett, is seated at the far right. (None of the women in our study was attending Brown in 1917–18.) Photograph courtesy of Brown University Archives.

New England Female Medical College as its medical school in 1873, the university claims to have been "the first university in the world to open all its departments to women."[31]

In 1889 Annie Louise MacKinnon (Fitch) and Mary Frances Winston (Newson), the first two women in our study who were granted bachelor's degrees by public universities, graduated from the University of Kansas and the University of Wisconsin, respectively. Annie MacKinnon was born in Ontario, Canada, grew up in Kansas, and attended the University of Kansas in Lawrence. The University of Kansas opened in 1866 with about fifty students, roughly half men and half women, although the first degree was not awarded until 1873. After her graduation in 1889, MacKinnon began her graduate work in mathematics at Kansas and became the third graduate student in mathematics in the university's history.[32] MacKinnon also taught in the high school in Lawrence 1890–92 and received her master's degree from Kansas in 1891. After one more year at Kansas, MacKinnon continued her graduate work at Cornell and received her PhD in 1894.

Mary Frances Winston, one of eight children of a physician in a small town in northern Illinois, enrolled at the University of Wisconsin in 1884. The University of Wisconsin in Madison was about eighty miles from Winston's hometown and had offered instruction at the postsecondary level since 1849 and degrees since 1854. The first women were admitted in 1863, three years before the university was designated

---

[31]Boston University, Office of the Provost, "Boston University, Past and Present," http://www.bu.edu/faculty/handbook/pastpresent/history.html.

[32]Price, *History of the Department of Mathematics of The University of Kansas 1866-1970*, 50.

a land-grant college. Winston taught one year in a country school and graduated from Wisconsin in 1889. Although Mary Winston wanted to remain at Wisconsin as a graduate student, she was denied a fellowship. She taught for two years at Downer College in Fox Lake, Wisconsin, before continuing her education at Bryn Mawr College and the University of Chicago.

As mentioned above, Kansas and Wisconsin were not only the first state universities to grant degrees to women in our group, but they were also the leading such universities. Not surprisingly, most of the twelve graduates of these two schools came from families involved with agriculture. Although the size of the school was not insignificant, especially in the case of Wisconsin, other factors appear to have contributed. All of the women who entered Kansas were living in the state at the time of their matriculation. Four of them obtained their master's degrees, and two of these also received their doctorates at Kansas. These two, Florence Black and Wealthy Babcock, spent their teaching careers at Kansas. The families of all but Mary Winston were living in Wisconsin at the time of the matriculation of the women who earned bachelor's degrees at Wisconsin. Four of the six who received bachelor's degrees from Wisconsin also received Wisconsin doctorates. Included among these six were the sisters Louise A. Wolf and Margarete C. Wolf (Hopkins).

Although neither Annie MacKinnon nor Mary Frances Winston, the first graduates of public universities, had been enrolled in normal courses, both taught before continuing their studies for the doctorate in mathematics. In contrast, Ruth Gentry graduated from the Indiana State Normal School (now Indiana State University) in 1880, twenty-eight years before it offered its first baccalaureate degrees. During the 1880s she taught at the academy and college in Deland, Florida. Resuming her education, she received a PhB from the University of Michigan in 1890 before enrolling in the graduate program at Bryn Mawr College, where she received that institution's first PhD in mathematics in 1894. This education and career pattern was not just a phenomenon of the nineteenth century. For example, Adrienne S. Rayl received a teaching diploma from the New Orleans Normal School in 1917 and taught in the public schools of New Orleans while earning her 1924 bachelor's degree in education in the course for teachers at Tulane University. Rayl remained a teacher while earning her master's degree in mathematics from Tulane in 1934 and her doctorate from the University of Chicago in 1939.

Many of the women went to college assuming that they would become teachers. Furthermore, many of the women religious taught before entering college as did some graduates of public institutions in the West and the South. While attending a normal school or teachers college as the only undergraduate experience was extremely uncommon for the women in this study, a small group, only about 5 percent, attended a normal school either before or after receiving a bachelor's degree. Only two women, Josephine Chanler and Marguerite Zeigel (Hedberg), received their undergraduate education exclusively at a teachers college. A dozen women attended normal school before starting a regular bachelor's program, while two others attended after receiving their bachelor's degrees. Most of the women in our study either took education courses as undergraduates or taught at the precollege level at some point before receiving the doctorate. Some who did not teach had other employment before or while completing their undergraduate education.

Table 3.2: Undergraduate institutions

Degree name with name used in the biographical entry, if different, in parentheses[33]

Baker University (KS): Armstrong 1917
Barnard College (NY): Seely 1911; McCormick (Torrance) 1931; Greenebaum (Epstein) 1935
Bates College (ME): O'Brien 1922
Baylor University (TX): McCoy 1925
Boston University (MA): Metcalf 1886; Simond 1927
Brown University (*Women's College) (RI): Stark, Torrey 1916; Stafford (Hirschfelder), E. C. Stokes 1923; Whelan (Sedgewick) 1925; Kenny (Landers) 1926
Bryn Mawr College (PA): Martin 1894; Darkow 1915; Litzinger 1920; Pillsbury (Baxter) 1927; Peters 1928
Catholic University (§Catholic Sisters College) (DC): Kelley 1914; Gough 1920; Fowler 1927
§Central Holiness University (IA): Boyce† 1918
College of Notre Dame of Maryland: Varnhorn 1936
§College of Saint Teresa (MN): Kloyda 1920; Schulte 1923
Colorado College: Whiton (McDonald) 1909
***Columbian University (DC):** McKelden (Dimick) 1899
Concordia College (MN): M. R. Anderson 1920; Rasmusen 1921
Cornell College (IA): Schmeiser (Barnes) 1926
Cornell University (NY): Van Benschoten 1894; Lester 1924
Dalhousie University (Nova Scotia, Canada): Baxter (A. B. Hill) 1891; Grant 1925
***Delta State Teachers College (MS):** Zeigel (Hedberg) 1928
Denison University (*Shepardson College) (OH): Hunt 1909
DePaul University (IL): Vaudreuil 1921
***Dominican College of San Rafael (CA):** Mayer 1927
Duke University (*Woman's College) (NC): Griffin (Reavis) 1928
Emmanuel College (MA): Burke 1926
Fordham University (NY): Morrison 1922; Garvin 1927
Georgetown College (KY): LeStourgeon 1909
***Georgia State College for Women:** Nelson 1926
Goucher College (MD): Miller 1907 (then Woman's College of Baltimore); Cohen, Sutton 1912; Barton 1913; Mullikin 1915; Mears 1917; Whelan 1918; Lehr 1919; Waters (Dean) 1924; Aitchison 1928
Grinnell College (IA): Harshbarger 1923
Guilford College (NC): Ragsdale 1892
***H. Sophie Newcomb College (LA):** Gibbens 1914
§Hedding College (IL): Bacon† 1886
***Heidelberg University (OH):** Bareis 1897
Hunter College (NY): Karl 1916; Guggenbuhl, Kramer 1922; Rees 1923; Griffin 1925; Tuller 1929; Becker (Mazur) 1930; Levin (Early) 1932; Manel (Greenfield) 1935
§Huron College (SD): Mauch 1919
***Indiana Normal School (PA):** Cowley† 1893
Indiana University: Hennel 1907; McCain 1908; Haseman 1910; Price 1926
***Iowa State College of Agriculture and Mechanic Arts:** Farnum 1909
***Kansas State Agricultural College:** Arnoldy 1929
Knox College (IL): Calkins 1916; Johnson (Yeaton) 1920
Loyola University of Chicago (IL): Galvin 1924
McPherson College (KS): Jacobs (Offermann) 1914
Miami University (OH): Leffler (Alden) 1930
Monmouth College (IL): McFarland 1916

[33]For the three individuals who earned two bachelor's degrees before undertaking graduate work, the first school is indicated by a † and the second by a ‡. Two other women, Madeleine Levin Early and Isabel Maddison, received bachelor's degrees after receiving their PhD's; these post-PhD degrees are not included below. The schools that are now closed are preceded by a §; those that have changed their names or have merged with other institutions are preceded by a *.

Mount Holyoke College (MA): Smith 1902; Wells 1904; Sanderson 1910; Alderton (Moore) 1914; R. L. Anderson 1922; Huke (Frink) 1924; Newton 1929; Modesitt (Reklis) 1931
New York University: Schlauch (Infeld) 1928; Fox (Delevie) 1931; Blanch 1932
Northwestern University (IL): Wishard (McMillan) 1935
Oberlin College (OH): Robinson (Roe) 1894; Sinclair 1900; Spencer 1928; Kohlmetz 1931
Ohio State University: Shover (Quinn) 1926
Ohio University: Bennett (Grennan) 1903
***Our Lady of the Lake College (TX):** M. L. Hill 1922
***Oxford College for Women (OH):** Taylor 1921
Pacific University (OR): Noble 1929
***Penn College (IA):** Boyce‡ 1921
***Radcliffe College (MA):** Hazlett 1912; Cooper 1913; M. M. Sullivan 1929
***Randolph-Macon Woman's College (VA):** Larew 1903; Chandler (Pixley) 1926
Realgymnasium der Schillerschule, (Frankfurt am Main, Germany): Bechtolsheim (*Abitur*) 1922
Reed College (OR): Stahl (Brady) 1924; Fry 1933
***Rice Institute (TX):** Hickey (Maria) 1926
Rockford College (IL): Rees (Bonner) 1934
***Saint Xavier College for Women (IL):** Haberzetle (M. H. Turner) 1934
***Shorter Female College (GA):** Hightower 1896
§Shurtleff College (IL): Terry (Nee) 1926
Smith College (MA): Peirce 1886; Dickerman 1894; Benedict 1895; Wood 1898; Rambo 1905; Sperry 1906; Barney 1908
***St. Benedict's College (KS):** M. H. Sullivan 1930
***St. Xavier College (OH):** Thuener 1920; Reilly 1923
Stanford University (CA): Weiss 1925; Manning (Little) 1933
Swarthmore College (PA): Gurney 1930; Stauffer (McKee) 1931
Syracuse University (NY): Bower 1925
Transylvania University (KY): Dale 1914
***Trinity College (DC):** Mangold 1910
Tulane University (LA): Rayl 1924
University of Arkansas: Hughes (Bushey) 1915
University of British Columbia (Canada): Humphreys 1932
***University of Buffalo (NY):** Montague 1927
University of California: Biddle (A. B. Andrews) 1908; Sznyter (Sagal) 1915; McFarland (Buck) 1917
***University of California, Southern Branch:** Beenken 1923
University of Chicago (IL): Collier 1911; Logsdon 1912; Killen (Huston) 1926
University of Cincinnati (OH): Eversull (B. E. Allen) 1921; Odoms (A. O. Gray) 1931
University of Colorado: Kendall 1912; Johnson (Rosenbaum) 1928
University of Edinburgh (Scotland): Pairman (Brown) (MA) 1917; M. C. Gray (MA) 1922
University of Georgia: Stith (Ketchum) 1924
University of Illinois: Burns (Glasgow) 1909
University of Iowa: Cronin 1903; Baker 1923
University of Kansas: MacKinnon (Fitch) 1889; Brewster (Owens) 1900; Black 1913; Babcock 1919; Hagen 1920; Shea 1927
University of Kentucky: Pence 1914
University of London (England): Maddison 1893; Coddington (E. C. Williams) 1896
University of Maine: Copeland 1904
University of Michigan: Gentry 1890; White 1893; Liu (Hsia) 1926
University of Minnesota: Carlson 1917; Wilder (Thornton) 1927
University of Missouri: Walker (Hull) 1903; Weeks (King) 1908; Wyant 1921; Anderson (Haynes) 1922
***University of Mount Allison College (New Brunswick, Canada):** Colpitts 1899
University of Nebraska: Rummons (Ballantine) 1916
University of Pittsburgh (PA): Taylor (Speer) 1926; Kanarik 1930
University of Rochester (NY): Heckel (Beaty) 1928
University of Saskatchewan (Canada): Hughes 1925; Ames (Widder) 1931

University of South Dakota: Johnson (Wheeler) 1903; Olson 1915
University of Texas: Lewis 1897; Horton (Porter) 1908
University of Toronto (Ontario, Canada): Cummings 1895
University of Washington: Pepper 1920; Griffiths 1921; Haller 1924
University of Wisconsin: Winston (Newson) 1889; Pengra (Crathorne) 1897; F. E. Allen 1900; L. Wolf 1931; M. Wolf (Hopkins) 1932; Bernstein 1934
Vassar College (NY): Ladd (Ladd-Franklin) 1869; Barnum 1881; Cowley‡ 1901; Morenus 1904; Thorndike (Cope) 1922; N. Cole 1924; Murray (Hopper) 1928
Wellesley College (MA): Edgerton (W. E. Merrill) 1883; H. A. Merrill 1886; G. Andrews, Bacon,‡ Bosworth (Focke) 1890; Vivian 1894; Young 1898; Worthington 1904; Curtis (Graustein) 1906; Anderton, Schoonmaker (Wilson) 1911; Blodgett (Adams), Weeks 1916; Wiggin (Casner), Russell 1921; Mason (Ballard) 1926; Hathaway (M. H. Williams) 1935
Wells College (NY): Howe 1908; Carroll (Rusk) 1920; Moody 1926
West Virginia University: Buchanan (M. B. Cole) 1906; B. M. Turner 1915
***Western College for Women (OH):** Jackson 1904; Tappan 1909; Stafford (Henriques) 1926
***Western Kentucky State Normal School and Teachers College:** Chanler 1927
Wilson College (PA): Johnson 1925
***Winthrop Normal and Industrial College of South Carolina:** R. W. Stokes 1911
***Woman's College of Baltimore (MD):** see Goucher College

CHAPTER 4

Graduate Education

The 228 women in this study received PhD's from thirty-four US and three European universities. These schools generally had strong programs, and the dissertation advisors were generally highly regarded in the mathematics community. The University of Chicago granted a fifth of the doctorates awarded to women in mathematics before 1940, more than twice as many as the next leading institution, Cornell University.

As noted in the introduction, Christine Ladd-Franklin was, in 1882, the first American woman to earn a doctorate in mathematics, although she did not receive the degree until 1926. After her graduation from Vassar College in 1869, Ladd taught for three years in girls' schools in Pennsylvania. While at her first position she wrote her family that she was engaged in "special investigations in Physics, and in mathematical research."[1] While at the second position she wrote that she planned to study with Professor Peirce in Cambridge [Massachusetts] or with Professor Chauvenet in St. Louis.[2] Instead, she studied with George C. Vose, a professor of mathematics at the nearby Washington and Jefferson College. Ladd described her teacher as "among the first ten or twelve mathematicians in the country [who] leaves his family to poverty and dirt, while he absorbs himself in Quarternions [*sic*]."[3]

In the early 1870s Christine Ladd began to contribute problems and solutions to the *Educational Times* of London, a "'monthly stamped journal of education, science and literature' to deal with general educational questions at school and university level" first published in 1847 by the College of Preceptors in London; she continued these contributions into the 1890s.[4] In the fall of 1872 Ladd was in Boston, where, according to letters to her family, she studied mathematics with W. E. Byerly and J. M. Peirce. After attending classes at Harvard, Ladd remained in the Boston area and began teaching again. In 1875 she published what was probably her first paper in an American journal, *The Analyst*, primarily a recreational journal that had started in Iowa a year earlier. By the spring of 1875 Christine Ladd was teaching in Union Springs, New York, a small town north of Ithaca, where the coeducational Cornell University was located, and where she studied biology in 1876. In March 1878, Christine Ladd wrote from Union Springs to J. J. Sylvester asking him whether Johns Hopkins would refuse her permission to listen to his

---

[1] Christine Ladd to her aunt, September 7, 1870, box 2, Christine Ladd Franklin and Fabian Franklin Papers, Rare Book and Manuscript Library, Columbia University.

[2] It is not clear whether Ladd in her diary entry of September 28, 1871 (1866–73 Diary, 18a, Christine Ladd-Franklin Diaries, Special Collections, Vassar College Libraries), is referring to James Mills Peirce, whose lectures she did eventually attend, or his father, Benjamin Peirce, both of whom were on the Harvard faculty in 1871.

[3] Entry in 1866–73 Diary, 19b, November 14, 1871, Christine Ladd-Franklin Diaries.

[4] Ivor Grattan-Guinness, "A Note on the *Educational Times* and *Mathematical Questions*," *Historia Mathematica* 19 (1992): 76.

lectures "on account of [her] sex?"[5] Sylvester strongly supported Ladd's application. He wrote Daniel Coit Gilman, the president of Johns Hopkins, that he "should rejoice to have her as a fellow worker among us."[6] The university permitted Ladd to attend classes on an unofficial and exceptional basis.[7]

While at Johns Hopkins she contributed three papers to the *American Journal of Mathematics* and wrote a dissertation that appeared in a volume containing works by her advisor, C. S. Peirce, and his students.[8] Her work was accepted by the department and her dissertation was highly praised by Peirce. Nonetheless, university trustees refused to award the degree because she was a woman. It was not until 1926 that Ladd-Franklin, at the age of seventy-eight, was finally awarded the PhD in mathematics she had earned forty-four years earlier.

Officially, therefore, the first PhD actually awarded to an American woman in mathematics was granted to Winifred Edgerton (Merrill) in 1886 by Columbia College.[9] Edgerton, an 1883 graduate of Wellesley College, began her graduate study at Columbia in 1884. The previous year Columbia's board of trustees had defeated President Barnard's plan for coeducation. Instead a Collegiate Course for Women was instituted that did not permit women to attend classes but did allow them to earn a certificate after a four-year course of examinations. When Edgerton applied for admission she "[did] not ask to be admitted to the College as a student, nor to join a class in the College, but to have access to the Observatory and its instruments, in order that she [might] learn the work of a practical astronomer."[10]

The Reverend Doctor Morgan Dix, a trustee of Columbia College who had been an opponent of coeducation in 1883, wrote in his diary describing the board of trustees meeting of June 7, 1886, "We recommended that the degree of B.A. in the Course for Women be conferred on all who shall pursue with success a course equivalent to that which secures the degree in the School of Arts.... I also moved that Winifred Edgerton have the degree of Ph.D. cum laude, which was unanimously agreed to. She fully deserves it."[11] Although the resolution that approved Edgerton's PhD refers to a course in practical astronomy and pure mathematics, Edgerton's dissertation was in mathematics.[12] The degree she received was the first degree of

[5]Christine Ladd to Professor Sylvester, March 27, 1878, MS 1, Daniel Coit Gilman Papers, Special Collections, Sheridan Libraries, The Johns Hopkins University.

[6]J. J. Sylvester to Daniel Coit Gilman, April 2, 1878. Reprinted in Karen Hunger Parshall, *James Joseph Sylvester: Life and Work in Letters* (Oxford: Clarendon Press, 1998), 188–89.

[7]A report of Ladd's acceptance to Johns Hopkins in the New York newspaper *The World* concludes: "She is, indeed, a very clever person—'for a girl'; and it is suggested by a rash and jealous mathematician of the stronger sex that 'she is a Ladd after all!'" (from *The Vassar Miscellany* 28 (October 1879): 202). See Parshall and Rowe, *The Emergence of the American Mathematical Research Community, 1876–1900*, 83–86, for a comprehensive discussion of Ladd's admission to and attendance at Johns Hopkins.

[8]"The Algebra of Logic," in *Studies in Logic by Members of the Johns Hopkins University*, ed. Charles S. Peirce (Boston: Little, Brown, 1883), 17–71.

[9]Columbia College together with its affiliates took the name Columbia University in 1896.

[10]"Excerpt from Trustees Minutes, February 4, 1894 [*sic*]," from 1933 speech by Dean Virginia C. Gildersleeve, Historical Biographical Files: Mrs. Winifred Edgerton [Merrill], box 215, folder 10. University Archives, Columbia University in the City of New York (hereafter cited as Winifred Edgerton Merrill File).

[11]"June 7 – Monday," from the diary of Rev. Morgan Dix, rector of Trinity Parish, Winifred Edgerton Merrill File.

[12]The title of her dissertation was "Multiple Integrals: a) Their Geometrical Interpretation in Cartesian Geometry, in Trilinears and Triplanars, in Tangentials, Quaternions, and in Modern Geometry; b) Their Analytical Interpretation in the Theory of Equations using Determinants,

any kind given to a woman by Columbia. Furthermore, prior to 1886 Columbia had granted only one other PhD in mathematics.

Edgerton's degree was among the first dozen doctorates granted to women in any field by universities in the United States. It followed by nine years the first, which was awarded to Helen Magill (White) in classics by Boston University in 1877. It appears that Edgerton's PhD in mathematics was the first recognized modern doctorate in the natural sciences or mathematics granted to a woman in the United States. By the turn of the century, women scientists were well represented among the new PhD's. Doctorates were first awarded in botany by Syracuse University and in zoology by the University of Michigan in 1888, in geology by the Johns Hopkins University in 1893, in astronomy by Yale University and in chemistry by the University of Pennsylvania and Yale in 1894, in physics by Cornell University in 1895, and in physiology by the University of Chicago in 1896. In all, about 230 doctorates were awarded to women in the nineteenth century in the United States. These include about 100 in the humanities, about 70 in the social sciences, and about 50 in the natural sciences and mathematics.[13]

By the early 1890s, several coeducational schools had graduate programs in mathematics, and a few traditionally male institutions admitted women at the graduate level only. Yale University is usually credited with being the first American institution to grant a PhD in mathematics in 1862 and was, with Columbia College and the University of Pennsylvania, one of the first of the men's schools to allow women to enroll in graduate courses early in the 1890s. Cornell had a flourishing mathematics program, and class rolls for the early 1890s show the presence of a significant number of women in both the advanced undergraduate and graduate courses. In particular, the records for the academic years 1892–93 and 1893–94 make frequent mention of five women who later earned PhD's in mathematics, including three women who earned doctoral degrees from Cornell between 1893 and 1895.[14] Bryn Mawr College was the only women's college to offer a PhD program in mathematics and granted its first doctorate in that field in 1894. Radcliffe College, the coordinate college for women associated with Harvard University, awarded doctorates to women later, but the classes were taught by Harvard faculty.

The belief that European education was important was typical of the American mathematical community during the late nineteenth and early twentieth centuries. The acknowledged center of mathematical research was Germany, particularly the universities in Göttingen and Berlin, and large numbers of American mathematicians either received German doctorates or studied in Germany before or after receiving American degrees. The participation of women in this movement was

---

Invariants, and Covariants as Instruments in the Investigation." She listed a second dissertation on her 1937 Owens questionnaire: "(a) Computation of the Orbit, the Comet of '83, Data Furnished by Harvard University; (b) Determination of Latitude and Longitude of New York City from Direct Observations." In a 1944 interview Merrill mentioned having practically had two dissertations, one in mathematics and one in astronomy. Roger Howson, "Interview with Winifred Edgerton Merrill, October 4, 1944," Winifred Edgerton Merrill File.

[13]Eells, "Earned Doctorates for Women in the Nineteenth Century," is the major source for this information, although we have made changes where we know Eells's material to be incomplete or in error.

[14]Ida M. Metcalf (PhD 1893), Anna L. Van Benschoten (BS 1894, PhD 1908), Annie MacKinnon (Fitch) (PhD 1894), and Agnes Baxter (Hill) (PhD 1895) earned their degrees at Cornell. Leona May Peirce (PhD 1899) earned her degree at Yale.

delayed owing to institutional bars to the admission of women to German universities. Their participation was facilitated in part by the introduction of a European fellowship granted annually by the Association of Collegiate Alumnae (ACA), the predecessor of the American Association of University Women. In an 1888 report of an ACA committee calling for the introduction of such a fellowship, the committee's chair, Christine Ladd-Franklin, wrote that

> there are plenty of women who are well educated; there are very few who are engaged in making additions to the world's stock of knowledge. We believe that this is not owing to any natural incapacity on their part. We believe that it is owing partly to their not having felt that it was expected of them, partly to the fact that the professors in many of their colleges are not themselves investigators, and hence cannot lead their students to become such, and partly to poverty, which compels them to do something for their own support immediately after leaving college.[15]

Ruth Gentry, who had studied at Bryn Mawr in 1890–91 and was awarded the ACA European fellowship for the following year, went to Berlin during the summer of 1891. From Berlin she wrote to "various prominent Professors of Mathematics elsewhere than in Berlin" asking for permission to attend their lectures; the only positive response she received was from Leipzig, "where the work in Mathematics was not exactly suited to [her] purpose."[16] When the university in Berlin opened for the new term, she spoke with Lazarus Fuchs who supported her request to attend lectures there. Although the rector of the university also gave his approval, and she attended lectures of Fuchs and Ludwig Schlesinger, Gentry was soon informed of an 1884 ruling by a former *Kultusminister* (minister of education and cultural affairs) strictly prohibiting exceptions to the exclusion of women. Gentry was permitted to hear lectures only through the end of the semester and left Germany but remained in Europe through the first semester of 1892–93, when she studied at the Sorbonne.

Also in 1891, Christine Ladd-Franklin accompanied her husband, the mathematician Fabian Franklin, on his sabbatical trip to Germany. At that time, Ladd-Franklin was also denied permission to enroll as a student at Göttingen. On May 15, 1892, Felix Klein wrote to Ladd-Franklin that

> not only our faculty but also our central office, the "executive administrative board," has expressed itself favorably for the women's movement. Indeed for the time being we mainly support only what will not satisfy you, the admission of female guest auditors (under control of the prorector [vice chancellor] and with the assumed assent of each individual docent). However, you must only consider this as a beginning; further development of the matter is not prejudiced [in the legal sense] thereby. Concerning this further development one will indeed judge best when we have had some experiences with the female guest auditors; besides, this further development will not be the matter of the

[15] *Report of Committee on Endowment of Fellowship*, 1. Association of Collegiate Alumnae, AAUW publications, Series II, no. 7, 1888.

[16] Ruth Gentry, "A Winter in Berlin," *The Lantern* (June 1892): 45–46.

> individual university but will depend on general considerations which partly lie outside our competence [legal authority].[17]

In the summer of 1893 Klein came to the United States to represent the German mathematical community at the International Mathematical Congress held in connection with the World's Columbian Exposition in Chicago and to look for students who might enroll at Göttingen as "female guest auditors." Klein was already aware of Mary Winston (Newson), a student at the University of Chicago during its inaugural year 1892–93, since Heinrich Maschke had written him in April of that year about her abilities and desire to study in Germany.[18] After having met with her, Klein agreed to sponsor her admission to the university. Later, writing about her student days, Newson reported that "it had been a dream of mine to go to Europe to study and when the announcement was made that the 'Association of Collegiate Alumnae' was offering a scholarship for a year's study in Europe, I made up my mind to apply."[19]

Figure 4.1: Das Pentagon, Göttingen: Felix Klein's students about 1893. Standing (l to r): Charles Jaccottet, Paul Heegard, and Gino Fano. Seated (l to r): Mary Winston (Newson) and Grace Chisholm (Young). Photograph courtesy of Sylvia Wiegand.

---

[17]Box 4, Christine Ladd Franklin and Fabian Franklin Papers, Rare Book and Manuscript Library, Columbia University. Translated from the German by Uta C. Merzbach.

[18]For excerpts from this letter see Parshall and Rowe, 241.

[19]Mary Newson, "My Student Days in Germany," 1. Paper read to a women's group at the Unitarian Universalist Church in St. Petersburg, Florida, ca. 1952. A typescript of this paper, letters from Winston to her family, 1893–95, transcribed by her daughter Caroline N. Beshers, and a biographical foreword to the letters were given to one of the authors by Beshers. The originals are in the Mary Frances Winston Papers, Sophia Smith Collection, Smith College (hereafter cited as Winston Papers).

Although Winston was not awarded the ACA European fellowship at that time, she was sent $500 by Christine Ladd-Franklin "toward her initial expenses."[20] As a consequence of Ladd-Franklin's generosity and Klein's support, in the fall of 1893 Winston sailed to Europe in order to attempt to enroll at the university in Göttingen. Also arriving in Göttingen that October were Grace Chisholm (Young), an Englishwoman who had attended Girton College at Cambridge and who wished to study mathematics, and Margaret Maltby, an American who had been a graduate student at the Massachusetts Institute of Technology and who wanted to study physics. It was not at first clear that they would be admitted to study at the university. In fact, five days after her arrival Mary Winston reported that her landlady was "always expressing her curiosity to know whether I can get in or not, but she always ends 'Aber ich glaube nicht.'"[21] However, by the end of October all three had received permission to attend lectures and all three eventually received PhD's from Göttingen, with Chisholm and Maltby becoming in 1895 the first two women to be granted doctorates from a university administered by the Prussian government.[22] Although Maltby's degree was in physics, Klein later wrote, "Mathematics had here rendered a pioneering service to the other disciplines. With it matters are, indeed, straightforward. In mathematics, deception as to whether real understanding is present or not, is least possible."[23]

Winston was the first of fourteen women in our group who studied mathematics at Göttingen in the years before World War I.[24] The only other one to receive a Göttingen PhD was Anne Lucy Bosworth (Focke), a student of David Hilbert, who received her degree in 1900. Most of these women did not hold fellowships to study at Göttingen, although four of the women, including three graduate students, arrived in Göttingen with ACA European fellowships. Although several women attended other European institutions during this period, some on ACA fellowships, the Göttingen contingent was by far the most significant, outnumbering all the others together. The only other European doctorates received by women in our study were granted after the end of World War I to Lulu Hofmann (Bechtolsheim)

[20]Caroline N. Beshers, "Foreword," 2, Winston Papers. Biographical foreword to letters of Mary Winston by her daughter.

[21]Mary Winston to Gene [Eugenie Winston], October 15, 1893 (p. 12 of transcription by Caroline N. Beshers), Winston Papers.

[22]A short description of the events leading to the granting of permission to the three women to attend lectures appears in Green and LaDuke, "Women in the American Mathematical Community: The Pre-1940 Ph.D.'s," 15. That description is taken from letters Mary Winston wrote to her family during that time. A more complete description appears in Tobies, "Zum Beginn des mathematischen Frauenstudiums in Preussen."

[23]Conveyed to the authors and translated by Uta C. Merzbach from Klein Nachlass XXII L (Personalia) 1913, Niederschsische Staats- und Universitätsbibliothek Göttingen, Georg-August-Universität Göttingen.

[24]They are Elizabeth Cowley, Charlotte Pengra Crathorne, Annie L. MacKinnon (Fitch), Anne Bosworth (Focke), Isabel Maddison, Emilie N. Martin, Helen A. Merrill, Eugenie Morenus, Mary Winston (Newson), Virginia Ragsdale, Clara E. Smith, Anna Van Benschoten, Anna Pell (Wheeler), and Ruth G. Wood. Although Ladd-Franklin was in Göttingen in 1891–92, she was not studying mathematics, but rather was working in the laboratory of the experimental psychologist G. E. Müller. For a general discussion of women studying in Germany during this period see Singer, *Adventures Abroad: North American Women at German-Speaking Universities, 1868–1915.* Later, in 1932–33, Margaret Gurney studied at Göttingen before completing her doctorate at Brown.

(Zürich 1927) and Myrtie Collier (Strasbourg 1930). As was the case for all American mathematicians, the incidence of women mathematicians studying abroad was much lower after World War I than before.

Although many of the early fellowships were designed for women who wanted to study abroad, it was also recognized that women who wanted to study in the United States might need financial assistance. In 1897 Christine Ladd-Franklin helped organize the Baltimore Association for the Promotion of the University Education of Women, an organization that provided fellowships to two of the women in our study. Because of this fellowship, in 1901 Virginia Ragsdale was able to return to her studies at Bryn Mawr after having taught three years in a private school. Likewise, in 1910–11 Clara L. Bacon was supported in her studies at Johns Hopkins by such a fellowship. In addition, both Bryn Mawr and Wellesley provided fellowships for their graduates, and fellowships specifically earmarked for women were available at the University of Pennsylvania.

As we noted above, thirty-four US and three European universities granted PhD's to women in our study. The University of Chicago leads by far, having granted forty-six doctorates to women in mathematics before 1940. The leading seven institutions granted more than 60 percent of all the degrees. Two tables serve to illustrate the importance of these institutions and of particular advisors. Table 4.1 lists all institutions that granted PhD's to women, while table 4.2 lists all advisors who directed the dissertations of more than five women.

Many of the institutions near the top of table 4.1 also rank high for overall PhD production in mathematics before 1940.[25] In particular, the University of Chicago leads by wide margins for both women and total number of degrees. It conferred 266 PhD's in mathematics through 1939, more than twice as many as any other school. However, two of the top four institutions granting degrees to women were important only for women. Bryn Mawr College admitted only women, and Catholic University, whose graduate program was coeducational, granted 70 percent of its pre-1940 PhD's in mathematics to women. On the other hand, Princeton University, which did not admit women into its graduate programs until the 1960s, ranks among the top ten schools granting mathematics PhD's before 1940.

Harvard University, which ranks second among all schools granting doctorates in mathematics before 1940, did not grant degrees to women until 1963. Harvard first allowed women to take graduate courses as early as 1894, but the degrees earned were officially awarded by Radcliffe College, even though the female students attended classes at Harvard and had Harvard faculty members as advisors. Radcliffe awarded its first doctorate in 1902 and its first in mathematics in 1917, to Mary Curtis (Graustein). Even though women could attend classes at Harvard, they could not be members of the Harvard Mathematical Club, which met in the common room of Conant Hall, a men's dormitory. "And no woman was permitted to enter Conant

[25]Table I, Number of PhD Degrees in Mathematics Conferred by American Universities, in Richardson, "The PhD Degree and Mathematical Research," 203, lists degrees by school for the period prior to 1934. We know that the list is not completely accurate since degrees earned by several women in our study are not included. Nonetheless, we will make some broad comparisons using Richardson's table augmented for the period 1935–39 from figures in table 8, Doctorate Production in the Ten Fields by 40 Leading Schools, by Five-Year Periods, in National Research Council, *Doctorate Production in United States Universities 1920–1962*, 20–26, and for other schools from the Dissertation Abstracts database.

Table 4.1: Doctoral institutions by decade, 1886–1939

	80s	90s	00s	10s	20s	30s	Total
University of Chicago			1	8	13	24	46
Cornell University		3	1	3	6	8	21
Bryn Mawr College		2	2	2	5	8	19
Catholic University				1	1	12	14
Yale University		3	5	1	1	3	13
Johns Hopkins University				5	5*	2	12
University of Illinois				3	1	8	12
Radcliffe College				1	2	6	9
Columbia University	1†		3	2	1	1	8
University of Wisconsin			2			5	7
University of California				2	2	3	7
University of Michigan				1	1	5	7
University of Pennsylvania			2	1	1	1	5
Ohio State University			1			4	5
University of Missouri				1	3	1	5
Brown University					1	4	5
Indiana University				2		1	3
Georg-August-Universität Göttingen		1	1				2
University of Kansas					2		2
University of Cincinnati					1	1	2
University of Minnesota					1	1	2
Stanford University					1	1	2
University of Colorado						2	2
Duke University						2	2
Massachusetts Institute of Technology						2	2
New York University						2	2
University of Pittsburgh						2	2
Syracuse University				1			1
University of Texas				1			1
University of Iowa					1		1
Rice Institute					1		1
Universität Zürich					1		1
Marquette University						1	1
Purdue University						1	1
Saint Louis University						1	1
Université de Strasbourg						1	1
University of Washington						1	1
Total degrees	1	9	18	35	51*	114	228
Total institutions	1	4	9	16	21	30	37

*Christine Ladd-Franklin's PhD, earned in 1882, was granted in 1926.
†Winifred Edgerton's 1886 PhD was awarded by Columbia College.

Hall."[26] In about 1930 the Harvard mathematics colloquium was organized and, since it met in a classroom building, women could attend.

Although Clark University did not grant degrees to women in our study and granted fewer than thirty to men before 1940, it granted about ten mathematics doctorates in the nineteenth century. This is comparable to Harvard and behind only Johns Hopkins, with about thirty, and Yale with about twenty. Three of these nineteenth-century Yale PhD's were earned by women, but none of the other three institutions admitted women to degree programs during that period. Leona May Peirce, who received a Yale PhD in 1899, was actually advised by William E. Story, a faculty member at Clark. The degree was granted by Yale, because Clark did not drop its formal barriers to granting doctorates to women until 1900, and, in fact, did not confer its first doctorate (in physiology) on a woman until 1908. Furthermore, in the late 1910s Clark stopped awarding doctorates in mathematics altogether and did not resume for another fifty years. Johns Hopkins did not formally accept women as graduate students until 1907, and it was not until 1911 that it granted PhD's to women without requiring special approval from the trustees. Clara L. Bacon was one of the women who received a PhD from Johns Hopkins in 1911, and during the next fifteen years mathematics was second only to psychology in the number of doctorates granted to women by the school.

Table 4.2: Leading advisors of women PhD's before 1940

Advisor	Women advised before 1940	Total advised before 1940	Women as % of total	Institution	Women PhD's before 1940	% of women PhD's advised
L. E. Dickson	18	67	27%	Chicago	46	39%
Virgil Snyder	14	40	35%	Cornell	21	67%
Aubrey Landry	13	19	68%	Catholic	14	93%
G. A. Bliss	12	52	23%	Chicago	46	26%
Frank Morley	8	49	16%	Johns Hopkins	12	67%
Charlotte A. Scott*	7	7	100%	Bryn Mawr	19	37%
Arthur B. Coble†	1 5	5 20	20% 25%	Johns Hopkins Illinois	12 12	8% 42%
Anna Pell Wheeler	6	6	100%	Bryn Mawr	19	32%
8 advisors	84	265	32%	6 schools	124	68%

*Although Scott served as her official advisor, Louise D. Cummings, a faculty member at Vassar, worked very closely with her colleague Henry Seely White, who had suggested the dissertation topic.

†Arthur B. Coble was on the faculty at the Johns Hopkins University 1904–18 and 1927–28 and at the University of Illinois 1918–27 and 1928–47.

In what follows, we will focus particular attention on the seven schools, Chicago, Cornell, Bryn Mawr, Catholic, Yale, Johns Hopkins, and Illinois, that led in the granting of doctorates to women in our study, and on the leading dissertation

[26] G. Baley Price, "A Lifetime in Mathematics 1925–1994," in *A Century of Mathematical Meetings*, ed. Bettye Anne Case (Providence, RI: American Mathematical Society, 1996), 187.

advisors. These institutions each granted more than ten doctorates to women and were among the earliest to award doctorates to women in our study; relatively few of the other US schools granted any doctorates to women in mathematics before the 1920s. Also, as we can see in table 4.2, the eight leading advisors were, among them, at six of the seven leading schools. They directed two-thirds of all the dissertations written by women at these six schools, indicating significant clustering of the women PhD students around these few advisors.

We start with the University of Chicago, which granted degrees to about 20 percent of the women in our study. Chicago opened its doors in 1892 as a coeducational institution with a commitment to research and with women in its graduate programs. Mary Winston (Newson), who had studied at Bryn Mawr the previous year, was a fellow by courtesy at Chicago during 1892–93. It also ran a summer quarter that brought in outside faculty and allowed graduate students with full-time jobs an opportunity to study while earning a living. Unlike Johns Hopkins, Clark, and Harvard, Chicago never placed any institutional barriers on the conferring of PhD's to women, and by 1900 the university had granted about thirty PhD's to women in a variety of fields. However, the first PhD in mathematics granted by Chicago to a woman was not conferred until 1908 when Mary Emily Sinclair earned a degree under the direction of Oskar Bolza. By the time Sinclair was awarded her PhD, the Chicago mathematics department had already granted twenty-nine PhD's to men.

Figure 4.2: Mathematics library in Eckhart Hall at the University of Chicago. Photograph courtesy of the University of Chicago Department of Mathematics.

From the beginning, the Chicago mathematics department sought, and quickly achieved, a place among the leading American research departments. The rapid rise of the Chicago department can be attributed primarily to the leadership of the first chair, E. H. Moore, and secondarily to the presence of the German mathematicians Oskar Bolza and Heinrich Maschke, whom Moore hired as original members of the department.[27] Bolza left Chicago in 1910, two years after Sinclair received her PhD. That year E. H. Moore became the advisor of the second woman to receive a Chicago doctorate, Anna J. Pell (Wheeler). Although he had only one other woman student, Mary E. Wells who received her degree in 1915, Moore's first doctoral student, L. E. Dickson (PhD 1896), served as advisor to eighteen women in our study, more than any other advisor. Pell Wheeler and two of Dickson's students, Mildred Lenora Sanderson and Olive C. Hazlett, are probably the three women in our study whose research in pure mathematics was most recognized by the American mathematical community. Another of Dickson's students, Mayme I. Logsdon, remained at Chicago as a member of the faculty and was the dissertation advisor for four students, one of whom was Anna Stafford (Henriques). Logsdon was the only female member of the department at a rank of at least associate professor until 1983 when Karen Uhlenbeck joined the University of Chicago faculty as a professor.

Figure 4.3: Leonard Eugene Dickson with an unidentified student. Photograph courtesy of the University of Chicago Department of Mathematics.

Dickson advised about one quarter of all the Chicago doctoral students before 1940. In a 1997 article Della Dumbaugh Fenster makes it clear that even though Dickson advised more students than any other faculty member before 1940, his role was not that of what we now call a mentor. Rather, she describes his interactions with his students as "almost impersonal."[28] One of Dickson's last students, Ivan Niven (PhD 1938), described him as "a very gruff person at that point in his life."[29]

[27]Moore's career and the early history of the department at Chicago are chronicled in Parshall, "Eliakim Hastings Moore and the Founding of a Mathematical Community in America, 1892–1902." See also Parshall and Rowe, 363–426.

[28]Fenster, "Role Modeling in Mathematics: The Case of Leonard Eugene Dickson (1874–1954)," 20.

[29]Donald J. Albers, G. L. Alexanderson, and Ivan Niven, "A Conversation with Ivan Niven," *The College Mathematics Journal* 22 (1991): 377.

On the other hand, a Chicago graduate student of the late 1920s reported that he was "good to his students, he kept his promises to them and backed them up."[30] We know of no evidence that Dickson treated his female students any differently from his male students.

Gilbert Ames Bliss (PhD Chicago 1900) joined the department following the death of Maschke in 1908 and served as advisor to twelve women between 1910 and 1939. Unlike Dickson, whose students often worked almost entirely on their own, Bliss scheduled time to see each of his doctoral students. Also, unlike Dickson, Bliss acted more as a mentor making sure, for example, that the graduate students could follow the talks of visiting lecturers.[31]

Figure 4.4: Described by Julia Bower in her Smithsonian questionnaire as "Snapshot of Abba Newton, Frances Baker, and me taken at Chicago in 1933." (Bower was a 1933 student of G. A. Bliss, Newton was a 1934 student of L. E. Dickson, and Baker was a 1933 student of E. P. Lane.) Photograph courtesy of Early Women Doctorates Collection, Mathematics, National Museum of American History, Smithsonian Institution.

Together Bliss and Dickson advised about two-thirds of the women who received PhD's at Chicago before 1940 but only advised about 40 percent of the men. During the first four decades of the twentieth century, Chicago produced about 20 percent of US PhD's in mathematics granted to women while granting about 17 percent of all mathematics doctorates during that period. Thus, the large number granted to women and men reflects not only a department held in high esteem and an institutional structure (most notably the summer quarter) that was conducive to graduate work for a wide range of students, but also an attitude that was welcoming to women and men alike. About 20 percent of all the women who earned doctorates in mathematics before 1940 took advantage of the summer quarter at Chicago.

[30]Duren, "Graduate Student at Chicago in the Twenties," 244.

[31]Bliss is discussed in Duren, "Graduate Student at Chicago," and in Albers, Alexanderson, and Niven, "A Conversation with Ivan Niven."

Cornell University was the second leading producer of women PhD's in the United States before 1940, with twenty-one women having received PhD's there between 1893 and 1939. Like Chicago, Cornell was institutionally receptive to women students. Cornell's first president, Andrew Dickson White, married Helen Magill, the first woman to receive a doctorate in the United States. Unlike Chicago, it did not have an extensive summer program of graduate courses. Although this situation changed in the 1920s when graduate courses were consistently offered during the summer, none of the women with Cornell doctorates took most of their coursework during summers. On the other hand, starting in 1885 there was an incentive for mathematics students to come to Cornell. At that time Cornell instituted the Erastus Brooks fellowship designated for graduate students in mathematics. Between 1890 and 1939, six women, five of whom received Cornell PhD's, held that fellowship. The sixth was Anna Helene Palmié, who held the fellowship 1890–91 and continued her graduate studies in mathematics at Chicago and Göttingen.

The first three women to get PhD's in mathematics at Cornell, Ida M. Metcalf (1893), Annie L. MacKinnon (Fitch) (1894), and Agnes S. Baxter (Hill) (1895), were students of James E. Oliver. Oliver had studied with Benjamin Peirce at Harvard and received a bachelor's degree in 1849 before working in the newly established Nautical Almanac Office in Cambridge for seventeen years. After studying in New York, at the School of Mines of Columbia College, and in Philadelphia, and after lecturing on thermodynamics at Harvard for a semester, Oliver accepted a position of assistant professor at Cornell and arrived there in 1871, three years after the school opened. In 1874 he became senior professor and chair of the department. He thought of himself as a teacher as well as a researcher and spent a year, 1889–90, "with the main purpose of seeing for himself the modes of teaching mathematics followed at the European universities."[32] During that year he formed a close association with Felix Klein of Göttingen that eventually resulted in a visit by Klein to Cornell after the 1893 Chicago Congress. This association was also a reason that a number of Cornell students went to Göttingen to study. For example, Annie MacKinnon, who had met Klein in Ithaca, went to Göttingen immediately after receiving her doctorate in 1894 and studied there for two years with fellowships.

It was an obvious loss to the department when Oliver died in the spring of 1895. Lucien A. Wait, who had done much of the administration of the department during Oliver's tenure as chair, took over the leadership of the department. The new departmental view was that a doctorate should not be granted without a significant period of study, preferably abroad, and then only in exceptional circumstances. These views are expressed by Wait in an 1897 letter concerning the final examination of Leona May Peirce.[33]

Wait had come directly to Cornell as an assistant professor after receiving his undergraduate degree at Harvard in 1870. Although he agreed with Oliver concerning the importance of study in Europe, apparently he disagreed with Oliver concerning Cornell's awarding of doctorates. In his 1897 letter about Peirce's examination Wait wrote:

[32]George William Hill, "James Edward Oliver," *Biographical Memoirs, National Academy of Sciences* 4 (1896): 69.

[33]Unsigned letter to Professor Hewett, June 24, 1897, file 12/5/656, box 18, Graduate School Records, Division of Rare and Manuscript Collections, Cornell University Library.

> In our graduate students, we are breeding our own future mathematical instructors and we desire to bring back all the very best European influences based on a thorough foundation laid at Cornell. To show you more fully what I mean, we gave Miss McKinnon [*sic*] the degree of PhD at the end of two years work. She then went to Göttingen and studied two years with Klein and at the end of that time she was not more than up to the German degree. Mr. Snyder after two years of graduate work with us could have complied with the faculty requirements for the degree here. He studied two years and a third in Germany before securing his doctorate.[34]

Neither Oliver nor Wait held a PhD, but their differing attitudes regarding the doctorate are shown in Wait's further remark that "the mathematical department of this University does not desire at the present time to give the degree of PhD except in rare cases."[35] Thus, it is not surprising that under Wait's leadership the production of PhD's in the department took a temporary halt, not resuming until 1901. Although eleven men received PhD's between 1901 and 1907, it wasn't until 1908 that the next woman, Anna Van Benschoten, was awarded the PhD under the direction of Virgil Snyder.

Perhaps the most significant cause of the resumption of the production of doctorates at Cornell was the presence of Snyder in the department. Snyder studied at Cornell 1890–92 and then in Göttingen, where he earned his PhD as a student of Klein. He returned to Cornell in 1895 and remained there for the next forty-three years. He officially directed thirteen of the twenty-one Cornell dissertations written by women and unofficially directed a fourteenth that was awarded in 1939, the year after he retired. There is no doubt that Snyder, who was responsible for somewhat fewer than half of the total number of PhD's conferred by the department during between 1900 and 1939, had a disproportionately large number of female thesis students. In fact, Snyder was one of the leading advisors of women PhD's in the country before 1940. Like Dickson and Bliss at Chicago, Snyder was a well-respected researcher who served as president of the AMS. Most of his students, both men and women, published their dissertations in the highly regarded *American Journal of Mathematics*, and, for students getting degrees about the same time, the dissertations were on similar topics of algebraic geometry.

Algebraic geometry was also a topic studied by many of the nineteen women who earned PhD's in mathematics at Bryn Mawr College. Although Bryn Mawr had a relatively small program, it was, nevertheless, one of the top five producers of women PhD's in mathematics in the period we are discussing. Furthermore, the graduate program offered fellowships that attracted other students as well, a dozen of whom received their PhD's at other schools. Most of the women who received PhD's at Bryn Mawr were students of Charlotte Angas Scott or Anna Pell Wheeler, both of whom had strong connections to the larger mathematical community.

At Bryn Mawr's founding in 1885, M. Carey Thomas, then dean of the college and later president, hired Charlotte Angas Scott to head her mathematics department. Scott was one of the few women in the world at the time with a doctorate in mathematics. She had been a student at Girton but received her doctorate in

[34]Ibid.
[35]Ibid.

1885 from the University of London by external examination, since Cambridge University did not grant advanced degrees to women until 1948. She was an algebraic geometer who worked in the area of higher singularities of algebraic curves. She was starred in the first edition of *American Men of Science*, a later list showing her to have been ranked thirteenth among all American mathematicians. She was the only woman vice president of the American Mathematical Society until 1976 when Mary W. Gray became the second; the first woman to be elected president was Julia Robinson, who served 1983–84.

Deeply involved in the education of all the graduate students at Bryn Mawr, Scott directed seven of the early PhD's. In 1896, two years after the first Bryn Mawr mathematics doctorate was granted, Scott started the Bryn Mawr College Mathematics Journal Club, which provided a forum for young graduate students, recent Bryn Mawr PhD's, and local faculty to report on their own and others' research. She headed the Bryn Mawr mathematics department from its inception until she retired in 1924. She remained in Bryn Mawr the year following her retirement in order to supervise her final student's doctoral dissertation.

Figure 4.5: Charlotte A. Scott and Anna Pell Wheeler, Bryn Mawr College faculty members 1885–1924 and 1918–1948, respectively. Photograph of Scott courtesy of Bryn Mawr College; photograph of Wheeler courtesy of Grace Shover Quinn.

Anna J. Pell, who had received a Chicago PhD in 1910, went to Bryn Mawr in 1918 as an associate professor and became head of the department in 1924 when Scott retired. The following year Pell married Arthur Leslie Wheeler, a classics professor at Bryn Mawr who was then joining the faculty at Princeton. She directed eight PhD's at Bryn Mawr, six of which were granted to women before 1940. Anna Pell Wheeler was an analyst, and it was this area in which her students worked. She was the first woman to give an AMS invited address (1923) and the first woman to give the AMS Colloquium lectures (1927). Wheeler was also responsible for bringing Emmy Noether to Bryn Mawr. Noether arrived in 1933, and the following year Bryn

Mawr provided postdoctoral support for three women to work with her. Two of the three were women in our study; the third was Olga Taussky (Taussky-Todd), a European who earned a doctorate from the University of Vienna in 1930. Noether died in the spring of 1935, just before her only American student, Ruth Stauffer (McKee), received her PhD.

The fourth largest producer of women PhD's in the United States before 1940 was the Catholic University of America. In 1917 Catholic granted its first two doctorates in mathematics, one of which was to Sister Mary Gervase (Kelley); three more were awarded before 1929, all to men. Between 1929 and 1939 thirteen women, twelve of whom were Catholic nuns, received PhD's in mathematics from Catholic.

After 1911, when the Catholic Sisters College opened with staff from Catholic University, the university became an attractive site for graduate work for women religious. Since the university emphasized graduate work and was a pontifical university, associated with no particular order, it appealed to the broad range of religious communities that desired graduate education for their teaching members. There were few alternatives for graduate study for women religious in a specifically Catholic residential setting, so the existence of the Sisters College as a residential college, providing living and study quarters, satisfied the desires of many communities to send their members to a congenial setting. Furthermore, Catholic University was the only Catholic institution with a membership in the prestigious Association of American Universities. Four women religious in our study received graduate degrees while residents of the college.[36]

All but two of the fourteen Catholic University doctorates in mathematics in the 1930s went to women. This proportion reflected wider opportunities for graduate study for men, as well as conditions at Catholic that were attractive to women religious. In biology, there were more women PhD's than men (about two-thirds were women), but this was not the case in the other natural sciences. Of about forty-five doctorates in the natural sciences in the 1930s at Catholic, about half went to women.

The key figure in mathematics at Catholic was Aubrey Landry, who directed the mathematics dissertations of all the women religious there until the early 1940s. Landry earned a bachelor's degree from Harvard University in 1900 and was associated with Catholic University most of the rest of his career. He was a teaching fellow at Catholic some of the time he was working towards his doctorate at nearby Johns Hopkins in Baltimore, where he received his PhD in 1907. His dissertation in algebraic geometry was directed by Frank Morley, who was himself one of the leading advisors of women. Landry remained at Catholic University until his retirement in 1952. In the thirty-two years between Landry's first student in 1917 and his last in 1949, his twenty-eight students accounted for all but six of the doctorates in mathematics conferred by Catholic University. Eighteen of his twenty-eight students were women; all but one were Catholic sisters. Landry was both helpful to and supportive of his students and had a ready supply of problems in algebraic geometry. Sister Helen Sullivan, one of his doctoral students, reported at the 1981 Smithsonian meeting that "it was very convenient for me to have a man who was a

[36]Sisters Mary de Lellis Gough, Mary Laetitia Hill, Mary Gervase Kelley, and M. Domitilla Thuener, received master's degrees; Sister Mary Gervase Kelley also received a PhD.

Figure 4.6: Mary C. Varnhorn (Catholic 1939) was the only lay woman to receive a PhD in mathematics at Catholic before 1940. Photograph courtesy of Trinity Washington University.

giant who was willing to take on the doctoral students and [who was] a very helpful, a very helpful, director of theses. You could hardly miss if you worked under Dr. Landry." Typical of the dissertations he supervised were those consisting of detailed descriptions of the triangles simultaneously inscribed and circumscribed to some special class of real plane curves. These specialized problems were outside the mainstream of research in algebraic geometry, but they were well suited to Landry's students, to whom the PhD usually represented a desired credential rather than the beginning of a research career.

Unlike the institutions we have already discussed, Yale University was not a newly established school nor did it admit women as undergraduates when, in 1895, Charlotte C. Barnum was awarded the first Yale mathematics PhD granted to a woman. Although Yale accounts for one-quarter of all mathematics PhD's granted to women before World War I, the number thereafter falls off sharply; only one woman received the PhD in mathematics from Yale in the 1920s, followed by three other doctorates in 1933 and 1934. The end of Yale's dominance as an educator of women mathematicians followed by a few years the last graduation from Smith College of a woman in our study. Of the nine women who were granted PhD's by Yale through 1911, four were graduates of Smith, two of Wellesley, and one each of Mount Holyoke, Vassar, and the University of Missouri.

James Pierpont supervised more than half of the pre-World War I degrees granted by Yale to women in our study.[37] Pierpont, who spent his entire career at Yale, joined the faculty in 1894 after receiving his doctorate from the University

[37]It is difficult to determine dissertation advisors at Yale during this period. More than at most other schools, the name of the advisor on table 4.3 is determined using internal evidence. In particular, internal evidence points to Pierpont as advisor of five, and possibly six, women in our study. In "James Pierpont–In Memoriam," *Bull. Amer. Math. Soc.* 45 (1939): 481–86, Oystein Ore lists Pierpont's students as reported by former Yale students. Ore's list includes five women, omitting Euphemia Worthington but including Leona M. Peirce. In addition, Pierpont was the first signature on the departmental recommendation for Ruth G. Wood (1901).

of Vienna. Like the advisors of most of the women in our study, he was very well regarded in the mathematical community, having had the distinction of being the first AMS Colloquium speaker when that series was inaugurated in 1896. He served on the council of the AMS 1899–1901 and was a vice president in 1905, during the period when he was advising women students at Yale.

Unlike Yale, Johns Hopkins did not officially allow women to enroll in its graduate school in the nineteenth century. Thus, Charlotte C. Barnum, Yale's first woman PhD in mathematics in 1895, attended lectures at Johns Hopkins starting in 1890; however, like Christine Ladd before her, she was not an official student. Barnum left in 1892 to attend Yale where she could be admitted and work towards a degree. Supporting Barnum's desire to be admitted to Johns Hopkins, Simon Newcomb, then on the faculty there and later to serve as the fourth president of the AMS, wrote to the university president supporting the admission of women to graduate courses in mathematics and astronomy:

> As a first trial step I would ask that the professors in the Department of Mathematics and Astronomy be permitted to invite qualified women to their courses during the pleasure of the university authorities. It would, perhaps, be well not to make any public announcement of this at first.
>
> If it is deemed desirable that such students should be regularly enrolled, then since (so far as I know) we have no statute prohibiting the admission of women, it would be done without any formal announcement of a new policy.[38]

Florence Bascom, like Christine Ladd a decade earlier, was admitted to graduate courses at Johns Hopkins as an exception. Unlike Ladd, however, Bascom was awarded a PhD in 1893 in geology. Also in 1893, the Johns Hopkins School of Medicine opened with women in its first class as the result of a condition set by a major woman benefactor to the new medical school.[39] Nonetheless, it was not until 1907 that the board of trustees unanimously approved the admission of women to graduate courses "provided there is no objection on the part of the instructor concerned."[40] In September 1907, Clara L. Bacon, then an associate professor at Woman's College of Baltimore, applied for admission to take a single mathematics course at Johns Hopkins. The following month her sister, Agnes L. Bacon, who later did statistical work at the Johns Hopkins School of Hygiene, applied for admission to Johns Hopkins to study physics. Both were admitted and Clara L. Bacon received her PhD in 1911, as one of the first four women who did so after the official admission of women. Clara Bacon was a student of Frank Morley, who was president of the AMS 1919–20 and a close associate of Charlotte A. Scott of Bryn Mawr. Like Scott, Morley was an English algebraic geometer. He was a Quaker and had come to the United States to teach at Haverford College, a Quaker school near Bryn Mawr. Morley directed one dissertation at Haverford and in 1900 moved to Johns Hopkins, where he directed forty-eight more doctoral students. He was the advisor of eight of

[38]Simon Newcomb to President D. C. Gilman, November 6, 1890, MS 137, Johns Hopkins University Collection, Special Collections, Sheridan Libraries, The Johns Hopkins University.

[39]See Rossiter, *Women Scientists in America: Struggles and Strategies to 1940*, 45–46, for a discussion of the circumstances surrounding the awarding of the degree to Bascom in this exceptional case and the admission of women to the medical school.

[40]*Johns Hopkins University Circular* n.s., 1908, no. 1: 18.

the twelve women who received PhD's in mathematics from Johns Hopkins before 1940. Two of them, Bacon and Florence Lewis, were faculty members at Goucher College, and all but one of Morley's other female advisees were students of Bacon and Lewis at Goucher. Two other Goucher graduates received PhD's from Johns Hopkins but did not study with Morley. Florence Lewis was the only woman with a Johns Hopkins degree who did not receive her bachelor's degree from a women's college.

The University of Illinois is the only public institution among those that granted more than ten degrees to women before 1940. It is also the only institution in that group, other than Bryn Mawr, whose mathematics department hired more than one woman faculty member. Two men, G. A. Miller and Arthur B. Coble, together were responsible for directing the dissertations of eight of the twelve women who were awarded doctorates by Illinois. G. A. Miller, an algebraist, came to Illinois in 1906 and served as vice president of the AMS in 1908 and as president of the MAA in 1921. He directed the dissertations of three women, including that of Elizabeth Bennett (Grennan) in 1910. Bennett's doctorate was only the second doctorate awarded by Illinois, the first having been granted in 1903. Miller's second student was Josephine Burns (Glasgow) in 1913 and his third was Beulah Armstrong in 1921. All of Miller's women students served on the faculty at Illinois at some time after receiving their doctorates. Coble was an algebraic geometer who was president of the AMS 1933–34. He came to Illinois from Johns Hopkins in 1918, having previously directed the dissertation of Bessie I. Miller in 1914. Two of Coble's students, Bessie I. Miller and Josephine Chanler (1933), also were on the faculty at Illinois.

The University of Illinois was one of many institutions that awarded doctorates in mathematics to women virtually from the beginning of its mathematics doctoral programs. At Purdue University the time span between the first PhD in mathematics in 1893 and the second in 1939 was considerably longer than that at Illinois, but the second doctorate went to a woman, Cleota Fry. In addition to these women who were among the first two to receive doctorates in mathematics from their schools, six women were the first, male or female, to be granted PhD's in mathematics from their institutions. Two others shared the honor of being the first to be awarded a doctorate in mathematics by their schools. For the most part, these firsts occurred at public institutions relatively early and at private institutions somewhat later. At public institutions the first PhD's in mathematics were awarded to Grace Bareis (Ohio State 1909), Cora Hennel (Indiana 1912), and Goldie Horton (Porter) (Texas 1916); Elizabeth Carlson was one of the first two awarded at Minnesota in 1924. At private institutions the first PhD's in mathematics were awarded to Rose Whelan (Sedgewick) (Brown 1929), Ruth W. Stokes (Duke 1931), and Joanna Meyer (Marquette 1931); Sister Mary Gervase (Kelley) was one of the first two awarded at Catholic in 1917. Furthermore, the doctorate that Winifred Edgerton (Merrill) received in 1886 was only the second awarded in mathematics by Columbia, the first having been awarded in 1880. Other schools where women were among the first two or three to receive PhD's were the University of Missouri, the University of Cincinnati, and Saint Louis University.

Even when women were not among the first few students to earn doctorates at a school, they were usually not very far behind. About 60 percent of all the US schools that granted doctorates to women by 1939 granted one to a woman

Figure 4.7: University of Illinois mathematics department Christmas party, about 1936. Virginia Modesitt (Reklis) is standing in the back row on the far right, Beulah Armstrong is seated second from the right in the middle row, and Arthur B. Coble is standing fifth from the left. (Other women on the faculty in 1936 were Ruth Mason (Ballard), Josephine Chanler, Olive C. Hazlett, and Echo Pepper.) Photograph courtesy of the University of Illinois at Urbana-Champaign.

within ten years of the first doctorate granted. Thus, at most schools that granted doctorates to women, the graduate education of women in mathematics did not lag far behind the graduate education of men.

Table 4.3: Doctoral institutions with advisors

Degree name with name used in the biographical entry, if different, in parentheses

Brown University: J. D. TAMARKIN: Whelan (Sedgewick) 1929; Gurney 1934; Bernstein, O'Brien, Torrance 1939

Bryn Mawr College: C. A. SCOTT: Gentry 1894; Maddison 1896; Ragsdale 1906; Cummings 1914; Haseman 1917; B. M. Turner 1920; Lehr 1925; J. HARKNESS: Martin 1901; A. P. WHEELER: Buchanan (Cole) 1922; M. C. Gray 1926; Guggenbuhl 1927; R. L. Anderson 1930; Hughes 1934; Ames (Widder) 1938; E. NOETHER: Stauffer (McKee) 1935; W. W. FLEXNER: Levin (Early) 1936; G. A. HEDLUND: Grant, Tuller 1937; H. W. BRINKMAN [OF SWARTHMORE COLLEGE]: Greenebaum (Epstein) 1938

Catholic University: A. E. LANDRY: Kelley 1917; Mangold 1929; Burke, Gough, Morrison, Vaudreuil 1931; Thuener 1932; Arnoldy 1933; M. H. Sullivan 1934; M. L. Hill 1935; Reilly 1936; Fowler, Galvin 1938; E. J. FINAN: Varnhorn 1939

Columbia University: J. H. VAN AMRINGE: Edgerton (W. E. Merrill) 1886 (then Columbia College); ADVISOR UNKNOWN: G. Andrews 1901; Coddington (E. C. Williams) 1905; Cowley 1908; E. KASNER: Seely 1915; Cronin 1917; Morenus 1922; Kramer 1930

Cornell University: J. E. OLIVER: Metcalf 1893; MacKinnon (Fitch) 1894; Baxter (A. B. Hill) 1895; V. SNYDER: Van Benschoten 1908; Owens 1910; Tappan 1914; Howe 1917; Torrey 1924; Farnum 1926; Schoonmaker (Wilson) 1927; Moody 1930; Carroll (Rusk) 1932; Adams (Infeld), Johnson 1933; Blanch, Montague 1935; C. F. CRAIG AND J. I. HUTCHINSON: Colpitts 1924;

W. A. Hurwitz: Dale 1924; Mears 1927; C. F. Roos [at AAAS from 1930] and D. C. Gillespie: Calkins 1932; V. Snyder [retired 1938] and R. J. Walker: Nelson 1939

Duke University: J. M. Thomas: R. W. Stokes 1931; Griffin (Reavis) 1933

Georg-August-Universität Göttingen: F. Klein: Winston (Newson) 1897; D. Hilbert: Bosworth (Focke) 1900

Indiana University: R. D. Carmichael: Hennel 1912; advisor unknown: McCain 1918; K. P. Williams and H. T. Davis: Price 1932

Johns Hopkins University: F. Morley: Bacon 1911; Lewis 1913; Young 1914; Cohen 1918; Sutton 1921; Whelan 1923; Barton 1926; Dean 1929; A. B. Coble: Miller 1914; C. S. Peirce: Ladd-Franklin 1926 (earned 1882); O. Zariski: Karl 1931; G. T. Whyburn: Aitchison 1933

Marquette University: H. P. Pettit: Mayer 1931

Massachusetts Institute of Technology: N. Wiener: Weeks 1930; D. J. Struik: Plass (M. H. Williams) 1939

New York University: R. Courant: Manel (Greenfield) 1939; D. A. Flanders: Griffin 1939

Ohio State University: H. W. Kuhn: Bareis 1909; H. Blumberg: Schmeiser (Barnes) 1931; C. C. MacDuffee: Shover (Quinn) 1931; T. Radó: Leffler (Alden) 1934; Kohlmetz 1937

Purdue University: H. K. Hughes: Fry 1939

Radcliffe College: advisor unknown: Curtis (Graustein) 1917; Blodgett (Adams) 1921; G. D. Birkhoff: Pairman (Brown) 1922; Cope 1932; J. L. Walsh: Russell 1932; O. D. Kellogg: M. M. Sullivan 1932; W. C. Graustein: Peters 1933; M. Morse: N. Cole 1934; L. V. Ahlfors: Wishard (McMillan) 1938

Rice Institute: L. R. Ford: Hickey (Maria) 1929

Saint Louis University: F. Regan: Garvin 1934

Stanford University: W. A. Manning: Weiss 1928; Manning (Little) 1937

Syracuse University: E. D. Roe: Roe 1918

Universität Zürich: E. G. Togliatti: Hofmann (Bechtolsheim) 1927

Université de Strasbourg: advisor unknown: Collier 1930

University of California: D. N. Lehmer: Biddle (A. B. Andrews) 1911; Alderton (Moore) 1921; McDonald 1932; Noble 1935; J. H. McDonald: Sznyter (Sagal) 1918; McFarland (Buck) 1920; Brady 1933

University of Chicago: O. Bolza: Sinclair 1908; E. H. Moore: Pell (Wheeler) 1910; Wells 1915; G. A. Bliss: White 1910; Larew 1916; LeStourgeon 1917; Hughes (Bushey) 1924; Stark 1926; Jackson, Johnson (Yeaton) 1928; Huke (Frink) 1930; Bower 1933; E. C. Stokes 1939; L. E. Dickson: Sanderson 1913; Hazlett 1915; Logsdon 1921; Ballantine 1923; Darkow, Hunt 1924; Pepper 1925; Griffiths 1927; Chandler (Pixley), Rees 1931; Mason (Ballard) 1932; Baker, Litzinger 1934; Humphreys 1935; M. R. Anderson, McFarland 1936; Mauch, Haberzetle (M. H. Turner) 1938; E. J. Wilczynski: Sperry 1916; Gibbens 1920; Kendall 1921; E. P. Lane: Beenken 1928; Hagen 1930; Olson 1932; Newton 1933; Rasmusen 1936; M. I. Logsdon: Stafford (Henriques) 1933; A. A. Albert: Killen (Huston) 1934; Rees (Bonner) 1937; Boyce 1938; G. A. Bliss and W. T. Reid: Wiggin (Casner) 1936; W. Bartky: Rayl 1939; G. A. Bliss and M. R. Hestenes: Landers 1939

University of Cincinnati: C. N. Moore: Eversull (B. E. Allen) 1924; Odoms (A. O. Gray) 1936

University of Colorado: A. J. Kempner: Beaty, Johnson (Rosenbaum) 1939

University of Illinois: G. A. Miller: Bennett (Grennan) 1910; Burns (Glasgow) 1913; Armstrong 1921; A. B. Coble: Jacobs (Offermann) 1919; Cooper, Harshbarger 1930; Taylor 1931; Chanler 1933; R. D. Carmichael: Ketchum 1934; H. R. Brahana: Terry (Nee) 1934; A. Emch: Pence 1937; H. Levy: Modesitt (Reklis) 1937

University of Iowa: E. W. Chittenden: McCoy 1929

University of Kansas: E. B. Stouffer: Babcock, Black 1926

University of Michigan: L. C. Karpinski: Benedict 1914; Schulte 1935; Kloyda 1936; W. B. Ford: Rambo 1920; L. A. Hopkins: Liu (Hsia) 1930; W. L. Ayres: Simond 1938; G. Y. Rainich: Baxter 1939

University of Minnesota: D. Jackson: Carlson 1924; Wilder (Thornton) 1933

University of Missouri: E. R. Hedrick: Weeks (King) 1915; G. E. Wahlin: Hightower 1927; Wyant 1929; L. Ingold: Anderson (Haynes) 1929; Zeigel (Hedberg) 1932

University of Pennsylvania: ADVISOR UNKNOWN: Vivian 1901; McKelden (Dimick) 1905; O. E. GLENN: Copeland 1913; R. L. MOORE [AT THE UNIVERSITY OF TEXAS FROM 1920]: Mullikin 1922; J. A. SHOHAT: Spencer 1936

University of Pittsburgh: M. M. CULVER: Kanarik 1934; F. A. FORAKER: Taylor (Speer) 1935

University of Texas: M. B. PORTER: Horton (Porter) 1916

University of Washington: R. M. WINGER: Haller 1934

University of Wisconsin: L. W. DOWLING: Pengra (Crathorne) 1901; F. E. Allen 1907; M. H. INGRAHAM: Stafford (Hirschfelder) 1930; L. Wolf, M. Wolf (Hopkins), 1935; T. L. BENNETT: Shea 1934; C. C. MACDUFFEE: Lester 1937

Yale University: ADVISOR UNKNOWN: Barnum 1895; Dickerman 1896; Wood 1901; W. E. STORY [OF CLARK UNIVERSITY]: Peirce 1899; J. P. PIERPONT: H. A. Merrill 1903; Smith 1904; Worthington 1908; Walker (Hull) 1909; Barney 1911; J. K. WHITTEMORE: Anderton 1925; O. ORE: Hopper, Becker (Mazur) 1934; Fox (Delevie) 1935

CHAPTER 5

Employment Issues

The first of the women in our study to receive a PhD did so in 1886, but the first instance of employment for this group was nearly 20 years earlier, when Christine Ladd taught in a high school in Utica, New York, in autumn 1867 before completing her undergraduate work at Vassar College in 1869. Employment for some continued into the 1990s, so we are considering an employment span of more than 125 years. While historical forces (especially the Great Depression and World War II) had an impact on these women's employment, social issues were more keenly felt.

Ruth Gentry, from a small town in central Indiana, was among the earliest to begin her career. Born in 1862, she began teaching school at age sixteen and later used her savings to pay for college. She graduated from the normal school in nearby Terre Haute at age eighteen, taught for several years in preparatory schools and college, and then graduated from the University of Michigan in 1890. This early work and schooling was a prelude to her holding fellowships, obtaining a doctorate from Bryn Mawr College, studying in Europe, and teaching at Vassar, all before 1900.

Indeed, it was rarely the case that the women went directly from high school to college to graduate school. About 17 percent worked full time one or more years before even beginning their undergraduate programs at the schools where they earned their bachelor's degrees. Almost all of these taught in elementary or high schools in the period between high school graduation and entrance into college. This was particularly common for the Catholic sisters, two-thirds of whom taught before beginning college work. A few women did clerical or other work, and about a dozen women first received some teacher training at a normal school before starting a regular bachelor's degree program.

The path from elementary school to graduate school taken by Nola Anderson Haynes, born in Missouri in 1897, was not unusual. She described the course of her entrance into the field of mathematics at the 1981 Smithsonian meeting.

> I come from the Middle West and my grandfather had been a pioneer in Missouri ... and was a landowner.... I went to a ... very small country school. My family were all good in mathematics, the men in the family particularly.... My father would buy corn and ship cattle.... He had me do a lot of the adding and figuring, and I learned how many bushels ... in a load and I liked it. I liked, of course, the arithmetic at school. Then I went to a very small town high school that was just a two-year high school when I started. But they added on the junior year and the senior year.... Then ... you took examinations to get certificates and you could go out and teach in the rural schools. So I did that and taught for four years at a rural school. And I

> was ambitious ... so I went to the University of Missouri and got my bachelor's degree. I taught two years in high school then, but I had known a man at the university who at that time was the president of a junior college ... [and] they needed a mathematics teacher there.... Since I had so much mathematics I was given permission to go there and teach in the junior college. Well, I liked it very much and I thought I will do junior college teaching now, but I better go back and get a master's degree.... One day ... the chairman of the department asked, "Miss Anderson, what are you going to do next year?" I said, "I guess go out and get a job in junior college," thinking I could very easily, of course. And he said, "Would you be interested in going on towards a PhD if you got a fellowship?" Well, that was an easy thing; ... I didn't have to think about getting a job, so I accepted it and went on and got my PhD.[1]

A few others taught while studying for their bachelor's degrees, so that nearly 20 percent had held full-time teaching positions before completing their undergraduate work. About half of the women began graduate studies immediately after graduating from college, although in several instances, they took courses during the summer or otherwise on a part-time basis while holding a full-time job. Furthermore, nearly two-thirds worked in the period between beginning and finishing graduate school, nearly all in full-time teaching positions.

Altogether, nearly 80 percent were employed full time at some point before receiving their doctorates.[2] Thus, it appears that because of financial considerations, opportunities available, or interest, most of these women who eventually earned PhD's in mathematics did so after some period of full-time employment. It is not easy to sort out to what extent employment before the doctorate was a means of financing education and to what extent it provided the motivation for obtaining the degree as a credential for college or university teaching. In any case, with few exceptions the jobs available after the doctorate were teaching jobs, usually in institutions that offered at most a bachelor's degree.

Some of the factors that would prove significant for women after receiving their doctorates were experienced even before those degrees were earned. For example, in 1931 Marjorie Heckel (Beaty) interrupted her graduate studies at Brown University to take a position as instructor at the University of South Dakota. Sally Krebs wrote in a 1987 paper about four professors emeriti at the university, that Heckel "came by train from New York and arrived at a university quite different from the way it is today. For one thing, she and ... Harry Lane were the only instructors the mathematical sciences department had. Also, it was a year of drought in South Dakota when [she] arrived so the campus was very dry and dusty with only a few buildings compared to how many there are today."[3] In her second year there she met Donald W. Beaty, a cattle feeder and farmer who was teaching some courses at the university. They married in 1933, and according to Krebs, "her marriage forced [her] to resign her position at the university because during the depression only one member of a household was allowed to hold a job." Krebs reported that

[1] Smithsonian meeting tapes.

[2] These numbers do not include the several who had assistantships or other part-time work.

[3] Sally Krebs, "A Century and a Half of Dedication," April 1987 (typescript).

the university needed another mathematics instructor during the year 1934–35, and Marjorie Beaty volunteered to teach one course for a semester without pay. After studying for two years at the University of Colorado she once again was given a regular faculty position at South Dakota. She received her PhD from Colorado in 1939 and in 1941 interrupted her career for fourteen years to rear two daughters. She returned, however, and eventually retired as professor after twenty more years as a faculty member.

The idea that only one member of a household should hold a job was not new at the time of the Depression. In an unsigned editorial in 1903 critical of a New York Board of Education by-law that a woman teacher who married had to resign, the writer noted, "Experience shows that, consciously or unconsciously, there lies behind all this opposition to married women as teachers some remnant of the spoils notion that a teaching post is a 'place,' and that a married woman ought to be 'supported' by her husband, while the 'place' goes to an unmarried girl dependent upon her own efforts."[4] As was the case with Marjorie Beaty, the Great Depression, marriage, and children were significant factors affecting the employment of many of the women in the study, before, while, or after they earned their doctorates.

About 90 percent of the women in our study found employment in the year following their doctorate.[5] The number of positions per person after the PhD ranged from none to at least ten, with over a quarter of the women holding exactly one job during their entire career. After receiving their doctorates, the 228 women had about 600 different jobs with nearly 350 different employers, mostly academic, but some in government and business. All but two or three were employed full time at some point in their lives. The most notable exception is Emily Coddington Williams, whose attorney father died when she was young and whose mother's family appears to have been very well-to-do. She attended private schools in New York, earned her undergraduate degree from the University of London and her master's degree and doctorate from Columbia University, the latter in 1905. After she received her doctorate, she attended mathematics meetings in New York City and international mathematical congresses in Rome (1908) and in Cambridge, England (1912). However, there is no evidence that she ever held a formal position or published mathematical research. A few years after receiving her PhD, she earned a law degree and was admitted to the New York Bar but never practiced. She married when she was forty-three, was active in the life of New York City and Newport, Rhode Island, published a play, two novels, and a lengthy genealogical sketch, and died in Paris in 1952 at the age of seventy-eight leaving an estate estimated at $12,000,000. Emily Coddington Williams is clearly not representative of the women in our study.

In general, the experiences of those who married and those who did not were quite different. Indeed, there are three distinct groups we can consider: those who remained single (or for whom marriage was irrelevant as far as employment was

[4] "Notes and News: Married Women as Teachers," *Educational Review* 25 (1903): 214.

[5] This number is comparable to the number given for persons with doctorates in mathematics who received the doctorate in 1934, at the height of the Great Depression, or who sought employment for 1934–35. See E. J. Moulton, "The Unemployment Situation for Ph.D.'s in Mathematics," *American Mathematical Monthly* 42 (1935): 143–44. In comparison, the final unemployment rate for 2005–06 doctoral recipients was 3.3 percent. See Polly Phipps, James W. Maxwell, and Colleen A. Rose, "2006 Annual Survey of the Mathematical Sciences (Second Report)," *Notices of the American Mathematical Society* 54 (2007): 877.

concerned); women religious; and those who married. The group of women religious warrants independent examination since some employment limitations did not affect these women.

Eighteen Catholic sisters received PhD's in mathematics in the United States before 1940, the majority in the 1930s. It appears that these degrees were generally obtained as part of a movement to upgrade the level of instruction in Catholic women's colleges. Many of these colleges were just emerging as full four-year colleges from preexisting academies and were facing increasing pressure to meet requirements for accreditation.

Whereas the most daunting task facing most new PhD's, especially in the 1930s, was getting a job, that was not the case for the women religious. As Sister Helen Sullivan reported at the 1981 Smithsonian meeting, "One of the advantages of going into a religious community is you don't have to seek your jobs; I was assigned. I came home [to Mount St. Scholastica College] with my degree and started teaching everything in the program that I could handle. I taught fifteen hours my first year and believe you me it wasn't easy."

Figure 5.1: Sister Mary Cleophas Garvin's calculus class at Notre Dame College, 1945. Photograph courtesy of the Archives of the Sisters of Notre Dame, Chardon, Ohio.

Typically the Catholic sisters were the mainstays of small departments, and some taught a variety of courses in addition to mathematics. For example, during her more than thirty years on the faculty at Trinity College in Washington, D.C., Sister Marie Cecilia Mangold was either the only member, or one of two members, of the mathematics department. In another instance, after receiving the PhD, Sister Catharine Frances Galvin returned to Siena College in Memphis, Tennessee, where her teaching assignments included chemistry, physics, and statistics, in addition to mathematics. Furthermore, she was bursar at both St. Agnes Academy and at Siena College and served as the superior of her community for many years.

The situation of Sister Catharine Frances was not unusual. In addition to exceedingly demanding teaching assignments, women religious holding doctorates in mathematics often provided academic and administrative leadership for their colleges. For example, in addition to spending more than thirty years as professor and chairman of the mathematics department at Regis College near Boston, Sister Leonarda Burke was, for twenty-five years, director of the Regis College Research Center and principal investigator for contractual work for the Air Force Research Center, Hanscom Field. As another example, after serving as chairman of the mathematics department at Nazareth College in Louisville from her arrival in 1937, Sister Mary Charlotte Fowler was president of the college from 1961 to 1969.

For the women who were not nuns, the more common profile is of a woman who taught full time before receiving her doctorate in mathematics, probably also having had some kind of scholarship or assistantship during her graduate studies. Afterwards she found a position teaching mathematics in a college or university where the department's focus was on teaching rather than research. If she was among the 50 percent of the women who received their PhD's in the 1930s and was not a nun, she may have had additional trouble finding a suitable position because of the Great Depression. Furthermore, if a woman married, anti-nepotism practices probably drastically curtailed her already limited professional opportunities.

Those who did not get jobs immediately and those whose initial postdoctoral jobs did not fully utilize their education illustrate some of the difficulties encountered. First, of course, the Great Depression affected nearly everyone, men as well as women, throughout much of the 1930s. It appears that at least six women in our study were without jobs immediately after getting the PhD primarily because of the Depression. Others found that their opportunities were not what they had expected. Rosella Kanarik (PhD Pittsburgh 1934) wrote in 1985, "I graduated during the depression. It was almost impossible for anyone, let alone a woman, to find a position in industry, college, or university. I was lucky to get into a high school to teach mathematics."[6] The first position for Ruth Peters, who received her Radcliffe College PhD in 1933, was as a personnel assistant doing job analysis for the Pennsylvania Emergency Relief Board.

Even for those who had finished their graduate work before the 1930s, the Depression sometimes made a significant impact on their careers. A particularly vivid account is offered by Elsie McFarland Buck (PhD California 1920). She later reported, "I was teaching and semi-starving at a very small college in Spokane.... At any rate, we were not getting paid very much money. I [taught] my whole nine months there for $360 and a box of apples and some kitchen cleanser donated by one of the students as part of his tuition. And I was sending out something like 300 or 400 letters of application all over the country. This was in 1932 ... when jobs were very few and far between."[7]

As Kanarik suggested above, just being a woman greatly limited job opportunities. In some cases there were formal barriers or institutional practices. Sometimes individual discriminatory attitudes were the issue. In other instances, the limitations reflected expectations about the professional woman's proper role in academia.

[6]Smithsonian questionnaire.

[7]Conversation between Mr. Don Haacke and Dr. Elsie Buck, November 21, 1980, OH-20, Special Collections Department, Albertsons Library, Boise State University.

Discrimination was prevalent enough to produce the following response on a 1926 questionnaire distributed by the Bureau of Vocation Information:

> Nothing but the most earnest conviction that she could never be satisfied without a PhD in mathematics would justify a woman's setting herself that end. It is a long, hard road and when the degree is obtained, she finds that all the calls for mathematics teachers are for men, and that when a woman is employed in one of the large universities she is practically always given long hours and freshmen work for *years*, with less pay than a man would receive for the same service. If all the women could fare as well as I have fared, I'd say "Go ahead," but alas! Such unexpected good luck does not come to many in a generation.[8]

Nonetheless, there were men who did support women in mathematics. In his positions as head of the mathematics department (1915–42), dean of the graduate school (1926–48), and secretary of the American Mathematical Society (1921–40), R. G. D. Richardson of Brown University was particularly influential in helping to place women looking for positions, especially those who had been students at Brown. In 1924, in a letter recommending Marian M. Torrey for a position at Smith College, he wrote, "The department feels that we have never had, among the twenty or more girls whom we have sent out to teach in colleges, any stronger candidate.... If Brown University would employ women, I would not hesitate to ask President Faunce to call her here at a good salary. She would do much better than many of the men whom we have at present on our staff."[9]

It is not surprising that Brown, a men's college, would not hire women. However, in a letter to Emily Chandler (Pixley) in 1927, Richardson writes that there is no opportunity for teaching at Brown's Women's College, because, "in general it is against the policy of our Women's College to employ women teachers."[10] Similarly, at some time no later than 1926, Frank Nelson Cole requested a recommendation from E. H. Moore at the University of Chicago for a man to fill a position at Columbia's coordinate college, Barnard.[11]

While most women's colleges did hire women faculty members, sometimes they preferred hiring men. Two years before Richardson recommended Marion Torrey for a job at Smith, Ruth Goulding Wood wrote to Richardson asking for recommendations for a young instructor in mathematics for the following year at Smith. She wrote that "the President is anxious to find a young man who has recently done his Doctors' thesis and who gives promise as a teacher."[12]

Men did not always think it was a wise idea for a women's college to hire only women faculty. For example, in 1947 C. R. Adams at Brown replied to Marion

[8]Hutchinson, *Women and the PhD*, 185–86. All we know of the respondent is that she had earned a PhD in mathematics between 1915 and 1924 and was an assistant professor at a college at the time of her response to the questionnaire.

[9]R. G. D. Richardson to Ruth G. Wood, February 13, 1924, Correspondence 1921–1925, R. G. D. Richardson Papers, Brown University Archives.

[10]R. G. D. Richardson to Emily Chandler, March 31, 1927, Richardson Correspondence 1926–1930.

[11]Undated letter, folder 2, box 1, E. H. Moore Papers, Special Collections Research Center, University of Chicago Archives.

[12]Ruth G. Wood to R. G. D. Richardson, November 24, 1921, Richardson Correspondence 1921–1925.

Stark at Wellesley College who had inquired about possible job candidates. He mentions a number of possibilities and then suggests they hire a man. "In any one of the group of fine women's colleges to which Wellesley, Smith, Mount Holyoke, and Bryn Mawr belong the department of mathematics is bound to deteriorate if a policy of women only on the staff is adhered to."[13] It should be noted that Rachel B. Adams (PhD Radcliffe 1921), the wife of C. R. Adams, had served as a tutor at Radcliffe for fifteen years but held no other formal position.

Other departments just did not want too many women on the faculty. After Louise Johnson (Rosenbaum) received her PhD from the University of Colorado in 1939 under the direction of the head of the department, Aubrey J. Kempner, "she knew that she had no future [at Colorado] since Kempner told her that there were enough women already in the department."[14] At that time two of the five members of the department were women; they were Claribel Kendall (PhD Chicago 1921), an associate professor, and Frances C. Stribic, an assistant professor who had completed all her work for a PhD at the University of Nebraska except for her dissertation.

Some correspondence reflects views that were overtly hostile to women. In a letter of 1920, the long-time head of the mathematics department at Ohio State University wrote regarding the possible placement of two women there. "Our dean (Math is in Engineering College) does not like girls. We have two girl teachers. Don't think we could get him to agree on another."[15] At the time, one of the "girls" was the forty-four-year-old Grace Bareis, who had earned her doctorate there in 1909 and was an assistant professor until her retirement in the mid-1940s.

As has been amply demonstrated in other accounts, women were not the only targets of discrimination. Frederick W. Owens, department head at Pennsylvania State College, sought candidates in a letter to H. S. Everett at Chicago in 1931. "We have an opening for an instructor for next year to supply for a man on leave of absence.... For this position we would prefer a Gentile and a man. While we have both Jews and women on our staff, we can not have too large a proportion of them."[16] Two years later in the midst of the Depression, Owens wrote to Richardson at Brown noting that they have an unexpected opening at Penn State and asking for suggestions. He writes that the salaries are reduced, probably about $1500 and one year only. "For this place, if filled as a full time appointment, I would prefer a man and a non-Jew."[17] It is noteworthy that Helen B. Owens, the wife of F. W. Owens, had earned a PhD in mathematics from Cornell in 1910 and was prohibited from taking a position at Penn State because of anti-nepotism practices.

Another frequently held prejudicial assumption was that women should teach only women or in women's colleges. In 1922 Julia Dale applied for financial support at Cornell University after having received her master's degree and having served as an instructor at the University of Missouri. A letter of recommendation contained

---

[13]Clarence Raymond Adams to Marion Stark, December 9, 1947, Richardson Correspondence 1947–1949.

[14]Jones and Thron, *A History of the Mathematics Departments of the University of Colorado*, 15.

[15]R. D. Bohannan to R. C. Archibald, 1920, Richardson Correspondence.

[16]F. W. Owens to H. S. Everett, July 16, 1931, folder 10, box 1, Mathematics Department Papers, Special Collections Research Center, University of Chicago Archives.

[17]F. W. Owens to R. G. D. Richardson, August 5, 1933, Richardson Correspondence 1931–1934.

Figure 5.2: Ohio State University Graduate Math Club and Faculty Club, April 15, 1929. Three of the five women who received PhD's from Ohio State are pictured here; Grace Shover (Quinn) and Grace Bareis, are seated on the left, and Mabel Schmeiser (Barnes), is partially hidden. Maude Hickey, younger sister of May Hickey Maria, is seated in front of Bareis and Schmeiser. Shover's advisor, C. C. MacDuffee, is standing on the right, and Schmeiser's advisor, Henry Blumberg, is seated on the left. Photograph courtesy of Grace Shover Quinn.

the following: "I consider Miss Dale to be a hardworking student with a good mind. Yet she is not an horrible example of what a female graduate student sometimes becomes. She unites a good scientific mind with sufficient youth and ordinary intelligence to make her influential with university women. I feel that in helping her in her education you will be not only advancing mathematics but furnishing some college with women students with a good teacher and an influential adviser."[18]

While many of the women were likely to face discrimination at some time in their careers, the experiences of those who married and those who did not were quite different. We will first consider issues relating to employment expectations and opportunities for those who married. These included finding employment for a two-person professional family, anti-nepotism practices (both formal and informal), child rearing, the perceived lack of a need to produce extra income, and societal and personal expectations.

Eighty-four of the 228 women in the study were married at some point in their life. Most were married once, seven were married twice, and one married three times.[19] All of the first and second marriages have ended, twenty of the first and two of the second by the death of the wife, fifty-three of the first and six of the

[18]W. D. A. Westfall to Professor J. H. Tanner, March 4, 1922, Julia Dale folder, box 61, Graduate School Records, Rare and Manuscript Collections, Cornell University.

[19]Six of the eight second marriages occurred after a divorce; the other second marriages and the third marriage occurred after the death of a husband.

second by the death of the husband, and eleven of the first by divorce. While 36 percent of the women who married had little or no employment while they were married, only about 4 percent of the single women, and none of the women religious, were never, or rarely, employed. Furthermore, the majority of those who were unemployed immediately after obtaining their doctorates were married. Initially the greatest difficulties were experienced by the approximately twenty women who married before or shortly after receiving the doctorate.

Of the women who married, 40 percent married men with PhD's in mathematics while another 21 percent married men with PhD's in other fields. Most of these husbands took academic positions; and the institutions in which they worked, their cultures, or their locations played a large role in whether the subsequent situation was favorable or not for the continued involvement of the woman in mathematics.

We can categorize roughly the nature of the careers of the women who married. Seven had continuous careers, without interruption, in which the role of the marriage seems to have been relatively insignificant. Ten had nearly continuous careers, although in some instances the marriage affected the professional situation adversely. Thirty-seven women had significant interruptions in their careers, frequently, but not always, because of family responsibilities. Finally, thirty of the women who married had little or no employment while they were married, even though almost all had professional positions at some point before the marriage.

For the seven who were both married and had continuous careers, the marriage and careers overlapped significantly for only three. They are Jewell Hughes Bushey, Evelyn Carroll Rusk, and Mary Kenny Landers. All three were married to fellow academics: Bushey and Landers to mathematicians, and Rusk to an art historian. Jewell Hughes married her Hunter College colleague of five years, Joseph Hobart Bushey, when she was thirty-nine years old and eleven years beyond her doctorate. Both were able to continue in their positions at Hunter. Mary Kenny and Aubrey Landers were fellow students at Brown when they first met. After receiving master's degrees they obtained teaching positions in New York City, she at Hunter and he first at Hunter and then at Brooklyn College. They married in 1933, and they both later resumed graduate work at the University of Chicago, in residence for a year and six summers to obtain their doctorates. They, too, were able to continue in their positions after marriage. Mary Landers, with three children, was the only mother among this group. Finally, Evelyn Carroll had just received her PhD and had been promoted to associate professor at Wells College, her alma mater, when she married William Sener Rusk, an art historian also on the faculty at Wells.

In the other instances of a continuous career it is obvious why the marriage was of little relevance. Mayme Logsdon married at nineteen and was widowed nine years later; her husband had two children, but they were largely in the care of other family members. Although she had taken a course for teachers and had taught during her marriage, her subsequent undergraduate work, graduate work, and mathematical career at the University of Chicago occurred long after she was widowed. In other cases, the marriage occurred relatively late in life, long after a career was established. Evelyn Wiggin Casner and Mina Rees were married in their fifties, both to non-academics, while Grace Murray Hopper was divorced relatively early in her career.

In some instances of a nearly continuous career the marriage affected the professional situation adversely. Emily Chandler and Henry Pixley were married in 1931, a few months after each received a PhD from the University of Chicago. Unable to find employment in the same city, they resumed their former positions for the next five years, except for a year when both were on leave for work with the National Recovery Administration. She was professor and department head at Saint Xavier College for Women in Chicago, and he was an instructor at the College of the City of Detroit, which soon became part of the newly created Wayne University. Finally, in 1936, Emily Pixley obtained a position at Wayne University, where her husband was then an assistant professor. She was hired as a special instructor with an hourly salary and worked both part time and full time. During the next dozen years the Pixleys had three children. In September 1947, Emily Pixley's position was designated "regular substitute assistant professor," and she had a teaching load of sixteen hours per week. Less than a year after that, in April of 1948, in a memo concerning "the university policy relative to employment of man and wife," it was noted that "while the quality of teaching will be lowered somewhat by making Mrs. [X] and Mrs. Pixley the first to go, university policy demands that this be done, and I have already made clear to those affected that this will require the termination of their services in June 1949."[20] Emily Pixley did not wait for the university anti-nepotism policy to be satisfied but found employment at the University of Detroit, from which she retired as professor emeritus after twenty-five years on the faculty there. In this case, the so-called two-body problem was initially an issue, but formal anti-nepotism practices and the demands of parenthood also played a role.

As observed above, a number of those who were married were married to fellow academics, frequently mathematicians. In some cases the husband took a position at an institution where there were limited opportunities for his wife. Eleanor Pairman (PhD Radcliffe 1922) married Bancroft Brown (PhD Harvard 1922) shortly after they both received their degrees. He took a position at Dartmouth College, an isolated men's college in New Hampshire. Appropriate professional opportunities simply did not exist for her because of the isolation and absence of schools that would have hired a woman.

Not all dislocations occurred immediately after the doctorate. Annie MacKinnon earned her PhD at Cornell University in 1894 and spent the next two years studying mathematics in Göttingen before taking a position at Wells College, a college for women in Aurora, New York. Five years later the thirty-three-year-old MacKinnon gave up her position as professor and registrar to marry Edward Fitch, a professor of Greek at Hamilton College, a school for men in Clinton, New York, nearly a hundred miles from Aurora.

While in 1901 Annie MacKinnon Fitch ended her professional career when she married, we have seen that thirty years later Emily Pixley and her husband continued jobs they had in different cities. The year after the Pixleys found their temporary solution to the two-body problem, another couple found a similar solution. Rose Whelan earned her doctorate in 1929 from Brown University. She had taught at the University of Rochester the year before receiving her degree and returned to that position. In 1932 she and Charles H. W. Sedgewick were married. For the first year and a half of their marriage she continued her position at Rochester, while he finished his doctorate in mathematics at Brown and held a full-time position at the University of Connecticut. Soon after the first of their four children was born

[20]Alfred L. Nelson, Mathematics Department, to Dean C. B. Hilberry, April 23, 1948, University Archives, Wayne State University.

the entire family lived full time in Connecticut. She was able to do some teaching as an instructor at the University of Connecticut and elsewhere, before the family moved to Washington, D.C. From age fifty-five to sixty-six she was instructor and then assistant professor at the University of Maryland.

In other instances, anti-nepotism practices were the primary factor affecting the woman's employment. Elizabeth Stafford (Hirschfelder) received her PhD in mathematics from the University of Wisconsin in 1930 and then taught for a year in Texas before marrying Ivan Sokolnikoff, who had just received his doctorate from Wisconsin. Both remained at Wisconsin in a variety of positions, he as instructor through professor and she in a number of irregular positions despite the fact that they had jointly published several significant mathematical papers and the classic text *Higher Mathematics for Engineers and Physicists.* It was clear that anti-nepotism sentiment at the university played a significant role in her options there. In 1932, Mark Ingraham wrote to the dean of his college that "although Mrs. Sokolnikoff is one of our best instructors and is better prepared than any other instructor to give advanced work, we have omitted her from the tentative budget for next year due to the fact that you do not feel it wise to retain the wife of a member of the Department on the staff."[21]

Anti-nepotism rules affected graduate students as well as faculty. In 1931, a year after going to the University of Illinois as a graduate student and teaching assistant, Gertrude Stith married Pierce W. Ketchum, a member of the mathematics faculty. Because of anti-nepotism practices, Gertrude Stith Ketchum was not permitted to continue as a teaching assistant after her marriage. However, some years later she was allowed to teach on a part-time basis because the department was in need of instructors.

Fifty-one of the eighty-four women who married had children. There were also stepchildren, adopted children, and a ward. Adopting children or having a ward was not limited to those who married. Ida Martha Metcalf (PhD Cornell 1893) advertised for and took a ward in 1915, while Mary E. Sinclair (PhD Chicago 1908) adopted two children as infants, one in 1914 and the other in 1915. In most cases, childrearing delayed the onset of a career, interrupted it, or made it seem impossible. Although several women established shortened careers later in their lives, for many the career effectively ended at marriage or after the birth of children.

In at least one instance, having a child brought an unwanted and abrupt end to a job. Jessie Jacobs (Offermann) earned her PhD from the University of Illinois in 1919. In 1920 she moved to a position as instructor in pure mathematics at the University of Texas, where she met Hermann Joseph Muller, who arrived the same year as associate professor of biology. Jacobs and Muller were married in June 1923, and she remained an instructor through the academic year 1923–24. Their son was born in November 1924 in Austin. According to Muller's biographer, "The birth ... led to a bitter blow for Jessie. The mathematics department terminated her appointment because her colleagues felt that a mother could not give full attention to classroom duties and remain a good mother. For Jessie it meant a permanent loss of her career as a teacher."[22]

---

[21]Mark Ingraham to Dean Sellery, May 19, 1932, folder Budget 1931–33, box 1: 1924–1937, Budget Files, Department of Mathematics, College of Letters and Science, University of Wisconsin-Madison Archives.

[22]Elof Axel Carlson, *Genes, Radiation, and Society: The Life and Work of H. J. Muller* (Ithaca: Cornell University Press, 1981), 133–34.

About 60 percent of the married women who had children had significant, although shortened, careers with the most common employment pattern being one in which there was an interruption during the childbearing and early childrearing years and then resumption of the mathematical career later. Mabel Schmeiser Barnes provides one such example. She received her PhD from Ohio State University in 1931. After spending the next four years, in the midst of the Depression, at a teachers college in Nebraska, at the Institute for Advanced Study at Princeton, and as a substitute mathematics teacher in New York City, she married John Landes Barnes, a recent mathematics PhD from Princeton. They moved to Massachusetts, where J. L. Barnes had a position as assistant professor at what was then Tufts College. Mabel Barnes "kept [her] hand in somewhat by marking papers and substituting for him when he was away, and by helping him edit the mathematics section of *Eshbach's Handbook of Engineering Fundamentals*, second edition."[23] Their two children were born before John Barnes took a leave from Tufts to do war work with Bell Labs. After World War II ended, Mabel Barnes was hired as an assistant professor at Tufts for 1946–47, partly because of the influx of veterans. She later noted that "nepotism and [her] being a woman were overlooked. Desperation again overcame prejudice."[24] Although she was rehired for the next year, they moved to California where her husband had taken a position as professor in engineering at UCLA. In 1950 Mabel Barnes resumed her career at Occidental College in Los Angeles, where she was hired as instructor and was promoted through the ranks before retiring as professor emeritus in 1971.

Audrey Wishard McMillan's career took a different turn after her years at home with her children. She earned a doctorate in 1938 from Radcliffe, after which she spent four years as an instructor at Vassar College. In 1942 she married Brockway McMillan, who had earned a PhD in mathematics from MIT. From 1942 to 1945 the McMillans were in Dahlgren, Virginia, where Brockway McMillan was with the US Naval Reserve. Audrey McMillan was an assistant to Hermann Weyl at the Institute for Advanced Study in Princeton in 1942–43 and worked for the US Navy's Bureau of Ordnance in Dahlgren during 1943–45. In 1946 Brockway McMillan began a long association with Bell Telephone Laboratories, where he spent most of the next thirty years. The McMillans had three children during the 1940s and early 1950s. Audrey McMillan wrote in 1997, "I do not think it possible to care for 3 small children and do mathematical research. I tried. Unfortunately, I did not realize that, at age 50, I might have been able to."[25] Instead, she turned her attention to elementary school teaching and consulting, which interested her because of the "new math." She later was an instructor at a day school, taught a course for elementary teachers at New York University, and served for a decade as a consultant to the Summit, New Jersey, board of education.

Without a doubt, World War II affected many of the women in the study. In several instances, women were called back into the academic workforce because of the demand for instructors during and after the war. These were often the same women who had been unable to hold academic jobs earlier because of anti-nepotism practices. We noted earlier the experience of Mabel S. Barnes who was hired at Tufts in 1946 because the college was willing to overlook its previous discriminatory

[23]Barnes, in "Centennial Reflections on Women in American Mathematics," 7.
[24]Ibid.
[25]Authors' questionnaire.

practices when it desperately needed more faculty members. Nola Haynes (PhD Missouri 1929) is another example of a women for whom World War II and the subsequent influx of veterans to college campuses was a positive influence. Haynes married in 1938 after having spent 1930 through 1938 as a faculty member at H. Sophie Newcomb College in New Orleans. At the 1981 Smithsonian meeting, she recounted the following:

> I left Newcomb College ... to get married to the chairman of the astronomy department of the University of Missouri. There was a very strict nepotism law and I was giving up my career for marriage, thinking I would never teach again. Then when the Second World War came along and they were bringing in all these various companies..., I was the first person called back into teaching. But at that time they didn't set any salaries, and they gave me an acting instructor's name or something like that; it was because of the nepotism law.... I thought it was a temporary sort of thing and then it went on and after the war the boys came back ... I was still an acting associate professor ... because of my husband. It was not until my husband retired that I became associate professor.

As we saw earlier in this chapter, marriage did not always hinder a woman's ability to pursue a career in mathematics. Several married women had continuous or nearly continuous careers, and many more had satisfying careers after significant interruptions. Whereas marriage often presented women with difficulties in finding or keeping a job, we know of no instance of a close female friendship having had a negative professional consequence. Moreover, we find that such friendships apparently contributed to the fulfillment of personal and professional lives of a number of women in our study. In at least two cases, a pair of these women mathematicians worked and lived together for much of their careers. Suzan Benedict and Susan Rambo were both graduates of Smith College, and both taught there for the major part of their careers. Benedict was ten years older than Rambo and joined the Smith faculty in 1906, two years before Rambo. Benedict earned a PhD from Michigan in 1914, as did Rambo in 1920. By 1918 they were sharing a house, and they continued to do so until Benedict's death in 1942. Similarly, Clara E. Smith and Lennie P. Copeland were colleagues at Wellesley College, traveled together, and were sharing a home in Wellesley at the time of Smith's death. Another pair, Wealthy Babcock and Florence Black, were close companions and colleagues for about fifty years. They both received their doctorates from the University of Kansas in 1926 and shared many interests including sports, camping, and riding. At least three other women in our study shared their lives with female mathematicians who were not in our study. For example, Harriet Montague and Mabel Montgomery were colleagues at the University of Buffalo, where Montgomery earned her PhD in 1953. They coauthored a textbook and shared a home in Buffalo until Montague's death, when Montgomery was listed as Montague's only survivor. Others lived with colleagues who were not mathematicians. Pauline Sperry and her friend Alice Tabor, who was in the German department at the University of California in Berkeley, were instrumental in establishing the Women's Faculty Club at Berkeley. Cleota Fry joined her Reed College classmate Vivian Johnson at Purdue University, completed her doctorate there, and remained on the mathematics faculty, while Johnson taught in the physics department. They traveled together and shared a house until Johnson's death, several years after they both had retired. These are just some of well over a dozen instances in which women in the study found close, sustaining female companionship as well as uninterrupted and successful professional lives.

CHAPTER 6

Career Patterns

We have considered some of the paths women took before getting the doctorate and noted that well over three-quarters of them had full-time employment before receiving the degree, almost always in teaching at the precollege or college level. We also noted a number of limitations that applied at various stages when they were seeking positions. These included overall economic conditions, assumptions about women's proper role in the workforce, and restrictions on—and hostile attitudes toward—the hiring of women, as well as special difficulties that were encountered by women who married. We will now examine in some detail overall employment patterns, focusing on employment after the PhD. We will consider the employment histories of some of the women and describe some of the institutions where they worked. We first look at employment in the year following the PhD and then consider the primary position for each woman, defined as that in which she worked the longest.

Of the 200 women who were employed in the first year after receiving the doctorate, all but five were associated with academic institutions, with nearly all in teaching positions at some level. The five include Caroline Seely who became in 1913, two years before receiving her PhD from Columbia, the first mathematician to be employed full time by the American Mathematical Society, as a clerical and editorial assistant. She remained with the AMS for twenty-two years. The others were Alta Odoms (Gray), who worked briefly for the AMS as an editorial assistant in the mid-1930s; Ruth Peters, who received her PhD in 1933 in the depth of the Depression and was a personnel assistant doing job analysis for the Pennsylvania Emergency Relief Board; Vivian Spencer, who in 1936 became a statistician for the US Bureau of Mines and the National Research Project (NRP) of the Works Progress Administration (WPA) in Philadelphia; and Marion Greenebaum (Epstein), who in 1938 was a research statistician with the International Statistical Bureau in New York City.

Of the remaining 195 women with jobs at academic institutions in their first postdoctoral year, about 20 were engaged as research assistants, editorial assistants, or postdoctoral fellows rather than as faculty members. Half of these had positions that could be described as research or postdoctoral positions. The latter included three National Research Council (NRC) fellows: Echo Pepper at Oxford University and Marie Weiss and Dorothy Manning (Little) at the University of Chicago, and two AAUW fellows: Olive C. Hazlett at Harvard University and Deborah May Hickey (Maria) at the Ludwig-Maximilians-Universität (LMU) in Munich. Mildred Sullivan was a research assistant at Harvard in the year following her doctorate and held an NRC fellowship the next year.

Thus, 175, or more than three-quarters of the entire group of women PhD's, took first jobs that involved teaching in an academic institution of some type. Of these, 107 took first positions in colleges; 48 were at schools that called themselves universities, although that designation did not necessarily imply the existence of a graduate program in mathematics; and 19 taught in high schools or preparatory schools. Of the 107 who taught in colleges, 87 taught at women's colleges or coordinate colleges, 14 taught at coeducational liberal arts colleges, and 8 taught at colleges for teachers, which were coeducational but had a small percentage of male students. Thus, nearly half of the first teaching jobs were at colleges for women. We will describe this situation in more detail later.

Over two-thirds of the women who held teaching jobs immediately after receiving the doctorate were known by the institution employing them, either as previous students at the undergraduate or graduate level or as previous employees. Over half of these returned to full-time positions held before the graduate work was completed. Well over a quarter of them took first jobs after the doctorate at the schools where they had taken their undergraduate degrees; of these, six were Wellesley College graduates. Some stayed for at least one year at the schools where they had completed their graduate work.

Generally, however, the schools that awarded doctorates to women in mathematics before 1940 did not become employers of these women. Twenty-seven women (11.8%) took full-time positions at one of these schools in the first year after receiving the doctorate, and only sixteen of the thirty-four US institutions are represented. All but four of these twenty-seven women received their doctorates before the 1930s, at which time the hiring of women in research-oriented institutions declined from earlier levels that were already very low. Of the twenty-seven women whose first position was at a doctoral institution, seventeen stayed and held their primary positions at the schools where they had earned their PhD's; the rest left after a year or two. In addition, two others later returned to teach at their doctoral schools.

When women were hired by schools that had awarded them doctorates, they typically did not fare well in terms of promotion or support for their research. Mary Taylor Speer's son reported that she "recounted that she left the University of Pittsburgh, and mathematics teaching, because men junior to her were promoted to Associate Professor while she was not, and because she was told that this was because the men had families to support and needed the money."[1] We will return to the consideration of women teaching at PhD-granting schools when we examine in more detail the employment and advancement of women who spent the major parts of their careers at such schools.

The number of different positions per person ranged from none to at least ten, but, as we indicated above, in almost all cases we can designate a primary position for each woman. We consider positions held before the doctorate as well as after and count for each woman the total number of years in the job of longest duration. Not only did women's colleges lead in the hiring of women for their first position after receiving their doctorate, but as we can see from figure 6.1 below, by some measures they also lead when we consider primary jobs.[2] While fewer women held their primary positions at women's colleges than at coeducational schools, they typically enjoyed jobs that lasted longer and where the advancement was

[1]Eugene R. Speer, Authors' questionnaire, 1990.

[2]In figures 6.1 and 6.2 we include coordinate colleges in the women's colleges category.

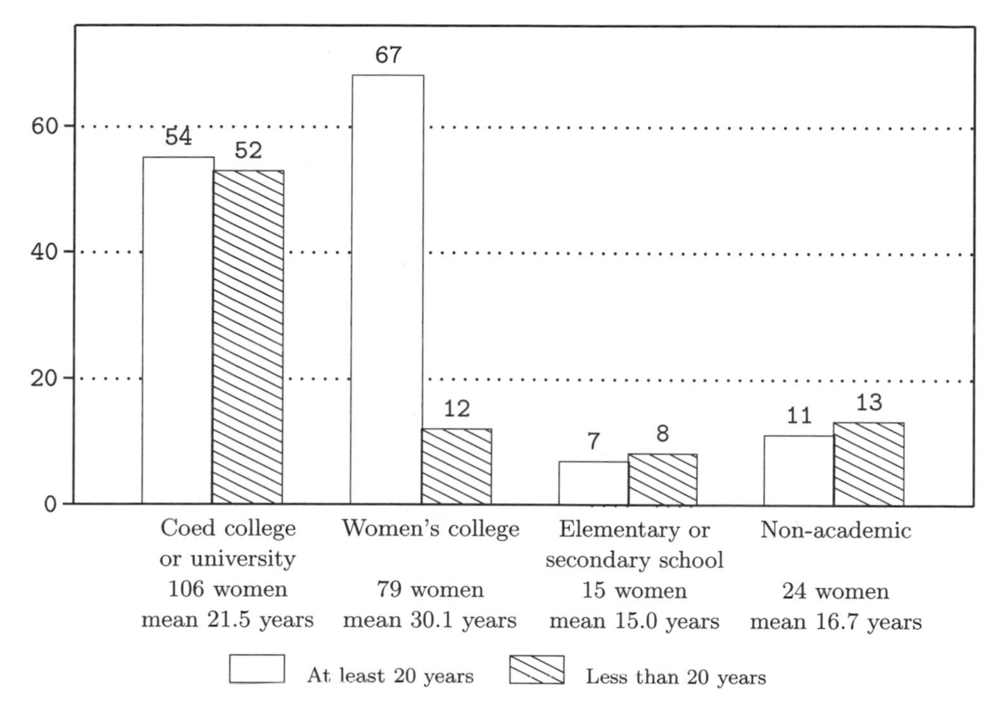

Figure 6.1: Primary employment for 224 of 228 women

significantly better than at other types of schools. We will explore some of these issues later.

Figure 6.2 makes obvious how employment at institutions of higher education differed among the groups of single women, married women, and women religious. Single women tended to hold their jobs for longer periods of time than married women did, and relatively few married women held jobs at women's colleges. Women religious had long tenures at colleges for women. Although not obvious from the bar graph, all of the colleges at which the nuns taught were run by their religious orders.

As we can see by table 6.1, with two exceptions women's colleges were the leading employers of women in the study by almost all measures. In addition, the schools that led in providing first positions after the doctorate were Hunter College with ten, Wellesley College with eight, and Vassar College and the University of Illinois with six each. As we noted earlier, women's colleges provided nearly half of the first jobs. Furthermore, these jobs tended to be stable; half of those whose first job was at a women's college remained at that college their entire careers.

With the two exceptions of the University of Illinois and Pennsylvania State College, these leading employers were women's colleges in the East. The five women's colleges, Bryn Mawr, Mount Holyoke, Smith, Vassar, and Wellesley, of the Seven Sisters were also prominent in the undergraduate education of the women in our study. Radcliffe and Barnard, the coordinate colleges of the Seven Sisters, rarely employed women or did so only as tutor or instructor. Moreover, the women's colleges Wellesley, Smith, and Hunter were among the largest educators of women

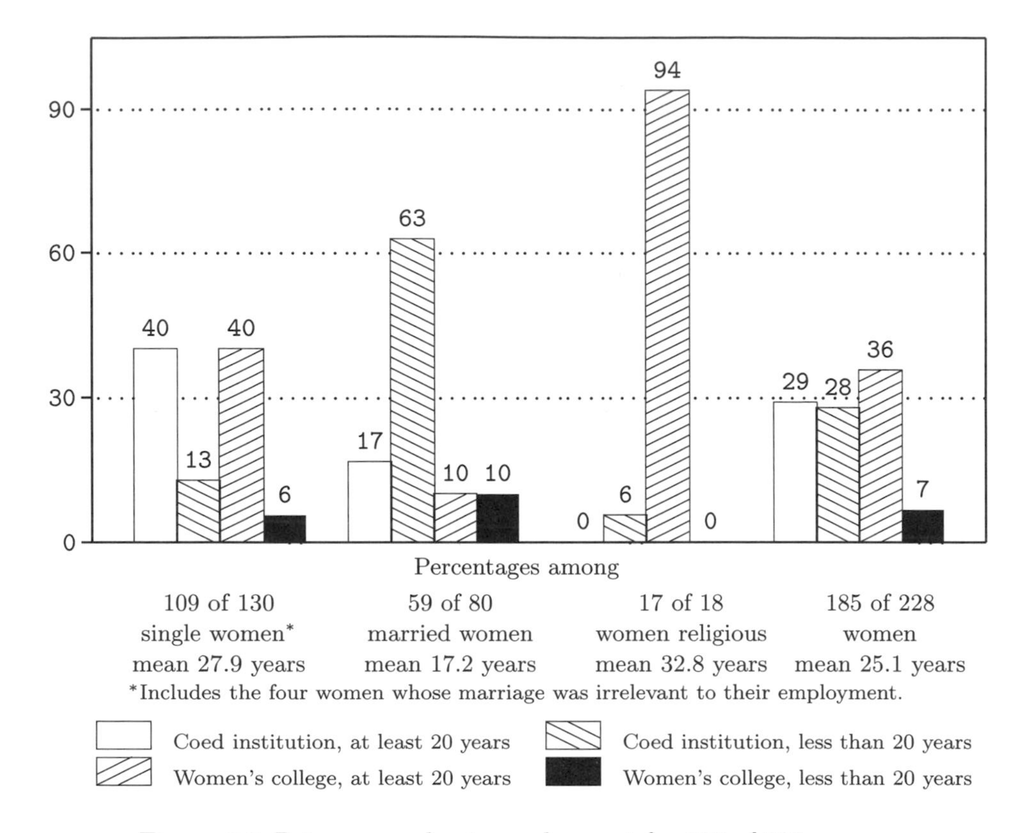

Figure 6.2: Primary academic employment for 185 of 228 women

in the late nineteenth and early twentieth centuries. Some of the women's colleges offered master's degrees, but as we indicated earlier, only Bryn Mawr had a doctoral program in mathematics. Most often, the mathematics departments at these women's colleges were all, or almost all, female.

Except for Hunter College, the leading women's institutions generally employed only women who were single. Just under a third of the women at these colleges had their primary jobs at schools where they had done their undergraduate work, and just over a third received their doctorates from the University of Chicago, a much larger proportion than the 20% of doctorates that Chicago awarded to women. Half of the women at these colleges earned their doctorates after they began their employment. The women who taught at these leading women's colleges tended to be engaged in a number of scholarly activities, occasionally mathematics research but more often in the writing of textbooks, expository articles, and reviews. Most were active in the professional mathematics organizations. There is frequent reference to their excellence as teachers. Almost all retired having advanced through the ranks to professor.

Hunter College of the City of New York was by far the leading academic employer of women in our study. As we saw earlier, it served women in the New York City area, was tuition-free, and by 1916 was the largest women's college in the country. Hunter's mathematics department was large, with seven regular faculty

Table 6.1: Leading academic employers

	Total number of women	Number who held primary job	Average number of years worked by women*	Total number of years worked by women†
Hunter College	20	9	16.8	335
Wellesley College	14	7	17.9	250
University of Illinois	15	6	12.0	180
Vassar College	14	7	12.9	180
Goucher College	5	4	26.0	130
Randolph-Macon Woman's College	6	3	20.9	125
Smith College	9	3	13.3	120
Bryn Mawr College‡	8	3	14.4	115
Pennsylvania State College (Univ.)	7	5	15.7	110
Mount Holyoke College	13	2	7.3	95

*These averages are based on data that has been rounded.
†These numbers are rounded to the nearest five years.
‡These numbers include Isabel Maddison who was employed at Bryn Mawr for thirty-one years, mainly in administrative positions, the last of which was recording dean and assistant to the president of the college. For the first fourteen years she also held faculty rank.

members in 1916; all were women and none had a doctorate. The Hunter department continued to grow, and, by the time the last woman in our study to get a Hunter degree graduated in 1935, there were sixteen regular faculty members, including seven women with PhD's in mathematics.

Nine women had their main positions at Hunter College, eight of them for 30 years or more. All but one of the nine were first hired between 1925 and 1931, and their teaching spanned a total of 50 years, from 1925 to 1975. Tomlinson Fort had joined the faculty at Hunter in 1924 and served as chair until he left three years later. Lao G. Simons, who held a doctorate in mathematics education from Teachers College in Columbia University and is not counted among our group, was chair from 1927 until her retirement in 1940, when she was replaced by Jewell Hughes Bushey.

An unusual feature of this group is that five of the nine were married before or while they were on the faculty at Hunter, and one of these married a fellow member of the mathematics department. One reason this situation was possible was the inauguration in 1929 of an enlightened president. At that time few of the women on the Hunter faculty were married and, as we have seen, it was not unusual to insist on the resignation of a woman upon marriage. The new president, James M. Kieran, replied to the question "Should a woman member of the staff resign when she got married?" by saying that "the only thing such a person had to do was

Table 6.2: Hunter College faculty - primary employment

	Baccalaureate		Doctorate		On Faculty
Mina Rees	Hunter	1923	Chicago	1931	1925–50* 1953–61
Laura Guggenbuhl	Hunter	1922	Bryn Mawr	1937	1926–72†
Marguerite Darkow	Bryn Mawr	1915	Chicago	1924	1928–59
Anna Marie Whelan	Goucher	1918	Johns Hopkins	1923	1928–65
R. Lucile Anderson	Mount Holyoke	1922	Bryn Mawr	1930	1928–65
Mary Kenny Landers	Brown (Women's College)	1926	Chicago	1939	1928–75
Jewell Hughes Bushey	Arkansas	1915	Chicago	1924	1930–66
Helen Schlauch Adams Infeld	New York Univ.	1928	Cornell	1933	1931-41
Annita Tuller	Hunter	1929	Bryn Mawr	1937	1937–68

*Rees was on leave 1929–31 and 1943–50.
†Guggenbuhl was also an instructor at Hunter 1922–23, the year after she received her bachelor's degree.

to file her married name for payroll purposes."[3] Jewell Hughes, who was recruited from the University of Arkansas in 1930, was a beneficiary of this policy. In 1935 she married J. Hobart Bushey, a mathematics department colleague. She was chairman of the mathematics department from 1940 until her retirement in 1966. This was a rare instance of a married couple within the same department. No doubt Hunter's location in New York City was a significant factor in providing a congenial setting for a two-person professional family.

Five of these Hunter faculty members, including Mina Rees, achieved the rank of professor. Rees held a sequence of influential positions at Hunter and then at the City University of New York (CUNY) after it was established in 1961. At that time CUNY consisted of three community colleges and four senior colleges, one of which was Hunter. Rees became professor and dean of graduate studies at CUNY, then provost of the graduate division, and finally the first president of the Graduate School.

Wellesley College followed Hunter as an employer. Seven women held their main positions at Wellesley with five of the seven serving more than thirty years and all serving at least twenty years. Furthermore, it was generally the case that the women joined the faculty some years before obtaining their doctorates. We will also see this pattern as we examine several of the other women's colleges.

Helen Merrill, an 1886 Wellesley graduate, joined the faculty there in 1893, having taught in private schools after her undergraduate work. By that time, Winifred Edgerton Merrill had already graduated from Wellesley and earned her Columbia PhD. Also by that time, five other women had graduated from Wellesley who would earn PhD's in mathematics early in the twentieth century. During Helen Merrill's

[3]Samuel White Patterson, *Hunter College: Eighty-five Years of Service* (New York: Lantern Press, 1955), 127.

Table 6.3: Wellesley College faculty - primary employment

	Baccalaureate		Doctorate		On Faculty
Helen A. Merrill	Wellesley	1886	Yale	1903	1893–1932
Roxanna H. Vivian	Wellesley	1894	Pennsylvania	1901	1901–27*
Mabel M. Young	Wellesley	1898	Johns Hopkins	1914	1906–41
Clara E. Smith	Mount Holyoke	1902	Yale	1904	1909–34
Lennie P. Copeland	Maine	1904	Pennsylvania	1913	1913–46
Marion E. Stark	Brown (Women's College)	1916	Chicago	1926	1919–60
Helen G. Russell	Wellesley	1921	Radcliffe	1932	1932–66

*Vivian was on leave 1906–09, 1918–24, and 1925–26.

first year teaching at Wellesley she also studied mathematics there. She later spent a year in graduate study at the University of Chicago and also had a leave of absence from the college in 1901–02 for study at Göttingen and for travel in England and Italy. In a letter of August 1902 to Professor Andrew Phillips of Yale, whom she had met two years earlier, she wrote, "I expect to come to Yale University this fall for work in Mathematics. There will, I suppose, be no formalities necessary for my admission, as I am a member of the Mathematics Department of Wellesley College."[4] Her Yale PhD was granted the following year. Merrill was chairman of the Wellesley department from 1916 until her retirement in 1932.

The last of this group to join the faculty was Helen Russell who earned her PhD at Radcliffe and was hired by Wellesley in 1932, the year Merrill retired. Russell retired in 1966, so for nearly three-quarters of a century the Wellesley faculty included women who had received their doctorates before 1940. Four of the seven had been Wellesley undergraduates, and all but one did their undergraduate and graduate work in the East. All achieved the rank of professor and all remained single.

We have noted earlier that members of the Wellesley faculty were active in the wider mathematical community and that Helen Merrill and Clara E. Smith, her close associate at Wellesley, both became vice presidents of the MAA. Indeed, when the MAA was founded in 1915, Helen Merrill, named one of the original associate editors of the *American Mathematical Monthly*, was the only woman among the initial sixty-two members of the executive council, the standing committees, and the section officers. All of the women in our study who were on the Wellesley faculty at that time were charter members of the association. All also remained active in a variety of scholarly activities and published some mathematical research, expository articles, textbooks, translations, and reviews. It was clear, too, that the department intended to maintain itself as something considerably more than just a teaching department. Several months after Clara Smith died in April 1943, Lennie Copeland wrote to R. G. D. Richardson at Brown asking for suggestions for a faculty member. She indicated that they were looking for an assistant professor or instructor who

[4]Helen A. Merrill to Prof. Andrew Phillips, August 3, 1902, Graduate School of Arts and Sciences, Yale University, Student Records (RU 262). Manuscripts and Archives, Yale University Library.

Figure 6.3: Wellesley College mathematics faculty, 1927–28. Front row: Marion E. Stark, Mary C. Graustein, Lennie P. Copeland; Back row: Helen A. Merrill, Clara E. Smith, (Jessie) Esther Comegys (PhD Radcliffe 1941), Mabel M. Young. (Graustein and Comegys taught at Wellesley for only short periods of time.) Photograph courtesy of Wellesley College Archives.

should be a young woman at most thirty-five who was interested in doing research work as well as teaching.[5]

The next most significant employers were Vassar College and the University of Illinois, two very different institutions. Vassar was a small women's college in Poughkeepsie, New York. Even though the department rarely consisted of more than five members during the first four decades of the twentieth century, seven women held their primary positions at Vassar. Ruth Gentry was the first on the Vassar mathematics faculty to hold a PhD. She attended lectures at the university in Berlin and at the Sorbonne, finished her doctoral work at Bryn Mawr in 1894, and then became part of a three-woman mathematics department at Vassar. Apparently illness was a factor in her resignation in 1902, two years after having been promoted to associate professor.

Vassar hired Henry Seely White as professor of mathematics in 1905. White, who had earned a doctorate from Göttingen in 1891, went to Vassar at age forty-four after more than a dozen years on the faculty, the last ten as professor, at Northwestern University. He had been one of the prime organizers of the Evanston Colloquium at which Felix Klein gave a series of lectures after the International

[5]Lennie Copeland to R. G. D. Richardson, September 30, 1943, Richardson Correspondence 1943–1945.

Table 6.4: Vassar College faculty - primary employment

	Baccalaureate		Doctorate		On Faculty
Ruth Gentry	Michigan	1890	Bryn Mawr	1894	1894–1902
Elizabeth B. Cowley	Vassar	1901	Columbia	1908	1902–30
Louise D. Cummings	Toronto	1895	Bryn Mawr	1914	1902–35
Mary E. Wells	Mount Holyoke	1904	Chicago	1915	1915–48
Audrey Wishard (McMillan)	Northwestern	1935	Radcliffe	1938	1938–42
Frances E. Baker	Iowa	1923	Chicago	1934	1942–68
Abba V. Newton	Mount Holyoke	1929	Chicago	1933	1944–67

Mathematical Congress held in conjunction with the 1893 World's Columbian Exposition in Chicago. White was to remain on the Vassar faculty for just over thirty years, the first twenty-two as the only full professor in the department. Early in his tenure there, in 1907–08, he was president of the AMS, an indication of his accomplishments and status in the profession. White seems to have been an encouraging and helpful presence. Elizabeth Cowley added in her dissertation vita, "To Professor Henry S. White, as head of my department at Vassar College, I am indebted for encouragement, criticism, and sympathy with my work."

Louise Duffield Cummings also received significant support from Henry S. White. She joined the Vassar mathematics faculty in 1902, immediately after receiving her master's degree from Toronto. Within the next few years she began graduate studies at Bryn Mawr and eventually earned a doctorate from there on a topic in combinatorics suggested by White. She continued to publish in the field and had two papers coauthored with White.

Mary Evelyn Wells, Frances Baker, and Abba Newton also enjoyed lengthy tenures at Vassar. Wells became chair of the department after White's retirement in 1936 and remained so until her retirement in 1948. Baker and Newton were hired in the early 1940s and were joined in the mid-to-late 1940s by three women who earned PhD's after 1939. All seven women listed in table 6.4 were unmarried at the time they were there; one, Audrey Wishard left after four years when she married Brockway McMillan. Others who taught at Vassar for a period of time but held their primary position elsewhere include Grace Murray Hopper, a Vassar graduate who was married at the time of her employment at Vassar, and Marie Weiss. Five others had limited experiences on the faculty at Vassar, all before receiving their doctorates. All of the women in our study who were on the faculty at Vassar for more than a few years reached the rank of professor before retirement.

The University of Illinois is unusual in that it is both the only academic institution, other than Bryn Mawr, that granted PhD's to women in the study and is among the leaders in employing them. Illinois had a large department serving many different constituencies. By some measures the Illinois department was among the top four mathematics graduate programs in the country throughout much of the first four decades of the twentieth century. By the end of 1939, the department had produced more PhD's in mathematics than any other state university. It also produced about three times as many master's degrees as PhD's and had a large

Table 6.5: University of Illinois faculty - primary employment

	Baccalaureate		Doctorate		On Faculty
Josephine Burns Glasgow	Illinois	1909	Illinois	1913	1913–15 1918–20 1924 (1st sem) 1925–26
Mary Gertrude Haseman	Indiana	1910	Bryn Mawr	1917	1920–27
Beulah Armstrong	Baker University	1917	Illinois	1921	1921–63
Olive C. Hazlett	Radcliffe	1912	Chicago	1915	1925–59*
Echo Pepper	Washington	1920	Chicago	1925	1928–65
Josephine Chanler	Western Kentucky State Teachers	1927	Illinois	1933	1932–71

*Hazlett was on disability leave 1944–59.

undergraduate program. In 1940 there were nearly sixty faculty members including twenty-seven with rank below instructor. Typically there were few faculty members at the top levels: the number of full professors ranged from three to five between 1920 and 1940 and the associate professors from one to five in the same period.[6]

Figure 6.4: Olive C. Hazlett. Photograph courtesy of LHM Institute.

Six women had their primary jobs at Illinois. Four were technically on the faculty for more than thirty years each; the other two were there for less than ten

[6]For a detailed description of the mathematics department at Illinois before 1940, see Stanford, "The History of the Mathematics Department at the University of Illinois."

years each. All but one remained single; Josephine Burns Glasgow married while on the faculty and taught a few years while her husband was on the faculty in entomology. Her career ended when they moved so he could take a position in New York State. Six others taught for short periods of time in the mathematics department after receiving their PhD's but held their primary positions elsewhere. Even though Illinois employed fifteen women in our group, most were there for very short periods, and only four of the fifteen were ever promoted beyond instructor. This latter group consisted of Beulah Armstrong and Josephine Chanler, both of whom remained after receiving their doctorates from Illinois, and Olive C. Hazlett and Echo Pepper. All eventually retired in their mid-to-late sixties after reaching the rank of associate professor.

Olive C. Hazlett was a distinguished researcher and was one of only two women in our study "starred" for mathematics in *American Men of Science*. She had a postdoctoral fellowship at Harvard the year after receiving her PhD from Chicago in 1915 and went to Illinois in 1925, taking a demotion in rank from associate professor at Mount Holyoke College to assistant professor at Illinois in order to be at an institution providing her with more time and better library facilities for research. While on the faculty, Hazlett held a Guggenheim fellowship for two years, was promoted to associate professor soon thereafter, directed one doctoral dissertation, and retired after spending her last years on disability leave.

Echo Pepper had studied with G. H. Hardy at Oxford University as a holder of a National Research Council fellowship in 1925–26, the year after she received her PhD from the University of Chicago. She went to the University of Illinois as instructor in 1928 after two years at Bryn Mawr. She was promoted to associate, a rank similar to senior instructor, in 1929 and to assistant professor in 1945. In the spring of 1960 Pepper considered moving to Southern Illinois University at a considerably higher salary than she was making at Illinois. The Illinois department chairman's recommended promotion to associate professor met with initial resistance from the dean after which several senior members of her department wrote in her behalf. She and her teaching were given high praise. Among the comments were: "I believe that Professor Pepper's promotion to an associate professorship is long overdue.... She succeeds in bringing the spirit of mathematical creativity into whatever course the department assigns her to teach."[7] Another wrote, "Other members of our staff, younger and of the opposite sex to that of Professor Pepper, have had immediate response to their requests for consideration."[8] She was promoted to associate professor in 1960 and remained in that rank until her retirement from the University of Illinois five years later at age sixty-eight.

Although Bessie I. Miller had her primary position elsewhere, we will also describe briefly her appointment at the University of Illinois. Miller received her doctorate from Johns Hopkins as a student of Arthur B. Coble in 1914 and remained at Johns Hopkins as a researcher the following year. The next year she became head of the mathematics and physics departments at Rockford College, a women's college in Illinois, where she was professor for thirteen years. It is unclear what prompted her to move to the University of Illinois as instructor in 1928. It is possible that she

---

[7]Harry Levy to M. M. Day [department head], May 2, 1960, Echo D. Pepper personnel file, Department of Mathematics, University of Illinois at Urbana-Champaign.

[8]Mahlon M. Day, Head, to Dean Lanier, May 2, 1960, Echo D. Pepper personnel file, Department of Mathematics, University of Illinois at Urbana-Champaign.

was encouraged by Coble, who had come to Illinois from Johns Hopkins in 1918 but returned to Hopkins in 1927. After only one year he came back to Illinois, the same year that Miller came from Rockford. At Illinois, Miller directed the master's thesis of Josephine Chanler and possibly others. She also published another research paper in her field, the first since her earlier work based on her dissertation. Miller died of an infection in February 1931, still an instructor at age forty-six. Although the University of Illinois employed a number of women, it appears that they did not necessarily fare very well in terms of salary or advancement.

Four women in the study had their primary employment at Goucher College, formerly Woman's College of Baltimore. As we have noted earlier, Goucher was a small college with a small mathematics department which, nevertheless, produced many women who eventually earned doctorates in mathematics. For most years between 1908 until 1924, the Goucher department consisted of two women, Clara Latimer Bacon and Florence Parthenia Lewis. In 1897 the thirty-one-year-old Clara Bacon was hired as an instructor at Woman's College of Baltimore, beginning a career there that was to last until 1934. At the time, five years after the first class graduated, the only other mathematician on the faculty was William H. Maltbie, who had earned his doctorate from Johns Hopkins two years earlier. Bacon was joined in 1908 by Lewis, who had previously studied mathematics and philosophy at Texas; philosophy at Bryn Mawr, the Sorbonne, and in Zurich; and mathematics at Radcliffe. Like Bacon she had begun her Johns Hopkins studies in 1907.

Table 6.6: Goucher College faculty - primary employment

	Baccalaureate		Doctorate		On Faculty
Clara L. Bacon	Wellesley	1890	Johns Hopkins	1911	1897–1934
Florence P. Lewis	Texas	1897	Johns Hopkins	1913	1908–47
Marian M. Torrey	Brown (Women's College)	1916	Cornell	1924	1925–59
Dorothy L. Bernstein	Wisconsin	1934	Brown	1939	1959–79

While Clara Bacon and Florence Lewis were on the Goucher mathematics faculty together, nine women graduated who later received PhD's in mathematics. Another graduated before Lewis joined the faculty. Of these ten, seven received their doctorates from the Johns Hopkins University. Bacon not only encouraged women to go to graduate school but assisted many financially so that they could do so. Marian Marsh Torrey joined the faculty in 1924, and Dorothy Bernstein came to Goucher as professor in 1959, having spent most of the previous sixteen years at the University of Rochester. Both Torrey and Bernstein later served as chair of the department.[9] None of these four women married.

It was rare when the schools that awarded doctorates to women in mathematics before 1940 became employers of these women. Furthermore, if they did hire them, normally the women did not advance through the ranks. Altogether, twenty-four women had their primary faculty position at one of these schools and attained the

[9] Some of the material about Goucher is based on Cockey, "Mathematics at Goucher: 1888–1979."

rank of at least assistant professor before retiring. Thus, just over 10 percent of our cohort spent the major part of their careers at schools that had awarded PhD's to the women in the study.

Table 6.7: Primary employment at schools awarding PhD's to women before 1940

	Name (PhD school, year)	On faculty	Final rank
Bryn Mawr	Marguerite Lehr (Bryn Mawr 1925)	1924–67	Professor
Bryn Mawr	Anna Pell Wheeler (Chicago 1910)	1918–48	Professor
California	Pauline Sperry (Chicago 1916)	1917–50*	Assoc. prof.
Chicago	Mayme Logsdon (Chicago 1921)	1921–46	Asst. prof.
Colorado	Claribel Kendall (Chicago 1921)	1913–57	Professor
Duke	Julia Dale (Cornell 1924)	1930–36	Asst. prof.
Illinois	Beulah Armstrong (Illinois 1921))	1921–63	Assoc. prof.
Illinois	Josephine Chanler (Illinois 1933)	1932–71	Assoc. prof.
Illinois	Olive C. Hazlett (Chicago 1915)	1925–59†	Assoc. prof.
Illinois	Echo D. Pepper (Chicago 1925)	1928–65	Assoc. prof.
Indiana	Cora Hennel (Indiana 1912)	1909–47	Professor
Kansas	Wealthy Babcock (Kansas 1926)	1920–66	Assoc. prof.
Kansas	Florence Black (Kansas 1926)	1918–60	Assoc. prof.
Minnesota	Elizabeth Carlson (Minnesota 1924)	1924–65	Professor
Minnesota	Gladys Gibbens (Chicago 1920)	1920–58	Assoc. prof.
Missouri	Nola Haynes (Missouri 1929)	1946–67‡	Assoc. prof.
New York Univ.	Fay Farnum (Cornell 1926)	1927–41	Asst. prof.
Ohio State	Grace Bareis (Ohio State 1909)	1908–46	Assoc. prof.
Purdue	Cleota Fry (Purdue 1939)	1939–77	Assoc. prof.
Syracuse	Nancy Cole (Radcliffe 1934)	1947–71	Assoc. prof.
Texas	Goldie Horton Porter (Texas 1916)	1913–57	Asst. prof.
Washington	Mary Haller (Washington 1934)	1929–65	Assoc. prof.
Wisconsin	Florence E. Allen (Wisconsin 1907)	1902–43	Asst. prof.
Wisconsin	Elizabeth Hirschfelder (Wisc. 1930)	1931–53§	Asst. prof.

*Sperry was dismissed during the McCarthy era when she refused to sign a loyalty oath. She was reinstated in 1952 as associate professor emeritus and received back pay.
†Hazlett was on disability leave 1944–59.
‡Nola Anderson (Haynes) also taught at Missouri in 1929–30, before her marriage, and one year, 1943–44, during World War II after her marriage.
§During the period 1931–47 while Hirschfelder was married to her first husband, the mathematician Ivan Sokolnikoff, she held various irregular faculty positions.

Among academic institutions, it was the case that women's colleges almost always proved the most supportive and rewarding setting, while those institutions that had granted PhD's to the women were generally the least supportive. The data for PhD-granting schools are almost identical for primary positions to those for first postdoctoral jobs. Just sixteen of these thirty-four US institutions hired and retained a total of 24 of the 228 women PhD's in faculty positions that eventually carried a rank above instructor. In all but two cases, these 24 also held their primary position at the school.[10] Of these 24, 15 were retained by the institution where they had done their graduate work. Once employed, only 5 attained the rank of professor before retirement. Two of these were at Bryn Mawr and the other 3 were at state universities in the Midwest and West (Colorado, Indiana, and Minnesota).

Looking more closely at the three at state universities, Kendall at the University of Colorado, Hennel at Indiana University, and Carlson at the University of Minnesota, we see many similarities. All were natives of the states where they were employed. Cora Hennel and Elizabeth Carlson spent their careers at the schools where they had done all of their undergraduate and graduate work. Hennel earned the first mathematics PhD at Indiana in 1912, and Elizabeth Carlson received one of the first two doctorates in mathematics, both awarded in 1924, at Minnesota. Claribel Kendall received her bachelor's and master's degrees from Colorado and taught as an instructor there for several years before receiving her doctorate from the University of Chicago in 1921, long before Colorado had an active doctoral program in mathematics.[11]

Although all three achieved the rank of professor before they retired, it appears that in at least two instances their advancement through the ranks was unusually slow. For example, in an essay on Claribel Kendall, Ruth Rebekka Struik describes Kendall's career advancement at Colorado and then indicates that one might compare it with that of George H. Light.[12] Doing so, we see that Kendall was an instructor from 1913 to 1921, when she received her doctorate from the University of Chicago at age thirty-two. She was then assistant professor 1922–28, associate professor 1928–44, and finally professor 1944–57. On the other hand, Light arrived at Colorado as assistant professor in 1916, just after receiving his PhD from Yale at about age forty. He was promoted to associate professor two years later and to professor two years after that. During this period there were just two professors in the department, and Kendall was not promoted until 1944, a year after Light had retired. Initial assumptions might be that Light was particularly distinguished to have jumped to full professor in just four years, whereas it took twenty-three years for Kimball. Neither, however, could be said to have been a particularly productive researcher. Kendall's dissertation was published in the *American Journal of Mathematics*, whereas a search through the major reviewing journals reveals that Light's dissertation apparently was not published except as a dissertation monograph. Light had one article in the *Monthly* in 1928, long after his last promotion. Neither directed any of the four PhD dissertations awarded by Colorado before 1950, and both directed master's theses. Thus, one can only speculate that issues other than professional competence played a role in this discrepancy.

While we will not try to make direct comparisons of Elizabeth Carlson to others in her department, we note that even though she was eventually promoted to professor, the promotion came when she was nearly sixty-seven, two years before

[10]Fay Farnum held her primary position at NYU but left and held another position elsewhere and so is not one of the 24 who were hired and retained. On the other hand, Ruth W. Stokes held her primary position at Winthrop College but was later hired and retained by Syracuse.

[11]Although a doctorate in mathematics had been awarded at Colorado in 1906, the next were not given until the 1930s. Much of the material about the mathematics department at Colorado is based on Jones and Thron, *A History of the Mathematics Departments of the University of Colorado.*

[12]Ruth Rebekka Struik, "Claribel Kendall (1889–1965)," in *Women of Mathematics: A Biobibliographic Sourcebook*, eds. Louise S. Grinstein and Paul J. Campbell, 92.

her retirement. Cora Hennel, on the other hand, had a considerably more positive experience at Indiana University. She and both of her sisters received bachelor's and master's degrees at Indiana, and all were on the faculty at Indiana at some point in their lives (in mathematics, English, and botany). Cora Hennel entered Indiana in 1903, continued her studies, and began teaching there immediately after receiving her bachelor's degree in 1907. She was a teaching fellow for two years after her graduation, earned her master's degree in 1908, and was appointed instructor of mathematics in 1909. At the suggestion of the head of the department, she continued her work for the PhD at Indiana and wrote her dissertation under the direction of R. D. Carmichael, a 1911 Princeton PhD. Hennel's PhD in June 1912 was the first in mathematics awarded by Indiana University. She remained an instructor until 1916, was assistant professor 1916–23, associate professor 1923–36, and became professor in 1936 more than a decade before her death at sixty-one. An obituary recorded that Hennel was "well known throughout the State, having addressed numerous Indiana University alumni groups, and various educational, professional and civic organizations." It noted further that "Dr. Hennel's chief contributions were in her excellent teaching and her influence on many students, both native and foreign."[13]

Two situations that deserve mention are those of Grace Bareis at Ohio State University and Florence E. Allen at the University of Wisconsin. Grace Bareis, a native of Ohio, graduated in 1897 from Heidelberg University in Tiffin, Ohio, was a graduate student at Bryn Mawr College, graduated from the Columbus (Ohio) Normal School, and was a teacher at a private school in Pennsylvania while resuming her graduate work in mathematics at nearby Bryn Mawr. In 1906 she returned to the Midwest to finish her graduate work in mathematics at Ohio State University. She was a fellow there 1906-08 and was appointed assistant professor in 1908. During her years at Ohio State, she directed master's theses, mainly in geometry, for at least fifteen graduate students. During World War II, she taught students in the Army Specialized Training Program at Ohio State, and for two years after her retirement in 1946 at age seventy, still as an assistant professor, she continued to teach because of the postwar shortage of instructors.

Perhaps the most extreme situation was that of Florence E. Allen, who received her PhD at Wisconsin and who served as instructor there for forty-three years before being promoted to assistant professor at the age of sixty-eight, two years before her retirement. Allen did all of her undergraduate and graduate work at Wisconsin and became instructor in 1902, the year after receiving her master's degree and five years before her PhD was completed. Anecdotal evidence suggests that she was always "Miss Allen" to the graduate students and that it was not commonly known that she had ever earned a doctorate. While she was not a steady or prolific publisher, three research papers by her did appear between 1914 and 1927, the last in the *American Journal of Mathematics*.

It is possible that Allen's lack of promotion resulted from anti-nepotism practices. Her brother, Charles E. Allen, received his PhD in 1904 from Wisconsin and was on the botany faculty there from 1901 to 1943, when he retired as professor emeritus. He had attained the rank of professor in 1909. His career was a distinguished one with many honors including election to the National Academy of Sciences. In response to an inquiry by Helen Owens concerning the outlook for women doctorates in mathematics, Florence Allen wrote in 1940:

[13] *Proceedings of the Indiana Academy of Science* 57 (1947): 3–4.

> Of course there will always be some women who should go in for a Ph.D. – some because it will be an actual necessity to qualify them for one of the occasional – very occasional – openings in college and university positions, some because of the leisure they may have to follow a congenial pursuit. But on the whole I see no great encouragement to be had from past experiences and observation. I do not believe that there is or will be a great future for any but a few in this field. At present, it seems to me, as I look about this campus, that in all strictly academic fields (not those special to women) that there is a decided drop in the number of women engaged. That may be peculiar to this economic phase, but I look for it to continue for some time to come.[14]

Altogether, six women spent their careers at the institution where they had earned both their undergraduate and graduate degrees. Besides Carlson, Hennel, and Allen, they include Mayme Irwin Logsdon (PhD Chicago 1921), and Florence Black and Wealthy Babcock (both PhD Kansas 1926). All but Logsdon were associated with the main public universities of the state in which they were born, and most of these universities had minimal graduate programs at the time of their degree and hiring.

Florence Black, from a cattle ranch in southwestern Kansas, and Wealthy Babcock, from a farm in northern Kansas, had similar educational and career paths and were close associates most of their lives. Both taught in high schools after completing their undergraduate work, and both earned their doctorates from the University of Kansas in 1926, having had the same dissertation advisor. Black and Babcock were the fourth and fifth, and the first since 1910, to receive PhD's in mathematics at Kansas. The next two doctorates at Kansas were awarded in 1930 and 1947. Although a number of master's degrees were awarded at Kansas, the pace of the graduate program didn't really accelerate until the 1950s. Black and Babcock were promoted to associate professor in 1940 and both retired at that rank, Black in 1960 and Babcock, who was six years younger, in 1966. Both were known as outstanding teachers and were formally recognized as such by the university.

Finally, Mayme Logsdon is the only one of these six, and one of the very few, to have been retained by one of the schools with a highly regarded graduate program in mathematics. Although the University of Chicago, which granted forty-six PhD's to women in our study, led by far in that category, its record has been dismal with respect to the employment of women in mathematics. Since 1892, and as of this writing, there have been only two women in the mathematics department at the level of associate or full professor. One was Mayme Logsdon, who directed four doctoral students in algebraic geometry, and who was an associate professor at the time of her retirement in 1946. The other was Karen Uhlenbeck, who was awarded a MacArthur fellowship and was appointed professor of mathematics at the University of Chicago in 1983. She was then elected to the National Academy of Sciences. After five years at Chicago, she resigned in order to accept a chaired professorship at the University of Texas.

---

[14]Florence E. Allen to Helen B. Owens, June 8, 1940, Series IV, Women in Mathematics and the Sciences, Helen Brewster Owens Papers, Schlesinger Library, Radcliffe Institute, Harvard University.

Figure 6.5: University of Chicago mathematics faculty members, mid-1920s: (l to r) E. H. Moore, G. A. Bliss, M. I. Logsdon, L. E. Dickson, W. D. MacMillan, F. R. Moulton, H. E. Slaught. Photograph courtesy of the University of Chicago Department of Mathematics.

A few other women spent many years on the faculty of schools that did not have active doctoral programs in mathematics when they joined the faculty, but were developed later. This was the situation at Penn State, which was Pennsylvania State College when several women joined the faculty, most at least two decades before mathematics doctorates were first granted there in the early 1950s. Indeed, Penn State was the only coeducational university besides the University of Illinois among the ten leading academic employers. Penn State employed at various times five women who held their primary positions there. Teresa Cohen (PhD Johns Hopkins 1918), Aline Huke (Frink) and Beatrice Hagen (both PhD Chicago 1930), and Ethel Moody (PhD Cornell 1930) joined the faculty as instructors in 1920, 1930, 1931, and 1933, respectively. As we noted in the introduction, Helen Owens (PhD Cornell 1910) was married to the department chairman at Penn State and was, thus, not able to assume a regular position until late in her career. She was appointed instructor in 1940 and retired nine years later as assistant professor. Frink was permitted to keep her position when she married in 1931; however, she resigned when she became pregnant and was not rehired into a regular position until 1947. Ethel Moody died in an automobile accident nine years after she was hired. She was still an instructor at the time. Cohen, Frink, and Hagen retired after having been promoted to professor.

Although it was not a leading employer of women, we mention the University of Kentucky because two women were on the faculty when the mathematics department granted its first PhD in 1930. Elizabeth LeStourgeon (PhD Chicago 1917) had arrived a decade earlier, at age forty, and retired as associate professor in her mid-sixties. Sallie Pence (PhD Illinois 1937) became an instructor in 1929. She had

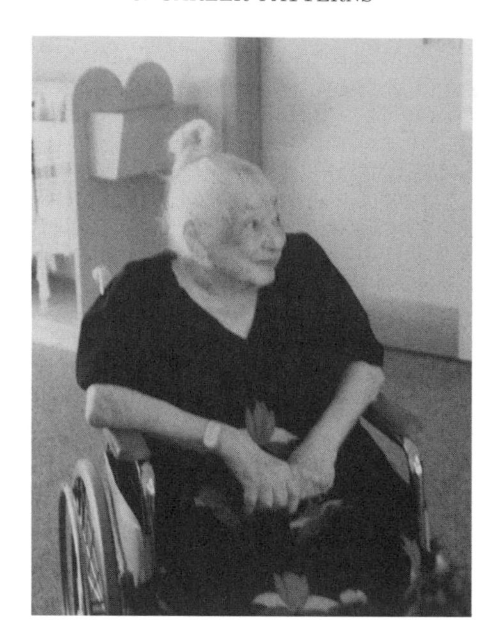

Figure 6.6: Teresa Cohen a week before her 100th birthday in February 1992. (As of mid-2008 Cohen, Elizabeth Hirschfelder, and Elizabeth Carlson were the only centenarians among the women in this study. Cohen and Hirschfelder died at age 100 and Carlson at age 104.) Photograph by James Leitzel, courtesy of Joan Leitzel.

earned bachelor's and master's degrees at Kentucky and had taught in high schools for many years before her appointment there. She earned her doctorate when she was forty-three and retired from Kentucky when she was nearly seventy, having been promoted to professor seven years earlier.

Two other women who held their primary positions at schools that had or were beginning doctoral programs in mathematics are Lois Griffiths (PhD Chicago 1927) and Grace Shover Quinn (PhD Ohio State 1931). Lois Griffiths was at Northwestern University from the time she received her doctorate until her retirement as associate professor in 1964. She published research articles during her first two decades at Northwestern, continuing well after that school granted its first mathematics doctorate in the late 1930s. Grace Quinn returned to full-time teaching after several years at home with children. She directed a number of master's theses at American University, mainly in the 1960s, and retired as professor in 1970 after fourteen years on the faculty. American University had begun to award doctorates in mathematics in the 1950s, with the first several in statistics, far from Quinn's area of algebra. None of the women in our study who taught at Penn State, Kentucky, Northwestern, or American directed doctoral dissertations.

Finally, two women who taught at schools with emerging doctoral programs in mathematics did direct PhD dissertations, although the job was not the primary position for one of them. Florence Mears (PhD Cornell 1927) and Dorothy Bernstein (PhD Brown 1939) held positions at George Washington University and the University of Rochester, respectively. Mears started her job at George Washington in 1929 and remained until her retirement as professor emeritus at age seventy, having directed two PhD dissertations when she was in her sixties. She then taught at Howard University for another ten years and was on the faculty there when the mathematics department started a doctoral program in the mid-1970s. Dorothy Bernstein was hired by Rochester as instructor in 1943, a few years before doctorates were awarded there. During the fourteen years she was on the faculty, she was on leave for two years, one as a member of the Institute for Advanced Study at Princeton and another as a visiting professor at the Institute for Numerical Analysis at UCLA. She directed three doctoral dissertations at Rochester and was promoted to full professor before moving to Goucher College for the last two decades of her career.

We have focused on academic careers in mathematics departments because they were so prevalent among the women in our study. About two dozen women, however, chose other directions, usually, but not always, in areas related to mathematics. In some cases, caretaker responsibilities for husband, children, or parents dictated certain paths outside the norm. These other types of careers included jobs with the federal government, in private industry, and in academic settings in fields outside of mathematics. Occasionally there were different types of employers over the working life of an individual.

An example of the last situation is provided by Irene Price (PhD Indiana 1932) who had three distinct careers. After spending about fifteen years, 1929–44, on the faculty of a state teacher's college, she worked as a statistician for the US Air Force for a dozen years at Wright-Patterson Air Force Base in Ohio and Holloman Air Force Base and White Sands Proving Ground in New Mexico. In this position she coauthored technical reports and articles in applied statistics for problems related to instrumentation and coordinate systems for missiles. Finally, she was self-employed as a real estate agent and owner of Price Realty from 1956 until her retirement in 1998 at age ninety-six.

Whereas the Great Depression was a generally negative factor in the careers of many of the women in our group, World War II sometimes opened opportunities. Even during World War I, there is indication of some war-related activity. Starting in about 1915, Elizabeth S. Dickerman (PhD Yale 1896) was primarily involved with literary pursuits, and during World War I she chose and translated war poems for a calendar, the proceeds of which were to go to wounded French soldiers.

One of the women in our study had an even closer connection to the war effort during World War I. Caroline E. Seely (PhD Columbia 1915), who had been working since 1913 as a clerk to Frank N. Cole, the secretary of the AMS, joined many prominent mathematicians who worked on ordnance with F. R. Moulton of the University of Chicago. Some, like Seely, worked as civilians and others, including Moulton who was a major, had military commissions.[15]

During World War II, several women, all with doctorates from the early 1930s, did special work as civilians in support of the war effort. For example, Grace Quinn (PhD Ohio State 1931) was first a cryptanalyst for the US Navy and then a research assistant for the National Defense Research Committee (NDRC). During and after the war Ruth W. Stokes (PhD Duke 1931) was an instructor of military cryptography and cryptanalysis. Ruth Peters (PhD Radcliffe 1933) contributed to the war effort by doing research on the theory of errors in bombing while serving as a technical aide in airborne fire control, guided missiles, and rockets for the Office

[15]"American Mathematicians in War Service," *American Mathematical Monthly* 26 (1919): 40-44.

of Scientific Research and Development (OSRD). Dorothy W. Weeks (PhD MIT 1930) and Marguerite Hedberg (PhD Missouri 1932) also worked as civilians for the OSRD, with Weeks serving as a technical aide in the liaison office.

At least four women, all with PhD's from the mid 1930s, served in the military. For one of them, Grace Murray Hopper (PhD Yale 1934), it was the beginning of a distinguished career that lasted nearly a lifetime. For the others, it was temporary. Henrietta Terry (Nee) (PhD Illinois 1934) had been teaching at the University of Illinois high school when she was commissioned a lieutenant (jg) in the US Naval Reserve. During the early years of World War II, Madeline Levin (Early) (PhD Bryn Mawr 1936) took a military leave of absence from her position at Hunter College, earned a bachelor's degree in nursing from New York University, and was appointed ensign in the Naval Reserve. She was eventually transferred to Guam where she served for nearly a year before being released from active duty. Caroline Lester (PhD Wisconsin 1937) was a lieutenant in the United States Coast Guard (Women's Reserve) stationed at the Naval Communications Annex in Washington, D.C., as a specialist in cryptanalysis.

Several women in the study, most notably Hopper, were actively engaged in activities related to the war, in some cases changing their careers entirely. After receiving her doctorate in 1934 Hopper returned to Vassar, her undergraduate institution, to teach. After the bombing of Pearl Harbor in December 1941, Hopper wanted to join the navy. Even though she was over-age for enlistment in the newly formed women's branch of the US Naval Reserve, the WAVES (Women Accepted for Volunteer Emergency Service), and she worked in a profession (mathematics teaching) that was considered crucial, Hopper took a leave of absence from Vassar and asked for a waiver to join the WAVES. During the summer of 1943 she taught an accelerated wartime calculus course as an assistant professor at Barnard College and joined the US Naval Reserve in December 1943. In July 1944, Lt. Grace Hopper reported for duty at the Bureau of Ordnance Computation Project housed at Harvard University and began work on the Mark I computer. This was the beginning of a distinguished career both in computer science and with the US Navy. In the former, she is noted as the creator and developer of the first compilers and as an advocate for making computers accessible to large numbers of people. Her navy career finally ended in 1986, when she was seventy-nine, the oldest officer on duty in the armed forces, and holding a rank of rear admiral. Her name is perhaps the most widely recognized of all the women in mathematics and computer science in the twentieth century, and she is among the most celebrated, having received nearly fifty honorary degrees, the Distinguished Service Medal from the Department of Defense, and the National Medal of Technology. In 1997, five years after her death, the USS *Hopper*, a guided missile destroyer bearing her name, was commissioned in recognition of her contributions to the navy.

Perhaps the most highly regarded within the mathematical, academic, and larger communities was Mina Rees, a woman whose work during World War II had a significant impact on the contributions of mathematics to the war effort and whose work afterwards was instrumental in developing support for basic research in science and mathematics. Soon after the war, her work was recognized with the President's Certificate of Merit in the United States and with the King's Medal for Service in the Cause of Freedom, awarded by the British government. Like Hopper, Rees (PhD Chicago 1931) started her career as a faculty member at her undergraduate institution, in this case Hunter College. In 1943 she took a leave of absence from Hunter to work for the newly established Applied Mathematics Panel (AMP) of the National Defense Research Committee; the AMP enlisted civilians to assist the military with weaponry and other war-related work. When Rees left the AMP in 1946 she became head of the mathematics branch of the Office of Naval Research (ONR) in Washington, D.C., and was deputy science director when she left the ONR in 1953. Later that year the AMS adopted a resolution recognizing the significance of her role during and just after the war.

After leaving the government, Mina Rees returned to Hunter College as professor of mathematics and dean of the faculty. Rees remained as professor and dean until 1961, when the City University of New York was founded and she became professor and dean of graduate studies. In the latter position she was instrumental in shaping the nature of graduate studies at CUNY. When CUNY formed the separate Graduate School, Rees became its first president and served as such until her retirement in 1972. In 1985 the library at the Graduate School and University Center of CUNY was dedicated as the Mina Rees Library. She served on the National Science Board 1964–70 and was president of the AAAS in 1971. Among her numerous other awards were the MAA's first award for distinguished service to mathematics in 1962 and the Public Welfare Medal of the National Academy of Sciences in 1983.

Gertrude Blanch (PhD Cornell 1935) also served the United States government during and after World War II. Unlike Hopper and Rees, Blanch became involved with the federal government as a result of the Great Depression. In 1938, unable to find an academic job, she joined the Mathematical Tables Project, which used mathematically untrained individuals to produce tables useful to mathematicians and scientists and was funded by the WPA. After the United States entered World War II, the mathematical staff of the Tables Project was transferred to the National Bureau of Standards, and the project was transformed from a works project to a professional organization concerned with computations. After the war, Blanch continued working with many of the same people, but at the Institute for Numerical Analysis at UCLA. Finally, in 1954, she returned to government service as a senior mathematician at the Aerospace Research Laboratories of Wright-Patterson Air Force Base. In 1964 she was given the Federal Woman's Award from the Civil Service Commission for her contributions.

Some of the women in our study who had non-academic careers split them between government and private industry; others remained in one sphere the entire time. Table 6.8 below includes those women whose main employment was non-academic, but the table does not indicate the broad range of employment held by many individual women. We list only one employer for each, the one with whom the woman remained the longest, and only one position, the last held with that employer.

Marion Epstein stayed at ETS, where she worked in the Test Development Division and in the College Board Programs Division, where she was named vice president in 1977. In addition to her work with ETS, Epstein served on local and statewide boards of education as well as in other positions relating to education starting in the mid-1950s and continuing for more than twenty years.

While most of the women who worked outside of academia held jobs at least somewhat related to the mathematical sciences, several worked in other areas. Most

Table 6.8: Instances of non-academic employment

Name (PhD date)	Employer, position
Ida M. Metcalf (1893)	New York City Finance Department, statistician
Charlotte C. Barnum (1895)	US Naval Observatory, computer
Leona M. Peirce (1899)	Peirce Music Company, manager and treasurer
Ida Barney (1911)	Yale Observatory, director of star mapping project
Caroline E. Seely (1915)	American Mathematical Society, assistant to the secretary
Sarah E. Cronin (1917)	Equitable Life Assurance Society, chief of policy change division
Nina Alderton Moore (1921)	National Bureau of Standards, mathematician
Flora D. Sutton (1921)	John C. Legg & Co. [brokerage firm], statistician
Marion C. Gray (1926)	Bell Telephone Laboratories, technical staff
Beatrice Aitchison (1933)	US Post Office, director of transportation rates and economics
Margaret Gurney (1934)	US Bureau of the Census, statistician
Grace M. Hopper (1934)	Sperry Rand Corporation, staff scientist
Gertrude Blanch (1935)	Wright-Patterson Air Force Base, senior mathematician
Ruth Stauffer McKee (1935)	Joint State Government Commission [Pennsylvania], analyst
Vivian E. Spencer (1936)	US Bureau of Mines, chief of commodity staff
Anna M. C. Grant (1937)	Airborne Instruments Laboratory, engineer
Marion G. Epstein (1938)	Educational Testing Service, vice president
Elizabeth Baxter (1939)	Jet Propulsion Laboratory, numerical analyst
Bella Kotkin Greenfield (1939)	RAND Corporation, technical staff

of these positions were in economics or statistics, but one woman, Ida Barney (PhD Yale 1911), was an astronomer for most of her career. Barney, after a decade of college teaching following her doctorate, and after having augmented her undergraduate astronomy courses with some graduate courses, joined the staff of the Yale Observatory as research assistant. Soon thereafter she became involved in a major star mapping project and eventually became director of the project. She received the Annie J. Cannon Prize from the American Astronomical Society in 1952. It was later reported that she received the prize "for completing one of the most intensive photographic mapping jobs ever undertaken by a single observatory."[16]

Both of the women who worked as economists, Beatrice Aitchison and Dorothy Stahl Brady, were connected with the federal government, and both received awards for that work. After receiving her doctorate, Aitchison (PhD Johns Hopkins 1933) had two one-year jobs, first teaching mathematics and then teaching statistics. Hoping that a master's degree in economics would help her secure a job teaching statistics in an economics department, she enrolled in such a program at the University of Oregon and was awarded the degree in 1937. While studying for this second

[16]"6 at Yale to Retire," *New York Times*, March 20, 1955.

master's degree, she became involved with transportation economics, which soon became her field of expertise. After teaching statistics for a year, Aitchison began her long association with the federal government with a position at the Interstate Commerce Commission. In 1951 she moved to the US Department of Commerce as director of the Transport Economics Division, and in 1953 she moved again, this time to the US Post Office, where she started as director of transportation research and retired as director of transportation rates and economics. In 1961 Aitchison was one of six recipients of the first annual Federal Woman's Award. In 1970, the year before she retired, she was honored with the Career Service Award by the National Civil Service League as one of the top ten government workers of the year, the only woman in the group.

In addition to working in the federal government, Dorothy Brady (PhD California 1933) also worked in an academic setting outside of mathematics. Brady married just after finishing her undergraduate work, earned a master's degree in mathematics, and worked at the National Bureau of Economic Research between brief teaching positions at Vassar and New York University. By the time she began her career as economist and statistician with the federal government in 1936, she had had a son and was divorced. Her first position with the government was as home economics specialist in the US Department of Agriculture's Bureau of Home Nutrition and Home Economics. She left that position in 1943 and for a year served as the chief of the Family Expenditure Section, Cost of Living Branch, Division of Prices and Cost of Living of the US Department of Labor's Bureau of Labor Statistics. Her next government position was also with the Cost of Living Branch, this time as chief of its Bureau of Statistics. She left that position in 1948, the same year she received the National Women's Press Club Award in Economics. That year Brady also temporarily returned to academia, teaching mathematics and statistics at the University of Illinois. While at Illinois, Brady was elected fellow of the American Statistical Association and fellow of the Econometric Association. She briefly returned to the Bureau of Labor Statistics before holding a professorship in economics for two years at the University of Chicago and ending her distinguished career as professor of economics for a dozen years at the Wharton School at the University of Pennsylvania.

Among the women in our study we have seen a wide range of careers, including the eclectic pursuits in mathematics, law, and literature of Emily Coddington Williams, and Grace Murray Hopper's rise to rear admiral in the US Navy. Others, too, had stimulating and fruitful careers outside of academia, most often with the federal government. But, for the most part, what we have seen are women who chose to teach mathematics at the college level and did so, often in the face of discrimination against women in general and married women in particular. Despite these obstacles, many, although clearly not all, of the women in our study succeeded in having satisfying and productive careers on college faculties. Examples of this satisfaction can be seen in the following two 1926 quotations from women who had earned doctorates in mathematics by 1915. The first is in response to the question "Would you advise a Ph.D. for the work you are doing?" while the second represents a more general remark.

> I should encourage women vigorously. My personal experience has been most fortunate. Apparently I have had only the knocks needed to act as spurs. I believe there is a better opportunity for

women in the profession than ever before, let one be only willing to surmount difficulties and seek real achievement.[17]

The freedom from monotony in the work in mathematics, the vision and grasp of fields of knowledge that may be interpreted through mathematics, the ideals of thought and of thinking, and the ability to interpret in conduct, relief from the turmoil of a crowded life, all these make the Ph.D. more valuable than any professional advantage to be derived from it.[18]

---

[17]Hutchinson, *Women and the Ph.D.*, 101.
[18]Ibid., 173.

CHAPTER 7

Scholarly and Professional Contributions

The women in our study made scholarly contributions in many areas and in many ways. They wrote research papers, expository papers, technical reports, chapters in books, and entire books and monographs in mathematics and in other fields; they gave talks at schools and at professional meetings; they served as editors and reviewers of professional journals; they directed dissertations; and they contributed to the American mathematical community through their participation in the various professional organizations.

We begin by looking at one aspect of the research records of the women in our study using the same criteria that R. G. D. Richardson did in 1936 when he examined, among other things, the publication records of those individuals who received PhD's by 1933.[1] For the purposes of his article, Richardson counted journal articles that had appeared on the American Mathematical Society's "Annual List of Papers Read Before the Society and Subsequently Published" and used other, unnamed, sources to find articles published before the period covered by these lists.[2] In table 7.1 below, we reproduce Richardson's table III with comparable data for women added to his data for all mathematicians who earned doctorates during specified periods of time.

While the percentage of mathematicians whose research records include no published papers appears high for both men and women, it is not unusual. In a 1939 article concerning the then current standards for the PhD degree, we find that among those who hold PhD's "the large majority never make another contribution to learning after their dissertations, and another large percentage are never again called upon to employ the methods of research in which they have been trained."[3]

At the beginning of the twentieth century C. J. Keyser of Columbia University described the conditions under which many men with mathematical training published very little, if any, mathematics. His description similarly details the conditions under which the vast majority of the women in our study worked.

> There is in our country a goodly number of men having the requisite ability and training, who, nevertheless, produce but little or nothing at all–a fact to be accounted for by the absence in their case of other essential conditions. In some cases library facilities are lacking.... Again, there are cases where able men have not

[1]Richardson, "The Ph.D. Degree and Mathematical Research."

[2]Fifty such lists were published. The first appeared in the *Bulletin of the New York Mathematical Society* in 1892. Each of the first four lists fits on a single page. Through 1925 no list was longer than ten pages, but in 1926 the list grew to sixteen pages. The longest list, twenty-one pages, appeared in 1932. Publication of the list was discontinued after the introduction of the *Mathematical Reviews* in 1940.

[3]I. L. Kandel, "The Ph.D. Degree," 235.

> the necessary leisure to engage successfully in investigation.... Again, scientific activity is not infrequently rendered impossible by the amount of administrative work which professors of notable administrative ability are called upon to perform.... Once more, it is very desirable, indeed it is really necessary, for men working in a branch of science to attend the meetings of scientific bodies, in order to meet their fellow-men, to take counsel of them, to create and share in a wholesome *esprit de corps*, to catch the inspiration and enthusiasm, and to gain the sustaining impulses which can come only from personal contact and co-operation. But our country is so vast, the distances so long, and traveling so expensive, that many mathematicians, owing to smallness of income, find themselves hopelessly condemned to a life of isolation, of which the result is a loss first of interest and then of power. It is in vain that one counsels such men to wake up and be strong and active, for their state of inactivity is less a defect of will than an effect of circumstances.[4]

Table 7.1: Papers published by Americans receiving PhD's by 1933*

	Persons taking degrees 1862–1933				Persons taking degrees 1895–1924			
	Number		Percent		Number		Percent	
	All†	Women	All†	Women	All†	Women	All†	Women
No papers	549	100	46%	63%	232	49	39%	59%
1 paper	227	39	19%	25%	109	22	18%	27%
2 papers	100	11	8%	7%	58	5	10%	6%
3–5 papers	131	6	11%	4%	66	4	11%	5%
6–10 papers	70	2	6%	1%	41	2	7%	2%
11–20 papers	69	1	6%	1%	50	1	8%	1%
21–30 papers	20		2%		17		3%	
More than 30 papers	22		2%		17		3%	
Total	1188	159			590	83		

*In this table we use Richardson's definition of a "paper" as described above.
†From Richardson, Table III, 209.

Nonetheless, while table 7.1 clearly indicates that women were less prolific than men when counting papers published after having been read at a meeting of the AMS, it also indicates that a significant proportion of women were reading papers at meetings and then publishing them. Furthermore, even using this definition of a research record, one woman, Olive C. Hazlett, placed well within the top 10 percent of all research mathematicians who earned degrees by 1933 and well within the top 14 percent of all research mathematicians who earned degrees between 1895 and 1924. Hazlett, who received her PhD in 1915, published fourteen papers between

[4]Keyser, "Mathematical Productivity in the United States," 353–55.

1914 and 1930 that were presented at meetings of the AMS. Since Richardson combines those who published between eleven and twenty such papers, we cannot be precise as to Hazlett's standing. Even so, it is not unreasonable to suppose she is near or among the top 10 percent of all mathematicians taking degrees in the restricted period 1895–1924.

Richardson chose the restricted period 1895–1924 "as being far enough in the past so that men have had an opportunity to get something into print, and not extending back far enough so that lack of publication facility and of stimulus enters into the calculation."[5] While he did not elaborate on what he meant by stimulus, it is likely that Richardson was referring to the growth of centers of mathematical research at various universities in the United States. In their book on the early American mathematical research community, Parshall and Rowe describe this growth as follows:

> The 1920s and 1930s witnessed further growth and consolidation within American mathematics, as many established programs in the East and throughout the Midwest expanded into solid departments. Besides Chicago, Harvard, and Princeton, publicly funded universities like Michigan, Illinois, Wisconsin, California, and Ohio State developed programs competitive with the older ones at institutions like Columbia, Yale, the Johns Hopkins, and Cornell. Meanwhile, the mathematics faculty at the Massachusetts Institute of Technology vied for a place alongside the nation's "Big Three" schools [Chicago, Harvard, and Princeton]. Outside this more élite coterie, however, teaching occupied twelve to eighteen hours per week, making it difficult for students trained in these strong programs to maintain their mathematical momentum once they entered the academic working force. Only on rare occasions might a mathematician at one of the more teaching-intensive institutions be given a lighter load as an inducement to do research.[6]

As we have seen, few women taught at schools that had significant graduate programs in mathematics. In fact, only nine of the women in our study achieved a professorial rank at any of the research-oriented institutions referred to above by Parshall and Rowe.[7] In addition, Mildred L. Sanderson at Wisconsin and Bessie I. Miller at Illinois were instructors but died shortly after taking those positions. Nonetheless, we find nine women who, by Richardson's criterion, "made ... substantial contributions to research as ... evidenced by the publication of three or more research articles."[8] Using Richardson's definition of a research article, Olive C. Hazlett published fourteen articles, Louise D. Cummings and Anna Pell Wheeler each published eight, Lois Griffiths published five, Mary Curtis Graustein and Caroline E. Seely each published four, and Elizabeth Carlson, Lennie P. Copeland, and

[5]Richardson, 210.

[6]Parshall and Rowe, *The Emergence of the American Mathematical Research Community*, 445.

[7]One of these nine, Florence E. Allen, was promoted to assistant professor when she was sixty-eight years old. Table 6.7 lists all twenty-four women whose primary employment was at an institution awarding doctorates to women before 1940 and who achieved a rank above that of instructor while there.

[8]Richardson, 210.

Marie J. Weiss each published three. Among those nine the only one who spent any part of her career at one of the institutions referred to above was Olive C. Hazlett, who went to Illinois ten years after receiving her doctorate.

Richardson's restriction to articles that had been presented at meetings of the AMS before publication turns out to be much more restrictive for the women in our study than it was for the larger population Richardson was considering. Richardson justified limiting the definition of research publications to those read before the AMS by noting that in using it he counted 78 percent "of the total publication of mathematical research for a typical year (1932)."[9] In comparison, during 1932 only five of the sixteen articles published by the women in our study were presented to the AMS. While 1932 is far from a typical year, over the entire period that Richardson looked at papers read before the AMS, 1890–1933, women in our study presented a significantly lower percentage, only about 53 percent, of their research papers at meetings of the AMS.[10] Furthermore, when we restrict ourselves to those women who received PhD's by 1933, we are considering only 70 percent of the women in our study. Of these, a much higher percentage, 29 percent compared to 15 percent for Richardson's group, had received degrees in the last four years considered (1930–33).

Therefore, in order to portray more accurately this aspect of the research records of the women in our study, we must expand our view of them. In order not to stray too far from Richardson's characterization, we will still not count monographs and articles in books but will continue to consider only articles in journals. However, our new definition will include both articles that appeared on the AMS "Annual List" and those that were reviewed in at least one of the mathematical reviewing journals, *JFM*, *Revue Semestrielle*, *Zentralblatt*, or *MR*.[11] Furthermore, we will not restrict the time period in which the articles appeared and will look at all of these publications regardless of date.

Using this new definition, we find forty-five women who received PhD's before 1940 and published three or more research articles as indicated in table 7.2 below. That is, during their lifetimes almost 20 percent of the women in our study published three or more articles that appeared on the AMS "Annual List" or were reviewed in at least one of the mathematical reviewing journals. We also find that almost 400 publications by American women with pre-1940 PhD's in mathematics appeared between 1877 and 1982. Since the reviewing journals include expository articles as well as research articles and include applications of mathematics to other areas, approximately 7 percent of the articles that are included might not fit the usual definition of research mathematics. On the other hand, about the same number of articles that are clearly mathematical research are not included because they either appeared in books, as opposed to journals, or were not reviewed. Table 7.2 indicates the numbers of papers published by the women in our study using this expanded definition.

[9]Richardson, 201.

[10]For this comparison we counted as a research publication any paper that was reviewed in at least one of the mathematical reviewing journals, *Jahrbuch über die Fortschritte der Mathematik* (*JFM*) (1868–1942), *Revue Semestrielle des Publications Mathématiques* (*Revue Semestrielle*) (1893–1934), and *Zentralblatt für Mathematik und ihre Grenzgebiete* (*Zentralblatt*) (first published in 1931).

[11]*Mathematical Reviews* (*MR*) was first published in 1940.

Table 7.2: Papers published by women mathematicians receiving PhD's 1886–1939*

	Number	Percent
No papers	76	33%
1 paper	71	31%
2 papers	36	16%
3 papers	17	7%
4 papers	8	4%
5 papers	10	4%
6–10 papers	6	3%
11–20 papers	3	1%
More than 20 papers	1	0%
Total	228	

*In this and the following tables, a "paper" refers to any publication that fits our expanded definition of a research publication, i.e., an article that appeared on the AMS "Annual List" or was reviewed in at least one of the following: *JFM*, *Revue Semestrielle*, *Zentrallblatt*, or *MR*.

The 19 percent of women who published three or more papers under our broader definition of research publications still represents a significantly smaller proportion than the 26 percent of Richardson's entire group for the period 1862–1933. We believe that an important reason for this lower percentage is that the women were not, for the most part, hired by the universities that provided stimulus for research. Unfortunately, Richardson's article does not include the employment data that would be needed to substantiate this assertion.

Nevertheless, it is instructive to look briefly at the careers of the ten women, listed in table 7.3 below, who published at least six research papers in mathematics. Three women, Gertrude Blanch, Marion C. Gray, and Bella Greenfield, worked outside academia. Most of their papers that we are counting as research publications were related to their positions in industry or government. The seven others had primarily academic careers.

Gertrude Blanch received her 1935 PhD in algebraic geometry at Cornell University under Virgil Snyder, and her first two publications were in the field of her dissertation. Most of her subsequent publications were concerned with numerical mathematics and the preparation of mathematical tables, reflecting her employment successively with the WPA-funded Mathematical Tables Project, the National Bureau of Standards, the Institute of Numerical Analysis, and the Aerospace Research Laboratories of Wright-Patterson Air Force Base. Blanch also wrote many reviews for the *Journal of the American Statistical Association* and for *Mathematical Reviews*. Her twenty-one research publications, which appeared between 1936 and 1969, constitute the most sustained and productive mathematical research career of any woman in our study.

Marion C. Gray, a native of Scotland, received her PhD from Bryn Mawr College as a student of Anna Pell Wheeler in 1926 with a dissertation on a boundary value problem in ordinary differential equations. After a brief period as university assistant in Edinburgh and then London, she returned to the United States

Table 7.3: Women publishing at least six papers

	PhD		Number	Time span
Gertrude Blanch	Cornell	1935	21	1936–69
Olive C. Hazlett	Chicago	1915	15	1914–30
Lois W. Griffiths	Chicago	1927	13	1925–46
Anna Pell Wheeler	Chicago	1910	11	1910–22
Louise D. Cummings	Bryn Mawr	1914	10	1913–33
Marion C. Gray	Bryn Mawr	1926	10	1925–53
Christine Ladd-Franklin	Johns Hopkins	1926 (1882)	7	1877–96
Elizabeth Hirschfelder	Wisconsin	1930	7	1930–39
Florence M. Mears	Cornell	1927	6	1928–48
Bella Greenfield	New York Univ.	1939	6	1940–64

where she worked for AT&T before joining the technical staff of Bell Telephone Laboratories in 1934. Her early publications were in differential equations and integral transforms; the later ones were mainly in the area of applied electrodynamics, reflecting her professional affiliation. In addition to her ten research publications between 1925 and 1953, she contributed 258 reviews, most in applied electrodynamics, to the first fourteen volumes of *Mathematical Reviews*.

Bella Manel (Greenfield) received her doctorate in 1939 from New York University under the direction of Richard Courant. She was at that time married to Max Shiffman, another student of Courant. Her first paper was coauthored with Shiffman and Courant, and the second was her dissertation; both were in the field of calculus of variations. In 1954 she began her non-academic career with what is now TRW Corporation. Her last four papers were published between 1962 and 1964, after her divorce from Shiffman and marriage to Emanuel Kotkin while she was employed by the RAND Corporation. Three of these four were joint publications with Richard Bellman and Robert Kalaba, while the fourth was joint with just Bellman. All four also appeared as RAND Corporation Research Memoranda. In addition to these four RAND technical reports, Bella Kotkin produced thirteen others, two of which were subsequently published, although not in journals. She remained with RAND until 1965 when she resigned because of the poor health of her husband and her mother. She married Moses A. Greenfield in 1984, three years after Emanuel Kotkin's death.

All of the seven women who published at least six research papers in mathematics and had academic careers received their doctorates by 1930. Again, Olive C. Hazlett was the only one of these seven who spent any part of her career above the rank of assistant professor at one of the schools in Parshall and Rowe's "élite coterie" of research institutions. Hazlett was also the only one of the seven to hold any postdoctoral fellowship. She held an Alice Freeman Palmer fellowship from Wellesley at Harvard in 1915–16, the year following her Chicago doctorate under L. E. Dickson, and published papers in 1916 and 1917. She also held a Guggenheim fellowship during 1928–30 and published papers in 1929 and 1930. Hazlett published fifteen research papers between 1914 and 1930, all in the area of linear associative algebras. During this period she taught successively at Bryn Mawr,

Mount Holyoke, and Illinois, and, as we noted earlier, she accepted a demotion from associate professor to assistant professor in order to move to Illinois. Even so, the majority of her publications appeared before she joined the faculty at Illinois. While she was at Mount Holyoke (1918–25) she was appointed associate editor of the *Transactions* of the AMS, and she served on the council of the AMS soon after her 1925 move to Illinois. She regained the rank of associate professor at Illinois in 1929 and directed a PhD dissertation in 1938. She was on disability leave from 1944 until her retirement as associate professor emeritus in 1959.

Lois W. Griffiths, like Hazlett, was a student of L. E. Dickson at Chicago. She received her degree in 1927 and spent her entire postdoctoral career at Northwestern University, reaching the rank of associate professor in 1938 and retiring at that rank in 1966. The first of her thirteen papers bears the same title as her University of Washington master's thesis in algebraic geometry. Her remaining twelve research publications were in number theory and appeared between 1928 and 1946. Three of the six publications that appeared after her promotion to associate professor were published in the *American Journal of Mathematics*.

Unlike Hazlett, who eventually moved from women's colleges to a research university, Anna Pell Wheeler and Louise D. Cummings spent their entire careers at women's colleges. Wheeler was probably more recognized as a mathematician than any other woman in our study. In 1921 she was "starred" in *American Men of Science*. She gave the prestigious AMS colloquium lectures in 1927 and published eleven research papers between 1910 and 1935, mainly in functional analysis. She also served on both the council and the board of trustees of the AMS and was an editor of the *Annals of Mathematics*. After studying at Göttingen, marrying Alexander Pell (her undergraduate mentor), and receiving her PhD under E. H. Moore from Chicago in 1910, she taught at Mount Holyoke and Bryn Mawr. Although her full-time teaching career at Bryn Mawr was interrupted for several years after her second marriage to classicist Arthur Wheeler and their subsequent move to Princeton, New Jersey, she directed eight PhD dissertations, all but two written by women in our study.[12]

Louise D. Cummings spent her entire career (1902–35) at Vassar College. She received her PhD from Bryn Mawr in 1914 at the age of forty-three. Of her ten research articles counted here, the first appeared the year before she was granted her doctorate, and the last was published two years before her retirement. One of her earliest papers was jointly authored with Henry Seely White, her senior colleague at Vassar and president of the AMS 1907–08; another was coauthored with White and Frank N. Cole of Columbia, secretary of the AMS 1896–1920.[13]

Although Christine Ladd-Franklin taught at Johns Hopkins and Columbia, and Elizabeth Hirschfelder taught at Wisconsin, neither served above the rank of assistant professor. Ladd-Franklin's seven research papers in mathematics all appeared in the nineteenth century. They were just a few of her numerous publications, over fifty of which were in physiological optics and philosophical logic, with about half of these appearing in the first three decades of the twentieth century. Despite her

[12]See table 4.3 for Wheeler's students before 1940. Her other doctoral students were Dorothy Maharam (Stone) in 1940 and Josephine M. Mitchell in 1942.

[13]Cummings, Cole, and White also published a monograph in the series *Memoirs of the National Academy of Sciences, U.S.A.* This paper did not appear in a journal, so it does not fit our definition of a research paper and, therefore, is not included among Cummings's ten research papers.

recognition as an important psychologist, Ladd-Franklin never held an academic position other than as a part-time lecturer in logic and psychology, first at Johns Hopkins and then, on an unpaid basis, at Columbia.

Hirschfelder, then Elizabeth Stafford, earned her doctorate at the University of Wisconsin under the direction of Mark Ingraham in 1930. She taught briefly at the university level both before and after receiving her PhD. In 1931 she returned to Wisconsin and married Ivan Sokolnikoff, then an assistant professor of mathematics. While married to Sokolnikoff she frequently taught in the Wisconsin mathematics department, although she did not hold a regular faculty position there. She did continue doing research, however, and coauthored the last five of her seven research papers with her husband. The Sokolnikoffs also coauthored the classic textbook, *Higher Mathematics for Engineers and Physicists* in 1934.[14] They divorced in 1947, and Elizabeth Sokolnikoff was appointed assistant professor that same year. In 1953 she married Joseph Hirschfelder, a professor of chemistry at Wisconsin, and resigned her position so she could accompany her new husband when he traveled for professional reasons.

Florence Mears was a student of W. A. Hurwitz at Cornell. Her 1927 dissertation was published in 1928 in the *Transactions* of the AMS and was the first of six articles that appeared over a span of twenty-one years, all of which concerned the summability of series. Her most frequently cited article, which appeared in 1937, was the first of her three in the *Annals of Mathematics*. Mears went to George Washington University as an assistant professor in 1929 and was promoted through the ranks, becoming full professor in 1944. Her first doctoral student graduated in 1958 when Mears was sixty-two years old, and her second doctoral student received his PhD two years later. Mears's students were preceded by only four at George Washington University, two in the 1930s and two in the early 1950s. After her retirement at age seventy, she taught an additional ten years at Howard University.

Hazlett, Wheeler, and Mears all served as advisors to doctoral students. At least four other women in our study also contributed to mathematical research by directing doctoral students in mathematics. Pauline Sperry directed five dissertations at the University of California at Berkeley, Mayme I. Logsdon directed four dissertations at the University of Chicago, and Dorothy L. Bernstein directed three dissertations at the University of Rochester. In addition, Harriet F. Montague directed one dissertation in mathematics at the University of Buffalo but was mainly involved with mathematics education and directed seven dissertations for EdD degrees. We look briefly at some of the scholarly contributions of Sperry, Logsdon, and Bernstein.

Pauline Sperry directed five dissertations, more than any woman in our study except Anna Pell Wheeler. Sperry, a 1916 PhD student of E. J. Wilczynski at the University of Chicago, published her dissertation in projective differential geometry and, in 1931, her only other paper that we classify as a research article. Her first PhD student received his degree in 1935 when Sperry was fifty years old, and her last fourteen years later. At least three of her students have themselves directed mathematics dissertations.

Mayme I. Logsdon received her PhD at the age of forty-one. Her 1921 dissertation in algebra was directed by L. E. Dickson at the University of Chicago. A

[14]A second edition appeared in 1941 and was reissued in 1962; a Spanish translation appeared in Argentina in 1959.

published version of her dissertation appeared in 1922 and a paper in algebraic geometry appeared in 1925; her next paper was on non-Euclidean geometries and appeared in 1938. Logsdon's first doctoral student was Anna Stafford (Henriques), who earned her doctorate in 1933. Her three other students received their doctorates between 1936 and 1938.

Dorothy Bernstein was one of three women to get 1939 PhD's under the direction of Jacob Tamarkin of Brown University.[15] After several postdoctoral appointments, Bernstein became an instructor at the University of Rochester in 1943. She had attained the rank of professor when she moved from Rochester to Goucher in 1959. Her most important publication was probably a 1950 monograph in Princeton's Annals of Mathematical Studies, *Existence Theorems in Partial Differential Equations*. Bernstein published five research articles, all in analysis, between 1941 and 1979, four of which appeared while she was on the faculty at Goucher. Three of Bernstein's articles were joint publications with Geraldine A. Coon, her first doctoral student at Rochester, with the last of these papers appearing in 1965 when both Bernstein and Coon were on the faculty at Goucher. Bernstein's last two students received their degrees shortly after she left Rochester for Goucher.

In contrast to all of the women mentioned above, Virginia Ragsdale, Mildred Sanderson, and Anna M. Mullikin published nothing after the PhD dissertation, but in each case the dissertation constituted a recognized contribution to the mathematical literature. Virginia Ragsdale, the third student of Charlotte A. Scott at Bryn Mawr, received her PhD in 1904. In her dissertation, published in 1906, Ragsdale proved that a then well-known class of real algebraic plane curves of even degree satisfies certain restrictions on the topological arrangement of their components. At the time it was not known whether all plane curves satisfied the hypothesis of her theorem. What became known as the "Ragsdale Conjecture," and was proved to be false in 1980, is that Ragsdale's conclusion applies to all real plane algebraic curves of even degree.[16] Ragsdale never explicitly made the conjecture but merely pointed out that if her hypotheses held universally, then so would her conclusion.

Mildred Sanderson received her PhD from Chicago in 1913, two years before Hazlett, as the first of L. E. Dickson's eighteen female doctoral students. The main result of her dissertation, which was often referred to as "Miss Sanderson's Theorem," is a technical, but foundational, result in the theory of invariants. This theorem addresses issues that arise from the fact that a polynomial function on a finite field does not uniquely determine its coefficients. E. T. Bell referred to this result "as one of the classics of the subject."[17] Sanderson became ill with tuberculosis and died in the fall of 1914.

Anna M. Mullikin was R. L. Moore's third student and received her PhD in 1922 from the University of Pennsylvania. She, too, had a theorem named after her. "Miss Mullikin's Theorem" states that any set that separates the plane cannot be expressed as a countable disjoint union of closed subsets, none of which separates the plane. Mullikin taught in high school in Philadelphia for thirty-seven years.

---

[15]The others were Katharine E. O'Brien and Esther McCormick (Torrance).

[16]O. Ja. Viro, "Curves of degree 7, curves of degree 8, and the Ragsdale conjecture," *Soviet Mathematics Doklady* 22 (1980): 566–70. In 1996 Viro and Ilia Itenberg published an expository article, "Patchworking algebraic curves disproves the Ragsdale conjecture," *Mathematical Intelligencer* 18 (4): 19–28.

[17]Bell, "Fifty Years of Algebra in America, 1888–1938," 22.

As might be expected, for those women with a single publication, that publication is most often the dissertation.[18] Altogether, considerably more than half of the 228 women in our study have research publications based on their dissertations, with most of the dissertations that were published appearing in prestigious journals.[19] The fields of the dissertations are skewed toward geometry.[20] Within geometry the field is skewed toward algebraic geometry, with 70 percent of the geometry dissertations in that one area. When we look at the distribution by field of all the research articles in table 7.4 we find that geometry is less heavily represented than analysis. This is less a consequence of widespread switching of fields than a reflection of the fact that half of the women who published most prolifically published primarily in analysis. Furthermore, almost one-quarter of the women who wrote dissertations in algebraic geometry were women religious, for whom pursuing research in mathematics was not a primary goal in seeking a doctoral degree. Because half of the women who published more than five papers each published exclusively or mainly in analysis, the number of research articles in analysis is greater than would be expected by looking at the number of dissertations in that field. In fact, 38 percent of all the research articles in analysis were published by just five women: Blanch, M. C. Gray, Greenfield, Mears, and Wheeler.

Table 7.4: Research fields

	Dissertations	Research articles
History of mathematics, philosophy, education	2%	3%
Logic, arithmetic, algebra, number theory	26%	23%
General and combinatorics	1%	3%
Probability and statistics	0%	3%
Analysis	29%	32%
Geometry (Algebraic geometry)	38% (27%)	31% (18%)
Mathematical physics and other applications	3%	3%
Computer science		1%

While the *Mathematical Reviews* and *Zentralblatt* reviewed some articles in computer science and applied statistics, they did not review all such articles nor did they review articles in other fields in which some women in our study worked. A few women who did not publish mathematics switched their intellectual interests away from science completely, but most who published in fields other than mathematics did so in one of the other sciences. Two exceptions were Emily Coddington Williams and Elizabeth Street Dickerman who published in the field of literature; Williams

[18]For more than 80 percent of the women with a single publication, the publication is based on the dissertation.

[19]More than 40 percent of the articles based on dissertations appeared in the *American Journal of Mathematics*. Another 40 percent appeared in four journals: 19 percent in the *Transactions* of the AMS, 10 percent in the *Duke Mathematical Journal*, 6 percent in the *Annals of Mathematics*, and 5 percent in the *Bulletin* of the AMS. The remaining twenty-five articles appeared in thirteen different journals.

[20]Except for the inclusion of computer science, which is one of the fields classified in *MR* and *Zentralblatt*, the classification scheme that we used is that used by the *JFM* in 1931.

published two novels and a play, and Dickerman published poetry and articles about poetry. Katharine O'Brien, while remaining a teacher of mathematics, published poetry that appeared in magazines, newspapers, and mathematical journals.

As we noted above, Christine Ladd-Franklin, who had wanted to study physics but was prevented from doing so because laboratories were closed to women, published extensively in the fields of logic and physiological optics. A volume containing her collected works on the theory of color vision appeared in 1929, when she was eighty-one years old, and was reprinted in 1973 in the series, *Classics in Psychology*. Her work in logic, the topic of her 1882 dissertation, appeared in philosophical journals and does not account for any of her seven mathematical research papers.

About ten years after receiving her 1911 PhD from Yale, Ida Barney was hired by the Yale Observatory. The star mapping project on which she worked resulted in a multiple-volume series of catalogues for which she served as author or coauthor of most volumes. She directed the project for its final ten years and, in connection with it, published many articles in the *Astronomical Journal*.

Perhaps the most unusual item published by one of the women in this study was a 1918 brochure, *Musical Autograms*, written by Winifred Edgerton Merrill and published by G. Schirmer, the leading sheet music publisher. The brochure consisted of "melodic outlines" based on the signatures of twenty famous men and set to music by Robert Russell Bennett, later her son-in-law and the orchestrator of such Broadway and movie musicals as "Oklahoma" and "The King and I."

A number of women worked for the federal government and published articles in areas related to their positions. As we noted earlier, Dorothy Brady spent most of her career with the US Department of Labor's Bureau of Labor Statistics and was later a professor of economics at the Wharton School of the University of Pennsylvania. Margaret Gurney spent most of her career at the US Bureau of the Census. Both Brady and Gurney published articles in the *Journal of the American Statistical Association*, and Gurney served as an editorial collaborator for that journal. Brady also published many articles and reviews in economics journals such as the *American Economic Review*, the *Journal of Political Economy*, and the *Journal of Economic History*, and served as book review editor of the last of these journals. Furthermore, she wrote many technical reports for the Bureau of Home Economics.

Brady was not the only woman who published technical reports. We have already indicated that many of the items written by Bella Kotkin (Greenfield) appeared as RAND Corporation Research Memoranda. During her many years of service in the government, Beatrice Aitchison was the author of technical reports issued by the Bureau of Transport Economics and Statistics of the Interstate Commerce Commission and by the US Post Office. Marian A. Wilder (Thornton) wrote reports while employed by the University of Minnesota Committee on Educational Research. Sister Leonarda Burke and Irene Price wrote reports for the Air Force. Sr. Leonarda's work for the Air Force was under the aegis of the Regis College Research Center, which she directed from 1953 until 1978. She involved her students in that research, describing their participation as follows. "I had what I called a research table right in my classroom, and if they wanted to I would let them check matrices or do something like that if I thought it would help them and advance them."[21] Sr. Leonarda was the only one of this particular group of women whose entire career was spent in academia.

Of the women who had academic careers but published few or no research articles after their dissertations, many published expository articles in such journals as the *American Mathematical Monthly*, *Mathematics Magazine*, the *Mathematics Teacher*, and *School Science and Mathematics*. Laura Guggenbuhl, who spent her entire career at Hunter College, published nine articles in the *Mathematics Teacher*, while Sr. M. Helen Sullivan, who spent her career at Mount St. Scholastica College in Atchison, Kansas, published six articles in various Catholic educational journals. Sr. Helen also ran an NSF-sponsored undergraduate research program for five years that resulted in eighteen papers published by her students.[22] Elizabeth Cowley taught at Vassar until her mid-fifties when she resigned to care for her elderly mother; she finished her career teaching in high school. While she was on the faculty at Vassar, she contributed nearly two dozen reviews of books written in French, German, Italian, and Spanish to the *Bulletin* of the AMS; she also served on the editorial board of the abstracting journal *Revue Semestrielle des Publications Mathématiques* and was a regular reviewer for that journal from 1908 until 1926. She published more than a dozen articles that appeared in the *Monthly* and in journals specializing in educational research or mathematics education. She also published two geometry textbooks and a book on public education in the United States.

The two most well-known and influential women in the study, Grace Hopper and Mina Rees, contributed in ways that cannot be measured by counting publications. Hopper, who served in the US Naval Reserve, and retired as rear admiral in 1986, worked on early computers and wrote extensively about computing and computers. She was an early advocate for the widespread use of computers and served on the editorial board of the *Communications of the Association for Computing Machinery*. Speaking at the 1981 Smithsonian Institution meeting about her career with computers and in the Navy, Hopper said that "without the mathematical background I couldn't have done it." Mina Rees, in her position as director of the mathematical sciences division of the Office of Naval Research, was instrumental in establishing policy for government funding of mathematical research. Later in her administrative positions at City University of New York (CUNY) she again was instrumental in setting policy. Most of her published articles concern graduate education or mathematics programs funded by the federal government. Much has been written about Hopper and Rees; for example, they are two of the four women profiled in a book about women scientists who served in or worked for the US Navy during World War II.[23]

In addition to publishing research or expository papers, the women in our study contributed to reviewing journals and published book reviews for the *American Mathematical Monthly*, the *Bulletin* of the AMS, and a variety of other publications. Among these, we have already mentioned M. C. Gray's extraordinary number of reviews for the *Mathematical Reviews* and Elizabeth Cowley's book reviews for the *Bulletin*. Cowley's reviews comprise more than half the total number published

[21]Smithsonian meeting tapes.

[22]Smithsonian questionnaire, 1981.

[23]Kathleen Broome Williams, *Improbable Warriors* (Annapolis, MD: Naval Institute Press, 2001).

in the *Bulletin* by women in our study. Laura Guggenbuhl, who published in the *Mathematics Teacher*, contributed a large number of reviews on the history of mathematics and on integral equations to the *Mathematical Reviews*; about three-quarters of these were reviews of works in Russian.

Several women also served as translators of mathematical and other works. Lulu Bechtolsheim, whose 1927 doctorate was from Universität Zürich, translated mathematics books from English to German, from Portuguese to English, and from Italian to German. Elizabeth Street Dickerman, whose 1896 degree was from Yale, translated poetry from the French. Mary Newson, whose 1897 degree was from Göttingen, translated Hilbert's famous 1900 "Mathematical Problems" paper for the *Bulletin* of the AMS. Mary E. Wells, whose 1915 degree was from Chicago, translated one of the papers of Vito Volterra that addresses fluctuation of animal populations. About forty items, including four in David Eugene Smith's *A Source-book in Mathematics*, were translated by the women in our study.

Not all of the scholarly contributions appeared in print. Women in our study served as editors of journals and gave numerous talks at meetings of professional organizations. In addition to Brady, Gurney, Hazlett, Hopper, and Wheeler, who were mentioned above as having served in editorial positions, five women served as associate editors of the *American Mathematical Monthly*, three held editorial positions for the *National Mathematics Magazine* or its successor, *Mathematics Magazine*, and two served on the editorial board of the Kappa Mu Epsilon journal, *Pentagon*. Furthermore, Echo Pepper served an important role as business manager of the *Pi Mu Epsilon Journal*. Ruth Stokes was the only woman in our study who served as a full editor of a mathematical journal when she served as the first editor-in-chief of the *Pi Mu Epsilon Journal* 1949–55.

Among the talks given by women in the study at meetings of professional organizations were nearly 300 talks at AMS meetings, almost two-thirds of which resulted in publications, and more than 150 talks at meetings of the MAA; more than 90 percent of the talks at the MAA were presented at section meetings. Whereas the talks at the AMS meetings tended to be research presentations and those for the MAA frequently concerned issues related to college teaching, the women also gave many talks for more general audiences, especially student groups.

When Helen Owens served as an associate editor of the *Monthly* (1935–38) she coedited the Mathematics Clubs section of the journal with her husband F. W. Owens. The column indicated speakers and topics for many of the talks that were given. While these data were probably never complete, and the space devoted to them actually decreased as the clubs proliferated, between 1918 and 1939 almost 200 talks given by women in our study are mentioned. Most of these talks were given when the women were faculty members, but others were given when they were undergraduate or graduate students.

As we have already mentioned, women participated in the national mathematics fraternities by serving as editors of journals. One of these national fraternities, Kappa Mu Epsilon, was founded by Kathryn Wyant. Soon after arriving at Northeastern State Teachers College in Tahlequah, Oklahoma, in 1930, Wyant began the conversion of its mathematics club into the first chapter of this national mathematics fraternity for four-year colleges. Wyant served as national president of the newly founded honor society 1931–35 and later served as its national historian 1939–41.

Sr. Helen Sullivan also served as national historian of KME (1943–47) and was awarded the KME George R. Mach Distinguished Service Award in 1991.

American women comprised a significant proportion of the members of the national mathematical organizations in the United States from almost the beginning of these organizations. The first women joined the New York Mathematical Society in 1891, about three years after it was founded and three years before it changed its name to the American Mathematical Society. By 1912 there were 50 women among the 668 members of the AMS, a significantly larger proportion of women than in European mathematical societies at the time.[24] However, by 1912, only one woman, Charlotte A. Scott, had been elected to a leadership position; it would be eight years before another woman, Florence P. Lewis, was elected to the council of the AMS. Olive C. Hazlett became the first woman to serve on an editorial board of an AMS journal when, in 1923, she became an associate editor of the *Transactions*.

More than 80 percent of the women in our study were members of the AMS at some point in their careers, and about two-thirds of these women members presented at least one talk to the AMS. However, only five of the women in our study, Olive C. Hazlett, Florence P. Lewis, Caroline Seely, Clara E. Smith, and Anna Pell Wheeler, served on an AMS editorial board, on the AMS council, or on the AMS board of trustees. In fact, these five plus Scott were the only women to be involved in the governance of the AMS and its journals during the entire first half of the twentieth century, and almost all of this involvement was during the 1920s. Furthermore, by 1950 only two women, Anna Pell Wheeler and Emmy Noether, had given invited addresses to the society; the next woman to give an AMS invited address was Olga Taussky-Todd in 1959.

Only one woman, Clara E. Smith, was active in both the AMS and the MAA on the national level before World War II. She served on the board of trustees of the AMS and of the MAA and was also vice president of the MAA, all in the 1920s. Although about two-thirds of the women in our study were members of both the AMS and MAA, fewer were members of the MAA than of the AMS. Despite this, more of the women took an active role in the association, possibly because of the emphasis on teaching rather than research, but also because of the opportunity to participate on the local level. Many women in our study participated in MAA sections, and, during the 1920s, some also participated in the governance of the MAA at the national level; these included Elizabeth Carlson, Elizabeth B. Cowley, and Helen A. Merrill. However, much of the early participation of women in the national MAA was on meeting arrangements and program committees, and it was not until the late 1930s that Helen B. Owens and Mary E. Sinclair appeared on national governing boards of the MAA.[25] In 1940 it seemed to W. D. Cairns, then secretary-treasurer of the MAA, that it was a new idea to appoint a woman as chair of a program committee. He wrote to the president, Walter B. Carver, asking "Why not ask Miss Weiss to act as chmn of the program committee for the La. meeting next Dec.? We have never had a woman to act in that capacity and she can certainly do the job with fine ability."[26] Marie J. Weiss did chair that committee,

[24]"Notes and News," *American Mathematical Monthly* 19 (1912): 84.

[25]During the twenty-five year period 1936–60 there was at least one woman on the board of trustees about two-thirds of the time.

[26]W. D. Cairns to Walter B. Carver, June 11, 1940, Mathematical Association of America Records 1916–present, Archives of American Mathematics, Center for American History, University of Texas at Austin.

but she was not the first woman to do so, as Lennie P. Copeland had chaired the program committee for the 1923 summer meeting at Vassar. It was not until 1978 that Dorothy Bernstein was elected the first woman president of the MAA.

Many of the women in our study made contributions, in print and otherwise, to the field of mathematics education. Some did this through the MAA but others through the National Council of Teachers of Mathematics or regional groups affiliated with the NCTM. Some, including non-members, wrote articles for the NCTM journals *Arithmetic Teacher* and *Mathematics Teacher*, some gave talks or other presentations to teachers' organizations, some served on committees of such organizations, and several, including Lennie Copeland, Elizabeth Cowley, Marie Litzinger, and Mary Winston Newson, served as officers of NCTM or an affiliated group. Ruth W. Stokes served on the NCTM board of directors 1944–47 while at Winthrop College and Syracuse University. Stokes had taught at various teachers colleges and, while at Syracuse, spent some time holding a joint position in mathematics and education. Eula Weeks (King), who earned her PhD in 1915, was vice president of NCTM 1922–23 and was a member of the board of directors 1923–26 while teaching in a high school in St. Louis. She also served on the National Committee on Mathematical Requirements. This committee, chaired by Dartmouth's J. W. Young, "included mathematicians E. H. Moore, Oswald Veblen, and David E. Smith in addition to several prominent teachers and administrators from the secondary school system."[27] The 1923 report of the committee was published by the MAA.[28]

While Richardson did not consider anything but journal articles that had been read before the AMS in his discussion of the state of the American mathematical profession in 1935, he concludes that "the rapidity with which a mathematical school of high distinction has been built up in America is one of the most striking phenomena in the history of science.... Building on the splendid foundations already laid, great forward movements are possible if the spirit of cooperation now animating mathematicians is fostered."[29] It is clear that he is not referring exclusively to the work of a few outstanding researchers but is congratulating the entire American mathematical community on its achievements. As we have seen, the achievements of the women in our study, not only through 1935 but well into the second half of the twentieth century, have been significant despite their initial almost complete exclusion from the professorial ranks at schools noted for their research programs in mathematics. Although women were not always included in the "spirit of cooperation now animating mathematicians" that Richardson saw in 1935, many American women mathematicians found ways to contribute to their profession.

[27]David Klein, "A Brief History of American K-12 Mathematics Education in the Twentieth Century," http://www.csun.edu/~vcmth00m/AHistory.html.

[28]National Committee on Mathematical Requirements, *The Reorganization of Mathematics in Secondary Education* (Oberlin, Ohio: Mathematical Association of America, 1923).

[29]Richardson, "The Ph.D. Degree and Mathematical Research," 215.

CHAPTER 8

Epilogue

As we have seen, in the period prior to 1940 significant numbers of women earned PhD's in mathematics and made contributions to the mathematical community. Of the 228 who earned PhD's before 1940 more than 200 were still living and almost three-quarters of these were employed when the United States entered World War II in December 1941. In chapter 6 we described in some detail ways in which several of these women contributed to the war effort by teaching war-related courses, by working in private industry, or by working for the military either as a civilian or on active duty. Others worked for programs run through the Office of Scientific Research and Development. Most, however, had careers in academia and remained in those jobs during the war.

The real impact of World War II was felt by those women entering the field of mathematics after the war. The preparation for war and the US entry into it resulted in a significant decrease of students in mathematics at every level and, in particular, at the graduate level. The total number of mathematics PhD's awarded to both men and women declined from a high of about one hundred in 1940 to fewer than thirty in 1945. Then, as those who had been involved in the war effort returned and the Servicemen's Readjustment Act of 1944, commonly known as the GI Bill, became available, enrollment at every level increased significantly. All fields felt the impact of this influx of veterans, and the proportion of women earning PhD's in all fields dropped from 15 percent in the 1920s and 1930s to 13 percent in the 1940s and then to 10 percent in the 1950s.

The proportion of women receiving doctorates in scientific fields dropped even more than the overall proportion, as can be seen in table 8.1. Furthermore, the figures in this table show that the percentage of women earning PhD's in mathematics dropped much more than in other scientific fields, from well over 14 percent in the 1920s and 1930s to 5 percent in the 1950s. The obvious explanation is that the proportion dropped because of the influx of students who were in graduate programs because of the GI Bill, and those students were predominantly male. This is a partial answer but does not explain why the drop was so much more dramatic in mathematics than in the other sciences.

As is well documented, the post-war period, especially the 1950s, was a time when there was great pressure for women to move out of the workplace and into the home. A detailed discussion of the implications of the war, especially for American women in science, appears in Margaret W. Rossiter's *Women Scientists in America: Before Affirmative Action 1940–1972*. Rossiter notes such influences as popular culture idealizing the image of woman as homemaker and helpmate, formal or informal quotas, reinstatement of anti-nepotism policies that had been relaxed during the war and immediate post-war period, and decreased opportunities for employment for women, especially, for example, as women's colleges hired more men and fewer

Table 8.1: Percentage of science and engineering doctorates granted to women by field and decade, 1920–1959*

	1920s	1930s	1940s	1950s
All Sciences and engineering	12.2%	11.0%	8.9%	6.7%
Physical sciences	7.6%	6.6%	5.0%	3.7%
Mathematics	14.5%	14.8%	10.7%	5.0%
Engineering	0.9%	0.7%	0.5%	0.3%
Life sciences	15.9%	15.1%	12.7%	9.1%
Social sciences	17.1%	15.8%	14.5%	11.0%

*Data extracted from National Research Council, *Climbing the Academic Ladder*, 20, table 2.1.

women for their faculties. An excellent discussion of the effect of the war on women earning degrees in mathematics in the 1940s and 1950s can be found in *Women Becoming Mathematicians: Creating a Professional Identity in Post-World War II America* by Margaret A. M. Murray.

In what follows we will illustrate the situation for women in mathematics and will make some tentative observations about underlying causes. However, we will leave a fuller investigation of the causes for others.

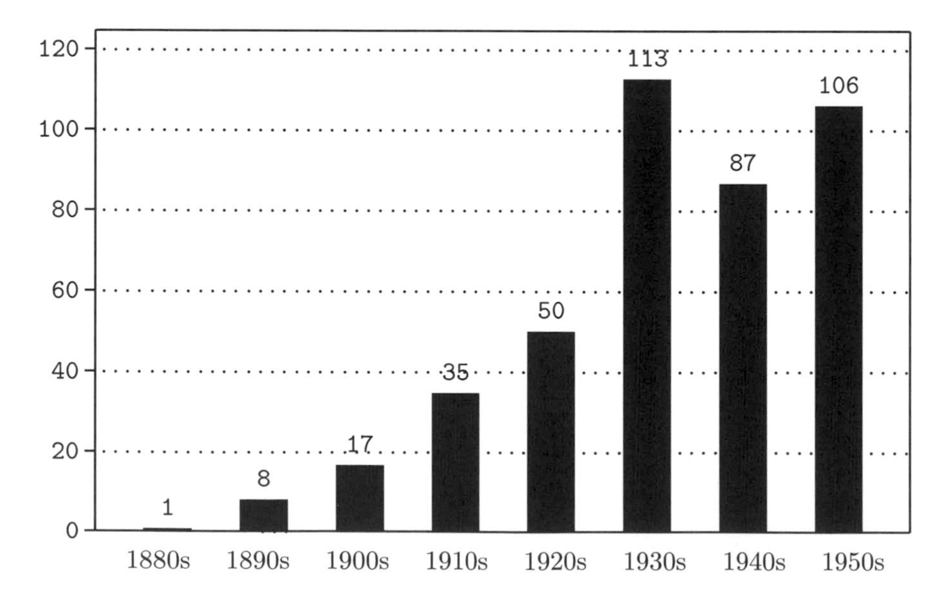

Figure 8.1: Mathematics PhD's to women in the United States by decade, 1886–1959

As we see in figure 8.1, the number of women receiving PhD's in mathematics in the United States had been approximately doubling each decade until the 1940s, when it decreased, largely because of the war. Furthermore, in the 1950s it had not yet recovered to the level of the 1930s. Looking at the number of degrees granted by five-year periods as shown in figure 8.2 gives a somewhat more dramatic view of the overall situation in mathematics. Comparing the five-year period 1930–34, when both the number and percentage of women was highest, with the span 1955–59, when the total number was highest but the percentage of women was lowest, we see that the number of women getting degrees in the later period decreased slightly from that in the earlier one, while the total number of PhD's in mathematics more than tripled.

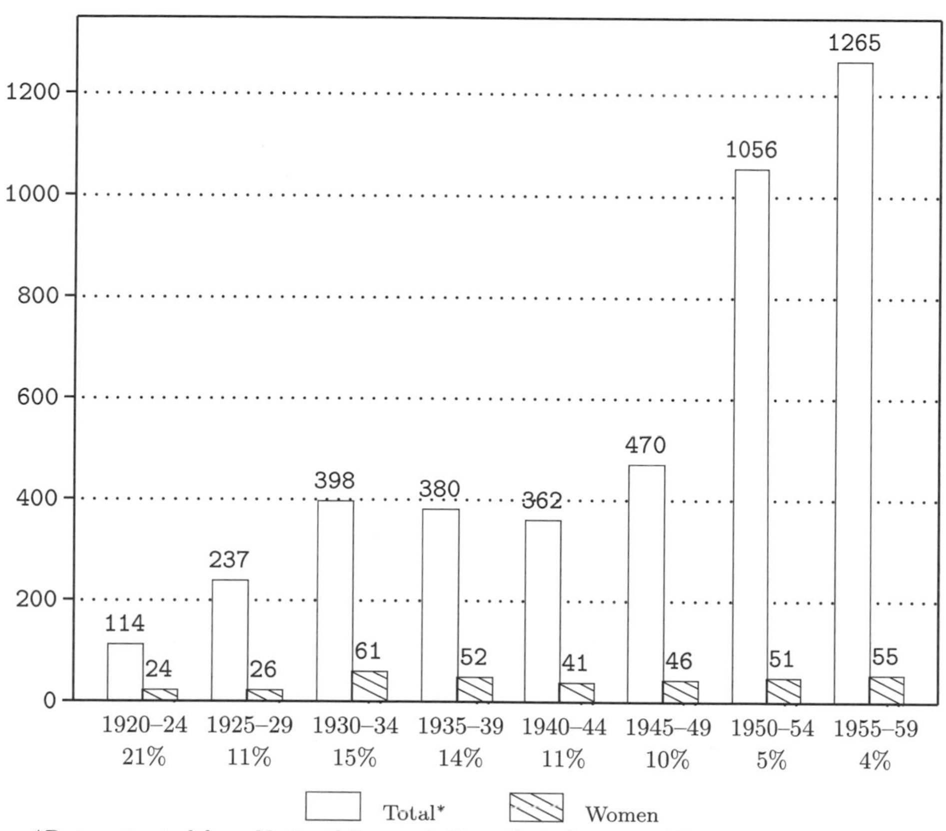

Figure 8.2: Mathematics PhD's by five-year periods, 1920–1959

As we saw in table 4.1, the five institutions that led in granting PhD's to women in mathematics for the entire pre-1940 period were Chicago, Cornell, Bryn Mawr, Catholic, and Yale. In table 8.2 below we see four of these five schools among the leading institutions of the 1930s. Only Yale University is no longer among the leaders, having granted most of its PhD's to women before 1910. The University of Chicago was also by far the leading granter of PhD's to both men and women in the first four decades of the twentieth century, as well as in the 1930s. In that decade Chicago was followed in overall production of mathematics PhD's by the University of Michigan, the University of Illinois, Harvard University, Cornell University, and Princeton University. Chicago, which had granted twenty-four PhD's to women in the 1930s, granted a few more to women in the early 1940s and then only three in the 1950s. Furthermore, while the number of women getting PhD's from Chicago dropped dramatically, the number of men increased slightly so that Chicago went from awarding 27 percent of the PhD's to women in the 1930s to less than 3 percent in the 1950s. Whereas the drop in the number and percentage of women who earned PhD's from Chicago is by far the most striking feature of our comparison of the 1930s and the 1950s, it is only the most extreme example of a general pattern. Because the decline is so precipitous at Chicago, we will examine the situation there in more detail.

Table 8.2: Leading PhD-granting institutions for women in mathematics, 1930s

PhD	1886–1939	1930s			1950s		
institution	women	total	women	% women	total	women	% women
Chicago	46	88	24	27%	102	3	3%
Catholic	14	14	12	93%	15	4	27%
Illinois	12	56	8	14%	92	5	5%
Cornell	21	41	8	20%	33	1	3%
Bryn Mawr	19	8	8	100%	1	1	100%

Many factors, some having nothing to do with the war, contributed to this sharp decline. Before the war many changes were occurring within the mathematical community. Margaret Murray discusses what she calls the growing polarization of teaching and research that increased during the 1930s, as partially exemplified by the creation of the Institute for Advanced Studies at Princeton. She notes, "Particularly during the years prior to 1930, even the most influential American mathematicians were not exclusively devoted to research or to teaching but, rather to a synthesis of the two into scholarship that placed mathematics into historical, critical, and cultural contexts."[1]

Among these influential mathematicians was E. H. Moore who headed the mathematics department at the University of Chicago until his retirement in 1931, a year and a half before his death in 1932. Under his leadership, the department welcomed women students to the graduate program and supported broad mathematical and pedagogical values. Moore had directed doctoral dissertations in several different areas of mathematics including group theory, algebraic geometry, number theory, geometry, and differential equations. While Moore was department head, some of his former students joined the faculty at Chicago and became leaders in their fields. One of these was Herbert Ellsworth Slaught, who had received his PhD in 1898. Slaught was promoted to full professor in 1913 and remained at Chicago until his retirement in 1931. This was despite the fact that a few years after he received his degree, "he decided to devote his life to the promotion and improvement of the teaching of mathematics rather than to a research career."[2] Thus, in the period before the 1930s, the Chicago department actively supported mathematics education at the school and college level.

Other changes in the mathematical community started in the late 1930s with the retirements and deaths of major figures of the first three or four decades of the

[1] Murray, *Women Becoming Mathematicians*, 7.

[2] G. A. Bliss, "Herbert Ellsworth Slaught–In Memoriam," *Bulletin of the American Mathematical Society* 43 (1937): 596.

century and subsequent changes in values within the mathematical community. A number of mathematicians who were advisors to a large number of women retired in the 1930s and shortly thereafter. These included the leading advisors for the pre-1940 women at Chicago: L. E. Dickson, who retired in 1939, and Gilbert Ames Bliss, who retired in 1941. Between them, Dickson and Bliss served as advisors for thirty of the Chicago women, half of them in the 1930s. Similarly, Virgil Snyder who directed fourteen dissertations of women at Cornell, with seven in the 1930s, retired in 1938, a year before his last student received her degree. At Bryn Mawr College, Charlotte Angas Scott directed seven, the last in 1925, and Anna Pell Wheeler directed eight, five in the period 1930–42, before her retirement in 1948. At Catholic University, Aubrey Landry advised thirteen of the fourteen women students before 1940, with eleven in the 1930s, and four of the six women students in the 1940s; he retired in 1952. Arthur Coble retired from Illinois in 1947, having directed the dissertations of five women, four in the 1930s, at Illinois, and an earlier one at Johns Hopkins.

In the 1930s, four of the five schools that led in the granting of PhD's to women were private institutions, and a total of twenty-nine schools awarded PhD's to women in mathematics. They include all of the leading producers of doctorates in the country except for Harvard and Princeton. Even though Harvard did not grant PhD's to women until 1963, the women who earned their degrees at Radcliffe College studied under Harvard faculty. Princeton, so prominent beginning in the 1930s, did not grant doctorates to women in mathematics until the early 1970s. Thus, women were unable to participate in one of the most fertile communities for mathematical research.

We are omitting a discussion of the 1940s since collegiate and graduate education was so disrupted by the war. However, by the 1950s many new schools were creating PhD programs. In the 1950s, forty-two institutions awarded doctorates to women in mathematics. Many of these were schools with new doctoral programs. The leading institutions, as measured by the total number of mathematics doctorates granted, were the University of California at Berkeley, New York University, Princeton, Michigan, and Chicago. In contrast to the 1930s, four of the leading five producers of women PhD's were public rather than private institutions. Comparing the schools listed in tables 8.2 and 8.3, we see that only the University of Illinois was among the top five schools granting PhD's to women in both the 1930s and the 1950s.

Table 8.3: Leading PhD-granting institutions for women in mathematics, 1950s

PhD	1886–1939	1930s			1950s		
institution	women	total	women	% women	total	women	% women
NYU	2	14	2	14%	120	10	8%
Brown	5	25	4	16%	70	6	9%
Michigan	7	59	4	7%	101	5	5%
Illinois	12	56	8	14%	92	5	5%
Minnesota	2	7	1	14%	48	5	10%

Although NYU was the leader in awarding PhD's in mathematics to women in the 1950s, it awarded only ten, or 8 percent of all the doctorates it awarded. Brown, which awarded six, 9 percent, of its PhD's in mathematics to women, was second in the 1950s. These two schools were among those that were contracted by the Applied Mathematics Panel to do work during the war, and these institutions and the degrees awarded reflected one trend that resulted from the war. Almost all of the advisors of the NYU 1950s women had fled Europe because of the Nazis. They included Richard Courant, who, having been born into a Jewish family, had been forced from Göttingen in 1933, and who created the graduate center at NYU. Others were Kurt Friedrichs, Peter Lax, and Lipman Bers. At Brown the applied mathematics program began in 1941 and became a formal PhD program in 1946, now the Division of Applied Mathematics. It was led by the German applied mathematician William Prager. All but one of the doctorates earned by women at Brown after the war and before 1960 were in applied mathematics, many related to problems that arose during the war.

By the 1950s the number and percent of women earning PhD's in mathematics were so low that women in mathematics were effectively invisible. In general, they were not earning degrees at the major institutions, and the employment outlook was bleak. It took until the early 1980s for the proportion of women receiving PhD's in mathematics to reach that of the first four decades of the twentieth century. Starting in 1991 the proportion of women among mathematics PhD's granted in the United States annually has been consistently above 20 percent. In 2002 the proportion reached 30 percent and did not change significantly during the following five years.

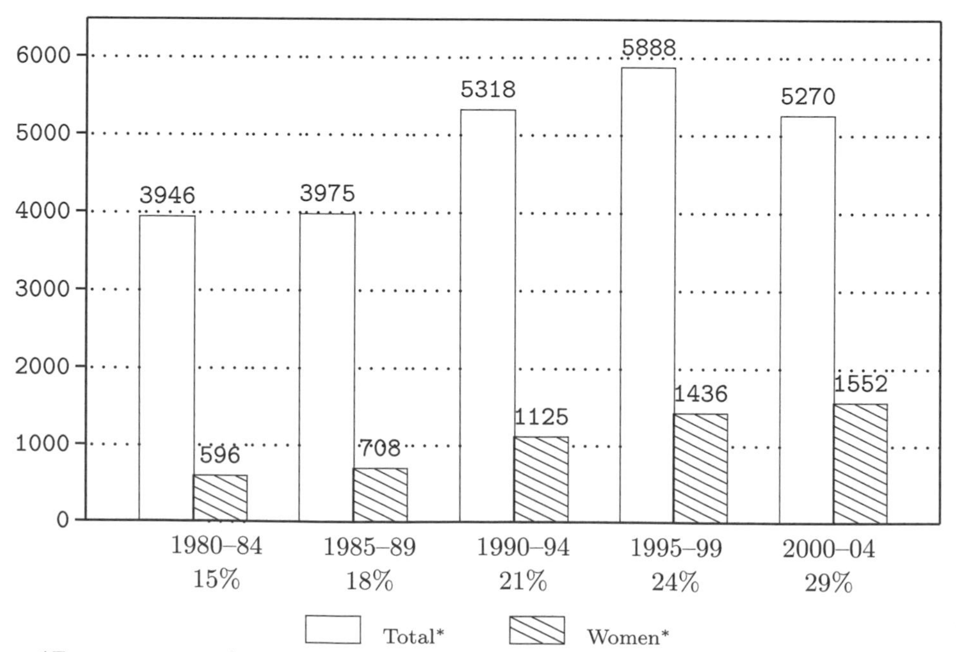

Figure 8.3: Mathematics PhD's by five-year periods, 1980–2004

Biographical Entries

Expanded versions of the biographical entries below may be found on the website hosted by the American Mathematical Society as indicated on the copyright page and the back cover of this book. Each expanded entry includes a list of works by the subject, which in most cases is complete, as well as more biographical detail. These entries also include works about the subjects as well as other sources used. This material is freely accessible and searchable.

ADAMS, Rachel (Blodgett). October 13, 1894–January 22, 1982.

Wellesley College (BA 1916), Radcliffe College (MA 1919, PhD 1921). PhD diss.: [Blodgett, R.] The determination of the coefficients in interpolation formulae; and A study of the approximate solution of integral equations.

Rachel Blodgett was born in Woburn, Massachusetts, the eldest of three children of Mabel Edith (Owen) and William Edward Blodgett. Her parents did not attend college; her father was treasurer of a company that manufactured tiles and was at one point mayor of Woburn. After attending public secondary schools Rachel Blodgett majored in mathematics at Wellesley, where several relatives of her father had studied in the early years of the college.

Following her graduation from Wellesley in 1916, Rachel Blodgett was employed 1916–18 as mathematics mistress at Miss Edgar's and Miss Cramp's School in Montreal. The next three years she did graduate work in analysis at Radcliffe, where she held an Edward Austin scholarship at least two years and a Mary E. Horton fellowship her last year, 1920–21. She earned her master's degree in 1919 and her doctorate in 1921. Before completing her PhD dissertation in analysis, she wrote two minor theses, one in algebraic geometry and one in dynamics. She was instructor of mathematics at Wellesley the year after she received her doctorate.

On August 17, 1922, Rachel Blodgett married Clarence Raymond Adams (1898–1965), a graduate of Brown, who had just completed his PhD in mathematics under George David Birkhoff at Harvard. They were in Rome and Göttingen 1922–23 while C. R. Adams was a traveling fellow from Harvard. Upon their return in 1923 they moved to Providence where he was to spend his career in the mathematics department at Brown. He retired as professor emeritus in 1965 and died later that year. There were no children of the marriage.

Rachel B. Adams was a member of the MAA for a year after her marriage and served as a tutor at Radcliffe from 1926 to 1941 along with **Florence Curtis Graustein**. In 1929 Adams published an article in the *American Journal of Mathematics* that was based in part on her PhD dissertation.

Adams was active in and served as an officer of a number of community organizations, including the First Baptist Church Women's Society, the Weavers Guild, a literary society, and an amateur theatrical group. During World War II she was registered in Washington with the National Roster of Scientific Personnel, although there is no evidence of her being asked to do scientific work at that time.

During their marriage Rachel and C. R. Adams traveled extensively in the United States and Europe. After her husband's death she continued her community work and served as treasurer for several years and on the board of directors of the Home for Aged Women (later the Tockwotton Home) in Providence. In 1980 she moved to the Tockwotton Home and died there at age eighty-seven in 1982. Following her death, a bequest from her estate was used to establish the Blodgett Fund at Wellesley, with the income to be used for scholarships.

AITCHISON, Beatrice. July 18, 1908–September 22, 1997.

Goucher College (BA 1928), Johns Hopkins University (MA 1931, PhD 1933), University of Oregon (MA 1937). PhD diss.: On mapping with functions of finite sections (directed by Gordon Thomas Whyburn).

Beatrice Aitchison was born in Portland, Oregon, the elder of two children of Bertha Beatrice (Williams) and Clyde Bruce Aitchison. Her mother was a musician and singer; her father was a lawyer who had a distinguished career serving on state and federal regulatory commissions and in private practice. Her father earned a PhD in economics in 1932 while a member of the Interstate Commerce Commission (ICC).

Beatrice Aitchison attended private school in Portland until 1916 when the family moved to Washington, D.C. In Washington she attended public schools and graduated with honors from Central High School shortly before her sixteenth birthday. She then attended Goucher College 1924–28. After graduating from Goucher in 1928, she worked for a year in New York City for the Metropolitan Life Insurance Company as its first female actuarial clerk. She took special courses while working in New York, and she passed the first four parts of the examinations of the American Institute of Actuaries and of the Actuarial Society of America in April 1929. She resumed her study of mathematics in October 1929 at the Johns Hopkins University. While there she held a scholarship and earned a master's degree in 1931 and a PhD in 1933. She immediately published two articles based on her work at Johns Hopkins.

Because of the Depression jobs were difficult to find in 1933. At the 1981 Smithsonian meeting Aitchison recalled, "I applied to 145 colleges and universities my last spring at Hopkins .... Someone whom I didn't know but knew my background held my application thinking she might have an opening at Westhampton College [the college for women] at the University of Richmond. Sure enough, [one of the women] got really sick and so on about three days notice I was down there teaching for one semester." After that semester as acting associate professor, she was lecturer in statistics at American University in Washington, D.C., 1934–35. She also wrote in her 1981 Smithsonian questionnaire that American University "invited [her] to 'help' with statistics for 'experience' and no pay." She did, however, get credit for a graduate course in history of economic thought.

Aitchison returned to Oregon in 1935 and took a University of Oregon extension course in Portland in principles of economics. In 1936 she was a field supervisor for the Oregon Historical Records Survey of the WPA in Portland. She then spent the academic year 1936–37 at the University of Oregon and earned a second master's degree, this one with honors in economics. When she returned to Washington, D.C., in 1937, she again lectured in statistics at American University and continued to do so through 1938–39; she also served as a member of the faculty of the graduate school of the US Department of Agriculture, and starting in 1938, she held temporary assignments at the ICC.

Aitchison returned to the University of Oregon as an economics instructor in 1939 but continued her temporary assignments with the ICC during the summers. At the beginning of 1942 Aitchison resigned her position in Oregon and began her long and distinguished career as a transportation economist with the US government. Her first position was with the ICC, where from 1942 to 1948 she advanced from junior to senior statistician and from 1948 until 1951 served as principal transportation economist. During her service with the ICC, Aitchison was again a lecturer in statistics at American University 1942–44. From 1942 until 1945, she was a consultant and head of traffic analysis for the Office of Defense

Transportation, and in 1950 she was a consultant to the director of military traffic of the Department of Defense.

Aitchison left the ICC in 1951 and worked until 1953 as director of the Transport Economics Division of the Office of Transportation at the US Department of Commerce. After this post was abolished in 1953, she worked for the US Post Office until she retired in July 1971. Aitchison held several positions while she was with the post office, and all involved directing research on transportation issues. She also consulted in that area after her retirement. From the mid-1940s to the 1970s, Aitchison conducted studies, testified before regulatory agencies, wrote technical reports for the government, and gave talks to various transportation groups and postmasters' conventions.

In 1961 Aitchison was one of six career women in government chosen to receive the first annual Federal Woman's Award given by the Civil Service Commission. As a consequence, she was a member of the Federal Woman's Award Study Group on Careers for Women, a small group of award recipients appointed by President Lyndon Johnson that successfully recommended that sex discrimination be added to the executive order banning discrimination in government employment. In 1970 Aitchison was honored by the National Civil Service League with the Career Service Award as one of the top ten government workers of the year, the only woman in the group.

In addition to consulting after her retirement, Aitchison spent significant time in volunteer work with the elderly. She was an active Episcopalian serving in a variety of positions; in 1963 she was the first woman elected to the vestry of Epiphany Episcopal Church in Washington, D.C. Among her interests were music, photography, sewing, and traveling. An obituary states that she had studied the mamba and the samba at Fred Astaire's dancing school and that "she also traveled widely, often by tramp steamer, to such locations as the South Pacific and Antarctica" (Richard Pearson, "Retired Postal Official Beatrice Aitchison Dies," *Washington Post*, September 29, 1997).

Aitchison was supportive of the schools where she had received her education. Summer grants for Goucher College junior faculty are available through the Beatrice Aitchison '28 Non-Tenured Faculty Professional Advancement Fund. In 1994 she established the Aitchison Public Service Fellowship at Johns Hopkins. Collections of her papers are deposited at the Schlesinger Library, Radcliffe Institute, Harvard University, and at the University of Oregon. In 1997 Beatrice Aitchison died at age eighty-nine in Washington, D.C.

ALDEN, Marjorie (Leffler). August 29, 1909–October 5, 2000.
Miami University (BA 1930), Ohio State University (MA 1932, PhD 1934). PhD diss.: [Leffler, M.] A lemma in potential theory (directed by Tibor Radó).

Marjorie Leffler was born in Kyle, Ohio, the eldest of four children of Stella Eugenia (Durbin) and William Homer Leffler. In 1910 her father was a commercial traveler; in 1920 the family farmed their own land in Butler County, Ohio.

Marjorie Leffler received her primary education in the local township school and her secondary education in the high school of Hamilton, Ohio. In 1930 she graduated from Miami University in neighboring Oxford, Ohio, magna cum laude, with honors in mathematics and with general honors. As an undergraduate Leffler was involved in many activities. These included several sports, a national literary society, student-faculty council, and the yearbook. She was elected to Phi Beta Kappa and to Mortar Board as well as to honor societies for education and journalism.

All of Leffler's graduate work was done at Ohio State University where she held a scholarship for two years. She earned a master's degree in 1932 and was an assistant in the university high school while completing her work for the PhD. Both her master's thesis and her doctoral dissertation were in analysis and were directed by Tibor Radó.

Marjorie Leffler received her PhD in 1934 and accepted a one-year position as tutor and instructor at Mills College in California. On December 20, 1934, she married Howard

Holston Alden (1908–1998), who had received a PhD in mathematics from Ohio State the previous year.

Marjorie Leffler Alden's subsequent academic positions were at the same institutions as her husband's, but at lower ranks. Howard H. Alden was head of the mathematics department at New Mexico Military Institute, a junior college in Roswell, from 1934 to 1943; was associate professor at the University of Wyoming 1945–46; and returned to Ohio State University as assistant professor in 1946. During World War II he was a chief engineer under Robert H. Goddard. During this period, Marjorie L. Alden was an instructor at New Mexico Military Institute 1936–43, instructor at the University of Wyoming 1945–46, and instructor at Ohio State starting in 1946. Howard Holston Alden was promoted to associate professor at Ohio State in 1948, while Marjorie Leffler Alden remained as instructor. Both resigned from the university in June 1956, after which it appears they left Ohio.

The Aldens lived for a time in Mesa, Arizona, before moving to California. For several years in the 1980s and 1990s they lived in a retirement home in La Mesa, California. Howard Alden died in 1998 at eighty-nine. After an illness of more than a decade, Marjorie Alden died in La Mesa at age ninety-one in 2000. Her ashes were scattered at sea.

ALLEN, Bess (Eversull). May 18, 1899–March 18, 1978.

University of Cincinnati (BA 1921, MA 1922, PhD 1924). PhD diss.: [Eversull, B. M.] The summability of the triple Fourier's series at points of discontinuity of the function developed (directed by Charles Napoleon Moore).

Bess Marie Eversull was born in Elmwood Place, a village adjacent to Cincinnati, Ohio, the first of two daughters of Olive (Magrew) and Warner Solomon Eversull. Warner Eversull was principal and later superintendent of a public school in the area. Bess Eversull attended Woodward High School in Cincinnati before enrolling at the University of Cincinnati where she received all of her formal postsecondary education. She held a scholarship 1917–19 and graduated in 1921 with a combined major in mathematics and English and a minor in French.

Eversull continued her studies at Cincinnati until she completed the work for her doctorate with a minor in physics. She held fellowships 1921–23 and in 1924, and assisted and taught intermittently. Both her 1922 master's thesis and her 1924 doctoral dissertation examined triple Fourier series, and each was published the year she received the degree, her master's thesis in the *Annals of Mathematics* and her PhD dissertation in the *Transactions* of the AMS. She was the first doctoral student of Charles N. Moore and the third person to earn a doctorate in mathematics at Cincinnati.

Bess Eversull was an instructor at Smith College 1924–27. In October 1927 she married Charles Easton Allen (1899–1975), and within a year they were living in Detroit, where by 1930 Charles Allen was a civil engineer working as a building inspector for the City of Detroit. Nearly twenty years after her marriage Allen indicated in a job application that she had no children and that she had done only volunteer work, mostly non-mathematical, since her marriage, except during World War II when she worked as a mathematician for the Jam Handy Organization, which made films for the armed services.

In February 1947 Allen resumed her professional work by taking a position as a regular substitute instructor at Wayne University (now Wayne State University) in Detroit. She continued there as instructor 1948–50, assistant professor 1950–59, associate professor 1959–64, and associate professor emeritus after her retirement in 1964. The department recommendation for her promotion to associate professor portrays her as a devoted teacher who spent much time helping students and who made sure that the essential work of the department got done.

At Wayne State Allen sponsored the campus chapter of a national music fraternity. She often attended MAA meetings and wrote several reviews for the *Monthly*. After her retirement in 1964, she taught for a few years in the late 1960s at the Detroit Institute of

Technology. Bess Allen died in Detroit, Michigan, in 1978, three years after her husband died and two months before her seventy-ninth birthday.

ALLEN, Florence E. October 4, 1876–December 31, 1960.
UNIVERSITY OF WISCONSIN (BL 1900, ML 1901, PHD 1907). PhD diss.: On the determination of cyclic involutions of order three (directed by Linnaeus Wayland Dowling).

Florence Eliza Allen was born in Horicon, Wisconsin, the second child of Eliza (North) and Charles Allen. Her father was a lawyer, who, it appears, died in 1890. In 1900, she, her mother, and her older brother were living in Madison, Wisconsin.

Florence Allen was a mathematics major as an undergraduate in the College of Letters and Science at the University of Wisconsin. She was active in a literary society for women promoting interest in the fine arts and served as secretary and president. She was also active in the self government association and the yearbook board. Florence Allen received her bachelor's degree in 1900. Her brother, Charles E. Allen, received his PhD in botany in 1904 from the University of Wisconsin and was on the faculty there from 1901 to 1943, when he retired as professor emeritus. His distinguished career included election to the National Academy of Sciences.

Florence Allen remained at Wisconsin as a resident graduate in mathematics and philosophy 1900–01. She received her master's degree in mathematics in 1901, and all of her subsequent positions were in mathematics at the University of Wisconsin. She was an assistant in 1901–02 and then became an instructor. She remained at that rank for forty-three years, even though she had obtained her doctorate with a dissertation in geometry in 1907 and published her dissertation in the *Quarterly Journal of Pure and Applied Mathematics* in 1914. She published two additional research articles: one in 1915 in the *Rendiconti del Circolo Matematico di Palermo* and the other in 1927 in the *American Journal of Mathematics*. She became an assistant professor in the fall of 1945, shortly before her sixty-ninth birthday. Her promotion came two years before she retired at the rank of assistant professor emeritus.

In response to an inquiry by Helen Owens concerning the outlook for women doctorates in mathematics, Allen wrote on June 8, 1940:

> Of course there will always be some women who should go in for a Ph.D. – some because it will be an actual necessity to qualify them for one of the occasional – very occasional – openings in college and university positions, some because of the leisure they may have to follow a congenial pursuit. But on the whole I see no great encouragement to be had from past experiences and observation. I do not believe that there is or will be a great future for any but a few in this field. At present, it seems to me, as I look about this campus, that in all strictly academic fields (not those special to women) that there is a decided drop in the number of women engaged. That may be peculiar to this economic phase, but I look for it to continue for some time to come (Helen Brewster Owens Papers, Schlesinger Library, Radcliffe Institute, Harvard University).

Florence E. Allen died at eighty-four in Madison, Wisconsin, in 1960. She was buried in Horicon, Wisconsin.

ANDERSON, Mae Ruth. May 31, 1899–March 28, 1948.
CONCORDIA COLLEGE (BA 1920), UNIVERSITY OF CHICAGO (MA 1923, PHD 1936). PhD diss.: Representation as a sum of multiples of polygonal numbers (directed by Leonard Eugene Dickson).

Mae Ruth Anderson was born near Westby, Wisconsin, the only daughter of Ida (Berg) and Norton A. Anderson. She had an older half-brother from her mother's first marriage. Norton Anderson was a merchant.

Sometime after 1910 Mae Ruth Anderson and her parents moved to Shelly in western Minnesota, where her father again owned a store. Anderson graduated from high school in Halstad, Minnesota, a few miles from Shelly, in 1916 before entering Concordia College in neighboring Moorhead, Minnesota.

Anderson was at Concordia, 1916–20, where she was a member of Concordia's major scholastic honor society and was on the staff of the student newspaper. A fellow student of Anderson's was **Ruth B. Rasmusen**, who was at Concordia 1918–21 and who received her PhD from Chicago in 1936, as did Anderson.

After her graduation from Concordia in 1920, Anderson taught in high school in Gayville, South Dakota, for a year before beginning her master's work in mathematics at the University of Chicago in the fall of 1921. After six consecutive quarters, she received the MA in March 1923. She taught at Waldorf Junior College and Academy in Forest City, Iowa, from 1923 to 1928 and joined the faculty at Concordia College as instructor in 1928. She was promoted to assistant professor in 1929. After her promotion she spent several summer quarters doing graduate work at the University of Chicago. A leave of absence from Concordia permitted her to be in residence at Chicago to finish her doctoral work from summer 1935 to summer 1936.

In 1937 Anderson was promoted from assistant to full professor and became head of the mathematics department at Concordia, positions she retained until her death. In addition to teaching mathematics, she taught some French and was noted for her tireless efforts to help students. Excerpts from Concordia College yearbooks of 1946 and 1947 convey a sense of her qualities: "Her obvious love of her subject, her patience and sense of humor instill a lasting respect in her students." "Her demands for work, thoroughness, concentration, and more work bring results."

Mae Ruth Anderson died of leukemia in Rochester, Minnesota, at age forty-eight in 1948. A bequest to the Trinity Lutheran Church in Moorhead reflects her involvement with that church.

ANDERSON, R. Lucile. October 22, 1900–February 18, 1999.

MOUNT HOLYOKE COLLEGE (BA 1922), BRYN MAWR COLLEGE (PHD 1930). PhD diss.: A problem in the simultaneous reduction of two quadratic forms in infinitely many variables (directed by **Anna Pell Wheeler**).

Rose Lucile Anderson was born in Jamestown, New York, the second of three children of Anna (Ebel), born in Germany, and Charles J. Anderson, born in Sweden. Both her parents became naturalized US citizens. Her father was the proprietor of a machine shop. Her older sister graduated from Mount Holyoke College, held a master's degree from Cornell University, and was a teacher of zoology at the University of Kentucky. Her younger brother attended Cornell and died in his thirties.

R. Lucile Anderson attended Jamestown public schools including the Jamestown high school from which she graduated in 1918. She immediately entered Mount Holyoke College and graduated in 1922 with senior honors in mathematics. Anderson went directly to Bryn Mawr College for graduate work in mathematics. She was a graduate scholar in mathematics 1922–23, a resident fellow 1923–24, and a fellow by courtesy and holder of the '86 fellowship from Mount Holyoke in 1924–25. In the spring of 1924 she was awarded the Mary E. Garrett European fellowship from Bryn Mawr, which she used for study at Cambridge University in England in 1925–26.

When Anderson returned from England she was hired as instructor of mathematics at the University of Rochester, where she remained for two years, 1926–28. In 1928 she began her long association with Hunter College. The following year, 1929–30, she was a graduate student and part-time instructor at Bryn Mawr College, where she completed her work for the PhD with a dissertation in analysis and a minor in physics.

In 1930 Anderson returned to Hunter, again as instructor until 1936, as assistant professor 1936–56, and as associate professor 1956–71. For her Mount Holyoke 50th reunion booklet Anderson reported, "I retired in 1971 after teaching mathematics at Hunter College of the City of New York for more years than I care to admit. The years have been spiced with some travel – Mexico, Alaska, Europe; some sports – golf, hiking, figure skating; some theater when tickets can be obtained without too much effort" (Mount Holyoke College Archives and Special Collections).

Anderson was still living in her apartment in New York City when she died at age ninety-eight in 1999. She was cremated and her ashes returned to her hometown of Jamestown, New York, where she had spent many of her summers.

ANDERTON, Ethel L. September 23, 1888–February 19, 1971.

WELLESLEY COLLEGE (BA 1911), YALE UNIVERSITY (MA 1922, PHD 1925). PhD diss.: Bioche curve pairs (directed by James Kelsey Whittemore).

Ethel Louise Anderton was born in Dover, New Hampshire, the daughter of Isabelle (Richardson) of New Hampshire, and Washington Anderton of England. Her father immigrated to the United States and came to Dover as a fabric colorist; he was superintendent of the print works from 1880 until his retirement in 1896, after which he was in real estate. For at least a dozen years in the early decades of the century, Isabelle Anderton was proprietor of a stationery and variety store in Dover. Their first child died as an infant, and Ethel was the second of three surviving children.

Ethel Anderton graduated from Wellesley College in 1911, after which she taught in high schools until 1924. She was teaching in Westfield, Massachusetts, in at least 1916 and 1917. The next year she began teaching in West Haven, Connecticut, and likely continued there until 1924. Also starting in 1918, she attended Yale University part time and earned her master's degree in 1922 and her doctorate in 1925. She received university scholarships part of the time she was at Yale and was elected to Sigma Xi.

Anderton became an instructor at Wellesley College in 1924 and remained for three years, after which she taught one year, 1927–28, at Smith College. While at Smith she was promoted to assistant professor and continued in that rank when she moved to Mount Holyoke College, where she remained for another three years. Her last college teaching job was in 1931–32 when she replaced a faculty member on leave at Goucher College.

In 1933, at age forty-five, she rejoined the faculty at the high school in West Haven, Connecticut, as teacher and head of the mathematics department. While in West Haven, she engaged in work on tests and measurements for the West Haven school system and was active in various educational organizations including the local NEA and the local teachers' credit union. After her retirement in 1954, she engaged in a tutoring business in college and high school mathematics. In her late seventies she was still tutoring an average of three hours a day and had served as an emergency substitute high school teacher.

In 1941 Anderton had a house built for herself in West Haven and was joined there by Elsie Schenker, a teacher friend, and, a few years later, by Schenker's sister, both of whom lived with Anderton for most of the rest of her life. Anderton's non-professional interests and activities included riding, gardening, and conservation; she was a member of the League of Women Voters, the Connecticut Forest and Park Association, and was president of a local literary club. Ethel Anderton died in New Haven, Connecticut, at age eighty-two in 1971.

ANDREWS, Annie Dale (Biddle). December 13, 1885–April 14, 1940.

UNIVERSITY OF CALIFORNIA (BA 1908, PHD 1911). PhD diss.: [Biddle, A. D.] Constructive theory of the unicursal plane quartic by synthetic methods (directed by Derrick Norman Lehmer).

Annie Dale Biddle was born in Hanford, California, the youngest of seven surviving children of Achsah A. (McQuiddy) and Samuel Edward Biddle. Her father was a grain dealer, later bank president and manager, and active in public service in Hanford.

Biddle attended the University of California in Berkeley. After receiving her bachelor's degree in 1908 she was a University Fellow in mathematics there in the academic years

1909–10 and 1910–11. Her minor subject was English literature, and a faculty member from that area served on the committee for her final examination for the PhD. Biddle's 1911 doctorate was the third granted by the California mathematics department, the first to a woman.

The year after completing her work for the PhD Biddle served as instructor of mathematics at the University of Washington in Seattle. She then returned to the Berkeley area and married, on October 7, 1912, William Samuel Andrews (1883–1952), a native of California and an attorney who had received his LLB degree from the University of California in 1906. A daughter was born in September the following year, and a son was born six years later.

Annie D. B. Andrews continued her association with the University of California mathematics department after her marriage, and taught while holding titles of teaching fellow (1914–16), assistant in mathematics (1916–17), associate in mathematics (1920–23), and instructor (1924–33). In 1933 Andrews was one of six junior faculty notified that they would not be reappointed because the department had decided to concentrate more on its graduate program. As this was during the Great Depression, the department considered whether the loss of the position would produce a significant hardship. Andrews, who was married to a practicing attorney, was the only one of the six who lost her job as of July 1, 1933. After an illness of two years Annie Andrews died in Oakland, California, in 1940 at fifty-four. Her husband noted soon after her death that "during the last few years of her life in addition to running her home and doing mathematical research she took an active interest in public affairs and charitable work" (Owens questionnaire 1940).

ANDREWS, Grace. May 30, 1869–July 27, 1951.

WELLESLEY COLLEGE (BS 1890), COLUMBIA UNIVERSITY (MA 1899, PHD 1901). PhD diss.: The primitive double minimal surface of the seventh class and its conjugate.

Grace Andrews was born in Brooklyn, New York, one of five children (four daughters and a son) of Susan Matthews (Hotchkiss) and Edward Gayer Andrews. Her father had studied at Cazenovia Seminary, graduated from Wesleyan University, and was serving as pastor of a church in Cooperstown at the time of his marriage in 1851. Later he was briefly president of Mansfield Female College in Ohio, was principal of Cazenovia Seminary, and in 1872 became a bishop of the Methodist Episcopal Church. From 1872 to 1880 the family was living in Des Moines, Iowa, and from 1880 to 1888 they were in Washington, D.C. In connection with her father's extended trip to India, Grace, her brother, and her mother spent about six months in Europe in 1877.

Grace Andrews was educated at the Mt. Vernon Seminary in Washington, D.C., before going to Wellesley College. She entered Wellesley in 1885 as a member of the class of 1889 and, having taken the five year course with music, graduated in 1890. She began graduate work at Columbia in October 1897 and earned a master's degree in 1899 and a PhD in 1901. From 1900 to 1902 Andrews served as an assistant in mathematics at Barnard College. In about 1904 she moved to her family's home in Brooklyn and remained there until shortly after her mother's death in 1915.

From 1905 to 1926 Grace Andrews was an accountant to the treasurer at Wesleyan University, work she was able to do while living in Brooklyn. From 1916 until her death she had responsible positions with the New York branch of the Women's Foreign Missionary Society of the Methodist Episcopal Church, serving at various times as home base secretary, receiving treasurer, treasurer, chairman of the finance committee, and director.

Andrews was a member of the Brooklyn Botanic Garden; the New York Wellesley Club; the Kosmos Club of Brooklyn, where she was treasurer and director; and the Women's University Club of New York. She liked to travel and spent about a month every summer out of the city, frequently in the nearby mountains. She was an avid reader and fond of doing double-crostic puzzles. Although she lived alone after the death of her mother, she had many nieces and nephews in the vicinity and was involved with her extended family. Grace Andrews died at eighty-two in 1951 in Margaretville, New York, near where she had frequently visited in the summers.

ARMSTRONG, Beulah. November 18, 1895–February 22, 1965.
BAKER UNIVERSITY (BA 1917), UNIVERSITY OF KANSAS (MA 1918), UNIVERSITY OF ILLINOIS (PHD 1921). PhD diss.: Mathematical induction in group theory (directed by George Abram Miller).

Beulah May Armstrong was born in Sterling, Kansas, the only daughter and third of five children of Lillie J. (Detter) and John Allen Armstrong. In 1900 the family was living in Enterprise township, Kansas, where her father was a farmer; in 1910 they were in Hutchinson, Kansas, and her father was in the hardware business.

The family remained in Hutchinson, and Armstrong graduated from Hutchinson High School. She then entered Baker University, a coeducational school affiliated with the Methodist Church, in Baldwin City in eastern Kansas. After her graduation in 1917, Armstrong held a scholarship for a year at the nearby University of Kansas. She received her master's degree in 1918 and continued her graduate work at the University of Illinois with a scholarship in 1918–19 and a fellowship in 1919–21. She received her PhD in 1921 from Illinois with a dissertation in group theory.

Armstrong remained at Illinois as instructor 1921–31, associate 1931–45, assistant professor 1945–59, and associate professor from 1959 until her retirement in 1963 as associate professor emeritus. In a 1983 interview with one of the authors, a former colleague, P. W. Ketchum, recalled that, "as a teacher she was tops." She was heavily involved with advising students who were planning to become mathematics teachers and taught many of the courses they were to take.

Armstrong was a charter member of the University of Illinois chapter of Sigma Delta Epsilon, graduate fraternity for women in science. She remained active at the local level and served as national treasurer 1946–47. She was also active in the Illinois chapter of Sigma Xi, serving as secretary and treasurer. She continued her interest in her college social sorority, Zeta Tau Alpha, and held advisory and corporate board positions for the chapter and offices in the alumnae association. After her death a memorial study in the chapter house at Illinois was established in her name.

Armstrong was described as athletic, and she indicated in 1937 that she was especially interested in reading, philanthropy, and traveling. She was a member of Wesley Methodist Church in Urbana, was active in the work of the Wesley Foundation, and in 1923 served as national president of Kappa Phi, a non-denominational group for Christian university women.

Beulah Armstrong died suddenly at age sixty-nine in Urbana, Illinois, in 1965. She was buried in Hutchinson, Kansas. Her estate included a bequest of $1,000 to Baker University.

ARNOLDY, Sister Mary Nicholas. March 7, 1893–September 28, 1985.
KANSAS STATE AGRICULTURAL COLLEGE (BS 1929), CATHOLIC UNIVERSITY OF AMERICA (MA 1930, PHD 1933). PhD diss. (1932): The reality of the double tangents of the rational symmetric quartic curve (directed by Aubrey Edward Landry).

Catherine Helen Arnoldy, born in Tipton, Kansas, was the eighth of nine children of Anna (Holz), born in Iowa, and Nicholas Arnoldy, who was born in Germany and had emigrated with his family when he was nine. Both parents attended school through the eighth grade. Her father was a storekeeper in Tipton.

Catherine Arnoldy attended parochial school in Tipton and did her high school work at the Nazareth Academy in Concordia, Kansas, before entering the Nazareth Convent of the Sisters of St. Joseph in 1910 and making her profession of vows in 1912.

In the seventeen years before she entered the Catholic University of America for graduate work, Sister Mary Nicholas had various teaching assignments in Kansas. During the 1920s she was doing her undergraduate work. From 1921 to 1923 she attended the Fort Hays Kansas State Normal School (now Fort Hays State University), and in 1923–24 she was among the first women to attend teachers' courses at Creighton University. She received her bachelor's degree from Kansas State Agricultural College (now Kansas State University) in 1929 after intermittent work there starting in 1924. She also took courses in music from Manhattanville College of the Sacred Heart, Pius X School of Liturgical Music, New York.

Sister Mary Nicholas entered Catholic University in 1929, received her master's degree in 1930, completed her doctoral work in September 1932, and was awarded her PhD the following year with minors in physics and education. She spent the next thirty-four years at Marymount College in Salina, Kansas. She was director of the mathematics department and, for several years, was the only mathematics instructor; at various times she was also the college registrar. A sibling, who was also a member of the Sisters of St. Joseph, taught biology at Marymount College.

In 1934 Sister Mary Nicholas presented a paper with the same title as her dissertation to the Kansas Section of the MAA. During the period 1949 through 1955 she presented three more papers to the section, two relating to degree requirements and one on magic circles.

After her retirement from Marymount College in 1966, two years before it became coeducational, Sister Mary Nicholas taught at Sacred Heart High School in Salina; Notre Dame High School in Concordia; and then worked part time at Central Catholic High School in Grand Island, Nebraska. She was known for her hard work, for her love of gardening, and for her mastery of tatting. She was described as very taciturn and as having a very dry sense of humor.

At age eighty Sister Mary Nicholas retired to Medaille Center in Salina. At ninety she moved back to the Nazareth Motherhouse in Concordia, where she died two years later in 1985. She had been a member of the Sisters of St. Joseph in Concordia for seventy-five years.

BABCOCK, Wealthy. November 18, 1895–April 10, 1990.

University of Kansas (BA 1919, MA 1922, PhD 1926). PhD diss.: On the geometry associated with certain determinants with linear elements (directed by Ellis Bagley Stouffer).

Wealthy Consuelo Babcock was born on a farm in Washington County, Kansas, near the Nebraska border. She was the second child of Ella (Kerr) and Cassius Lincoln Babcock; her brother was four years older. In 1900 Cassius Babcock was a farmer; in 1910 he was a carpenter.

Wealthy Babcock graduated in 1913 from Washington County High School and taught one year at Liberty School followed by a year at Lone Mound, both one-room country schools in Washington County. She then attended the University of Kansas where she was a member of the women's basketball team, an early indication of her interest in sports. After receiving her BA in 1919, she taught for a year at Neodesha High School in southeastern Kansas.

Babcock then returned to the University of Kansas, where she was to spend the rest of her career. She was an instructor starting in 1920, while also earning her graduate degrees: a master's in 1922 and a doctorate with a minor in physics in 1926. She was promoted to assistant professor in 1926 and to associate professor in 1940; she retired at that rank in 1966. During her tenure on the Kansas faculty Babcock regularly attended meetings of the Kansas Section of the MAA.

According to G. Baley Price's 1975 history of the department, she was known as an outstanding teacher. She was the department's librarian for more than thirty years, and this contribution was acknowledged in 1966 by the naming of the Wealthy Babcock Mathematics Library. She was particularly active in the KU Alumni Association's activities, for which she received the Fred Ellsworth Medallion, the highest award for service, in 1977.

A number of scholarships have been given in her name, and a floor of a residence hall is named the Wealthy Babcock House.

Babcock's career paralleled in many ways that of **Florence Black**, a close friend and colleague at Kansas, with whom she shared many interests. Black, too, received a PhD from Kansas in 1926 and was on the faculty for her entire career. They were both avid sports fans, faithfully attending university football and basketball games. Both Babcock and Black were interested in camping and the out-of-doors. Price notes that Babcock owned a stable of horses that she and Black rode. Both were elected to the University of Kansas Women's Hall of Fame in 1973. The College of Liberal Arts and Sciences reported in 2005 that the Florence Black and Wealthy Babcock Professorship in Mathematics was newly established.

Wealthy Babcock died at ninety-four in Lawrence, Kansas, in 1990. She was cremated with inurnment in the cemetery on the campus of the university.

BACON, Clara L. August 13, 1866–April 14, 1948.

HEDDING COLLEGE (PHB 1886), WELLESLEY COLLEGE (BA 1890), UNIVERSITY OF CHICAGO (MA 1904), JOHNS HOPKINS UNIVERSITY (PHD 1911). PhD diss.: The Cartesian oval and the elliptic functions ρ and σ (directed by Frank Morley).

Clara Latimer Bacon was born in Hillsgrove, Illinois, the daughter of Louisa (Latimer) and Larkin Crouch Bacon. Her father was a farmer and dealt in livestock. Her parents had married, each for a second time, in 1864, and Clara was the eldest of four children of this marriage. Clara Bacon attended primary school at the Hillsgrove county school and secondary school at the North Abingdon High School. After receiving her PhB in 1886 from Hedding College in Abingdon, Illinois, Bacon taught for the year 1886–87 at a private school in Dover, Kansas. She enrolled at Wellesley College in 1887 and received her second bachelor's degree in 1890.

After graduating from Wellesley, Bacon taught in a private school in Litchfield, Kentucky, 1890–91. She taught mathematics and German and served as librarian at Hedding College 1891–93. She was principal of North Abingdon High School 1893–95 and preceptress and teacher of mathematics at Grand Prairie Seminary, a secondary school in Onarga, Illinois, 1895–97. In 1897 Bacon was hired as an instructor at Woman's College of Baltimore (Goucher College after 1910), beginning a career there that was to last until 1934. During the summers of 1901 to 1904, Bacon studied at the University of Chicago. She received her master's degree in 1904 and was promoted to associate professor the following year. She also studied at Chicago during the summers of 1907 and 1908.

In 1907 the trustees of the Johns Hopkins University voted to allow women to be admitted to graduate courses without special permission. In September that year, Bacon and **Florence P. Lewis**, who later taught at Goucher, successfully applied to study mathematics. During 1909–11 Bacon was a fellow by courtesy at Johns Hopkins, and during the second of these two years she held a fellowship from the Baltimore Association for the Promotion of the University Education of Women, an organization that **Christine Ladd-Franklin** had helped to organize in 1897. In October 1910 Bacon applied for candidacy for the PhD degree in mathematics with subordinate subjects education and philosophy; she received her PhD from Johns Hopkins in 1911, the first year Johns Hopkins granted PhD's to women without special approval of the trustees.

In 1914 Bacon was promoted to professor; she retired in 1934 as professor emeritus. Her only other teaching position was during one summer at Hunter College. During her years at Goucher, students, faculty colleagues, and administrators all valued her teaching, organization, and general humanity. While Bacon was on the Goucher mathematics faculty ten women graduated who later received PhD's in mathematics. Bacon not only encouraged women to go to graduate school but assisted many financially so that they could do so. Two years before her retirement Bacon provided Goucher with funds so that the department could purchase a calculating machine.

Bacon was a member of the executive committee and then chairman of the Maryland-District of Columbia-Virginia Section of the MAA. She was president of the Baltimore chapter of the AAUP and of the Baltimore branch of AAUW and served for many years on the College Entrance Examination Board. She was also a member of the Equal Suffrage League of Baltimore, the Foreign Policy Association, the League of Women Voters, and the Daughters of the American Revolution. In 1937 she reported that she was a member of the Methodist Episcopal Church and of several peace organizations. Bacon shared her house and traveled with her sister Agnes, who had worked in the field of biometry at the School of Hygiene and Public Health at Johns Hopkins. After Bacon's retirement she fulfilled a promise to Agnes, who died in 1930, to travel again, and she visited England, Italy, Egypt, and Palestine during a fifteen-month trip abroad.

Bacon died in Baltimore at age eighty-one in 1948 and is buried near Abingdon, Illinois. One of the residential houses at Goucher, Bacon House, is named after her, and one of Goucher's endowed scholarships is named for Clara and Agnes Bacon.

BAKER, Frances E. December 19, 1902–April 4, 1995.

University of Iowa (BA 1923, MS 1925), University of Chicago (PhD 1934). PhD diss.: A contribution to the Waring problem for cubic functions (directed by Leonard Eugene Dickson).

Frances Ellen Baker was born in Anna, Illinois, the elder of two daughters of Katherine (Riedelbauch) and Richard Philip Baker. Frances Baker's mother earned a diploma in music from Illinois Wesleyan University and was head of the music department at Lamar College (now closed) in Lamar, Missouri, 1898–1901, the last three years that R. P. Baker was president there. R. P. Baker was born in England, studied at Oxford University, received a bachelor's degree from the University of London, and moved to Texas in 1888. During the next sixteen years he lived in Texas, Illinois, Oregon, Wisconsin, and Missouri: he practiced law; studied mathematics; taught music, mathematics, and science; was a college president; and was an administrator at a private school. He also constructed mathematical models for use in teaching, many of which are at the National Museum of American History, Smithsonian Institution, and at the University of Arizona.

R. P. Baker was head chemist at a printing company in Chicago 1904–05. In 1905 he joined the mathematics department of the University of Iowa in Iowa City (officially the State University of Iowa) as instructor and in 1910 received his PhD in mathematics from the University of Chicago. Katherine Baker continued to give private lessons in piano and voice and with her husband established and played with the first chamber music ensemble in Iowa City. R. P. Baker was associate professor at the time of his death in 1937.

Frances Baker received her elementary education, including piano and German, from her parents. She attended public high school in Iowa City, graduated as valedictorian in 1919, and entered the University of Iowa, where she studied Latin, Greek, and mathematics. She was elected to Phi Beta Kappa her junior year and graduated magna cum laude in 1923. She then studied mathematics and physics and received her master's degree in 1925 with a thesis written under the direction of her father. During 1924–25 she held a graduate fellowship and taught a freshman mathematics course one semester.

Baker was an instructor of mathematics and physics and head of the department at Tabor College in Iowa from 1925 until the college closed in 1927. She held a similar position at Jefferson City Junior College in Missouri 1927–28 and then enrolled at the University of Iowa for one semester to earn a teaching certificate. She attended the University of Chicago during the spring and summer quarters of 1929 and was instructor of mathematics and head of the department at the Creston, Iowa, Junior College and Senior High School 1929–31.

Frances Baker returned to the University of Chicago in 1931, held a fellowship 1932–33, and received her PhD in March 1934. She lived at home in Iowa City until she became instructor at Vassar College spring 1935. In fall 1935 she went to Mount Holyoke College and was instructor 1935–40 and assistant professor 1935–42. She had a leave in the fall semester 1938, having been injured in an automobile accident.

Gladys E. Baker, Frances Baker's sister, received her doctorate in botany from Washington University in St. Louis in 1935, taught at Hunter College, and joined the faculty at Vassar College in 1941. Frances Baker joined her sister there and was associate professor 1942–51, professor 1951–68, and emeritus professor after she retired in 1968; she chaired the department 1948–50 and 1951–52. She had three research leaves while at Vassar: at the University of Wisconsin part of 1945–46, at Princeton University 1952–53, and at the University of North Carolina 1960–61.

In the 1940s Baker gave talks on the mathematical models her father had constructed forty years earlier. Starting a few years before her retirement, she became a frequent book reviewer for the library journal *Choice* and later read college mathematics books for Recordings for the Blind. Baker was a member of the Protestant Episcopal church.

Gladys Baker left Vassar in 1961 to teach at the University of Hawaii, Manoa. In January 1974, she moved to Sun City, Arizona, and Frances Baker moved to a separate home there that autumn. Frances Baker was living in a care facility in Peoria, Arizona, at the time of her death at age ninety-two in 1995. Gladys Baker died at ninety-nine in 2007.

BALLANTINE, Constance (Rummons). August 15, 1896–January 14, 1974.
UNIVERSITY OF NEBRASKA (BA 1916, MA 1919), UNIVERSITY OF CHICAGO (PHD 1923). PhD diss.: Modular invariants of a binary group with composite modulus (directed by Leonard Eugene Dickson).

Constance Juliet Rummons was born in Lincoln, Nebraska, the daughter of Clara (Schroeder) and Nestor Rummons. Her mother was a homemaker and legal secretary; her father was a lawyer. A year or two after their 1895 marriage the family moved from Lincoln to Elwood, Nebraska, and in 1901 to Hobart, Oklahoma. She had two younger sisters; one died in infancy and the other became a teacher of Latin, Greek, and English in a Chicago high school after her education at the University of Chicago. Census records indicate that her parents were living apart in Hobart, Oklahoma, in 1910 and that they were divorced by 1920, when Clara Rummons and her daughters were living in Lincoln, Nebraska.

Constance Rummons entered the University of Nebraska in 1912 and received her bachelor of arts degree at the midwinter commencement of 1915–16. She also took some teachers' courses and in the summer session of 1916 earned the teachers' college diploma and the university teachers' certificate. She was a graduate student in 1916–17 and an assistant in philosophy 1917–19 before receiving her master's degree in May 1919 in philosophy and English literature. Rummons was a graduate student, presumably in mathematics, at Nebraska 1919–21 and was also an assistant instructor in mathematics 1920–21, having studied at the University of Chicago in the summer of 1920. She returned to Chicago for summer 1922 and the three quarters of the academic year 1922–23. She received her PhD from Chicago in 1923.

During this period she met John Perry Ballantine, a fellow graduate student in mathematics at Chicago; they were married on September 21, 1922. J. P. Ballantine was born in 1896 in Rahuri, India, where his father was a medical missionary. He received a bachelor's degree from Harvard in 1918 and his PhD from the University of Chicago in 1923.

In 1923 the Ballantines moved to New York City, where J. P. Ballantine was instructor at Columbia University 1923–26. Constance Ballantine was an instructor at Washington Square College, New York University, 1924–25. In August 1926 the first of their three children, a daughter, was born. That same year they moved to Seattle where J. P. Ballantine began his work at the University of Washington, where he remained, except during World War II when he served with the US Army at military schools in Europe, until he retired in 1966. Their two sons were born in 1927 and 1929. All the children subsequently had professional careers. Their daughter became an epidemiologist, their elder son a computer programmer, and their younger son a mathematician.

Within two years of moving to Seattle, Constance Ballantine was given the title of associate at the University of Washington. She held the title, which appears to have carried no specific duties, at least until 1950, and taught some correspondence courses. According to her younger son, anti-nepotism practices, especially during the Depression, played an inhibiting role in her pursuit of professional work. She wrote a review for the *Monthly* in 1939, and when the children were in school she was active in the PTA. J. P. Ballantine died in March 1970, and Constance R. Ballantine died in Seattle in January 1974, at age seventy-seven.

BALLARD, Ruth (Mason). April 24, 1906–June 7, 1980.

WELLESLEY COLLEGE (BA 1926), UNIVERSITY OF CHICAGO (MS 1928, PHD 1932). PhD diss.: [Mason, R. G.] Studies in the Waring problem (directed by Leonard Eugene Dickson).

Ruth Glidden Mason was born in Chicago, Illinois, the younger of two children of Bertha Hall (Sickles) and Fred Bonfoy Mason. Her mother was in one of the earliest classes at Michael Reese Hospital School of Nursing in Chicago. Her father was an 1890 graduate of Oberlin College and became an insurance agent in Chicago. Her brother attended Yale University and the University of Michigan.

Ruth Mason attended the Knickerbocker grammar school and Robert Waller High School, both public schools in Chicago. After her high school graduation at age sixteen, she studied one year at the University of Chicago before completing her undergraduate work in 1926 after three years at Wellesley College.

Mason immediately entered the University of Chicago where, after graduate study during the academic year 1926–27 and the spring and summer quarters of 1928, she received her master's degree in 1928 with a thesis written under the direction of **Mayme Logsdon**. During 1930–31 she was both a Horton-Hallowell fellow (from Wellesley) and a graduate fellow at Chicago. She studied one semester with D. N. Lehmer at the University of California in 1930 before finishing the work for her PhD at Chicago in 1932.

Ruth Mason returned to Berkeley to do some postdoctoral work 1932–34 and taught at the College of the Holy Names in Oakland, California, in 1933. She then returned to the East and taught at Wellesley in the spring semester of 1935. She was an instructor at Hood College in Frederick, Maryland, 1935–36 before moving back to Illinois where, except for travel, she remained. She taught at the University of Illinois in Urbana 1936–38, as assistant the first year and instructor the second. She then joined the faculty at Wright Junior College in Chicago and remained at Wright for five years. She was chairman of the mathematics section of the Illinois Association of Junior Colleges 1939–40.

On April 20, 1940, Ruth Mason married Foster K. Ballard (1899–1978), a graduate of the University of Chicago and chemist for the US Customs Service. She remained at Wright until 1943, when the navy took over the building in which classes met. Ballard was on leave of absence, but was teaching again in July 1943, this time in a Navy V-12 program for sailors, housed at the Illinois Institute of Technology in Chicago. She continued teaching in this program until 1945.

The Ballards' son was born in May 1946. For the next few years Ruth Ballard's activities focused on her son's activities and her church, in which she continued to teach Sunday school and serve as a member of the vestry. She also maintained her involvement with the Women's Mathematics Club of Chicago and Vicinity.

From 1955 until her retirement in 1972, Ballard taught full time as assistant professor and then associate professor at the University of Illinois at Navy Pier (moved and renamed the University of Illinois at Chicago Circle in 1965, now the University of Illinois at Chicago). During this period she was program director in 1962–63 and president in 1963–64 of the Women's Mathematics Club. Ruth Ballard had been living in Wilmette when she died at seventy-four in 1980 in Chicago.

BAREIS, Grace M. December 19, 1875–June 15, 1962.

HEIDELBERG UNIVERSITY (BA 1897), OHIO STATE UNIVERSITY (PHD 1909). PhD diss.: Imprimitive substitution groups of degree sixteen (directed by Harry Waldo Kuhn).

Grace Marie Bareis was born in Canal Winchester, Ohio, the first of two daughters of Amanda Jane (Schoch) and George Frederick Bareis. Her father conducted a successful lumber business in Canal Winchester. He was a student of history and archaeology and published a history of the local township in 1902.

After attending the public schools in Canal Winchester, Grace Bareis studied at Heidelberg University (Heidelberg College after 1926) in Tiffin, Ohio. She took the classical course and participated in a literary society and an oratorical association before graduating in 1897 as valedictorian of her class. Bareis was a graduate student at Bryn Mawr College 1897–99 and then attended the Columbus Normal School in Ohio 1899–1900. She was a teacher at Miss Roney's School in Bala, Pennsylvania, from 1900 until 1906 while continuing her graduate work in mathematics at nearby Bryn Mawr College during 1902–06.

In 1906 Bareis returned to the Midwest to finish her graduate work in mathematics at Ohio State University. She was a fellow there 1906–08, was appointed assistant professor in 1908, and was the first recipient of a doctorate in mathematics from Ohio State when she received her PhD in 1909. After nearly forty years on the faculty she retired as assistant professor emeritus in 1946. During her years at Ohio State, she directed master's theses, mainly in geometry, for at least fifteen graduate students, of whom at least eight were women. At various times she was also a faculty member in the education college, advisor for arts college women, and group leader for freshman week. She spent the summer of 1920 at Columbia University. During World War II, she taught students in the Army Specialized Training Program at Ohio State, and for two years after her retirement she continued to teach because of the postwar shortage of instructors.

In 1948 Bareis gave a gift of $2000 to Ohio State University to help commemorate its seventy-fifth anniversary. She asked that the interest on her donation be used to fund the Grace M. Bareis Mathematical Prize to be awarded annually after a written contest. During her lifetime Bareis remained active in her college alumni organizations and from 1935 to 1956 was a member of the board of trustees of Heidelberg College. Heidelberg gave Grace Bareis an honorary degree of Doctor of Pedagogy at the hundredth anniversary of the college in 1950, and the Bareis Hall of Science was built there in 1964 to honor her and her father, who had also served on the board of trustees. Grace Bareis had made a sizable gift towards its construction.

Bareis attended many national meetings of the MAA and attended more than thirty annual meetings of the Ohio Section from its inception in 1916 through 1955, nine years after her retirement. She was active in church activities: she was a member of the Evangelical and Reformed Church at Canal Winchester, belonged to the Missionary Society, participated in the business women's Sunday school class, and was an affiliate member of the Indianola Methodist Church. She also volunteered with the Red Cross and belonged to the DAR as well as numerous civic organizations and historical societies. She maintained her home, and spent the summers, in Canal Winchester. At one point she described her hobbies as gardening and farming; she managed a farm near Brice, Ohio.

Grace Bareis lived for twenty-six years with a friend, her former student and later mathematics department colleague, Margaret Eloise Jones (1895–1979), in Columbus. Bareis died in 1962 at age eighty-six in Columbus after a lengthy illness. Bareis's estate was valued at more than $500,000, with the largest single bequest to Margaret Jones. Other bequests included several to various church and mission organizations and to educational institutions, including one to Ohio State to aid women students majoring in mathematics.

BARNES, Mabel (Schmeiser). July 29, 1905–February 22, 1993.
CORNELL COLLEGE (BA 1926), UNIVERSITY OF WISCONSIN (MA 1928), OHIO STATE UNIVERSITY (PHD 1931). PhD diss.: [Schmeiser, M. F.] Some properties of arbitrary functions concerning approach to a straight line (directed by Henry Blumberg).

Mabel Frances Schmeiser was the second of six children of Christena M. (Wehmeyer) and Emanuel Schmeiser. Her mother was educated through the eighth grade, and her father, a farmer, received some elementary education. All of the children were born in Wapello, Iowa, and all attended college.

Mabel Schmeiser began her education in a one-room country school and graduated from the high school in Wapello in 1922. She then attended Cornell College and graduated in 1926. She spent the academic year 1926–27 at the University of Wisconsin on a mathematics department fellowship. She received her master's degree from Wisconsin in 1928 after having entered Ohio State University for further graduate work in 1927. She was at Ohio State 1927–30 as a graduate assistant the first two years and a university fellow the last year; she also attended the summers of 1930 and 1931.

In 1930 Schmeiser accepted a temporary position as professor and acting chairman of the mathematics department at Nebraska State Normal School and Teachers College (now Wayne State College) in Wayne, filling in for someone on leave of absence. During her first year she finished her dissertation and received her PhD from Ohio State in 1931. Her predecessor returned to resume the position in 1933. Because jobs were hard to find during the Depression, Schmeiser applied to the recently opened Institute for Advanced Study at Princeton, where she was accepted and spent the following academic year. At the end of that year she took an examination to become a substitute mathematics teacher in New York City and was hired at Wadleigh High School in Manhattan for 1934–35.

On July 3, 1935, Mabel Schmeiser and John Landes Barnes were married. He was born in New Jersey in 1906, had two degrees in electrical engineering from MIT, and two in mathematics from Princeton, a 1930 MA and a 1934 PhD. They first lived in Massachusetts, where J. L. Barnes had a position as assistant professor at what was then Tufts College. Mabel Barnes helped her husband grade papers and edit the mathematics entry for a handbook for engineers; moreover, she substituted for him as necessary. John Barnes remained at Tufts until 1942, during which time he was promoted to professor of mathematics and served as chairman of applied mathematics as well as acting chairman of electrical engineering. During that period their two children were born, a son in 1936 and a daughter in 1940.

During the war John Barnes was on leave from Tufts and did war work with Bell Telephone Laboratories, first in New York and then in New Jersey. They returned to Tufts after the war, and Mabel Barnes was hired as an assistant professor for 1946–47, partly because of the influx of veterans. Although she was rehired by Tufts for the next year, they moved to California where her husband had taken a position as professor in engineering at UCLA. He remained there until his retirement in 1974. In 1957 John L. Barnes founded Systems Corporation of America. He served as president until his death in La Jolla in 1976 shortly before his seventieth birthday.

In 1950 Mabel Barnes resumed her career by taking a position at Occidental College in Los Angeles, where she was instructor 1950–51, lecturer 1951–52, assistant professor 1952–56, associate professor 1956–64, professor 1964–71, and then professor emeritus. From 1955 until 1961 she served as one of the editors of *Pentagon*, the journal of Kappa Mu Epsilon.

Barnes liked to travel, especially with her young grandsons. Her main activity during retirement was working on conservation with the Sierra Club and with the Desert Protective Society. She also was a volunteer at a local hospital for many years. Mabel Barnes died in Los Angeles in 1993 at age eighty-seven.

BARNEY, Ida. November 6, 1886–March 7, 1982.

Smith College (BA 1908), Yale University (PhD 1911). PhD diss.: Line and surface integrals (directed by James Pelham Pierpont).

Ida Barney was born in New Haven, Connecticut, the first of two daughters of Ida (Bushnell) and Samuel Eben Barney. Her father graduated from the Sheffield Scientific School of Yale University and was on the civil engineering faculty at Yale from 1882 until his retirement in 1924. Her sister, Elizabeth Hunt Barney, graduated from Smith College in 1914 and was secretary to the dean, and later to the registrar, of the Yale Divinity School.

Ida Barney attended New Haven High School 1900–04, Smith College 1904–08, and Yale Graduate School 1908–11. She was a mathematics major at Smith, where her studies included a course in general astronomy and one in the theory and use of transit instruments.

After Barney received her PhD from Yale, she spent the next ten years on various mathematics faculties, starting as professor of mathematics at Rollins College in Winter Park, Florida, 1911–12. She was at Smith College for six years, as an assistant 1912–13, as instructor 1913–17, and as assistant professor 1920–21. Between her two periods of time at Smith, she was professor first at Lake Erie College in Painesville, Ohio, 1917–19, and then at Meredith College in Raleigh, North Carolina, 1919–20. All but Rollins College were women's colleges at the time.

Barney spent the rest of her life in or around New Haven, where she and her sister lived with their parents until their parents' deaths in the early 1940s. After that Ida Barney and her sister lived in the family home until her sister's death in 1958.

In 1921 Barney enrolled in a graduate course in astronomy at Yale, and the following year she was appointed research assistant at the Yale Observatory. Early in her tenure at the observatory she became involved in a major star mapping project under the direction of Frank Schlesinger. This project, cataloguing positions and proper motions of stars, was done with the participation of the Lick Observatory of the University of California, the US Naval Observatory, and the Royal Observatories at Greenwich and the Cape of Good Hope. Barney was coauthor with Schlesinger, and sometimes others, of twelve volumes of catalogues that appeared in the *Transactions of the Astronomical Observatory of Yale University*. The first six of these volumes, appearing 1925 though 1934, were produced before IBM punch-card machines were used to facilitate computations.

After Frank Schlesinger's retirement in 1941, Barney was named director of the project. Between 1945 and 1950 she authored another eight volumes of catalogues. In 1949 she was promoted to research associate in astronomy, with the rank of associate professor, for a five-year term. When the project was completed in 1950 it resulted in what is known as the Yale Photographic Zone catalogues. In 1951 through 1954 Barney authored a supplementary volume and two revised catalogues. She also coauthored three new catalogues, one in 1954 and two in 1959, four years after her retirement.

Barney was recognized by her colleagues for her achievements. In 1945 the American Astronomical Society inaugurated a Council, and the following year Barney was elected to a three-year term. She was a fellow of the Royal Astronomical Society of London. In 1952, Barney was awarded the Annie J. Cannon Prize, established in 1934 and given no more often than every three years, by the American Astronomical Society. At the time of her retirement in 1955, it was reported that she received the prize "for completing one of the most intensive photographic mapping jobs ever undertaken by a single observatory" ("6 at Yale to Retire," *New York Times*, March 20, 1955).

Ida Barney continued to live in the family house in New Haven for several years. She had no immediate survivors at the time of her death in Hamden, Connecticut, at age ninety-five in 1982.

BARNUM, Charlotte C. May 17, 1860–March 27, 1934.

VASSAR COLLEGE (BA 1881), YALE UNIVERSITY (PHD 1895). PhD diss.: Functions having lines or surfaces of discontinuity.

Charlotte Cynthia Barnum was born in Phillipston, Massachusetts, the third of four children of Charlotte (Betts) and Samuel Weed Barnum, who received a BA from Yale and a BD from Yale Divinity School. Both of her brothers graduated from Yale University, and her sister graduated from Vassar College and attended the Yale graduate school. Samuel Weed Barnum was pastor of Congregational churches in Connecticut and Massachusetts before moving to New Haven in 1865. He was engaged in various editorial projects, including the 1847 revision of *Webster's Dictionary*.

Charlotte Barnum attended New Haven public schools, including Hillhouse High School. She enrolled at Vassar in 1877 and graduated in 1881. She taught at Betts Academy, a boys' preparatory school in Stamford, Connecticut, 1881–82; at Bradford Academy in Massachusetts; and at Hillhouse High School in 1883 and 1885–86. She did computing for the Yale Observatory 1883–85 and, also in 1885, computed angles of crystals for a revision of James Dwight Dana's *System of Mineralogy*. She worked in 1886–90, and again in 1897, as assistant editor for *Webster's International Dictionary* and its Australian supplement. During the academic year 1889–90 Barnum was a teacher of astronomy at Smith College.

In 1890 Charlotte Barnum requested that she be allowed to take courses at the Johns Hopkins University. However, Johns Hopkins did not then officially admit women, even for graduate work. Barnum was supported in her request by Simon Newcomb, professor of mathematics and astronomy at Johns Hopkins, who wrote to President Daniel Coit Gilman in November 1890 making an argument for allowing qualified women to study mathematics and astronomy at the graduate level at the university. He asked that Barnum be admitted to classes of Thomas Craig and Fabian Franklin without setting any precedents. In December the executive committee of the board of trustees voted that Barnum be allowed to attend Dr. Craig's lectures without enrollment and without a charge for tuition. Consequently, she studied mathematical astronomy and physics at Johns Hopkins 1890–92. Wishing to earn a degree, however, Barnum entered Yale University in 1892, the first year women were officially admitted to the graduate school there. She remained at Yale for three years and received her PhD in 1895. It is unclear from the record who directed her dissertation. In 1893 she had attended the International Mathematical Congress in Chicago, one of the four women among forty-five attendees. The following year she became a member of the AMS, about three years after the first women joined.

Barnum was an instructor at Carleton College in Northfield, Minnesota, 1895–96. She worked in applied areas of mathematics and in editorial work most of the rest of her career. She joined the American Institute of Actuaries in 1898 and worked as an actuarial computer for the Massachusetts Mutual Life Insurance Company, Springfield, Massachusetts, in 1898 and for the Fidelity Mutual Life Insurance Company, Philadelphia, 1900–01. From 1901 to 1913 she was in Washington, D.C.: as a computer for the US Naval Observatory in 1901 and for the tidal division of the US Coast and Geodetic Survey 1901–08, and as editorial assistant in the biological survey for the US Department of Agriculture in 1908–13. She also worked on books by the Carnegie Institution of Washington.

From 1914 until 1918 Barnum was for the most part in New Haven and engaged mainly in editorial work: for the Yale Peruvian Expeditions 1914–15, for the Yale University secretary's office 1915–16, and for the Yale University Press in 1915 and 1918–19. She was also in New York City some of the time from 1917; in about January 1917 she was working there as a consulting actuary, and during much of 1919 she worked there on the *Chronicles of America* series for the Yale University Press and edited a book on labor unrest. She taught in 1922 in the Scoville and the Columbia preparatory schools in New York and in 1923 in the Walnut Hill School in Natick, Massachusetts.

During her life Charlotte Barnum was involved in various social and charitable activities. When she was living in New York City she was a member of the Broadway Tabernacle, a Congregational church known for its social activism. She died of meningitis at age seventy-three in Middletown, Connecticut, in 1934 and is buried in New Haven.

BARTON, Helen. August 9, 1891–March 19, 1971.

GOUCHER COLLEGE (BA 1913), JOHNS HOPKINS UNIVERSITY (MA 1922, PHD 1926). PhD diss.: Some applications of the generalized Kronecker symbol (directed by Frank Morley).

Martha Helen Barton was born in Baltimore, Maryland, the daughter of Mary Irene (Eichelberger) and James Sheridan Barton, a merchant. She had an older brother and a younger sister.

Helen Barton received her primary and secondary education in the public schools of Baltimore and graduated with honors from Western High School. She attended Goucher College (Woman's College of Baltimore until 1910) as a Western High School alumnae scholar 1909–11 and graduated in 1913. While Barton was a student at Goucher, two of her mathematics instructors, **Clara L. Bacon** and **Florence P. Lewis**, were awarded PhD's at Johns Hopkins University. The year after her graduation Barton served as an assistant in physics at Goucher and the following year, 1914–15, did graduate work in mathematics and physics as a Goucher alumnae fellow at Johns Hopkins. Her sister, Vola Price Barton, graduated from Goucher with a major in mathematics in 1915 and earned a PhD in physics from Johns Hopkins in 1923; she was a member of the physics faculty at Goucher.

From 1915 until 1919 Helen Barton served as head of the department of chemistry and physics at Salem College, a private college for women in Winston-Salem, North Carolina. The following two years she was an instructor in mathematics at Wellesley College. She continued her graduate work at Johns Hopkins during the summers of 1920 and 1921 and did some graduate work at Harvard University during the winter of 1921 while also teaching at Wellesley. She received the master's degree in June 1922 from Johns Hopkins.

Barton left her position at Wellesley in 1921 and moved to Albion College in Michigan, where she was associate professor of mathematics and dean of women 1921–25. In 1925 she resumed her full-time graduate work at Johns Hopkins and was a university scholar for the year 1925–26. The doctorate was awarded in June 1926 with physics as her first subordinate subject and applied mathematics as the second. Barton spent the following year as professor and head of the department at Alabama College, then a college for women in Montevallo, Alabama; it is now the coeducational University of Montevallo.

In 1927 Helen Barton joined the faculty of the North Carolina College for Women (later Woman's College of the University of North Carolina and now the University of North Carolina at Greensboro), where she was to remain until her retirement. She began her career there as associate professor and acting head of the department, and the following year, 1928, became professor and head of the department. She retired as professor emeritus at age sixty-eight in 1960 but continued teaching part time.

In addition to teaching and chairing the mathematics department, Barton was active both on campus and in various professional groups including the faculty science club at Woman's College, the campus chapter of Phi Beta Kappa, AAUW, the North Carolina Education Association, and the North Carolina Academy of Science. In 1960 the faculty lounge in the McIver classroom building at Woman's College was named in her honor.

Helen Barton died in Greensboro at age seventy-nine in 1971 after a long illness and was buried in Baltimore. She left $5000 to the University of North Carolina at Greensboro to establish the Helen Barton Mathematics Scholarship Fund. A second bequest of $500 was made to the Helen Barton Professorship Fund, which was established in 1962 and helps support the Helen Barton Excellence Professorship.

BAXTER, Elizabeth (Pillsbury). June 8, 1906–April 14, 1966.

Bryn Mawr College (BA 1927), University of Michigan (MA 1935, PhD 1939). PhD diss.: On the geometry of the Dirac equations (directed by George Yuri Rainich).

Margaret Elizabeth Pillsbury was born in Ann Arbor, Michigan, the elder of two children of Margaret May (Milbank) and Walter Bowers Pillsbury. Her mother earned a BA degree from the University of Michigan. Her father, who had received a BA from the University of Nebraska and a PhD from Cornell, was a faculty member in the department of philosophy and psychology at the University of Michigan. He became the chairman of the psychology department when the disciplines were separated in the 1920s and remained on the faculty at Michigan until his retirement. His interests lay primarily in reasoning and attention; he was starred in the fifth edition of *American Men of Science* and was a member of the National Academy of Sciences.

After Elizabeth Pillsbury's graduation from Ann Arbor High School, she entered Bryn Mawr College. While there she held a James E. Rhoads junior scholarship and three scholarships her senior year: the Charles S. Hinchman memorial scholarship, the Elizabeth S. Shippen scholarship in science, and the Elizabeth Wilson White memorial scholarship. Pillsbury graduated summa cum laude in mathematics and chemistry in 1927.

Elizabeth Pillsbury first enrolled at the University of Michigan in the summer of 1927. She also studied at Göttingen in 1927 and in Berlin in 1928. From 1928 to 1932 she was a graduate student in chemistry at the University of California in Berkeley. Also during this period she married, in 1930, Warren Phelps Baxter. Baxter was born in Montreal in 1901 and received a PhD in chemistry from the California Institute of Technology in 1928. In 1930 he was a chemist for Shell Development, the research division of Shell Oil Company, in Emeryville, near Berkeley, California. In 1949 he was employed by National Technical Laboratories in South Pasadena. Elizabeth and Warren Baxter had a daughter, born in 1934, and a son, born in 1939.

In summer 1935 Baxter enrolled as a candidate for a master's degree in mathematics at the University of Michigan and received the degree by the end of the year. She continued her work there in the academic year 1935–36, the summers of 1936 and 1937, and part time in the second semester of 1937–38. She completed her dissertation in 1938 and received the PhD from Michigan in February 1939.

From 1945 to 1947 Elizabeth Baxter worked as a research analyst for the Douglas Aircraft Company. In 1947 she was hired by the Jet Propulsion Laboratory at the California Institute of Technology. Baxter was a theoretical analyst 1947–50 and a research engineer after 1950, working mainly in the area of numerical analysis. Having worked as an engineer for eighteen years, Elizabeth Baxter died of cancer at fifty-nine in April 1966 in Pasadena, California. Warren P. Baxter had died two months earlier.

BEATY, Marjorie (Heckel). January 21, 1906–July 18, 2002.

University of Rochester (BA 1928, MA 1929), University of Colorado (PhD 1939). PhD diss.: On the complex roots of algebraic equations (directed by Aubrey Kempner).

Marjorie Louise Heckel was born in Buffalo, New York, the elder of two children of Josephine Mary (Fisher) and Henry George Heckel, the manager of a furniture store. Marjorie Heckel attended grade school and a year of high school in Buffalo. Her family then moved to Rochester, New York, where she graduated from East High School. She attended the University of Rochester on a state scholarship and received her bachelor's degree in 1928. She remained for a year after her graduation to complete her master's degree. Heckel then went to Brown University to continue her graduate studies. She was an assistant there for two years, 1929–31, but had to leave for financial reasons.

Heckel took a position as instructor at the University of South Dakota in fall 1931. On March 30, 1933, she married Donald W. Beaty (1903–1979), a cattle feeder and farmer who was also teaching at the university, and Marjorie Beaty was forced to resign her position because of anti-nepotism policies. During 1934–35 the mathematics department was short-handed and Beaty volunteered to teach one course for a semester without pay. In the 1940s Donald Beaty served as a Republican state senator for South Dakota.

Marjorie Beaty studied at the University of Colorado during 1935–38; she was a university fellow and teaching assistant her first year, a research fellow her second year, and a teaching assistant her third year. Beaty and **Louise Johnson (Rosenbaum)** both earned PhD's in 1939 as students of Aubrey Kempner. These were the third and fourth PhD's given in mathematics at Colorado and were the only mathematics PhD's received by women there before 1960.

Beaty returned to the mathematics department at the University of South Dakota as acting head in 1938–39; she was assistant professor 1938–41 and served as acting dean of women for one semester in 1941. She took an extended leave to raise her two daughters, born in 1943 and in 1944, and returned to the department as assistant professor in 1955. She was promoted to associate professor in 1958 and to professor in 1961, about the time that two papers that she had coauthored were published. She retired in 1976 but taught at nearby Yankton College for a number of years after her retirement.

Beaty was a recipient of an Outstanding Educator of America award in 1974. In conjunction with her speech at the 1982 summer commencement, her daughters donated funds to the university to set up the Dr. Marjorie H. Beaty mathematics scholarship. The university's Phi Beta Kappa chapter also gives a Marjorie H. Beaty award to a senior who plans to pursue graduate work there. After her retirement, Beaty served on the board of directors of the university's America's Shrine to Music Museum, helped to create the Emeritus Club Shakespeare Garden, and was chair of the university's Centennial of South Dakota. She and her late husband were awarded the University of South Dakota Foundation's Inman Award in 1998 in recognition of their contributions to the university.

In about 1940 Beaty reported that she was Congregationalist and that her favorite recreation was hunting. From 1950 to 1976 she served as the secretary to the Clay County Republican Central Committee. She was a founding member of the Eta Chapter of South Dakota of Delta Kappa Gamma, a professional honor society for women educators, and received that society's state achievement award in 1985. She also belonged to the P.E.O. Sisterhood, a society promoting educational opportunities for women, and served as president of the local Vermillion chapter. She joined the AMS in 1929 as a nominee of Brown University and maintained her membership until her death in Vermillion, South Dakota, in 2002 at age ninety-six.

BECHTOLSHEIM, Lulu (Hofmann). May 27, 1902–August 29, 1989.

Realgymnasium der Schillerschule (Frankfurt am Main) (*Abitur* 1922), Universität Zürich (PhD 1927). PhD diss.: [Hofmann, L.] Über einige spezielle Strahlenkongruenzen, die mit analytischen Funktionen zusammenhängen (directed by Eugenio Giuseppe Togliatti).

Lulu Hofmann was born in New York City, the middle daughter of German parents Clara (Olshauson) and Otto Hofmann. Her mother attended a Lyzeum, a secondary school for girls, and became a homemaker. Her father attended a Gymnasium, the classical German secondary school, and was a stockbroker. By 1910 the family was living in Frankfort, Germany, where all three sisters attended a Gymnasium. Her older sister also attended an academy of art and became a painter; her younger sister studied at universities in Brussels and Paris and became a certified translator.

Hofmann was raised bilingually in German and English and received her complete education in Germany and Switzerland. She attended the Realgymnasium der Schillerschule in Frankfurt and passed her *Abitur* in 1922. She then attended the Universität Freiburg 1922–23 and the Universität Zürich 1923–26. During 1924–26 she also studied part time at the Eidgenössiche Technische Hochschule (ETH) in Zurich. Starting in 1926, Hofmann assisted Hermann Weyl, who was then at ETH, in translating German manuscripts into

English. Hofmann received her doctorate from the Universität Zürich in 1927, having also studied philosophy under Edmund Husserl and Martin Heidegger.

Lulu Hofmann returned to the United States in January 1927 and gave an address of Springfield, Ohio, when she applied for membership in the AMS that spring. She was at Columbia University as an assistant in mathematics 1927–28 and as a lecturer 1928–29; she was an instructor at Barnard College 1929–37. While at Columbia and Barnard, she worked closely with Edward Kasner and continued to assist Weyl, who came to the Institute for Advanced Study in 1933, with translations.

In November 1936 Lulu Hofmann married Baron Wilhelm Alfred von Bechtolsheim (1881–1968). He was born in Upper Bavaria and had been a Commander in the Imperial German Navy before immigrating to the United States in 1926. He later was a pharmaceutical salesman. They had no children, although he had children by a previous marriage.

During the summer session of 1937 Lulu von Bechtolsheim was an instructor at Hunter College. That fall she became an instructor at Queens College where she remained until 1940. By 1944 she and her husband had moved to California, where, during the winter quarter of 1943–44, she was a lecturer in mathematics at Stanford University; she was also an acting assistant professor at Stanford in summer 1947. Lulu Bechtolsheim, as she was then known, spent the largest part of her career at the University of Redlands, east of Los Angeles. She was assistant professor 1944–50, associate professor 1950–56, and professor 1956–61, before her retirement in 1961. In addition to her role as a mathematics professor, she also taught astronomy, Italian, French, and German at Redlands. She also knew some Spanish, Portuguese, and Greek. Starting in the mid-1950s, and continuing for about twenty years, Bechtolsheim translated mathematical manuscripts from Portuguese into English, from English into German, and from Italian into German.

Bechtolsheim was interested in poetry. From 1972 she was a member of the Anthroposophical Society, an organization promoting the spiritual philosophy based on the works of Rudolf Steiner, and, from 1981, the Christian Community, inspired by his work as well. Lulu Bechtolsheim died in 1989 at age eighty-seven in Redlands, California.

BEENKEN, May M. October 22, 1901–December 21, 1988.
UNIVERSITY OF CALIFORNIA, SOUTHERN BRANCH (BEd 1923); UNIVERSITY OF CHICAGO (MA 1926, PhD 1928). PhD diss.: Surfaces in five-dimensional space (directed by Ernest Preston Lane).

May Margaret Beenken was the daughter of Sophie (Kirn), born in Pennsylvania, and Henry Beenken, a native of Germany. She was born in Philadelphia, Pennsylvania, and had at least three older siblings.

May Beenken attended public elementary school and three years of high school in Philadelphia before the family moved to southern California. After one year there she graduated from the Manual Arts High School of Los Angeles in 1919. She then was a member of the first class to receive bachelor's degrees at the Southern Branch of the University of California (now UCLA); she received a bachelor of education degree in 1923.

Beenken taught in the public schools of Los Angeles 1923–24 and was an associate in mathematics at the Southern Branch of the University of California 1924–25. She entered the University of Chicago in the summer of 1925 and was in residence there, except for summer quarter 1927, until June 1928, when she received her doctorate with a dissertation in projective differential geometry. She was a university fellow her last two years at Chicago.

After receiving her PhD, Beenken joined the faculty at Oshkosh Teachers College (now University of Wisconsin Oshkosh) as head of the mathematics department and remained there until 1947. She served as secretary of the Wisconsin Section of the MAA in 1934–35 and as chairman in 1943–44. She was active in other state organizations as well; she was chairman in 1936–37 of the mathematics section of the Wisconsin Education Association and was a member in the late 1930s of a state committee on the uses and appreciation of mathematics. In 1943–44 she was in charge of mathematics for the US Air Force 96th column training detachment at the college. From 1944 to 1947 she directed the division of preprofessional education at Oshkosh.

Beenken was a lecturer at UCLA in the summer of 1945, and in 1947 returned to southern California as associate professor of mathematics and department chairman at Immaculate Heart College in Los Angeles; she was promoted to professor in 1951. During the summer of 1953 she attended an eight-week conference on collegiate mathematics in Boulder, Colorado, that was sponsored by the National Science Foundation. She received grants from the NSF for work at the University of Oregon in summer 1954 and at UCLA in summer 1959 and from the Social Science Research Council for work at Stanford in summer 1957. She retired from Immaculate Heart College as emeritus professor in 1969.

May Beenken died at age eighty-seven in 1988 in Los Angeles. In the memorial note that appeared in the *Immaculate Heart College Alumni Newsletter* after her death, a former student and colleague recalled that she was a superb teacher and a creative person who had turned to painting after her retirement.

BENEDICT, Suzan R. November 29, 1873–April 8, 1942.

SMITH COLLEGE (BS 1895), COLUMBIA UNIVERSITY (MA 1906), UNIVERSITY OF MICHIGAN (PHD 1914). PhD diss.: A comparative study of the early treatises introducing into Europe the Hindu art of reckoning (directed by Louis Charles Karpinski).

Suzan Rose Benedict was born in Norwalk, Ohio, the youngest of seven children of Harriett (or Harriott) Melvina (Deaver) and David DeForest Benedict. The family lived in Norwalk, where David D. Benedict practiced medicine. During the Civil War he had served as a surgeon for the Union army.

Suzan Benedict attended the high school in Norwalk before entering Smith College in 1891. She graduated in 1895 with a chemistry major and minors in mathematics, German, and physics. She then returned to Norwalk and taught mathematics there for the next decade.

In the fall of 1905 Benedict entered Teachers College, Columbia University. Three of her courses were taken with David Eugene Smith, and she wrote her 1906 master's thesis, "The Development of Algebraic Symbolism from Paciuolo to Newton," under his direction. In 1906 Benedict joined the three-person mathematics department at Smith and remained there the rest of her career. She was hired for the year 1906–07 as an assistant in mathematics and reappointed as instructor. Benedict resumed her graduate work by studying at the University of Michigan in the summers of 1911 through 1913. She took a leave from Smith in 1913–14 to finish her course work in mathematics and her dissertation in the history of mathematics with L. C. Karpinski.

Benedict returned to Smith in 1914 as associate professor, became professor in 1921, and was department chairman 1928–34. While on leave the first semester of 1927–28, it is probable that she did the research for her 1929 article on Francesco Ghaligai, a sixteenth-century Florentine who wrote a popular introductory mathematics text. Benedict credited D. E. Smith with teaching her to love the sixteenth-century books that she read whenever she could find the time and the books.

In the late 1930s Benedict described her chief interest as teaching. She was an Episcopalian and a member of the Daughters of the American Revolution. From at least 1918 Benedict shared a home in Northampton with **Susan M. Rambo**. Rambo, who had graduated from Smith in 1905, returned as assistant in mathematics in 1908, received her PhD from Michigan in 1920, and was promoted through the ranks in the mathematics department until her retirement in 1948.

Benedict retired at sixty-eight as professor emeritus in February 1942. At that time she indicated in the *Smith Alumnae Quarterly* that she had expected to go south but changed her plans when war was declared and would "stay at home to be near the Red Cross and other opportunities for service." She died suddenly of a heart attack in Northampton,

Massachusetts, in April and was buried in Norwalk, Ohio. In the minute read to the faculty after her death, it was noted that she had continued her research in the history of mathematics and "it is owing to her interest that the College library possesses a collection of rare books dealing with that field," and that "she was impatient with those who suggested that any branch of science was too difficult for girls to attempt" (Smith College Archives). The Suzan R. Benedict Prize, to be given to a member of the sophomore class who has done outstanding work in differential and integral calculus, was established by the college president, former members of the mathematics club, and others. Her home was bequeathed to Susan Rambo. Upon Rambo's death it was to go to Smith to be sold with the proceeds to be used for a scholarship fund. Rambo died in 1977.

BERNSTEIN, Dorothy L. April 11, 1914–February 5, 1988.

UNIVERSITY OF WISCONSIN (BA 1934, MA 1934), BROWN UNIVERSITY (PHD 1939). PhD diss.: The double Laplace integral (directed by Jacob David Tamarkin).

Dorothy Lewis Bernstein was born in Chicago, Illinois, the eldest of four surviving daughters and one son of Tillie (Loyev, later Lewis) and Jacob Louis Bernstein. Both parents were born in Russia and emigrated early in the century. Her father had a dairy business, first in Chicago and later in Milwaukee, Wisconsin. All of the surviving children earned advanced degrees: one MD and four PhD's.

Bernstein attended public primary and secondary schools in Milwaukee and graduated from high school as valedictorian in 1930 at age sixteen. She then entered the University of Wisconsin in Madison, where she majored in mathematics. By October 1932, when she was just over eighteen, the department voted that she be permitted to follow a program of advanced independent study. Bernstein received both her bachelor's and master's degrees in 1934. She was first in a class of two thousand, and her bachelor's degree was awarded summa cum laude. She held a scholarship her last year as an undergraduate, and remained at Wisconsin as a fellow for the year 1934–35, doing further graduate work and teaching.

Bernstein continued her graduate studies at Brown University, where for two years, 1935–37, she held a scholarship and taught at Pembroke College, Brown's coordinate women's college. She completed her dissertation and received her PhD from Brown in 1939. She was also an instructor at Mount Holyoke College from 1937 to 1940. It appears that she then moved back to Milwaukee.

In 1941 Bernstein returned to the University of Wisconsin for a year as an instructor. In June 1942 she worked on theoretical problems in probability at the Statistical Laboratory of the University of California, Berkeley, where she also taught a graduate course in probability theory in the mathematics department. After leaving the University of California she was unemployed for several months before joining the faculty at the University of Rochester as an instructor in fall 1943. In 1946 she was promoted to assistant professor, in 1951 to associate professor, and in 1957 to professor. She spent three years as acting chairman of the department. She also directed three PhD dissertations, including that of her later collaborator and close friend, Geraldine Coon, in 1950.

While at Rochester she undertook a study of the then current state of knowledge of existence theorems in partial differential equations. As she explained in a 1978 contribution to an AWM panel discussion, "some of the proofs could be used as basis for the computational solutions of non-linear problems that were just being tackled by high-speed digital computers" ("Women mathematicians before 1950," *AWM Newsletter* 9, no 4 (1979): 11). Her 1950 book, *Existence Theorems in Partial Differential Equations* (Princeton University Press), was the result of this undertaking. She spent the year 1950–51 as a member of the Institute for Advanced Study and the year 1957–58 as a visiting professor at the Institute for Numerical Analysis of the University of California, Los Angeles. During the 1950s she wrote more than fifty reviews for *Mathematical Reviews.*

In 1959 Bernstein went to Goucher College as professor and remained there until she retired in 1979 as professor emeritus. She was chairman of the department from 1960 until

1970 and director of the computer center from 1962 until 1967. In 1966–67, she was visiting professor of applied mathematics at Brown. While at Goucher, Bernstein became active in the uses of the computer in education and in the spring of 1971 was part of a group that founded the Maryland Association for the Educational Uses of Computers. At her retirement Goucher presented her with an award for distinguished service to the college. In 1981 Towson State University awarded her an honorary Doctor of Humane Letters (LHD). She was a fellow of the AAAS. Bernstein was particularly active in the MAA, at both the section level and the national level. Her contributions were recognized when she was elected first vice president for 1972–73 and president of the MAA to serve 1979–80, the first woman in this role in the MAA's history.

At the time of her retirement from Goucher College in 1979, a pair of articles appeared in the *Goucher Quarterly*, "Bernstein on Coon" and "Coon on Bernstein." Geraldine Coon, Bernstein's 1950 doctoral student, joined the faculty at Goucher in 1964 and retired the same year as Bernstein. After describing Bernstein's mathematical activities and work at Goucher, Coon added that "whenever possible, she indulges in her favorite hobbies of gardening, canning, and freezing. She intends to maintain the famous Bernstein Box at the Preakness, where annually the laws of probability and statistics fall into complete disarray" ("Coon on Bernstein," *Goucher Quarterly* 58, no 1 (1979): 17). Coon remained at Goucher an additional year before she returned to her home on the Pawcatuck River in Connecticut; Bernstein and Coon shared the home until Bernstein's death.

For several years after her retirement Dorothy L. Bernstein maintained an affiliation with Brown University. She died in Providence, Rhode Island, in February 1988 at the age of seventy-three and was buried in Milwaukee.

BLACK, Florence. November 22, 1889–September 13, 1974.

UNIVERSITY OF KANSAS (BA 1913, MA 1921, PHD 1926). PhD diss.: A reduced system of differential equations for the invariants of ternary forms (directed by Ellis Bagley Stouffer).

Florence Lucile Black was born in Meade County, Kansas, the fourth of five children of Mary Ella (Winslow) and Moses Black. Moses Black was a land surveyor and rancher. Her brother, Ernest Bateman Black, became a well-known consulting engineer in Kansas City who in 1942 was president of the American Society of Civil Engineers.

Florence Black grew up on a cattle ranch in southwestern Kansas. She attended the University of Kansas in Lawrence and graduated in 1913. After her graduation, she taught for two years at the high school in Anthony, Kansas, and for three years at a high school in Wichita. In 1918 she returned to the university as instructor in mathematics and remained on the faculty until her retirement. Initially she also took graduate courses and received her master's degree in 1921 and her PhD with a minor in physics in 1926, the same year that **Wealthy Babcock** earned her doctorate at Kansas.

Having earned her PhD, Black was promoted to assistant professor in 1926. As was the case with her colleague and friend Wealthy Babcock, she was promoted to associate professor in 1940. She retired as associate professor emeritus in 1960.

G. Baley Price notes in his 1976 history of the Kansas mathematics department that "many will remember Professor Black best as an outstanding teacher" (p. 747). Upon her retirement the Florence Black Excellence in Teaching Award was established to provide a gift to the best teacher among the mathematics department's first-year assistant instructors. The university also awards a Florence Black scholarship, and in its 2005 annual report the College of Liberal Arts and Sciences announced the establishment of the Florence Black and Wealthy Babcock Professorship in Mathematics. Black's other contributions to the University of Kansas included her service as secretary of the college faculty for nineteen years and as faculty advisor of the women's pep club from its founding until 1952. She was also a faithful attendee at meetings of the Kansas Section of the MAA.

A paper read by Price at Black's memorial service noted her many extracurricular interests: horseback riding, camping, swimming, tennis, and traveling. He also quoted a letter from former Kansas chancellor Deane W. Malott, and later president emeritus of Cornell University. Malott wrote: "I remember once offering her camping rights in the center of the Cornell campus, should she and Wealthy desire to come" (Price 1976, 747). She and Wealthy Babcock were known, too, for their extraordinary attendance at football and basketball games at the University of Kansas.

Among her travels was a trip at age eighty to the Mediterranean, Tanzania, Uganda, and Kenya. In 1973 Black was elected to the Faculty Women Hall of Fame by the Commission on the Status of Women. Florence Black died in Lawrence, Kansas, at age eighty-four in 1974. She was cremated and her ashes were buried in the cemetery on the University of Kansas campus.

BLANCH, Gertrude. February 2, 1897–January 1, 1996.

New York University (BS 1932), Cornell University (MS 1934, PhD 1935). PhD diss.: Properties of the Veneroni transformation in S_4 (directed by Virgil Snyder).

Gertrude Blanch was born Gittel Kaimowitz in Kolno, Poland. She was the last of seven children of Dora (Blanc) and Wolfe Kaimowitz (also "Kamovitz" and "Kamowitz"). Her father had emigrated from Poland to the United States early in the century, and he was joined by his wife, Gittel, and another daughter in 1907. She attended public elementary and secondary school in Brooklyn, New York, and graduated from the Eastern District High School at the beginning of 1914, the year her father died. By this time Gittel had Americanized her name to Gertrude. She then took a clerical job to support her mother.

Gertrude Kaimowitz became a naturalized US citizen in 1921. After her mother died in 1927, she began taking night courses at the Washington Square College of New York University. She decided to leave her job working for a hat dealer, but her employer offered to pay her tuition if she would remain. She accepted the offer and graduated with a major in mathematics summa cum laude in 1932. In February 1932 she legally changed her name to Gertrude Blanch, an Americanization of Blanc, her mother's name.

Blanch entered Cornell University in September 1932 and received her master's degree in February 1934. She held a graduate scholarship her final year at Cornell and earned her PhD in 1935 after writing a dissertation in geometry and with minors algebra and analysis. Blanch then returned to New York City and spent a year as a tutor at Hunter College replacing a faculty member on leave. She then took a job as a bookkeeper in Manhattan and took an evening course in relativity at Brooklyn College. The instructor of the course was Arnold N. Lowan, who invited her to join the Mathematical Tables Project, a WPA project that he had been asked to head. Blanch worked for the project from the beginning of 1938 until 1942 and was a tutor in the evening session of Brooklyn College 1940–42. In 1942 many of the professional staff of the project became employees of the National Bureau of Standards (NBS), contracted to the Applied Mathematics Panel of the National Defense Research Committee. The group continued to work in New York and was disbanded after the war. In 1948 Blanch went to California, still as an employee of NBS, as assistant director for computing at the Institute for Numerical Analysis (INA) located on the campus of UCLA. During the McCarthy era, she was one of the professional staff of the Institute who was investigated by a loyalty board of the Department of Commerce, which oversaw the NBS. Although previously Blanch had been denied security clearance, presumably because her sister was a member of the Communist Party, she was allowed to continue her job at the Institute. Nonetheless, the Institute was under attack and it closed in June 1954. Blanch left at the end of 1953 and worked for the ElectroData Corporation in Pasadena.

At the end of 1954 Blanch became a senior mathematician at the Aerospace Research Laboratories at the Wright Air Development Center (later Division) of Wright-Patterson Air Force Base in Dayton, Ohio. At this time she was granted security clearance. She remained at Wright-Patterson until her retirement in 1967. The Air Force gave her its Exceptional Service Award in 1963 and a Senior Citizens Award the following year. Also during this period she continued her work on Mathieu functions and published extensively in this area. At the base she taught mathematics to officers needing training in aerodynamics. She was active in the Dayton branch of the AAUW and was higher education chairman in at least 1963. On March 3, 1964, Blanch received the Federal Woman's Award.

Even in retirement Blanch continued her involvement with mathematics. Until 1970 she was an Air Force consultant through a contract with Ohio State University. She then returned to California where she was writing a book on numerical analysis for people who had computational experience but did not understand the numerical processes involved. The unpublished manuscript is in her papers at the Charles Babbage Institute. Gertrude Blanch died in San Diego on New Year's Day 1996, a month before her ninety-ninth birthday.

BONNER, Harriet (Rees). July 16, 1914–June 28, 1978.

Rockford College (BA 1934), University of Chicago (MS 1935, PhD 1937). PhD diss.: [Rees, H.] Ideals in cubic and certain quartic fields (directed by Abraham Adrian Albert).

Harriet Rees was born in Mt. Morris, Illinois, the daughter of Wilhelmina (Mina) Marie (Middour) and Vernon Victor Rees. Her only sibling, a brother, died as an infant. She grew up on a farm near Mt. Morris, where her father was farming in 1920. In 1930 her father worked in the bindery of a printing plant.

Rees received her elementary and secondary education in the public schools of Mt. Morris and attended nearby Rockford College from 1930 to 1934. At Rockford she was a member of the classical club, an honor society, and the orchestra, and received a BA with honors in mathematics and a diploma in piano in 1934. She began her graduate work at the University of Chicago in October of 1934 and continued her work there until June 1937 when she received her doctorate. She was a Talcott fellow from Rockford College 1934–35. At Chicago she studied under **Mayme Irwin Logsdon** among others, and wrote her master's thesis in number theory and her PhD dissertation in a related area in abstract algebra.

Rees was appointed instructor of mathematics at the University of Utah in 1937 and remained there a year and a half before resigning in December 1938. She was married on January 1, 1939, to James Fredrick Bonner (1910–1996), son of the chairman of the chemistry department at Utah. He had received his PhD from the California Institute of Technology in 1934 and became a distinguished molecular biologist. He was on the faculty at Caltech from 1935 until 1981 and was elected a member of the National Academy of Sciences. The Bonners were divorced in 1966, and James Bonner remarried the following year.

The Bonners had two children, both born in Los Angeles: a daughter born in 1948, and a son born in 1950. Both children later earned PhD's. Before their children were born the Bonners were serious climbers. In October 1947 they were both listed as having climbed fifteen peaks listed by the Desert Peaks Section of the Sierra Club (http://angeles.sierraclub.org/dps/dpsemblem.htm).

During the 1940s and 1950s Harriet Bonner worked with her husband and his colleagues, receiving thanks for her mathematical analyses in several publications in biology. She coauthored one paper with James Bonner and another with A. J. Haagen-Smit, an authority on air pollution who was later elected a member of the National Academy of Sciences.

Harriet Bonner taught mathematics at the continuation high school in Pasadena from 1968 until 1975. She was also a serious pianist and studied piano and voice at Pasadena City College. She developed symptoms of Alzheimer's disease and was moved to San Francisco where she died in 1978, shortly before her sixty-fourth birthday.

BOWER, Julia Wells. December 27, 1903–February 19, 1999.

SYRACUSE UNIVERSITY (BA 1925, MA 1926), UNIVERSITY OF CHICAGO (PHD 1933). PhD diss.: The problem of Lagrange with finite side conditions (directed by Gilbert Ames Bliss).

Julia Wells Bower was born in Reading, Pennsylvania, the second of three daughters of Maud Estella (Weightman) and Andrew Park Bower. Her parents had attended elementary school, her father through fifth grade. He later supplemented this education with night classes and a year at a business college. Her mother was a housewife; her father was a cigar maker and became a labor leader and mediator. He was national vice president of the Cigar Makers' International Union of America for many years and then president in the 1940s; he was also vice president of the Pennsylvania Federation of Labor for forty years. He held offices in other state and local labor associations and was involved in many charitable and religious activities at the community level. His political affiliation was Socialist. Julia Bower's older sister held a master's degree in child development from Pennsylvania State College. Her younger sister died in childhood.

Julia Bower attended the public elementary school and high school in Reading. After graduating from high school as salutatorian she went to Syracuse University as holder of a Reading High School alumni scholarship during her first year. After receiving her bachelor's degree in 1925, she continued at Syracuse one more year as a teaching fellow and received her master's degree at the end of that year.

In 1926 Bower was appointed to a one-year temporary instructorship at Vassar College. The next three years she was instructor at Sweet Briar College in Virginia. She also was instructor in the summer school at Syracuse the summers of 1927 through 1931.

Bower began her graduate work for the PhD at the University of Chicago in October 1930. She remained there, except in summers, through June 1933, when she received her degree with a dissertation in the calculus of variations. She held a fellowship 1931–32 and 1932–33 and taught a freshman course during the first quarter of 1932–33.

After receiving her doctorate, Bower went to Connecticut College for Women in New London as instructor in the fall of 1933. The college became coeducational and adopted its present name, Connecticut College, in 1969, shortly before her retirement. Bower was instructor 1933–38, assistant professor 1938–42, associate professor 1942–53, professor 1953–72, and professor emeritus after 1972. She served as department chairman 1943–69. She taught a broad range of courses including one for non-majors in which she used her 1965 text, *Introduction to Mathematical Thought.* Her second book, *Mathematics: A Creative Art*, was published the year after she retired. From 1961 to 1967 she served as director and as instructor in NSF in-service institutes for high school teachers that had weekly sessions during the school year.

In the summers of 1943–46 Bower was director and instructor of a training course for engineering aides for the research division of United Aircraft Corporation. She published two articles in the *Mathematics Teacher* relating to her work during World War II. During the summers of 1950 and 1954 she was visiting professor at Washington Square College of New York University and Boston University, respectively. She was visiting scholar at the University of Miami 1962–63. She served as secretary of the Connecticut Valley Section of the Association of Teachers of Mathematics in New England 1941–43, as vice president 1949–50, as president 1950–51, and as director 1951–52.

Bower was active in Pan-Hellenic and sorority work from the early 1920s to the early 1940s. She was a Baptist, a member of the League of Women Voters, and a local officer in AAUW, AAUP, and Phi Beta Kappa. Her hobbies were reading and swimming.

Some time after her retirement Julia Wells Bower moved to Florida. She died in Orange City in 1999 at age ninety-five. Before her death, an anonymous donor established the Julia Wells Bower prize for mathematics in her honor at Connecticut College, which also presents an annual Julia Wells Bower lecture.

BOYCE, Fannie W. March 16, 1897–February 13, 1986.
Central Holiness University (BA 1918), Penn College (BA 1921), University of Wisconsin (MA 1928), University of Chicago (PhD 1938). PhD diss.: Certain types of nilpotent algebras (directed by Abraham Adrian Albert).

Fannie Wilson Boyce was born near Lentner, in Shelby County, Missouri, and moved to Colorado with her younger sister and her parents, Mary Virginia (King) and George Wesley Boyce, when she was four. Her father was a farmer in Missouri in 1900 and later a grocer. Her sister was also born in Missouri, earned a BA, taught mathematics, married, and was a county treasurer for sixteen years.

The family was living in Colorado Springs in 1900, and Fannie Boyce attended grade school in Colorado 1903–10. Her family moved to University Park, Iowa, and she did her high school and first college work there. She indicated in her PhD dissertation vita that she was at the academy of John Fletcher College 1910–14 and received her BA from John Fletcher College in 1918. The college was called Central Holiness University at the time she was there; it adopted the name John Fletcher College in 1925. After her graduation, Boyce taught in Iowa high schools for two years; she taught mathematics in Leon 1918–19 and mathematics and Latin in Rolfe 1919–20. She attended Penn College (now William Penn University), a Quaker college in Oskaloosa, Iowa, for one year to earn a second bachelor's degree in 1921.

Boyce taught at two schools during the next six years. In both she taught mathematics at the academy level and the first two years of Greek at the college level. In 1921–22 she was at Olivet University (now Olivet Nazarene University) in Illinois, and in 1922–27 she was at Marion College in Indiana. Boyce studied at the University of Wisconsin during the academic year 1927–28 and received a master's degree in 1928. She taught at the high school in Platteville, Wisconsin, 1928–29 and at the high school and junior college in LaSalle, Illinois, the following year, before being hired by Wheaton College in Illinois in 1930.

Boyce remained at Wheaton as assistant professor 1930–41, associate professor 1941–46, and professor 1946–62. She pursued her graduate work at the University of Chicago by taking leaves of absence in the academic years 1933–34 and 1936–37 and by taking courses during four summers. She received her doctorate in August 1938. Boyce's parents lived with her when she was in Illinois. Her father died in 1942. In March 1953 Boyce offered to resign since she was unable to carry her full load because of the time she needed to care for her mother, who was injured nearly a year earlier. Her mother died about three years later.

After her retirement from Wheaton College in 1962, Boyce took a position as associate professor at Olivet Nazarene College, where she taught from 1963 to 1970. She was then at Owosso College in Michigan in 1970. She returned to Wheaton, where she was a member of the Wheaton Bible Church and tutored students in her home until she was in her eighties. Fannie Boyce was a resident of a heath care center in Lombard, Illinois, at the time of her death at eighty-eight in 1986. It was noted in her obituary that she was an accomplished artist, working with oils.

BRADY, Dorothy (Stahl). June 14, 1903–April 17, 1977.
Reed College (BA 1924), Cornell University (MA 1926), University of California (PhD 1933). PhD diss.: On the solutions of the homogeneous linear integral equation (directed by John Hector McDonald).

Dorothy Elizabeth Stahl was born in Elk River, Minnesota, the eldest of two daughters and two sons of Agnes M. (Roche) and Henry V. Stahl. In 1900 Agnes Roche and Henry Stahl were both teachers. In 1910 the family was living in Bayfield, Wisconsin, where Henry Stahl was a high school teacher. By 1920 they were living in Portland, Oregon, and Henry Stahl was working as an insurance agent.

Dorothy Stahl attended Lincoln High School in Portland and studied mathematics and physics at Reed College, where she was a student assistant and from which she graduated as a mathematics major in 1924. On June 14, 1924, she married Robert Alexander Brady (1901–1963) who had graduated from Reed College in 1923 and remained there as a teaching assistant in history 1923–24. He then worked as a principal of a high school in Florence, Oregon, 1924–25.

Robert and Dorothy Brady attended graduate school at Cornell University 1925–26, and she received her master's degree in mathematics in 1926. She was instructor of mathematics at Vassar College the following year when Robert Brady was an assistant professor at Hunter College. In 1927 she worked at the National Bureau of Economic Research as a research assistant on a study of labor statistics before studying and teaching mathematics and economic statistics at New York University 1927–29. Robert Brady taught at NYU those two years. He received his PhD in economics from Columbia University in 1929 and joined the University of California faculty. Dorothy Brady continued her study of mathematics at the University of Berlin during the winter of 1930–31 when Robert Brady had a fellowship to do research in Germany. She then attended the University of California 1931–33 and received a PhD in mathematics in 1933. The Bradys' son was born in December 1933. They later divorced and Robert Brady remarried in 1936. Their son earned a master's degree in electrical engineering from MIT in 1958, is based in Oslo, Norway, and writes about Nordic skiing and living abroad.

In 1936 Dorothy S. Brady began a long career as an economist and statistician working for the federal government and in academic institutions. From 1936 to 1943 she worked in the area of family expenditures for the Bureau of Home Economics of the US Department of Agriculture in Washington, D.C.; she was associate economist 1936–38, home economics specialist 1938–42, and senior statistician 1942–43. During 1937–39 she also studied statistics at the US Department of Agriculture Graduate School. In 1943 she moved to the US Department of Labor where she held administrative positions in the cost of living division of the Bureau of Labor Statistics, including chief of the division 1944–48. She was a lecturer for a graduate course at American University in spring 1946.

Brady assumed two half-time professorial positions in economics at the University of Illinois in 1948. She taught in the Department of Economics and did research in the Bureau of Economic and Business Research in the College of Commerce. In 1951 she became a consultant on costs and standards of living for the Bureau of Labor Statistics. Because of this commitment, she took a leave of absence for most of second semester 1950–51. She resigned her Illinois positions effective August 31, 1951, and returned to full-time government service in Washington to work on the revision of the consumer price index. In 1953 she was chief of the division of prices and cost of living at the Bureau of Labor Statistics and was responsible for price indexes, consumer expenditure surveys, and other cost of living surveys.

In 1956 Brady returned to an academic setting as professor of economics at the University of Chicago. Two years later she moved to the Wharton School of the University of Pennsylvania, where she was research professor of economics. She was chairman of the graduate group in economic history 1964–70 and retired as professor emeritus in 1970 but remained a consultant there until 1974. She was a consultant to the Bureau of Labor Statistics 1956–60 and a consultant to the Social Security Administration, Office of Research and Statistics, 1964–71. She was book review editor of the *Journal of Economic History* 1969–74.

In 1948 Brady received the National Women's Press Club Award in Economics for her work in developing the city worker's family budget. In 1950 she was made a fellow of the American Statistical Association and the following year she was made a fellow of the Econometric Society. Dorothy Brady died in 1977 at the home she shared with her friend and fellow economist, Eleanor M. Snyder, in Pine Hill, New York. Brady was seventy-three at the time of her death. There was no funeral and her body was donated to science.

BROWN, Eleanor (Pairman). June 8, 1896–September 14, 1973.

University of Edinburgh (MA 1917), Radcliffe College (PhD 1922). PhD diss.: [Pairman, E.] Expansion theorems for solutions of a Fredholm linear, homogeneous, integral equation of the second kind, with kernel of special nonsymmetric type (directed by George David Birkhoff).

Eleanor Pairman was born in Broomieknowe, Lasswade, Midlothian, Scotland, the youngest of four daughters of Helen (Dunlop) and John Pairman, solicitor, Supreme Courts of Scotland. Her father died when she was quite young, apparently before 1901.

Eleanor "Nora" Pairman attended Lasswade Higher Grade School 1903–08 and George Watson's Ladies College 1908–14 before enrolling in the University of Edinburgh. In 1917 she received her MA, then the first Scottish university degree, with first class honours in mathematics and natural philosophy. The same year she was awarded the Vans Dunlop scholarship in mathematics, a three-year scholarship, awarded by competitive examination, which could be used for study at any university.

Soon after her graduation, Pairman joined the staff of Karl Pearson's laboratory in the Department of Applied Statistics at University College London. She read two papers at meetings of the Edinburgh Mathematical Society early in 1918. In 1919 she and Pearson coauthored an article that appeared in *Biometrika*, a journal that Pearson edited. In the same issue in which their article appeared, Pearson published an editorial in which he corrected a previous article and noted, "As the problem is an exceedingly important one the writer asked Miss Eleanor Pairman to revise his work" (12:267). Also in 1919, Pairman produced the first volume of tables in the *Tracts for Computers* series edited by Pearson.

Eleanor Pairman arrived in New York from London on October 12, 1919, to begin her studies at Radcliffe College. She finished her dissertation in analysis in the fall of 1921 and received her doctorate in 1922. On August 10, 1922, Eleanor Pairman and Bancroft Huntington Brown, a fellow graduate student, were married at Roselea, the Pairman home in Broomieknowe, Scotland. B. H. Brown was born in 1894 in Hyde Park, Massachusetts, and received a bachelor's degree in 1916 and a master's degree in 1917 from Brown University. He then served in the US Army, and after his discharge in 1919 he taught for two years as instructor at Harvard University. He continued his graduate work in mathematics there and received his doctorate in June 1922, the same time as Eleanor Pairman.

After receiving their PhD's, Bancroft and Eleanor Brown moved to Hanover, New Hampshire, where B. H. Brown joined the faculty of Dartmouth College, then a men's school with an all-male faculty. He remained at Dartmouth his entire career; he was hired as instructor in 1922 and was B. P. Cheney professor when he retired forty years later.

The Browns had four children, a son born in 1923 and daughters born in 1925, 1935, and 1937; the second daughter died in infancy. Soon after her first two children were born, Eleanor P. Brown published a paper with Rudolph E. Langer, another 1922 student of G. D. Birkhoff. The Browns' son majored in mathematics and classics at Dartmouth and earned a doctorate from Union Theological Seminary. He held a number of positions, was a participant at peace events throughout the world, and has published extensively in classics and on biblical themes. The older daughter earned a PhD in language and literature from Rutgers University and was a university lecturer in English for several years before her death at fifty-four; the younger daughter studied at Brown University and became a medical editor and transcriptionist.

Eleanor Brown began to learn Braille in about 1950 and later learned the Nemeth Code for mathematical notation. She made transcriptions using household items and her sewing machine to create geometric diagrams and mathematical symbols. Examples of her work included a freshman mathematics text and a reference book on group theory. Eleanor P. Brown was also a part-time instructor of mathematics at Dartmouth from September 1955 until June 1959. Her younger daughter reported that teaching was what made her mother truly happy. After a lengthy illness, Eleanor P. Brown died in nearby White River

Junction, Vermont, at age seventy-seven in 1973. Bancroft H. Brown died the following year.

BUCK, Elsie (McFarland). June 3, 1897–January 11, 1984.

UNIVERSITY OF CALIFORNIA (BA 1917, MA 1918, PHD 1920). PhD diss.: [McFarland, E. J.] On a special quartic curve (directed by John Hector McDonald).

Elsie Jeannette McFarland was born in St. Louis, Missouri, the only child of Lillian (Hope) and Francis W. McFarland. In 1900 they lived in San Francisco, California, where Frank McFarland was a clerk in a pension office. In 1910 the family was living in Covina, in southern California, and Francis McFarland was a high school teacher. In 1920 they were in Berkeley and he was a clerk.

Elsie McFarland attended high school in Covina and Pasadena, California, before doing all of her undergraduate and graduate work at the University of California in Berkeley. After graduating in 1917 she began her graduate work and received her master's degree in 1918. She was a university fellow in mathematics 1918–19 and an assistant in mathematics 1919–20 before receiving her doctorate in 1920 with minor subject physics.

The next two years McFarland taught mathematics and physics at high schools in California; she was at Maryville High School 1920–21 and at Newman High School 1921–22. During the years 1922–25 McFarland taught as a part-time instructor at three schools in the San Francisco area: the University of California 1922–25, Dominican College in San Rafael 1922–24, and Mills College 1924–25.

In 1925 McFarland moved to the University of Oklahoma as an instructor and was promoted to assistant professor in 1927. She resigned her position at Oklahoma in 1931 for personal reasons and was a visiting PhD at the University of Chicago in the summer of 1931. McFarland then moved to Spokane, Washington, and taught at Spokane University for nine months.

In 1932 McFarland joined the first faculty of eight at the new Boise (Idaho) Junior College as a teacher of mathematics and German. After two years the possibility that the school would close caused McFarland to accept a job at a combined agricultural high school and junior college, Jones County Junior College, in Ellisville, Mississippi. She was there 1934–37 before accepting an invitation from the new president of the still surviving Boise Junior College to return and teach mathematics and German. Initially McFarland taught all the mathematics at the school and later was the ranking member in mathematics within the Physical Science Division. Her teaching of veterans after World War II was among the most satisfying of her career.

When McFarland originally moved to Boise, her parents accompanied her. In 1947 she married Roy M. Buck, a civil engineer, who died in 1952. She retired from the college as professor emeritus in 1968, three years after it was granted four-year status and renamed Boise College (now Boise State University). She taught an occasional course after her retirement.

Elsie M. Buck played clarinet in the college orchestra, built a room in her home to accommodate her grand piano, and made frequent trips to San Francisco for the opera. She belonged to the faculty women's bridge club for nearly fifty years and was a member of the Episcopal church. She was fond of animals, especially cats, loved to garden, and owned a small cabin outside of Boise where she spent time in the summer. Elsie McFarland Buck died in Boise at age eighty-six in 1984.

BURKE, Sister Leonarda. November 24, 1900–March 1, 1998.

EMMANUEL COLLEGE (BA 1926), BOSTON COLLEGE (MA 1928), CATHOLIC UNIVERSITY OF AMERICA (PHD 1931). PhD diss.: On a case of the triangles in-and-circumscribed to a rational quartic curve with a line of symmetry (directed by Aubrey Edward Landry).

Ethel Louise Burke was born in Boston, Massachusetts, the daughter of Caroline Dorcas (Leonard) and James Henry Burke. Her parents were educated in the Boston public school system. They probably had five children survive to adulthood of eight born; of these Ethel was the youngest. Her father was chief telegrapher for Western Union in Boston before his death in his late forties.

Ethel Burke attended public primary and secondary schools in Boston. After her high school graduation in 1918 she attended the Boston Normal School for three years, 1918–21, to obtain her certification to teach in the Boston school system, and during the summers of 1918–23 she worked as playground instructor for the Boston public schools. She taught in a public elementary school in Boston 1921–23.

In 1923 Burke entered the Congregation of Sisters of Saint Joseph in Boston. During 1924–26 Sister Mary Leonarda Burke studied at Emmanuel College and received her bachelor's degree as a mathematics major in 1926. She then studied part time at Boston College while teaching mathematics, French, and history at the secondary level in the Boston diocesan schools. The first year she took courses on weekends; the second year she took courses after classes finished at her school. Sister Leonarda earned her master's degree in 1928 with a thesis on calculus in high school mathematics. From 1928 to 1931 she was in residence at the Catholic University of America, where she studied mathematics, education, and physics. She was among a group of five students, four women religious and one lay man, who earned their PhD's in 1931 under the direction of Aubrey E. Landry.

In 1931 Sister Leonarda began her long association with Regis College, a Catholic college for women in Weston, near Boston. She served at Regis College in a variety of capacities: as teacher, administrator, and as researcher. From 1931 to 1964, as professor of mathematics and chairman of the department, she taught a full complex of courses including computer programming in the later years. She engaged in postdoctoral research at Catholic University in the summer of 1940 and studied modern mathematical techniques in teaching at the University of Chicago in the summer of 1950. From 1963 to 1965 she served as director of the computer center at Regis.

For twenty-five years, 1953–78, Sister Leonarda was director of the Regis College Research Center and principal investigator for contractual work for the Air Force Research Center, Hanscom Field. Her work involved mathematical analysis primarily associated with geophysics, and she authored several technical reports in this area. After 1978 she held the title Research Analyst, Institutional Research, at Regis. She also held the rank professor emeritus after her retirement from teaching in 1964.

Sister Leonarda Burke received the Outstanding Educator Award from Regis College in 1975. She was given a special achievement award in acknowledgment of twenty-five years service in support of the research mission of the Geophysics Laboratory of the Air Force on November 15, 1978. From 1958 she made tape recordings for the blind of books from a variety of fields, including mathematics. Sister Leonarda Burke died at age ninety-seven in Framingham, Massachusetts, in 1998.

BUSHEY, Jewell (Hughes). March 13, 1896–May 5, 1989.

University of Arkansas (BA 1915), University of Missouri (MA 1916), University of Chicago (PhD 1924). PhD diss.: [Hughes, J. C.] A problem of the calculus of variations in which one end-point is variable on a one parameter family of curves (directed by Gilbert Ames Bliss).

Jewell Constance Hughes was the fifth of six children of Cora A. (Stanley) and James R. Hughes. She was born in Fayetteville, Arkansas, where her father was a merchant in 1900, an operator of a sawmill in 1910, and later a timberman and sawmill operator.

Jewell Hughes attended public elementary school 1901–08 in Fayetteville and Fayetteville High School 1908–11. She entered the University of Arkansas in 1911 at age fifteen and graduated in 1915. She immediately received a scholarship at the University of Missouri, where she studied mathematics 1915–16 and received her master's degree in 1916.

Hughes taught for two years at the Columbia, Missouri, high school before returning to the University of Arkansas in 1918 as instructor of mathematics. She remained at Arkansas until 1930: as instructor 1918–24, assistant professor 1924–27, and associate professor 1927–30. She was granted a leave of absence for the academic year 1923–24 for study at the University of Chicago. Hughes studied first in the summer quarters 1920–23 and then in the following four quarters; she was a teaching fellow 1923–24. She wrote her dissertation in the calculus of variations before receiving her PhD magna cum laude in August 1924.

While continuing to teach at the University of Arkansas, Hughes did further study at the University of California in summer 1925 and at the University of Chicago summer 1927. Later postdoctoral study included a course at Columbia University during 1934 and work at the University of Chicago in summer 1938.

In 1930 Hughes joined the faculty at Hunter College as assistant professor and remained there the rest of her career: as assistant professor 1930–39, as associate professor 1939–51, as professor 1951–66, and as professor emeritus after her retirement at age seventy. On June 24, 1935, Jewell Hughes and Joseph Hobart Bushey, a colleague at Hunter College, were married. J. Hobart Bushey (1903–1976) had received his BS from the Johns Hopkins University in 1924 and his MA and PhD in mathematics from the University of Michigan in 1928 and 1930, respectively. He had also joined the mathematics department at Hunter College as an assistant professor in 1930. He was promoted to associate professor in 1936 and to professor in 1953. He retired at age seventy.

In 1940 Jewell H. Bushey was elected chairman of the mathematics department. In addition to her teaching and administrative work, Bushey was extremely active in college service and in organizations at the college. She was chairman of the executive committee of the faculty 1943–50 and president of the Hunter chapter of AAUP 1938–40. She was also president for several years of the Hunter Phi Beta Kappa chapter and was director at times of the local Pi Mu Epsilon chapter. During the summer of 1959 she directed an NSF institute for junior and senior high school teachers.

Jewell Hughes Bushey was active in a number of professional organizations at the national level, as well. Bushey served the MAA in various ways: she was chairman of the Metropolitan New York Section 1944–45, was national second vice president 1951–52, and was governor from the Metropolitan New York Section 1957–60. She was also an advocate for faculty and was active at the national level of the American Association of University Professors (AAUP). She was a member of the council 1942–44 and of its executive committee 1942–43, a vice president 1944–46, a member of the committee on the economic status of the profession 1948, and on the board of the *Bulletin* of the AAUP in 1951.

In June 1949 Jewell Bushey was awarded an honorary LLD degree from the University of Arkansas, the second woman to receive that degree from Arkansas. The Jewell Hughes Bushey scholarship for proficiency in mathematics was established by the Hunter College Pi Mu Epsilon chapter in 1966, the year she retired.

After J. Hobart Bushey retired from Hunter in 1972, the Busheys moved to Jewell Bushey's home town of Fayetteville, Arkansas. J. Hobart Bushey died four years later, shortly before his seventy-third birthday. Jewell Bushey continued to live in Fayetteville, Arkansas, until her death there in 1989 at age ninety-three. The Jewell C. Hughes Bushey papers are in the special collections division of the University of Arkansas libraries.

CALKINS, Helen. October 20, 1893–June 17, 1970.

KNOX COLLEGE (BA 1916), COLUMBIA UNIVERSITY (MA 1921), CORNELL UNIVERSITY (PHD 1932). PhD diss. (1931): Some implicit functional theorems (directed by Charles Frederick Roos [at AAAS from 1930] and David Clinton Gillespie).

Helen Calkins was born in Quincy, Illinois, the first of two daughters of Anna Burns (Schermerhorn) and Addison Niles Calkins. By 1900 her father was superintendent of Electric Wheelworks in Quincy.

Helen Calkins attended Quincy High School 1908–12 and Knox College in Galesburg, Illinois, 1912–16. The year after her college graduation with special honor in mathematics, she taught mathematics in junior high school in Quincy; the following year she taught mathematics in the senior high school in Jacksonville, Illinois. She then returned to Knox College, where she was instructor in mathematics 1918–20.

Calkins began her graduate work in mathematics as a university scholar at Columbia University in 1920–21. While there she studied history of mathematics and the teaching of mathematics with David Eugene Smith, fundamental concepts with Edward Kasner, and differential equations with W. B. Fite. She received her master's degree at the end of that academic year.

Calkins returned to Knox College as instructor in 1921. After one year she was promoted to assistant professor and after one year at that rank she took a leave of absence to return to Columbia. While at Knox she developed a course in the history of mathematics inspired by her earlier work at Columbia with Smith. She again held a scholarship at Columbia in 1923–24, when she studied theoretical physics, mathematics, and the philosophy of mathematics. She continued her leave from Knox in 1924–25, when she traveled in the United States and Cuba. She attended Columbia part time during 1925–26 and traveled in Europe in the summer and fall of 1926. Calkins served as instructor at the University of Nebraska, substituting for someone on leave of absence, during the second semester 1926–27. By the summer of 1927 she had completed residency requirements for the PhD degree at Columbia.

Calkins was hired at Sweet Briar College in Virginia for the year 1927–28 as professor of mathematics and acting head of the mathematics department, while **Eugenie M. Morenus** was on leave in England. In a letter of January 13, 1930, the president of Sweet Briar summarized the quality of Calkins' work to a Chicago teacher's agency. "We were so much impressed with Miss Calkins' ability that it was with deep regret that we allowed her to leave. There is a lucidity of mind, an ability to present any matter with which she is dealing, a gentleness and charm of personality, and an all-round sanity that will go far in academic usefulness" (Faculty Files, Sweet Briar College Archives).

In the summer of 1928 Calkins enrolled at Cornell and took classes through the summer of 1930 as a scholarship holder. Her intended advisor, Charles F. Roos, left Cornell for a position with the AAAS in 1930, at the same time that Calkins took a job as professor and acting head of the department of Pennsylvania College for Women, in Pittsburgh. Roos was asked to stay on as her advisor, and he agreed to do so as long as someone else at Cornell assumed joint responsibility with him and suggested D. C. Gillespie. In her dissertation, which was approved in October 1931, she acknowledges both Roos and W. A. Hurwitz, whom she later listed as a personal friend; the degree was awarded in February 1932.

Except for brief periods, Calkins spent the remainder of her career as professor and head of the department at Pennsylvania College for Women, where she was usually the only member of the mathematics department. In the summer of 1941, Calkins worked as a statistician in engineering defense training at Pennsylvania State College, and during 1943–44 she taught mathematics to preflight cadets at the University of Minnesota. Having spent the year 1956–57 on leave, Calkins retired as professor emeritus in 1957, two years after the school changed its name to Chatham College.

After her retirement, Calkins moved back to Quincy, Illinois, where she was a member of the Quincy country club and the First United Presbyterian Church. Helen Calkins moved to the Good Samaritan Home in Quincy about four years before her death at age seventy-six in 1970.

CARLSON, Elizabeth. October 2, 1896–November 1, 2000.

UNIVERSITY OF MINNESOTA (BA 1917, MA 1918, PHD 1924). PhD diss.: On the convergence of certain methods of closest approximation (directed by Dunham Jackson).

Sally Elizabeth Carlson was born in Minneapolis, Minnesota, the third of five surviving children of six born to Alice (Alise) (Johnson) and Carl Emil Carlson. Both of her parents were born in Sweden and emigrated several years before their marriage in 1891. Her mother had an elementary school education and her father no formal education, although he was literate in Swedish and English. He was a stone mason in Minneapolis. Her eldest brother was a salesman, while her next oldest brother was an electrical engineer. Her sister served forty-six years as a missionary to Venezuela, and her youngest brother was a lawyer.

In 1913 Carlson graduated as valedictorian from South High School in Minneapolis. Even though her parents were not supportive of her desire for a college education, she attended the University of Minnesota and received her bachelor's degree in 1917 and her master's degree in 1918; she then taught at McIntosh High School in northern Minnesota for nine months. She was an instructor of mathematics and physics at Knox College in Illinois during 1919–20. Carlson returned to the University of Minnesota in 1920 where she was a teaching assistant in mathematics her first year and remained an assistant until she received her PhD four years later. She and Carey Morgan Jensen (a male) received the first two PhD's in mathematics at Minnesota in 1924, both as students of Dunham Jackson in analysis, and both with minors in physics.

Elizabeth Carlson remained at Minnesota until her retirement as professor emeritus in June 1965. She was instructor 1924–28, assistant professor 1928–50, associate professor 1950–63, and professor 1963–65, a pattern very similar to that of **Gladys Gibbens**, her slightly older colleague. In the fall of 1965, just after her retirement, she taught at Macalaster College as a visiting professor. During her career at Minnesota she was advisor for a number of MA candidates and won a Distinguished Teacher Award in 1962. Carlson published three papers based on her dissertation. Her interests then turned to teaching.

Carlson was an associate editor of the *American Mathematical Monthly* 1927–31 and served on the executive committee of the Minnesota Section of the MAA 1961–62. She was an active member of the Evangelical Free Church of America and helped start the Central Evangelical Free Church in downtown Minneapolis, where she taught Sunday school, was deaconess, and sang in the choir and in a mixed quartet. She also served as faculty advisor for the Minnesota Christian Fellowship and for the Inter-Varsity Christian Fellowship chapter at the University of Minnesota, and conducted Bible studies in her home and elsewhere. Carlson traveled extensively throughout the world and retained her fluency in Swedish.

A former colleague of Carlson reported to one of the authors in 1998 that "she was strong and vigorous right up to retirement and beyond. . . . I recall meeting her in downtown Minneapolis in 1984, when she was eighty-eight. She said she was doing an errand for one of the older ladies in her retirement home."

Elizabeth Carlson died at a nursing home in Minneapolis at age 104 in November 2000. The following fall the mathematics library at the University of Minnesota mounted an exhibit, "Elizabeth Carlson, notable alumna."

CASNER, Evelyn (Wiggin). March 1, 1900–November 5, 1964.

Wellesley College (BA 1921), Brown University (MA 1924), University of Chicago (PhD 1936). PhD diss.: [Wiggin, E. P.] A boundary value problem of the calculus of variations (directed by Gilbert Ames Bliss and William Thomas Reid).

Evelyn Prescott Wiggin was born in Stratham, New Hampshire, the elder of two daughters of Margaret Prescott (Green) and George Herbert Wiggin, a farmer. Her mother died after giving birth to their second daughter in 1905. Her father remarried in 1908.

Evelyn Wiggin received her early education at Robinson Seminary, a public secondary school, in Exeter, New Hampshire. She entered Wellesley College in 1917 and while there was a Durant scholar. After her graduation as a mathematics major in 1921, she taught mathematics in the high school in Massena, New York, 1921–22. In March 1922, at the suggestion of **Helen A. Merrill** and **Clara E. Smith** of Wellesley, R. G. D. Richardson invited Wiggin to apply for a position as a graduate assistant at Brown University. She entered Brown that fall as a student and Richardson's assistant. She taught remedial courses in the Women's College and completed her work for the master's degree in 1924.

Wiggin was an instructor of mathematics at Hood College in Frederick, Maryland, 1924–27. While there, Richardson indicated that she might be able to take a course for credit in absentia at Brown or go to Chicago some summer. She left Hood in 1927 to enter the University of Chicago for full-time graduate work, where she was a teaching fellow her second year, 1928–29. In 1929 Wiggin joined the faculty as associate professor at Randolph-Macon Woman's College (now the coeducational Randolph College) in Lynchburg, Virginia. Except for leaves, she remained at Randolph-Macon the rest of her career. Wiggin returned to the University of Chicago and was in residence 1935–36, when she finished her dissertation in the calculus of variations; she received the PhD in 1936. She returned to Randolph-Macon and was promoted to full professor in 1941.

Wiggin was a lecturer at Wellesley College during 1942–43 and taught at Emory University in summer 1949. In December of 1951 she attended a computing laboratory seminar at the Oak Ridge Institute of Nuclear Studies. In the summers of 1953 and 1954 she attended National Science Foundation summer institutes in mathematics for college teachers. During 1954–55 she was on leave and studying at the University of Chicago.

On June 20, 1956, Evelyn Wiggin married Sidney Casner, a retired lawyer, in Chicago. Casner was born in 1890, grew up in Chicago, had degrees from the University of Illinois and Kent College of Law, and attended the University of Chicago Law School. He practiced law, taught social sciences in high schools, and taught in law schools, all in Chicago. He was in Europe July 1945 to August 1947, where he served as a military judge and taught at army university centers. After her marriage, they lived in Lynchburg and Evelyn Wiggin Casner continued in her position at Randolph-Macon. She developed an experimental course for social science students based on work done at Stanford on a summer grant from the Social Sciences Research Council. During the late 1950s and early 1960s she taught in seven NSF summer institutes for high school teachers.

Casner was president of the college faculty club and local chapters of Sigma Xi and AAUP. She was president of the Lynchburg Consumer Cooperative 1932–40, and over many years served as president 1958–62, trail supervisor, and council member of the local Appalachian Trail Club. Evelyn Wiggin Casner was a widow at the time of her death in Charlottesville, Virginia, at age sixty-four in 1964.

CHANLER, Josephine H. April 7, 1906–December 23, 1992.
WESTERN KENTUCKY STATE NORMAL SCHOOL AND TEACHERS COLLEGE (BA 1927), UNIVERSITY OF ILLINOIS (MA 1930, PHD 1933). PhD diss.: Poristic double binary forms (directed by Arthur Byron Coble).

Josephine Hughes Chanler was the only child of Louisa Castle (Hughes) and James H. Chanler. A few weeks after her birth in St. Louis, Missouri, she and her mother returned to Bowling Green, Kentucky, where her mother's family lived. Her father was away from the family after that time, although her parents corresponded until at least 1918. Her mother taught in nearby Franklin, Kentucky, so Josephine and her mother typically were in Franklin in the winters and in Bowling Green in the summers.

Josephine Chanler contracted polio as a child and had surgery at various times in her life as a consequence. She attended schools in both Franklin and Bowling Green; she started high school at age eleven in Franklin and then attended a half year of high school in Bowling Green before she and her mother went to Jacksonville, Florida, to stay with family. She graduated in 1923, at age seventeen, from Duval High School in Jacksonville.

Chanler credited her grandfather with starting to teach her mathematics at an early age and her mother with encouraging her to go to college and later to do graduate work. Thus, after her high school education they returned to Bowling Green where she entered Western Kentucky State Normal School and Teachers College (now Western Kentucky

University). Immediately after her graduation in 1927 she taught mathematics for two years in the high school in Bowling Green.

Chanler's mother mortgaged the house she had inherited in Bowling Green, and they moved to Illinois in 1929 so Josephine Chanler could begin her graduate work at the University of Illinois. Although her undergraduate degree was not from a fully accredited school, she earned her master's degree at the end of one year. Her thesis was written under the direction of **Bessie Irving Miller**, who died suddenly a few months after Chanler received her master's degree. Chanler was a fellow in mathematics 1930–31 and then an assistant in mathematics while finishing the work for her doctorate in 1933 with a minor in physics. Her advisor, Arthur B. Coble, had directed Miller's 1914 dissertation from Johns Hopkins University. During the six years after she received her PhD, Chanler published five papers in algebraic geometry.

Chanler remained at Illinois for the rest of her career, although she had an opportunity to take a position as assistant professor at a women's college in New England in the mid-1930s. She was assistant until 1937, instructor 1937–39, associate 1939–45, assistant professor 1945–61, and associate professor 1961–71. Chanler was known as a very fine teacher and directed one doctoral dissertation in algebraic geometry in 1941. In the summers during World War II she taught in programs for servicemen. During the summers of 1959 through 1970 she was an instructor at National Science Foundation institutes for high school teachers at Saint Louis University.

Josephine Chanler and her mother lived together until her mother's death in 1948. After that her office mate, **Beulah Armstrong**, who loved to drive, took Chanler on trips nearly every summer; they camped in the West and often went to mathematics meetings.

Chanler's religious interests were serious and eclectic; in a 1983 interview with one of the authors, she described herself as "a Southern Presbyterian–Calvinist–Catholic with leanings toward Judaism and Islam and a very high regard for Buddhism, Mormons, and some Hindu gods." At the time of her death, she was a member of a Catholic church.

Chanler maintained an apartment in Champaign, Illinois, for many years after her retirement. There she kept in touch with numerous former students and tutored without charge countless students in mathematics. She was in a health care facility for about five years before her death in Urbana at age eighty-six in 1992.

COHEN, Teresa. February 14, 1892–August 10, 1992.
GOUCHER COLLEGE (BA 1912), JOHNS HOPKINS UNIVERSITY (MA 1915, PHD 1918). PhD diss.: Investigations on the plane quartic (directed by Frank Morley).

Teresa Cohen was born in Baltimore, Maryland, the daughter of Rebecca (Sinsheimer) and Benjamin Cohen, a wholesale shoe merchant. She was the eldest of four children, one son and three daughters. Cohen graduated from Friends School in Baltimore in 1909 before entering Goucher College from which she graduated in 1912 as a member of Phi Beta Kappa. Her two sisters also graduated from Goucher; one became a social worker, and the other an attorney. Her brother was ordained a rabbi and spent most of his career engaged in academic work.

In 1912 Teresa Cohen entered the Johns Hopkins University with a resident fellowship for the year 1912–13. She received her master's degree in 1915 with a thesis in algebraic geometry. She was made a university scholar at Johns Hopkins for the year 1915–16 and was a fellow 1917–18. Her 1918 doctoral dissertation was also in algebraic geometry.

After receiving her PhD, Cohen went to the University of Illinois for the year 1918–19 as an instructor. She also taught courses at Johns Hopkins during the summers of 1918, 1919, and 1920. In September 1920 Teresa Cohen became the first woman on the mathematics faculty at Pennslyvania State College (now University). She said in a 1979 interview, "I used to stay out of the way and hope that if I wasn't a nuisance they'd get some more women" (Barbara Hale, "Matriarch of Mathematics," *Penn Stater*, Jan/Feb

1976, 1). Over the years a number of women joined Cohen and stayed at Penn State; these include **Aline Huke Frink**, **Beatrice Hagen**, and **Helen Owens**.

Cohen remained on the faculty of Penn State until she was forced to retire at age seventy. She was promoted to assistant professor in 1921, to associate professor in 1938, and to professor in 1945. After her retirement, Cohen, who had emeritus status, taught some classes and worked as an unpaid volunteer tutoring students who had problems with mathematics. She continued to tutor on a regular basis for twenty-four years until a serious accident in early 1986 prevented her from doing so. After several months in the hospital and in rehabilitation in Pennsylvania, her family moved her to a nursing home in Pikesville, Maryland, where she lived until her final illness.

Both before and after the accident, Cohen was recognized for her contributions to Penn State and its mathematics department. In November 1979 she received the Lion's Paw Medal for extraordinary service to the university, and in 1982 the department created the Teresa Cohen Service Award. After Cohen's accident her niece and nephew contributed to the establishment in 1987 of the Teresa Cohen Tutorial Endowment Fund in the mathematics department; four years later the department established the Sperling-Cohen tutoring program that bears her name along with that of an alumnus who contributed generously to the program.

Cohen was active in Sigma Delta Epsilon and was national treasurer from 1954 until 1963, a year after her retirement from teaching. At the time of her retirement, the *Sigma Delta Epsilon News* wrote about her love of music that was evidenced by her attendance at concerts and her playing of the violin. Cohen started her violin lessons at age ten and continued them until she was eighty-five; she performed a violin sonatina at a concert held during the mathematics meetings at Penn State in September 1937. She spent her summers at a home in the Baltimore area that she maintained at least through 1979. Teresa Cohen died in Baltimore, Maryland, in 1992, almost six months after celebrating her one hundredth birthday. Her papers are in the archives of Penn State University.

COLE, Margaret (Buchanan). August 18, 1885–September 10, 1959.

West Virginia University (BA 1906), Bryn Mawr College (PhD 1922). PhD diss.: [Buchanan, M.] Systems of two linear integral equations with two parameters and symmetrizable kernels (directed by **Anna Johnson Pell [Wheeler]**).

Margaret Buchanan was born in Clinton, Pennsylvania, the second of three children of Sarah (Wiley) and Aaron Moore Buchanan. Her father graduated from Washington and Jefferson College and from Western Theological Seminary in Pittsburgh, Pennsylvania. He was a Presbyterian minister who moved from Pennsylvania to Morgantown, West Virginia, in 1886. He received the DD degree from Washington and Jefferson College in 1899.

From 1898 to 1902 Margaret Buchanan attended the preparatory school of West Virginia University at Morgantown before entering that university in 1902. While there she was one of a handful of women in her class. After her graduation from West Virginia University in 1906, Buchanan taught mathematics for a year at Marshall College (then a state normal school, now Marshall University) in Huntington, West Virginia. She returned to West Virginia University in 1907 as a graduate student. She was an assistant in Greek and mathematics and taught in the university's preparatory school 1907–09. She was also an instructor of mathematics at the university 1909–12 before entering Bryn Mawr College for graduate study in 1912. She studied at Bryn Mawr College 1912–13, taught mathematics at the Baldwin School in Bryn Mawr 1913–14, taught mathematics in the high school at Parkersburg, West Virginia, 1914–15, and was instructor of mathematics at West Virginia University 1915–18. In the summer of 1918 she was a temporary office assistant on the scientific staff of the state geologist preparing the statistics and coal maps for the West Virginia geological survey of Webster County.

Buchanan returned to her graduate studies at Bryn Mawr College as a resident fellow in 1918–19 and as a graduate scholar in 1919–20, having attended the University of

Chicago in the summer of 1919. She was awarded a President M. Carey Thomas European fellowship in 1918–19 for use the following year. She postponed that fellowship and was a resident scholar 1919–20. She used the postponed fellowship and an Association of Collegiate Alumnae European fellowship in 1920–21 to study at the Sorbonne in Paris. In 1921–22 she finished her work for the PhD from Bryn Mawr as a fellow by courtesy, the first doctoral student of Anna J. Pell (Wheeler). Her doctorate, with a minor in physics, was awarded in June 1922.

Buchanan returned to the faculty of West Virginia University as assistant professor and was promoted to associate professor in 1925. She resigned her position to marry, on February 14, 1929, Harry Outen Cole (1874–1950). Cole was a native of Morgantown, an 1898 graduate of West Virginia University, and a construction engineer who had played a role in the design of the Panama Canal. He had been widowed in 1921 and from about that time was president of Cole Bros. Construction Company in Morgantown. In 1938 Margaret B. Cole again joined the faculty at West Virginia University and was instructor 1938–41, assistant professor 1941–52, associate professor 1952–55, and associate professor emeritus after her retirement in 1955 at age sixty-nine.

Cole became the first woman president of West Virginia University's alumni association in June 1939. In 1956 she joined the alumni association's emeritus club, for alumni of fifty or more years, and was vice president in 1958–59. Margaret Buchanan Cole died in Morgantown in 1959 at age seventy-four.

COLE, Nancy. October 15, 1902–July 7, 1991.

VASSAR COLLEGE (BA 1924), RADCLIFFE COLLEGE (MA 1929, PHD 1934). PhD diss.: The index form associated with an extremaloid (directed by Marston Morse).

Nancy Cole was born in Boston, Massachusetts, the only child of Gertrude (Groce) and Albert Henry Cole. Ancestors of her mother had come to Plymouth on the Mayflower in 1620; those of her father had come to Plymouth in 1630. Her mother was a normal school graduate, and her father, who was in the grocery business, was a high school graduate.

Cole graduated from high school in Plymouth, Massachusetts, in 1919 and studied there as a postgraduate an additional year before entering Jackson College, the coordinate women's college associated with Tufts College (now University) at that time. She was at Jackson College with a scholarship 1920–22 and then attended Vassar College for two years before receiving her bachelor's degree from Vassar in 1924.

From 1924 to 1926, Cole taught at the Oxford School, a preparatory school in Hartford, Connecticut. In 1926–27 she studied at Radcliffe College, and the following year she taught at Vassar as an instructor. During 1928–29 Cole again studied at Radcliffe while also working as a tutor there. She received her master's degree in 1929 and continued as a student at Radcliffe with scholarship assistance through 1931.

Cole was an instructor at Wells College 1931–32. In February 1933 she was at home in Plymouth writing her thesis and working on Marston Morse's colloquium lectures on the calculus of variations in the large. She did the typing, filled in the formulas, and checked the mathematics for Morse, who was also her dissertation advisor. During 1932–34 Cole finished her graduate requirements and received her PhD from Radcliffe in 1934. In 1933 she took a position as instructor at Sweet Briar College in Virginia, where, except for leaves, she remained for a decade. She was acting department head at Sweet Briar in 1934–35 and in 1941–42.

Cole began a leave from Sweet Briar in February 1943 to take a wartime appointment as visiting assistant professor and then visiting associate professor at Kenyon College in Gambier, Ohio. Her first full year at Kenyon, she was lecturer and teacher in mathematics in the Pre-Meteorology Program for the Army Air Force, and was a teacher in the Student Army Training Program from February until July 1944.

Nancy Cole resigned from Sweet Briar in May 1944 and, after leaving Kenyon, was assistant professor at Connecticut College for Women 1944–47. In 1947 she took a position as assistant professor at Syracuse University, where she was to remain the rest of her career. At Syracuse, Cole was assistant professor 1947–52, associate professor 1952–71, and associate professor emeritus after her retirement in 1971. In addition to her standard undergraduate teaching, Cole taught undergraduate courses in finance, statics, and dynamics, and some graduate courses, including calculus of variations. In March 1980 she was called back to take over an advanced calculus class when a colleague died suddenly. She reported in 1985 that it was "a stimulating experience after nine years of retirement!" (Smithsonian questionnaire).

Cole was a reader for the College Entrance Exam Board 1938–41 and was on the promotion and services committee of the Association of Mathematics Teachers of New York State 1958–65. She also held various offices with the Syracuse chapters of Phi Beta Kappa, Sigma Xi, and Sigma Delta Epsilon, and, for the twenty-year period before her retirement, was the permanent faculty secretary of the local chapter of Pi Mu Epsilon. She traveled extensively and attended the International Congress of Mathematicians in Oslo (1936), Cambridge, Massachusetts (1950), Edinburgh (1958), Stockholm (1962), Nice (1970), Vancouver, British Columbia (1974), Helsinki (1978), and Berkeley (1986). In 1983 she traveled to the People's Republic of China as a member of a delegation of American university mathematics education specialists organized by People-to-People International.

From 1950 Nancy Cole was affiliated with the Foundation for Christian Living, now the Peale Center for Christian Living, in Pawling, New York. She maintained a summer home in Plymouth, Massachusetts, where she was a member of the Antiquarian Society of Plymouth and of the Pilgrim Society. She died in Plymouth at the age of eighty-eight in 1991.

COLLIER, Myrtie. September 14, 1877–June 25, 1974.

University of Chicago (BS 1911), Université de Strasbourg (PhD 1930). PhD diss.: Sur quelques points de la théorie des fonctions entières ou méromorphes d'ordre fini; Groupes primitifs simplement transitifs de degreé vingt et un.

Myrtie Collier was born in Carthage, Missouri, the third of four children of Angeline (Willson) and Ervin Collier, a farmer. In the 1900 census Myrtie Collier and her younger sister were both listed as teachers.

After obtaining her bachelor's degree at age thirty-three in 1911 from the University of Chicago, Collier moved to California to become a mathematics teacher and chairman of the department at the Los Angeles State Normal School, which in 1919 became the Southern Branch, University of California, a two-year institution. It added a four-year teachers curriculum in 1922 and a four-year letters and science curriculum in 1923, and awarded its first bachelor's degree in 1923. While teaching at the normal school, Collier was doing educational experiments and publishing articles in the journal *School Science and Mathematics*. She published one more article in this journal after she received her PhD in 1930.

Collier remained at the Southern Branch of the University of California as chairman of the department through 1920 and as assistant professor from 1920 until 1928, a year after the school's name changed to the University of California at Los Angeles. She took a year's leave of absence 1920–21 and was, for some time in 1920, an associate in the mathematics department at the University of Chicago. In 1928 she left UCLA to pursue further graduate study and received her PhD cum laude from the University of Strasbourg in 1930, when she was fifty-two. She returned to Los Angeles, where she served as department chairman at Immaculate Heart College from 1930 until her retirement in 1947 as professor emeritus.

Collier was, by her own account, a great traveler. She considered herself a student of the history of science and anthropology, and she enjoyed music. Myrtie Collier maintained her residence in Los Angeles until her death there at age ninety-six in 1974.

COLPITTS, Julia T. February 22, 1875–August 8, 1936.
University of Mount Allison College (BA 1899), Cornell University (MA 1900, PhD 1924). PhD diss.: Entire functions defined by series of the form $\sum f(n)\frac{z^n}{n!}$ (directed by Clyde Firman Craig and John Irwin Hutchinson).

Julia Trueman Colpitts was born in Pointe de Bute, New Brunswick, Canada, the third of eight children of Celia Eliza (Trueman) and James Wallace Colpitts. The family was from a long line of farm families in New Brunswick. Several of the children went into teaching or scientific careers. Edwin, the eldest, was a distinguished communications engineer with the American Bell Telephone Company. Elmer, four years younger than Julia, earned a PhD in mathematics from Cornell University in 1906, and Mary, the youngest, was an instructor of mathematics at the University of Wisconsin when she died of tuberculosis at age twenty-eight in 1920.

Julia Colpitts attended the public high school in Pointe de Bute and the normal school in Fredericton, New Brunswick. She graduated from the University of Mount Allison College in 1899 with honors in mathematics. The next year she was at Cornell University where she earned her master's degree in 1900. Colpitts then joined the faculty at Iowa State College of Agriculture and Mechanic Arts (now Iowa State University of Science and Technology) in Ames as an instructor. She was promoted to assistant professor in 1906 and to associate professor in 1913. She coauthored an analytic geometry textbook with Maria Roberts, her mathematics department colleague; the book first appeared in 1918, with a second edition in 1926.

Colpitts studied at the University of Chicago during the summer of 1912, at Columbia University during the summer of 1915, and again at Chicago during the summer of 1919. In addition, she took many graduate courses at Iowa State for which she received no credit since she was a member of the faculty. She had also taught graduate courses by the time she applied to Cornell in February 1922. While on leave at Cornell 1922–23 Colpitts took courses and worked on her dissertation with C. F. Craig. She returned to Iowa State at the end of the year. Craig was on leave in Europe during the second semester 1923–24 and was replaced as chairman of her doctoral committee by J. I. Hutchinson. At age forty-nine in 1924 Colpitts received her PhD from Cornell with major analysis, first minor geometry, and second minor philosophy. In the summer of 1929 she did further studies at the University of Chicago.

Colpitts was active in the MAA and served as chairman of the Iowa Section. She was also national secretary in 1926 and president in 1927 of Sigma Delta Epsilon, the women's scientific fraternity, and was a member of the Iowa Academy of Science. She attended the International Congress of Mathematicians in Toronto in 1924, in Bologna in 1928, and in Oslo in 1936. Julia Colpitts was scheduled to depart from Hamburg on August 6, 1936, after the Oslo congress but became ill and died on August 8 in Southampton, England. She was sixty-one at the time of her death. The Julia Colpitts Memorial Fund was established in her memory by her brother Edwin H. Colpitts with the income to be used to purchase books in mathematics for the Mount Allison library.

COOPER, Elizabeth M. January 19, 1891–May 17, 1967.
Radcliffe College (BA 1913), Bryn Mawr College (MA 1923), University of Illinois (PhD 1930). PhD diss.: Perspective elliptic curves (directed by Arthur Byron Coble).

Elizabeth Morgan Cooper was born in Syracuse, New York, the eldest of six children of Jessie Marian (Bagg) and Henry R. Cooper. Her parents had three daughters followed by three sons. It appears that all of the children later graduated from either Radcliffe or Harvard. Her father was manager of the caustic soda department of the Solvay Process Company for more than twenty years before his retirement. Her parents were active in the social life of Syracuse.

Cooper attended the Orchard School and the Goodyear Burlingame School, both private schools in Syracuse, before entering Radcliffe College in 1909, where she held a scholarship all four years. She graduated magna cum laude in mathematics in 1913.

During the next seventeen years Cooper taught in private preparatory schools and earned her graduate degrees. She first taught at the Baldwin School in Bryn Mawr, Pennsylvania 1913–27. During 1920–23, she took graduate courses at Bryn Mawr College and received a master's degree in 1923, having been a graduate scholar her last year there. During 1924–25 she had a leave of absence from the Baldwin School and read mathematics at Newnham College, Cambridge University, England, on a fellowship from Radcliffe College. She studied at Radcliffe 1927–28. Cooper entered the University of Illinois in 1928, served as an assistant in 1929–30, and wrote her dissertation in geometry before receiving her doctorate in 1930 with a minor in physics.

After receiving her PhD, Cooper continued to work in private schools for the next seven years. She was principal of the Buckingham School in Cambridge, Massachusetts, from 1930 to 1935. She was then head of the mathematics department of the Chapin School in New York before becoming supervisor of mathematics and teacher training at Hunter College and chairman of the mathematics department at Hunter College High School in 1937. She remained chairman of the department at Hunter High School until she retired in 1958. While at Chapin, Cooper wrote, but did not publish, a textbook that she called "Algebra for Anne." Later she worked on a text called "Mathematics Made Plain" and on an article explaining some of her philosophy of teaching called "Mathematics for the Bright Student."

After her retirement Cooper returned to Syracuse, where she taught some undergraduate mathematics courses at Syracuse University and participated in teacher training seminars. Among her contributions as a life member of the American Friends Service Committee was her service as director in summer 1939 of Sky Island Hostel for European refugees at Nyack, New York. She was also a charter member and chairman of the management committee of a cooperative residence club in New York, which she described as a hostel for elderly refugees. In 1948 it became Newark House of the New Jersey Fellowship Fund for the Aged, and Cooper served as its president. Elizabeth Cooper died in Syracuse, New York, at age seventy-six in 1967. Her papers are at the Schlesinger Library, Radcliffe Institute, Harvard University.

COPE, Frances (Thorndike). August 19, 1902–May 14, 1982.

VASSAR COLLEGE (BA 1922), COLUMBIA UNIVERSITY (MA 1925), RADCLIFFE COLLEGE (PHD 1932). PhD diss.: Formal solutions of irregular linear differential equations (directed by George David Birkhoff).

Elizabeth Frances Thorndike, known as Frances, was born in New York City, the eldest child of Elizabeth (Moulton) and Edward L. Thorndike. She came from a distinguished academic family. Her mother had taken the four-year classical course at Boston University, while her father earned a PhD from Columbia in 1898 and became one of the most influential educational psychologists of the first four decades of the twentieth century. He was on the faculty at Teachers College, Columbia University, 1904–40.

In 1904, two years after Frances Thorndike's birth, a second daughter was born but lived less than two weeks. Three sons followed and all later earned PhD's: two in physics, and one in psychology.

Frances Thorndike first attended the Horace Mann School, a private school associated with Teachers College, Columbia University. Thorndike also went to the public school in Montrose, New York, and then to Drum Hill High School in Peekskill, New York. She attended Vassar College and, according to her father's biographer, impressed psychologist Margaret Washburne as the most brilliant student she had ever had. After Thorndike's graduation from Vassar in 1922, she worked as an engineering assistant at the American Telephone and Telegraph Company in New York City 1922–24 and again 1925–27, after taking the year 1924–25 to earn a master's degree in mathematics from Columbia University.

Frances Thorndike became instructor of physics at Vassar for the year 1927–28 and studied at Radcliffe College in 1928–29 as a Vassar alumnae fellow. On June 29, 1929, Frances Thorndike married Thomas Freeman Cope. Cope was born in Texas in 1900, earned a bachelor's degree in 1923 and a master's degree in 1925 from Tulane University, and received his PhD in mathematics from the University of Chicago in 1927. He was at Harvard for the year 1928–29 as a National Research Council fellow. They remained in Cambridge the year after their marriage, he as instructor of mathematics at Harvard.

In 1930 the Copes moved to Ohio, where T. Freeman Cope was head of the mathematics department at Marietta College until 1937. Frances T. Cope completed her work for the PhD at Radcliffe as a student of George David Birkhoff in 1932. The publications resulting from her dissertation continue to be cited in mathematics and physics literature. The Copes had three children, a son born in 1930 and daughters born in 1934 and in 1935. The older daughter earned a PhD from Harvard in psychotherapy.

Frances Cope was an instructor of mathematics at Vassar College during second semester 1935–36. In 1937 the family moved to New York, and T. Freeman Cope joined the faculty at Queens College, where he remained until his retirement in 1970. Frances T. Cope served as instructor at Queens in 1941 and at Adelphi College 1941–43.

Frances Cope was living in Montrose, New York, at the time of her death at age seventy-nine in 1982. T. Freeman Cope died in Rochester, Minnesota, two years later.

COPELAND, Lennie P. March 30, 1881–January 11, 1951.

UNIVERSITY OF MAINE (BS 1904), WELLESLEY COLLEGE (MA 1911), UNIVERSITY OF PENNSYLVANIA (PHD 1913). PhD diss.: On the theory of invariants of n-lines (directed by Oliver Edmunds Glenn).

Lennie Phoebe Copeland was born in Brewer, Maine, the only child of Emma (Stinchfield) and Lemuel Copeland. Her parents married in 1867 and by 1900 were living in Bangor, Maine, where her father was engaged in manufacturing.

Copeland graduated from the Bangor high school before going to the University of Maine, from which she graduated in 1904. Copeland was elected a member of Phi Kappa Phi; in 1921 she was named an alumna member of Phi Beta Kappa. She also belonged to Alpha Omicron Pi social sorority.

Copeland taught mathematics from 1905 to 1910 in the Bangor high school. She then went to Wellesley College, where she earned a master's degree in 1911. She continued her graduate work at the University of Pennsylvania as a Bennett fellow and a Pepper fellow 1911–13. She received her doctorate in 1913 with a dissertation in invariant theory.

Immediately after receiving her doctorate, Copeland was hired by Wellesley and spent the rest of her career there. She was instructor 1913–20, assistant professor 1920–28, associate professor 1928–37, professor 1937–46, and professor emeritus after her retirement in 1946. Among Copeland's teaching responsibilities was a course she developed in the history of mathematics. In connection with this interest she collected rare books, particularly on mathematical recreations, for the Treasure Room of the Wellesley library and wrote a descriptive catalogue of the rare mathematical books in the library.

Copeland was a charter member of the MAA, often attended national meetings of the association, and served as chair of the program committee for the annual meeting in 1923. She was the first woman president of the New England Association of Teachers of Mathematics 1925–27.

In the late 1930s, Copeland described herself as a Republican and as Congregational. She was on the Service Fund of the Christian Association and was treasurer at one time. Copeland's many interests included attending plays, reading, driving, hiking, and traveling. She was a member of the Appalachian Mountain Club and served as a counselor in natural history for the club. Her travels included trips in the 1930s to Europe; Alaska;

Labrador; Egypt, central Africa, and the Near East; Scandinavia; Mexico; and Puerto Rico. One trip, which included countries in Europe, Africa, and the Middle East, lasted from mid-February 1935 until early June 1935 and was made with **Clara Eliza Smith**, her close friend who had retired from the mathematics department at Wellesley the summer before. Smith and Copeland shared a home in Wellesley until Smith's death in May 1943. At the time of Copeland's retirement, it was noted in the Wellesley alumnae magazine "that the courage shown by Miss Copeland, when after Miss Smith's death she picked up the shattered plans of her life and rebuilt them beautifully, has made a profound impression on all of us" ("Lennie P. Copeland," *Wellesley Magazine* (July 1946): 319).

Following her retirement Copeland made her home in St. Petersburg, Florida, with Carol S. Scott, whom she had known at Wellesley. After Copeland was awarded an honorary DSc degree by the University of Maine in 1948, Scott reported that the "citation especially stressed her making Mathematics 'a joyous adventure' to her years of classes" (Alumni Records Collection; Folder: Copeland, Lennie Phoebe; University of Pennsylvania Archives). Lennie P. Copeland died in St. Petersburg in January 1951 at age sixty-nine. The funeral was held in St. Petersburg and was followed by burial in Brewer, Maine.

COWLEY, Elizabeth B. May 22, 1874–April 13, 1945.

INDIANA NORMAL SCHOOL (BS 1893), VASSAR COLLEGE (BA 1901, MA 1902), COLUMBIA UNIVERSITY (PHD 1908). PhD diss.: Plane curves of the eighth order with two real four-fold points having distinct tangents and with no other point singularities.

Elizabeth Buchanan Cowley was born in Allegheny, Pennsylvania, the eldest and, by 1900, the only surviving child of five born to Mary Junkin (Buchanan) and John Cowley. Mary Buchanan attended the Pittsburgh Female College (now Chatham College), from which she received a Mistress of English Literature (or Language) degree in 1870. She was elected supervisor of playgrounds and vacation schools and was a member of the school board in Pittsburgh. In 1900 John Cowley was an auditor; he was deceased by 1914.

Elizabeth Cowley attended the public schools of Allegheny after which she studied for two years before receiving a BS degree in 1893 from Indiana Normal School (now Indiana University of Pennsylvania). She taught for the next four years in public schools in Pennsylvania and was given a life diploma by the state.

Cowley entered Vassar College in September 1897 and received her BA degree four years later. She was awarded the graduate scholarship in mathematics and astronomy for the next year and received her MA degree from Vassar in 1902. She was then appointed instructor in mathematics at Vassar and continued at that rank until 1913.

Cowley attended the summer sessions at the University of Chicago 1903–05, where she had twelve weeks of mathematics and physics each year. She began work at Columbia University during the second semester of the academic year 1905–06 and remained for summer school in 1906 and the next academic year. Her doctorate was awarded by Columbia in 1908.

Cowley remained at Vassar, except for leaves of absence, until 1929. She was assistant professor 1913–16 and associate professor 1916–29. While at Vassar she continued her studies, was active in a number of mathematical societies, and published regularly. She reported in 1914 that she had studied at the universities of Göttingen and Munich as well as those indicated above. She belonged to several foreign mathematical societies and was a charter member of the MAA. She was particularly active in the MAA during its first decade. She was elected to serve a three-year term on the MAA council beginning January 1918; in 1920 the MAA incorporated and the council became a board of trustees. Cowley contributed articles and book reviews to mathematical, educational, and popular publications. She also served on the editorial board of the abstracting journal *Revue Semestrielle des Publications Mathématiques* and was a regular reviewer for the journal from 1908 until 1926. Seventeen of her two dozen book reviews in the *Bulletin* of the AMS were of books written in French, German, Italian, or Spanish.

Cowley had a leave of absence from Vassar 1926–29 and was promoted to professor while on leave. She resigned her position at Vassar in June 1929 to return to Pittsburgh to be with her mother. She immediately took a teaching position in a Pittsburgh high school and continued teaching until 1938, shortly after her mother's death. She was active in local professional organizations: she was sponsor for various activities of the Pittsburgh public high schools, she became president of the mathematics section of the Pennsylvania State Education Association in 1934, and she was president of the Association of Teachers of Mathematics of Western Pennsylvania.

Cowley's mathematical interests were broad. During the 1920s she published and spoke on historical topics. During that period she also studied crystallography because she had read some work on phases of atomic structure that interested her. Later her publications and talks were usually in the field of mathematics education; among these was a book about public education in the United States that appeared in 1941. Some of her poetry was also published.

At some time after her retirement, Elizabeth Cowley moved to Fort Lauderdale, Florida. She suffered a stroke in 1941 that left her unable to speak or use her right side. Even so, using her left hand she kept up her writing and correspondence and continued to embroider, weave, and sew. Elizabeth Cowley died in Fort Lauderdale, Florida, at age seventy in 1945.

CRATHORNE, Charlotte Elvira (Pengra). May 30, 1875–February 9, 1916.
University of Wisconsin (BA 1897, PhD 1901). PhD diss.: [Pengra, C. E.] On functions connected with special Riemann surfaces, in particular those for which $p = 3$, 4, and 5 (directed by Linnaeus Wayland Dowling).

Charlotte Elvira Pengra was born in Juda, Wisconsin, the eldest of five children of Mary Ellen (Preston) and Winfield Sherman Pengra. In 1880 the family was living in Jefferson, Wisconsin, near Madison, and her father was a farmer; in 1900 they lived in Madison and he was a landlord. All of the Pengra children received bachelor's degrees from the University of Wisconsin.

Charlotte Pengra graduated from the Madison, Wisconsin, high school before attending the University of Wisconsin, from which she received her BA in 1897, the same year that Wisconsin awarded its first PhD in mathematics. She then taught in high schools in Fox Lake and Sparta, both in Wisconsin. After teaching for two years, she received a fellowship from the University of Wisconsin and completed her work for the PhD two years later, as the third Wisconsin PhD in mathematics. After receiving her doctorate, Pengra was in charge of the mathematics department of a high school in Elgin, Illinois, for three years.

On June 21, 1904, in Madison, Charlotte Pengra married the English-born Arthur Robert Crathorne (1873–1946), who had been a graduate student and instructor in mathematics at Wisconsin. He had come to the United States with his family as a young boy, received a bachelor's degree from the University of Illinois in 1898, and had been at Wisconsin since 1900.

The Crathornes spent the years 1904 through 1907 in Germany studying mathematics at Göttingen, where A. R. Crathorne received his PhD in 1907. Their daughter was born in November 1906. On their return from Göttingen, A. R. Crathorne took a position at the University of Illinois. The Crathornes had two more children born in 1909 and in 1911. Charlotte Crathorne's daughters report that she was a collaborator in the preparation of *College Algebra* by H. L. Rietz and A. R. Crathorne in 1909, although she is not mentioned in the acknowledgements in the preface to the book.

After an illness of about three years, Charlotte Crathorne died of breast cancer at her parents' home in Brodhead, Wisconsin, in 1916 at age forty. She was buried in Juda, Wisconsin.

CRONIN, Sarah Elizabeth. July 1, 1877–October 2, 1958.
University of Iowa (BS 1903, MS 1905), Columbia University (PhD 1917). PhD diss.: Geometric properties completely characterizing all the curves in a plane along which the constrained motion is such that the pressure is proportional to the normal component of the acting force (directed by Edward Kasner).

Sarah Elizabeth Cronin was the youngest of nine (living) children of Deborah and Cornelius Cronin, a farmer in northeast Iowa. Her parents had emigrated from Ireland in the mid-nineteenth century before their marriage. She was born in Iowa, probably in Allamakee County in the northeast corner of the state.

According to census records, in 1900 Sarah Cronin was a school teacher and living in Cherokee County, in western Iowa, with her parents. She received her undergraduate degree in 1903 from the University of Iowa (officially the State University of Iowa) in Iowa City, and in April of that year the university awarded her a scholarship in mathematics to continue her studies. However, the following fall she was teaching mathematics in the Creston, Iowa, high school. After a year she returned to the University of Iowa as a fellow in mathematics and received her master's degree in 1905, with a thesis on the mathematical theory of diffusion.

Cronin was instructor of mathematics at Iowa State College of Agriculture and Mechanic Arts (now Iowa State University of Science and Technology) in Ames 1905–07. She was then instructor at the University of Iowa 1907–15, although she apparently had a leave of absence 1913–15, when she was a graduate student at Columbia University. Cronin returned to the University of Iowa as assistant professor in 1915 and received her PhD from Columbia in 1917. In 1918 she went to the Equitable Life Assurance Society in New York (now the AXA Equitable Life Insurance Company) as chief of the policy change division. The following year she passed the first two parts of the associate examinations of the Actuarial Society of America. Cronin passed the third and fourth parts in 1922 and 1923 and was then enrolled as an associate of the society. She kept her membership in the society through 1940, two years before her retirement.

Sarah E. Cronin moved from Forest Hills, Queens, to Daytona Beach, Florida, when she retired in 1942. She was eighty-one when she died in Daytona Beach in 1958.

CUMMINGS, Louise D. November 21, 1870–May 9, 1947.
University of Toronto (BA 1895, MA 1902), Bryn Mawr College (PhD 1914). PhD diss.: On a method of comparison for triple-systems (directed by Charlotte Angas Scott).

Louise Duffield Cummings was the daughter of Anne (Morison) and James Cummings, born in Scotland and Ireland, respectively. Louise Cummings was born in Hamilton, Ontario, Canada, where, at the time of the 1881 Canadian census, her father's occupation was described as collector. At that time, five children in the household were listed, three sons and two daughters, ranging in age from ten to twenty-three. Her eldest brother earned a BA from the University of Toronto and became a physician in Wayne, Michigan. Her sister was a pianist who taught at one time at a private high school in Toronto. Her youngest brother graduated in medicine from Toronto and became a surgeon who was noted for his pioneering work with X-rays for diagnosis and treatment of medical problems.

Louise Cummings attended public schools and the Hamilton Collegiate Institute in Hamilton, Ontario. After attending the University of Toronto 1889–90, she was away for a year but returned as a second-year student in 1891 and remained as a third-year student 1892–93. She was not in attendance 1893–94 but was a fourth-year student in 1894–95. She won the William Mulock scholarship in mathematics her second year, and half of the physics scholarship her third year. Cummings received her bachelor's degree from Toronto in 1895 with first class honors in mathematics.

Louise Cummings engaged in graduate study at four different schools during most of the next five years. Although she was not listed as a graduate student, she continued her study at Toronto in 1895–96. She held a Bennett fellowship, one of two given annually to women in the graduate school (then called the department of philosophy) for study at the University of Pennsylvania 1896–97 and supervised examinations at Toronto in 1897. In 1897–98 she was a student at the University of Chicago, and in 1898–99 she was a resident fellow in mathematics at Bryn Mawr College. She held a fellowship by courtesy in mathematics during the second semester of 1899–1900 at Bryn Mawr. Cummings studied at the Ontario Normal College in Hamilton 1900–01 and then returned to Toronto to continue her studies and to teach mathematics at St. Margaret's College for Ladies, a private high school there. Her master's degree was awarded by the University of Toronto in 1902.

In 1902 Cummings joined the faculty at Vassar College in Poughkeepsie, New York, where she remained until her retirement more than thirty-three years later. While there Cummings was instructor 1902–15, assistant professor 1915–19, associate professor 1919–27, professor 1927–35, and professor emeritus after her retirement. She was on leave for the second semesters of 1931–32 and 1934–35.

Louise Cummings continued her studies at Bryn Mawr College at various times while on the faculty at Vassar. She was a graduate scholar during the first semester 1905–06 and during the second semester 1912–13, when she was in residence at Bryn Mawr for the academic year. She received her doctorate from Bryn Mawr in 1914 with subjects mathematics, applied mathematics, and physics. In her dissertation vita Cummings acknowledged her indebtedness to Charlotte A. Scott of Bryn Mawr, "for helpful criticism and unfailing interest in the preparation" of her dissertation and to Henry Seely White of Vassar for suggesting the subject. She published a dozen papers over twenty years, and some of her work was mentioned by E. T. Bell in his article on algebra in the *Semicentennial Addresses of the American Mathematical Society.*

Cummings was given an honorary DSc degree by the University of Toronto at the university's centennial celebration in 1927. Louise Cummings was at the home of a niece when she died in 1947 at the age of seventy-six in Wayne, Michigan. She was buried in the Hamilton, Ontario, cemetery.

DALE, Julia. October 16, 1893–January 13, 1936.

TRANSYLVANIA UNIVERSITY (BA 1914), UNIVERSITY OF MISSOURI (MA 1921), CORNELL UNIVERSITY (PHD 1924). PhD diss.: Some properties of the exponential mean (directed by Wallie Abraham Hurwitz).

Julia May Dale was born in Shelbyville, Kentucky, the second of six children of Ida May (Todd) and James Harrison Dale. Her father was a farmer who had served as Kentucky's secretary of agriculture. He moved with the family to Alabama in 1916 and became a prominent breeder of shorthorn cattle there. Both parents were graduates of Eminence College, a small coeducational college in Kentucky that closed in 1895.

Julia Dale graduated from Shelbyville High School in 1910. She then attended Transylvania University in Lexington, Kentucky, where she was a class secretary and was an officer in the local YWCA and the literary club. After graduating cum laude from Transylvania in 1914, Dale taught for a year in the high school in Aberdeen, Mississippi. In 1916–18 she taught at William Woods College, then a junior college for women, in Fulton, Missouri. After a year on a graduate scholarship at the University of Missouri, Dale spent 1919–22 there as an instructor in mathematics; she received her master's degree in 1921.

Dale then attended Cornell University: as a graduate scholar in mathematics 1922–23 and as an Erastus Brooks fellow in mathematics 1923–24. After receiving her PhD in 1924, having written a dissertation in analysis and with minors in geometry and physics,

Dale was an instructor for a year at the University of Oklahoma. While there she also volunteered at nearby Native American reservations.

From 1925 until 1930, Dale was head of the mathematics department at Delta State Teachers College (now Delta State University) in Cleveland, Mississippi. She was active in the MAA and was secretary-treasurer of the Louisiana-Mississippi Section in 1929. She was an avid bridge player, did needlepoint, was interested in athletics, and helped coach the women's basketball team at Delta State. In 1930 Dale moved to Duke University, where she was an assistant professor and head of the women's division of mathematics at Duke University. While at Duke, Dale taught a variety of classes including some in the graduate school.

Julia Dale suffered from chronic kidney disease and died in Durham, North Carolina, in 1936 at age forty-two; her remains are interred in Greensboro, Alabama. Friends and relatives established the Julia Dale Memorial Fund at Duke in 1938. The income from this fund supports the Julia Dale Prize awarded annually to one or more mathematics majors on the basis of excellence in mathematics.

DARKOW, Marguerite D. November 12, 1893–December 9, 1992.
BRYN MAWR COLLEGE (BA 1915), UNIVERSITY OF CHICAGO (MA 1923, PHD 1924). PhD diss.: Arithmetics of certain algebras of generalized quaternions (directed by Leonard Eugene Dickson).

Marguerite Daisy Darkow was the second of three daughters of Flora (Singer) and Martin Darkow. Her mother was born in Vienna and studied piano from a well-known teacher there. Her father was born in Riga, then in Russia, and attended college in Vienna. Her parents immigrated to the United States in about 1891, and Marguerite was born in Philadelphia two years later. All three daughters earned higher degrees: the eldest a PhD in Greek literature and the youngest an LLB. Both Flora and Martin Darkow taught piano, but, for the most part, he earned his living as a music and theatre critic for a German-language newspaper. In September 1917, after the United States had declared war on Germany, Martin Darkow, who was then managing editor of the newspaper, was indicted and tried for treason and for violation of the Espionage Act for making false statements to help the enemy. Darkow was acquitted on the treason charge but was convicted on the other charges. He was never imprisoned and was pardoned by President Woodrow Wilson in June 1920.

Marguerite Darkow attended the E. Spencer Miller School in Philadelphia 1899–1907 and the Philadelphia High School for Girls 1907–11 before entering Bryn Mawr College in 1911. She was First Bryn Mawr Matriculation Scholar for Pennsylvania and the Southern States 1911–12, Simon Muir scholar 1911–15, James E. Rhoads junior scholar 1913–14, and Maria L. Eastman Brooke Hall memorial scholar 1914–15. She graduated in 1915 in mathematics and physics and led the list of honor students. Darkow was awarded a Bryn Mawr European fellowship for 1915–16 but delayed using the fellowship because of the war in Europe.

Darkow taught mathematics and science 1915–16 at Tudor Hall, a private school for girls in Indianapolis. During summer 1916 she studied at the University of Pennsylvania and entered Johns Hopkins University that fall as a graduate student and fellowship holder in physics and mathematics. After a year there, she did statistical work for the children's bureau of the US Department of Labor summer 1917 and taught mathematics and science at Rogers Hall in Lowell, Massachusetts, 1917–18. The following year she did astronomical research at the Leander McCormick Observatory at the University of Virginia. During several summers she worked as both a private tutor and a tutor for fall college entrance exams at tutoring camps in New England. From 1917, and continuing into the 1920s, Darkow's father was often mentioned in the news making it difficult for her to pursue a normal career.

Darkow began working in the actuarial department of the Provident Life Insurance Company in Philadelphia in 1919 or 1920 and passed section A, part I, of the Actuarial

Society of America examinations in May 1920. She passed part II of the associate examination in May 1921 and part III in May 1922. Darkow left her position at Provident in 1922 to enter graduate school at the University of Chicago with a fellowship. She earned her master's degree in 1923, continued another year with a fellowship, and received her PhD in August 1924. In 1924–25 she traveled in Europe and finally used the European fellowship awarded by Bryn Mawr in 1915 to study at the Sorbonne in Paris.

Darkow was appointed assistant professor at Indiana University in Bloomington in 1925 and was let go in 1927. Correspondence at Indiana indicates that the university president maintained that the original appointment was for one year; Darkow contested that assertion and believed that a report to the president that she was an admirer of H. L. Mencken contributed to the dismissal. The following year she was an instructor at Pennsylvania State College, and in 1928 Darkow began her career at Hunter College in New York. She was instructor 1928–30, assistant professor 1928–42, associate professor 1942–56, professor 1956–59, and professor emeritus after her retirement in 1959. She continued her study of mathematics at the University of Chicago in summer 1938 and studied statistics at the University of Michigan the following two summers. She presented papers and published articles sporadically into the 1950s.

Darkow was interested in birding, travel, flowers, photography, reading, and corresponding with friends. She knew many languages and traveled frequently. She regularly contributed to charities and was a member of the American Civil Liberties Union and antinuclear organizations. In about 1950 Darkow became close friends with Edna Sheinhart (1923–1995), who had graduated from Hunter in the 1940s and was a research engineer. After Darkow's retirement they shared an apartment in New York, traveled together, and then shared a home in Berkshire County, Massachusetts. Sheinhart had bought a farm in Sheffield, and the renovated barn became their permanent home in 1972. In about 1982 Darkow was diagnosed with Alzheimer's disease but remained in her home and was cared for by Sheinhart and a staff of care givers. Marguerite Darkow died in her home in 1992. She was ninety-nine. A large tract of land was donated to the Massachusetts Audubon Society in her name and forms the basis of the Lime Kiln Farm Wildlife Sanctuary.

DEAN, Mildred (Waters). February 17, 1904–September 11, 1981.

GOUCHER COLLEGE (BA 1924), JOHNS HOPKINS UNIVERSITY (MA 1927, PHD 1929). PhD diss.: Studies in inversive geometry, with reference to a special set of six points (directed by Frank Morley).

Mildred Caroline Somerset Waters was born in Baltimore, Maryland, the second of two daughters of Caroline F. and Robert F. Waters, the proprietor of a feed store. She had her early education in public and private schools in Baltimore before doing her undergraduate work at nearby Goucher College.

Waters graduated from Goucher in 1924 and was a graduate student in chemistry at the Johns Hopkins University in Baltimore the following year. She transferred to the mathematics department in the fall of 1925 and continued her work in residence at Johns Hopkins the next two years. Her master's essay was the beginning of the work in inversive geometry that she was to continue for her PhD dissertation.

In the fall of 1927 Mildred Waters married Charles Earle Dean (1898–1993), who had received his PhD in physics that year from Johns Hopkins. Since her husband's work as a technical writer with AT&T required a move to New York, Mildred Dean discontinued her course work at Johns Hopkins. She continued her graduate work by taking a full-time program at Columbia from the summer of 1927 through the spring of 1928. She also continued to work on her dissertation, mainly on her own, with occasional visits to Baltimore for discussions with her advisor, Frank Morley. She took her oral examination at Johns Hopkins in January 1929 and received her doctorate in February with subordinate subjects chemistry and geological physics.

From 1929 until 1963 Charles E. Dean worked as a technical writer and editor for the Hazeltine Corporation in Little Neck, Queens, New York. During that period the Deans raised a family, and Mildred Dean had various teaching positions in the New York area. A son was born in 1929, and a daughter in 1932. Mildred Dean taught part time at Brooklyn College 1931–33, at Queens College 1942–47, and at Adelphi College 1947–48. She was also a temporary instructor at Hunter College and taught at Hofstra University.

In 1963 the Deans moved to the Washington, D.C., area, where Mildred Dean participated in the Forum for Professionals and Executives, a study group, and volunteered for Travellers Aid in Union Station and in the accounting office for the American Cancer Society. Having suffered from heart and lung ailments for some time, Mildred Waters Dean died in Takoma Park, Maryland, at seventy-seven in 1981.

DELEVIE, Jeanette (Fox). January 13, 1912–July 29, 1997.

New York University (BS 1931), Yale University (PhD 1935). PhD diss.: [Fox, J.] Existence of a Euclid algorithm in quadratic fields (directed by Oystein Ore).

Jeanette Fox was born in New York, New York, the second of four children of Sophie (Levy) and Jacob Fox. Her mother was born in New York; her father was born in Russia, immigrated to the United States as a young man, and was naturalized in 1905. According to census records, her father was in the grocery business in 1910 and was an electric broker in 1920. Later he was president of the Owners and Tenants Electric Company. Her older sister was born in 1906 and earned a PhD in psychology; her younger brother was born in about 1914 and her younger sister in about 1916. At some point her parents were divorced.

Jeanette Fox graduated from Erasmus Hall High School in Brooklyn and lived at home while attending the Washington Square College of New York University, where she majored in mathematics and minored in physics. During her junior year, 1929–30, she studied at the University of Munich. Fox graduated from NYU in 1931 and studied at the University of Berlin 1931–32. She was a fellow in mathematics at Yale University 1932–35 and received her doctorate in 1935.

In 1935 Jeanette Fox married Albert Sol Keston (1911–1992). Albert Keston received his doctorate in chemistry from Yale in 1935. He taught chemistry at the College of the City of New York (CCNY) before joining the faculty at the New York University College of Medicine. Later he was at Mount Sinai School of Medicine.

During the academic year 1938–39 Jeanette Fox Keston taught at Brooklyn College during the day as a substitute, that is, on an hourly wage. She was also an instructor in the evening session 1938–41. The first of her two children, a daughter, was born in January 1942. She again taught during the day as a substitute from January to June 1943. Her son was born in January 1944. The Kestons were separated when the children were young and divorced in 1950. Their daughter received a PhD in experimental psychology in 1967.

Jeanette Fox Keston was a temporary lecturer in a veterans' program at Hunter College 1946–48 and remained in a temporary position at Hunter the following year. The next two years, 1949–51, she was an assistant professor at Jersey City Junior College. Keston again taught at Brooklyn College starting in 1953; she was a substitute 1953–56 and a lecturer 1956–57. She then began her association with the mathematics faculty at CCNY. She remained until 1971, by which time it had become City College, CUNY. She was an instructor September 1957–December 1961, an assistant professor January 1961–December 1966, and an associate professor from January 1967 until her retirement in 1971. In the early 1960s she coauthored two mathematics research papers with others at City College. In addition to her undergraduate teaching responsibilities, Keston taught the course in abstract algebra in the graduate school of education. During Saturdays of the academic year 1963–64 she also taught an NSF-sponsored course in number theory for high school students. For many years she served as chairman of the departmental committee on teacher preparation.

While traveling in Europe, Jeanette Keston met, and later married, Solomon Delevie (1907–1992), a native of Pennsylvania who was living in San Francisco. They appear to have married at the end of the 1969–70 academic year, and Jeanette Delevie moved to California while she was on a sabbatical leave 1970–71. She was granted a waiver from the condition of returning after a sabbatical and retired. In 1997, at the age of eighty-five, Jeanette Delevie died in San Francisco.

DICKERMAN, Elizabeth Street. November 13, 1872–April 24, 1965.

SMITH COLLEGE (BA 1894), YALE UNIVERSITY (PHD 1896). PhD diss.: Curves of the first and second degree in $x\ y\ z$ where $x\ y\ z$ are conics having two points in common.

Elizabeth Street Dickerman was born in West Haven, Connecticut, the eldest daughter of Elizabeth Mansfield (Street) and George Sherwood Dickerman. Her father was a Congregational minister who earned a BA from Yale and a BD from the Yale Divinity School. She had three siblings: a brother born in 1874 who received a BA from Yale, a PhD from Halle, and was a professor of Greek; a sister born in 1879 who graduated from Smith College, attended the Yale School of Music, and later married; and a brother born in 1885.

Elizabeth Street Dickerman graduated from high school in Amherst, Massachusetts, in 1890 before entering nearby Smith College, where her chief subjects of undergraduate study were mathematics, philosophy, social science, French, German, and music before her graduation in 1894. She entered Yale University that autumn, held a scholarship there her second year, and received her PhD in 1896.

In the decade after receiving her PhD, Dickerman was primarily occupied with private teaching. She also attended mathematics courses at Yale in 1896–97 and 1899–1900. She traveled in Europe in 1897 and in 1902 and was fluent in French, German, and Italian. The next several years she had formal teaching positions. She taught mathematics at Greenwich Academy 1904–05, substituted as professor of mathematics in the College for Women of Western Reserve University in Cleveland, Ohio, 1906–07, and taught mathematics and psychology at Ingleside School in New Milford, Connecticut, 1907–13.

After 1914 Dickerman was mainly engaged in literary work and some private teaching while living in New Haven. She served as assistant to the editor of the *Yale Review* in 1917. Her literary work includes translations of poetry from the French, books of poetical reminiscences of her travels, and various other articles. In 1924 she conducted a Round Table on Industrial Relations in New Haven in connection with the AAUW.

Dickerman reported in 1936 that she was devoting time to the care of her father who was seriously ill. Her primary residence before his death in 1937 was her family home in New Haven; after his death she moved to an apartment there. From the mid-1920s she also maintained a summer home on Squirrel Island, Maine, where her maternal grandparents had also had a home. During her last few years she lived most of the time at the home of her sister in Greenwich, returning to her apartment in New Haven for a few weeks each spring and fall. Elizabeth Dickerman died in Greenwich, Connecticut, in 1965; she was ninety-two.

DIMICK, Alice (McKelden). December 6, 1878–July 13, 1956.

COLUMBIAN UNIVERSITY (BA 1899), UNIVERSITY OF PENNSYLVANIA (MA 1900, PHD 1905). PhD diss.: [McKelden, A. M.] Groups of order 2^m that contain cyclic sub-groups of order 2^{m-1}, 2^{m-2}, and 2^{m-3}.

Alice Madeleine (Elsie) McKelden was born in Washington, D.C., the seventh of nine children of Alice Maria (McIntosh) and William Blagrove McKelden. Her father worked for the US Treasury Department and was cashier of the Treasury at the time of his death. Her mother was a teacher in the public schools of Washington before her marriage. There were seven daughters and two sons in the family.

Alice McKelden graduated from Central High School in Washington, D.C., in 1895. She then received by competitive examination a four-year scholarship to Columbian (now George Washington) University. She received medals in both mathematics and Greek and graduated from Columbian in 1899. She entered the University of Pennsylvania as a University scholar in mathematics in October 1899 and studied both mathematics and Greek before receiving her master's degree in 1900. McKelden studied at Harvard in the summer session 1901 and at Johns Hopkins in 1901–02. In 1902–03 and 1903–04 McKelden was a Moore fellow and a Bennett fellow, respectively, at the University of Pennsylvania. She finished her work for the PhD, with a dissertation in the theory of finite groups, in 1905.

After receiving her doctorate, McKelden taught in at least one preparatory school and at the Philadelphia High School for Girls. On June 12, 1907, Alice McKelden married Chester Edward Dimick. Dimick was born in New Hampshire in 1880, received a BA in 1901 from Harvard and an MA in 1907 from the University of Pennsylvania, where he taught 1903–06. In 1906 Chester Dimick became an instructor of mathematics on the *USS Chase* with the US Revenue Cutter Service, later the US Coast Guard. From the time of their marriage, Chester E. Dimick was a member of the mathematics department faculty at what later became the United States Coast Guard Academy. He joined the faculty there as a civilian instructor, and they lived in Baltimore when the academy was in Arundel Cove, Maryland. In 1910 the academy moved to New London, Connecticut, where they lived for the next thirty-five years. Chester Dimick was later commissioned a commander and was a captain at the time of his retirement; he served as professor and dean. There were no children of the marriage.

Alice M. Dimick occasionally served as a substitute teacher in the high school in New London and as a private tutor, but was mainly involved with various social, church, and civic volunteer activities. She was president of the New London Unit of the League of Coast Guard Women, vice chairman of the New London Unit of the Coast Guard League, and president of the New London Tuesday Book Club. Dimick also served on the national level with the Girls' Friendly Society, an international religious, educational, service, and creative arts group, affiliated with the Episcopal Church. She served as national secretary in the late 1930s, on the national board of directors, and as national president.

After Chester Dimick's retirement as head of the mathematics department at the US Coast Guard Academy in 1945, the Dimicks moved to Tryon, North Carolina. Chester Dimick died in January 1956, and Alice Dimick died in Tryon at age seventy-seven six months later. They are buried at Arlington National Cemetery.

EARLY, Madeline (Levin). April 1, 1912–January 20, 2001.

Hunter College (BA 1932), Bryn Mawr College (MA 1933, PhD 1936), New York University (Bellevue Schools of Nursing) (BS 1945). PhD diss.: [Levin, M.] An extension of the Lefschetz intersection theory (directed by William Welch Flexner).

Madeline Levin was born in Brooklyn, New York, the youngest of seven children of Dora (Siegal) and Hyman Levin. Both of her parents were born in Grodno, Russia (Lithuania), and emigrated from there in the early 1890s. Her father was a house painter.

Madeline Levin attended public schools in New York City for her primary, secondary, and collegiate education. She attended Hunter College as a scholar of the University of the State of New York before graduating magna cum laude as a mathematics major at age twenty. After her graduation in 1932, Levin went immediately to graduate school at Bryn Mawr College. She was a college scholar 1932–34 and a college fellow in the fall of 1934. A paper in algebraic topology coauthored by Levin and William W. Flexner, her advisor at Bryn Mawr, appeared in 1934. Flexner joined the faculty at Cornell University in 1934, and Levin spent the spring semester of 1935 at Cornell as a fellow from Bryn Mawr. She received a master's degree in 1933 and her doctorate in 1936 from Bryn Mawr.

In the fall of 1935, at age twenty-three and a year before completing the work for her doctorate, Levin joined the faculty of her alma mater, Hunter College, as tutor. She remained as tutor through 1938 and then was appointed instructor. It was a lonely time because of the heavy work load and because she was so much younger than the other faculty. She did postdoctoral study at the University of Chicago in the summers of 1937 and 1938.

During the early years of World War II, Levin took a military leave of absence to attend the Bellevue Schools of Nursing, which had recently affiliated with New York University in order to allow students to do additional course work at, and earn a bachelor's degree in nursing from, NYU. She was appointed ensign in the Naval Reserve in March 1945 and in April was directed to the US Naval Hospital, St. Albans, on Long Island. The following year she was transferred to Guam where she served for nearly a year before being released from active duty in March 1947. Then, using the GI Bill, Levin did postdoctoral study at the University of Michigan during 1947–48. It was there that she met Harold C. Early, who had graduated from Michigan State University in 1939 with a major in mathematics and who did graduate work at the University of Michigan where he became a member of the electrical engineering faculty. Levin returned to Hunter in 1948 but formally resigned her position as instructor in February 1949. She married Harold Early on April 8, 1949, and moved back to Ann Arbor, Michigan. The Earlys had one son born in July 1951. They divorced in July 1958.

In 1956 Madeline Early joined the faculty of Eastern Michigan University in nearby Ypsilanti, where she taught for the next nineteen years. She was assistant professor 1956–59, associate professor 1959–67, and professor from 1967 until her retirement in 1975, when she became professor emeritus. In the late 1970s, Early sold her house in Ann Arbor and moved into an apartment in a retirement community, also in Ann Arbor. She stayed active and took a group trip once or twice a year until her eighties. Early died in 2001 at age eighty-eight and, as a veteran, was buried at Fort Custer National Cemetery in Augusta, Michigan.

EPSTEIN, Marion (Greenebaum). June 14, 1915–.

BARNARD COLLEGE (BA 1935), BRYN MAWR COLLEGE (MA 1936, PHD 1938). PhD diss.: [Greenebaum, M.] The non-existence of integral normal bases in certain algebraic fields (directed by Heinrich Wilhelm Brinkmann [Swarthmore College]).

Marion Belle Greenebaum was born in Brooklyn, New York, the third daughter of Anna (Rheinhold), born in Germany, and Milton Greenebaum, born in Maryland. Her mother, who attended the Normal College of the City of New York (now Hunter College), was a teacher; her father, who attended the New York City public schools, was an importer and exporter. Marion Greenebaum's sisters both earned bachelor's degrees from Barnard College, attended graduate school, and became licensed social workers.

Marion Greenebaum attended Erasmus Hall High School in Brooklyn 1927–31 and was awarded the Julia M. Dennis memorial scholarship when she graduated. She entered Barnard College at age sixteen, graduated in 1935, and enrolled at Bryn Mawr College. Greenebaum earned her master's degree in 1936 and her doctorate in 1938 at age twenty-three. She held a graduate scholarship 1935–37 and a resident fellowship 1937–38. She became interested in abstract algebra from Nathan Jacobson, who was at Bryn Mawr her first year, and then worked with H. W. Brinkmann of nearby Swarthmore College.

Greenebaum worked as a research statistician with the International Statistical Bureau in New York City from June 1938 until June 1939. On June 25, 1939, she married Jess Epstein, a research engineer who worked for RCA Manufacturing Company in New Jersey. Epstein, born in Cincinnati, Ohio, in 1907, received his bachelor's and master's degrees from the University of Cincinnati and was an instructor of physics at the Cincinnati

College of Pharmacy 1934–35. He went to RCA in 1935 and remained with the manufacturing company until 1942, when he moved to the newly opened RCA Laboratories in Princeton, New Jersey; he remained there until his retirement in 1973.

The Epsteins had three children: a son born in 1941 and daughters born in 1943 and in 1948. All of the children later earned master's degrees. In 1944 Marion G. Epstein returned to work, part time, with the Test Development Division of the Educational Testing Service (ETS) in Princeton. From 1944 until her daughter was born in 1948, she worked as an associate examiner in mathematics about three hours a day mainly on actuarial exams. She resumed part-time work in 1954 as a professional associate in mathematics. In 1962 she began full-time work as senior examiner in mathematics. In this position she was responsible for graduate record exams, advanced placement exams, upper level college board tests, the original development of CLEP general mathematics examinations, development of computer assisted assembly of tests, and was project director for actuarial examinations. She became assistant director of the division in 1967 and associate director in 1968.

In 1973 Epstein moved to the College Board Programs Division of ETS, where she continued her administrative work. She was director of development and analysis 1973–75, administrative director of professional services 1975–77, and vice president 1977–80. After July 1980 she was senior advisor to the senior vice president for programs. Epstein was engaged in curriculum writing for the School Mathematics Study Group (SMSG) during the summers of 1959 and 1960. She was also a consultant for test development workshops with teachers of mathematics for the West African Examinations Council in Ghana and Nigeria in spring 1964 and for the University of London School Examinations Department in spring 1967.

Epstein was on the Princeton Township Board of Education 1955–66: as vice president 1958–61 and president 1961–63. She was on the New Jersey State Board of Education 1966–77 and the New Jersey Board of Higher Education 1973–77; on the former she was vice president 1975–77, chair of the legal committee 1975–77, and chair of the affirmative action committee 1974–77. In 1980 she became a trustee of Kean College of New Jersey (now Kean University) and was appointed to the advisory council to the Princeton University mathematics department.

In the 1950s, Epstein was active in the Princeton chapter of the League of Women Voters and was vice president 1950–54 and president 1954–56. She was a trustee of the Princeton Jewish Center and in 1979 became a member of the National Education Committee of the American Jewish Committee. She was elected secretary of Community Without Walls, a nonprofit association of Princeton-area residents supporting senior citizens, when its first governing body was elected in 1995. Jess Epstein died on December 30, 1998, and Marion Greenebaum Epstein continued to live in their home in Princeton.

FARNUM, Fay. August 24, 1888–March 11, 1977.

IOWA STATE COLLEGE OF AGRICULTURE AND MECHANIC ARTS (BS 1909), CORNELL UNIVERSITY (MA 1915, PHD 1926). PhD diss.: On triadic Cremona nets of plane curves (directed by Virgil Snyder).

Eugenia Fae Farnum, born in Spencer, Iowa, was the second of four children of Josephine (Jacobs) and George Edwin Farnum. Her father was a farmer in northwestern Iowa at the time of the 1900 census.

Farnum graduated with a degree in general science from Iowa State College of Agriculture and Mechanic Arts (now Iowa State University of Science and Technology) in 1909. Between 1909 and 1914 she taught in high schools in Lyons, LeMars, and Ames, Iowa. She attended Cornell University 1914–15 and received her master's degree in 1915 having written her thesis using the name Fae Farnum. By the late teens she used only the name Fay Farnum.

Farnum then returned to Iowa where she was an instructor at Iowa State 1915–24. She attended the University of Chicago during the summers of 1921 and 1922 and returned to study at Cornell during 1924–26. During her final year at Cornell, Farnum was an instructor; she registered only for thesis supervision and taught two classes per semester. She received her doctorate in 1926 with major subject geometry, first minor mathematical analysis, and second minor physics. Her youngest sibling, a sister, received a PhD in biochemistry the same year from the University of Chicago.

Soon after receiving her doctorate Farnum began a lengthy period of teaching at Washington Square College, New York University. She served as instructor 1926–28 and then as assistant professor. While at NYU Farnum usually taught at least one graduate course a semester. During 1939–40, she took a leave of absence from NYU to study at the Physics and Mathematics Institute in Copenhagen. She reported to her father in Ames, Iowa, on the April 1940 Nazi invasion of Denmark and described her trip back to the United States via Berlin, Munich, the Brenner Pass, Milan, and Genoa, where she took the USS *Manhattan* to New York.

Farnum returned to NYU after her leave and taught for at least one more year, 1940–41; both the undergraduate and graduate bulletins list her as scheduled to teach 1941–42. In 1943 she returned to the faculty of Iowa State, primarily to help the department meet the extra teaching required by the naval and army students. She was an assistant professor from 1943 to 1949, when she retired, and taught again as assistant professor 1955–57.

Farnum was living at least part time in Tucson, Arizona, by about 1952. In 1957 she was appointed assistant professor at the University of Arizona and that year, at age sixty-nine, rejoined the MAA, having let her membership lapse in the early 1940s. She died in Tucson at age eighty-eight in 1977.

FITCH, Annie (MacKinnon). June 1, 1868–September 12, 1940.

UNIVERSITY OF KANSAS (BS 1889, MS 1891), CORNELL UNIVERSITY (PHD 1894). PhD diss.: [MacKinnon, A. L.] Concomitant binary forms in terms of the roots (directed by James Edward Oliver).

Annie Louise MacKinnon was born in Woodstock, Ontario, Canada, the daughter of Annie Louise (Gilbert) and Malcolm MacKinnon, natives of Ontario. The family was living in Kansas by 1870, and according to census records from 1870 and 1880 there were at least five children. In 1870 her father was a real estate agent, and in 1880 he was a hardware merchant. She lived in Concordia, Kansas, during her childhood and graduated from Concordia High School.

MacKinnon then enrolled at the University of Kansas in Lawrence and graduated in 1889 as one of the first two women in our study to graduate from a public university; the other was **Mary Frances Winston (Newson)** at the University of Wisconsin. MacKinnon continued at Kansas as a graduate student, the third graduate student in mathematics in the university's history, and received her master's degree in 1891. She also taught in the high school in Lawrence 1890–92 and continued her studies at the University of Kansas.

In October 1892 Annie MacKinnon enrolled at Cornell University. She was an Erastus Brooks fellow 1893–94 and received her PhD in 1894 with major subject theory of functions, first minor quantics and statics, and second minor mathematical physics. MacKinnon then spent two years studying mathematics at Göttingen, as an Association of Collegiate Alumnae European fellow in 1894–95 and as a Women's Education Association of Boston European fellow in 1895–96. Also at Göttingen during this period was Edward Fitch, an American student of classical philology whom MacKinnon later married.

After returning from Göttingen in 1896, MacKinnon was appointed professor of mathematics at Wells College, a small college for women in Aurora, New York. She was the only mathematics teacher on the faculty and was also registrar her final year, 1900–01. On July 3, 1901, in Lakeside, Ohio, Annie MacKinnon married Edward Fitch (1864–1946). Fitch had received his PhD at Göttingen in 1896, before returning to his position teaching

Greek at his alma mater, Hamilton College, a school for men in Clinton, New York, nearly a hundred miles from Wells College. They had no children. Except for the year 1932–33 when he was professor at the American School of Classical Studies at Athens, Edward Fitch remained at Hamilton until he retired in 1934 as professor and dean emeritus.

Annie Fitch was a member of the League of Women Voters and was on the book committee of a nearby town library. Although she reported to the AAUW in the late 1920s that she continued mathematical investigations after her marriage, she did not publish after 1898. She was a charter member of the MAA and maintained that membership until her death at age seventy-two. She died in 1940 in Clinton, New York, after a long illness, having a few months earlier written to a friend, "It seems to me worthwhile that some women are intelligent about things mathematical even if their own accomplishments are not great" (Whitman, "Women in the American Mathematical Society before 1900," pt. 2, 9).

FOCKE, Anne (Bosworth). September 29, 1868–May 15, 1907.

WELLESLEY COLLEGE (BS 1890), GEORG-AUGUSTS-UNIVERSITÄT GÖTTINGEN (PHD 1900). PhD diss.: [Bosworth, A. L.] Begründung einer vom Parallelenaxiome unabhängigen Streckenrechnung (directed by David Hilbert).

Anne Lucy Bosworth was born in Woonsocket, Rhode Island, the only surviving child of Ellen (Metcalf) and Alfred Bosworth. It appears that her mother was widowed some time in the early 1870s and in 1874 was working at the public library in Woonsocket. In 1880 Ellen Bosworth was described as a librarian, and she, her sister, and Anne were living with Anne's maternal grandmother in Woonsocket.

After attending public schools in Woonsocket, Anne Bosworth entered Wellesley College in 1886 and graduated in 1890, in the same class as **Grace Andrews** and **Clara Latimer Bacon**. Bosworth spent the two years after her graduation as a high school teacher in Amesbury, Massachusetts. In 1892 she moved to Rhode Island College of Agricultural and Mechanical Arts (now University of Rhode Island) in Kingston as professor of mathematics, the same year that the college first offered instruction at the postsecondary level. During the summer quarters of 1894, 1896, and 1897, she studied at the University of Chicago.

In April 1898 Anne Bosworth was granted a year's leave from Rhode Island College for study in Göttingen. A 1978 letter from her granddaughter to one of the authors relates a family story about her time there. In spring 1899 Bosworth "was 'summoned' to tea" with Hilbert, whose lectures on Euclidean geometry she had attended, and was asked when she would take her doctoral exams. "She said she had not any such intention, had not even thought of a dissertation topic! [Hilbert] said, 'But your dissertation is finished'(!). It appeared she had done a special exercise for him, and it was considered an entirely original approach & acceptable as a thesis. So instead of spending her summer travelling in Italy, Greece, etc. she remained at Göttingen & took her exams & and passed with honor." Bosworth took her oral examination on July 31, 1899, just over a year after she had arrived in Germany, and, accompanied by her mother, returned to the United States in August 1899. The PhD from Göttingen was awarded in 1900. She became a member of the AMS at about that time.

Bosworth then returned to Rhode Island College and remained for the next two years, living with her mother in Kingston. While in Göttingen, however, she had met Theodore Moses Focke, who studied mathematics and physics there 1896–98 and received his PhD in 1898. On August 7, 1901, they were married in Kingston, Rhode Island. Theodore Focke was born on January 3, 1871, in Massillon, Ohio, and was an 1892 graduate of the Case School of Applied Science (later Case Institute of Technology, now Case Western Reserve University) in Cleveland. After his studies abroad, he returned to Case Institute of Technology, where he was an instructor of mathematics and civil engineering at the time of his marriage.

Anne Bosworth Focke left her position in Rhode Island and moved to Cleveland, where she did not teach but did help her husband with grading. They had three children, a daughter born in 1902 and sons born in 1904 and 1906.

On May 15 1907, Anne Bosworth Focke died at age thirty-eight of pneumonia, when her children were not quite five, three, and one. She was buried in Massillon, Ohio, her husband's birthplace.

FOWLER, Sister Mary Charlotte. August 26, 1899–August 26, 1997.

Catholic University of America (Catholic Sisters College) (BA 1927), Catholic University of America (MA 1935, PhD 1938). PhD diss. (1937): The discriminant of the sextic of double point parameters of the plane rational quartic curve (directed by Aubrey Edward Landry).

Josephine Fowler was born in Mechanicsville, Maryland, the eldest of eleven children of Charlotte (Burch) and Thomas Henry Fowler. Her father was a merchant and later a farmer. Josephine Fowler attended public school until 1916 and then attended St. Mary's Academy in Leonardtown, Maryland, 1916–18, where she was awarded the gold medal for mathematics when she graduated. On September 24, 1918, she entered the novitiate of the Congregation of the Sisters of Charity of Nazareth, in Nazareth, Kentucky, and took the religious name Sister Mary Charlotte Fowler.

Sister Mary Charlotte taught religion, mathematics, physics, English, and history primarily in high schools from 1920 to 1933. In the same period she was doing course work, including five summers at the University of Kentucky in Lexington and one summer at Notre Dame. Her first teaching assignment was at St. Vincent's Academy in Union County, Kentucky, 1920–21. Others were in Henderson, Kentucky, 1921–23 and in Yazoo City, Mississippi, 1923–26. She was in residence as a student at Catholic Sisters College of the Catholic University of America 1926–27 and received her bachelor's degree in 1927. She then taught in a high school in Mt. Vernon, Ohio, 1930–32 and in summer school at Nazareth Junior College in Kentucky in the summers 1927–30.

Sister Mary Charlotte was an assistant instructor in mathematics at Nazareth College in Nazareth, Kentucky, 1932–33 after which she was on leave for study at Catholic University. She was in residence at Catholic from 1933 until 1937 and earned her master's degree in 1935. When she left Catholic in 1937 she began teaching mathematics and physics at the Nazareth College campus in Louisville, Kentucky. In 1938 she received her PhD with minors in physics and education, having written a dissertation in algebraic geometry.

Sister Mary Charlotte Fowler was professor of mathematics and physics at Nazareth College in Louisville (now Spalding University) until her retirement in 1969. She was chairman of the department from 1950 until she was appointed president of the college in 1961. During her tenure as president, Fowler is credited with many positive changes and innovations. These included overseeing a period of growth of about fifty percent in the student population and the first capital campaign in the college's history. Graduate programs were added, and science, library, and dormitory buildings were constructed. The Fowler Fellowships, which allowed anyone sixty-five or older to take classes tuition-free, were created as was the Center for Continuing Education for the non-traditional student.

Sister Mary Charlotte was active in the community and was inducted into the Honorable Order of the Kentucky Colonels in 1967. Among hobbies she enjoyed were reading, crafts, photography, stamp collecting, collection of unusual recipes, and weaving, which continued well into her retirement. In the early 1970s she was living in St. Vincent's Home, an orphanage for boys, in Roanoke, Virginia. At some point, after the orphanage disbanded in 1975, she was living at the Motherhouse in Nazareth, Kentucky. Sister Mary Charlotte died on her ninety-eighth birthday in Louisville, Kentucky, in 1997.

FRINK, Aline (Huke). March 2, 1904–March 14, 2000.

Mount Holyoke College (BA 1924), University of Chicago (MA 1927, PhD 1930). PhD diss.: [Huke, A.] An historical and critical study of the fundamental lemma in the calculus of variations (directed by Gilbert Ames Bliss).

Aline Huke was born in Torrington, Connecticut, the elder of two daughters of Mary Evelyn (Feustel) and Allen Johnson Huke. It appears that her parents divorced soon after 1905. Her father had remarried by 1910 and was later in the building supply business. Her mother had studied languages for two years as a special student at Mount Holyoke College in South Hadley, Massachusetts, before her marriage. She returned to the area and in 1920 was a public school teacher in South Hadley Falls. Huke's sister died in 1927 while a graduate student in microbiology at Yale.

Huke received her elementary and secondary education in the public schools of South Hadley and graduated from the high school in 1920. She graduated magna cum laude from Mount Holyoke College in 1924. Huke taught physics and algebra 1924–26 at the high school in Cobleskill, New York, and in summer 1926 took courses in pedagogy at the New York College for Teachers in Albany (now SUNY at Albany). She began her graduate work at the University of Chicago in October 1926, remained through August 1927, and received a master's degree with a thesis in mathematical physics in 1927. She was a resident fellow in mathematics at Bryn Mawr College 1927–28 and returned to Chicago where she held a graduate service scholarship for four consecutive quarters. She was instructor of mathematics at Mount Holyoke 1929–30. She completed her dissertation in the calculus of variations that year and received her PhD from Chicago in 1930.

Aline Huke then went to Pennsylvania State College (now University) as an instructor of mathematics. On June 3, 1931, she married Orrin Frink, Jr., who had joined the mathematics faculty at Penn State in 1928 as an assistant professor. Frink, born in 1901 in Brooklyn, New York, had earned his BA in 1922, MA in 1923, and PhD in 1926, all from Columbia University. He was assistant professor 1928–29, associate professor 1929–33, and professor 1933–69, and was head of the department 1949–60. He was at the Institute for Advanced Study 1940–41 and was a Fulbright lecturer in Dublin, Ireland, 1960–61 and 1965–66. During the latter visit Aline Frink lectured for a month at University College, Dublin, filling in for the analysis lecturer who was ill.

Aline Frink later recalled that she was able to continue to teach after her marriage because of the efforts of **Helen B. Owens**, wife of F. W. Owens, then chairman of the Penn State mathematics department. However, she soon gave up her full-time position. Their three sons were born in 1932, 1939, and 1941, and their daughter was born in 1945. The Frinks' eldest son earned a PhD in Slavic languages and literature from Harvard and pursued an academic career; the second son received an MFA degree from Yale and became an architect; the third son was a mathematics major at Penn State and became a computer analyst; and their daughter received an MFA degree from the University of North Carolina at Greensboro and studied at the Slade School of Art in London.

While her children were young, Aline Frink published two mathematics papers. She taught part time 1936–37 and occasionally, some part time and some full time, through 1944 before resuming her career as a full-time faculty member in 1947. She was assistant professor until 1952, associate professor until 1962, and professor until her retirement in 1969 as professor emeritus. While at Penn State, she directed several master's theses and published her last research paper in 1960.

Having long had an interest in languages, Aline Frink began to study Russian in the late 1940s, and her translation from Russian of a book on the calculus of variations was published in 1962. She began the study of Chinese shortly before her retirement. After the Frinks retired they moved to a house in Kennebunkport, Maine, that had been in her family since 1747. She had a strong interest in music, bird watching, local history, and her grandchildren. Orrin Frink died in 1988. Aline H. Frink was ninety-six when she died in Kennebunk, Maine, in 2000. In 2006 her youngest son gave Penn State $100,000 in his parents' memory to create the Orrin and Aline Frink Trustee Matching Scholarship in the mathematics department.

FRY, Cleota G. December 30, 1910–July 1, 2001.
REED COLLEGE (BA 1933), PURDUE UNIVERSITY (MS 1936, PHD 1939). PhD diss.: Asymptotic developments of certain integral functions (directed by Howard Kibble Hughes).

Cleota Gage Fry was born in Shoshone, Idaho, the eldest of four children of Coral (Gage) and Holmes L. Fry, who was a machinist. Fry's early and secondary schooling was in Portland, Oregon, where she graduated from Roosevelt High School in 1929.

Cleota Fry attended Reed College, also in Portland, and graduated, probably as a physics major, in 1933. She then moved to Indiana, where her Reed College friend Vivian Annabelle Johnson was a graduate student in physics at Purdue University in West Lafayette. Fry at first sat in on mathematics and physics graduate courses while working part time on a research project to discover the properties that make a violin good; she also tutored. Fry earned a master's degree in physics in February 1936 and continued in that department as a graduate student until the following fall, when she switched to mathematics. In 1939 Fry received her doctorate in mathematics with a minor in physics, having written a dissertation in analysis. Her PhD was the second in mathematics given by Purdue, the first having been awarded in 1893.

Fry remained at Purdue for her entire career, teaching at various times both mathematics and physics. She was an assistant instructor of mathematics 1939–40 and an instructor of physics 1941–45. During the war years she taught elementary and intermediate physics and was in charge of the physics laboratory. She also organized intensive physics courses for high school students. The rest of her career at Purdue she was in the mathematics department: as instructor 1945–47, assistant professor 1947–55, and associate professor 1955–77. Her teaching consisted of mathematics courses from the freshman through the graduate level. In the period 1952–61 she was also assistant to the dean in the School of Science. Fry coauthored a college mathematics textbook that was published in 1952. She was secretary to the faculty of the School of Science, Education, and Humanities for at least three years in the early 1950s.

Fry and Vivian Johnson built a house together in West Lafayette. Johnson had earned a PhD in physics from Purdue in 1937 and became a member of the physics department there; she retired as professor in 1979 and died in 1985. She and Fry traveled every summer and also took a seven-month trip around the world. Cleota Fry died at ninety in 2001 in Lafayette, Indiana. There is a Cleota Gage Fry Scholarship, a need-based scholarship at Reed College, as the result of an annuity purchased by Fry.

GALVIN, Sister Catharine Francis. November 24, 1885–December 2, 1955.
LOYOLA UNIVERSITY OF CHICAGO (PHB 1924), UNIVERSITY OF NOTRE DAME (MS 1930), CATHOLIC UNIVERSITY OF AMERICA (PHD 1938). PhD diss.: Two geometrical representations of the symmetric correspondence $C_{n,n}$, with their interrelations (directed by Aubrey Edward Landry).

Joanna Galvin was born in Spalding, Nebraska, the daughter of Mary (Lawless) and Eugene Galvin, Nebraska pioneers, each of whom had emigrated from Ireland several years before their marriage. Her father was a farmer, and by 1900 there were five surviving children of six born.

Joanna Galvin taught in public schools in Nebraska before becoming a member of the Dominican Sisters of St. Catharine of Siena in Kentucky. She entered the order in March 1914, and was professed in August 1915. All of Sister Catharine Francis Galvin's ministry was in academies and colleges sponsored by the Dominican Sisters. For the first six years, until June 1921, she was a teacher at Immaculate Conception Academy in Hastings, Nebraska. She taught at Holy Rosary Academy in Louisville, Kentucky, 1921–22 and at St. Agnes Academy and College in Memphis, Tennessee, in 1922–24. During the summers 1918–21 she attended undergraduate courses at Loyola University in Chicago and received a PhB degree from Loyola in 1924 when she was thirty-eight. Sister Catharine Francis taught again at Immaculate Conception in Nebraska 1924–26 before returning to St. Agnes. While at St. Agnes she was professor in the mathematics department and was, for two years, dean of the college. She remained there until the fall of 1933.

Sister Catharine Francis Galvin continued her studies by taking mathematics classes at the University of Notre Dame during the summer sessions 1926–28 and in 1930. She received a master's degree from Notre Dame in August 1930, graduating magna cum laude. In 1933 she entered the Catholic University of America for further graduate work. She was in residence at Catholic 1933–35 and 1936–37. During 1935–36 she returned to Holy Rosary Academy in Louisville as a high school teacher. Although she returned to St. Agnes College in Memphis in 1937, she also was a registered student at Catholic 1937–38. She received her PhD in 1938, with minors in chemistry and physics, and remained at St. Agnes College (renamed Siena College in 1939) except for a year at Holy Rosary Academy 1945–46. Over the years her teaching assignments included chemistry, physics, and statistics, as well as mathematics. Sister Catharine Francis was also bursar at the academy and the college in Memphis and served as the superior of her community for many years.

Sister Catharine Francis Galvin died in 1955 at age seventy in Memphis, Tennessee. She was buried in the cemetery at the motherhouse of the Dominican Sisters of St. Catharine of Siena in Kentucky.

GARVIN, Sister Mary Cleophas. December 11, 1899–January 16, 1990.

Fordham University (BS 1927), Saint Louis University (MS 1931, PhD 1934). PhD diss.: On the convergence of a generalized series and the relation of its coefficients to those of the corresponding power series (directed by Francis Regan).

Sister Mary Cleophas Garvin was born Linetta Anna Garvin in Vickery, Ohio, the daughter of Odelia Margaret (Wilhelm) and Austin Edward Garvin. Her father was a meat dealer in 1900, a salesman in a meat market in 1910, and foreman in an auto factory in 1920. She had one older sister and four younger brothers. Garvin attended grade school at St. Ann's and high school at Notre Dame Academy, both in Toledo, Ohio. She graduated from high school in May 1917 and entered the Congregation of the Sisters of Notre Dame in Cleveland, Ohio, on July 2, 1917. Her investment was on January 2, 1918, and her final vows five years later. Her religious name, Mary Cleophas, was chosen in memory of her youngest brother who died at age four when she was eleven.

Garvin was a full-time instructor in mathematics and physics, as well as religion, history, and Latin, at Notre Dame High School in Cleveland 1919–29 and a part-time student completing her undergraduate work in science, mathematics, and education 1919–27. She had extension courses at Notre Dame College and John Carroll University in Cleveland and summer courses at St. Xavier College (now Xavier University) in Cincinnati, Catholic Sisters College of the Catholic University of America, University of Notre Dame, and Fordham University, which awarded her a bachelor's degree in 1927. In the summers of 1928 and 1929 she had courses in French and education at Fordham and in education at Columbia University. In autumn 1929 she entered Saint Louis University for full-time graduate study. She spent the next five years there and earned a master's degree in 1931 and a PhD in 1934 with studies in mathematics, physics, and astronomy. She was the second person to receive a PhD in mathematics from Saint Louis University.

Immediately after receiving her doctorate, Sister Mary Cleophas returned to the Cleveland area to serve as professor of mathematics and physics and head of the science department at Notre Dame College, then primarily for women. Garvin remained as department head until her retirement in about 1976. She taught until 1975, was chairman of the computer center 1975–76, and was associated with the computer center until 1977. She was also superior at the college 1947–53 and 1958–60.

Sister Mary Cleophas continued to take or audit courses. These included courses in religion and physics at John Carroll University and courses in mathematics the summers of 1959 through 1962 at Oklahoma State University, the University of Kansas, and Bowdoin College. She audited other courses at the University of Oklahoma and at Seton Hall University. Among her contributions at Notre Dame College were special classes for parents to learn the "new math" with their children. In connection with her interest in computer science, in the late 1960s and early 1970s Sister Mary Cleophas ran grant-supported computer-related workshops and institutes, especially for teachers. The Garvin Computing Center at Notre Dame College is named after her.

Garvin's main intellectual interest outside of mathematics was music, and she attended concerts of the Cleveland Orchestra whenever possible. She was also concerned with, and tried to help, anyone who was needy: students who needed financial aid, priests who had left the ministry, the poor. She organized the distribution of food, clothing, and other necessities. When she retired she founded the Cleveland Co-workers of Mother Teresa, a group that would continue her assistance to the poor. Sister Mary Cleophas received the Vincent P. Haas Memorial Award for "putting Gospel values in action" and the Mayor's Award for volunteerism in 1985 After her retirement Sister Mary Cleophas remained at the college as archivist before moving to the Provincial House in Chardon, Ohio, in 1985. She died there in 1990 shortly after a celebration of her ninetieth birthday.

GENTRY, Ruth. February 22, 1862–October 18, 1917.
UNIVERSITY OF MICHIGAN (PHB 1890), BRYN MAWR COLLEGE (PHD 1894). PhD diss.: On the forms of plane quartic curves (directed by Charlotte Angas Scott).

Ruth Ellen Gentry was born in Stilesville, Indiana, the daughter of Lucretia and Jeremiah Gentry. Her father was a farmer and stock trader and, later, dry goods merchant. She had at least two older siblings, a brother and a sister. She attended the public schools of Stilesville and began teaching school at age sixteen.

Gentry graduated from the Indiana State Normal School in Terre Haute in 1880. For the next ten years she taught in preparatory schools and colleges and completed her undergraduate education. She was admitted to the University of Michigan in 1885. She taught in Deland, Florida, 1886–88. The school was called DeLand Academy and College her first year, Deland University her second year, and John B. Stetson University after she left. During 1886–87 college preparatory classes were the highest level taught, but the following year three students were enrolled in the college department. Gentry then returned to the University of Michigan and graduated with the PhB degree in 1890.

That fall Ruth Gentry began her graduate studies at Bryn Mawr College and held the fellowship in mathematics in 1890–91. She was awarded the European fellowship of the Association of Collegiate Alumnae for the year 1891–92. Gentry first went to Berlin and from there requested admission to the lectures of several prominent European mathematicians. She was refused everywhere except Leipzig, where the work being done was not related to her mathematical interests. However, with the backing of Lazarus Fuchs and the rector of the university she was permitted to attend lectures at the University of Berlin. Nonetheless, upon the discovery of an 1884 ruling that prohibited exceptions to the exclusion of women, Gentry was permitted to attend classes only through the end of the semester. She remained in Europe through the first semester of 1892–93 and attended lectures at the Sorbonne before returning to the United States in February 1893.

Gentry again held the fellowship in mathematics at Bryn Mawr in 1892–93 and was a fellow by courtesy there in 1893–94. She passed her examinations for the PhD degree in June 1894 and was Charlotte A. Scott's first PhD student. Because of a two-year delay in the printing of her dissertation, she is often listed as sharing that honor with **Isabel Maddison** who received her degree in 1896.

Gentry began teaching at Vassar College in 1894. She was instructor 1894–1900 and associate professor 1900–02. In 1901–02 she played cornet in the Vassar symphony orchestra. She left Vassar in 1902 and took a position as associate principal and head teacher of mathematics at Miss Gleim's School for Girls, a private school in Pittsburgh, where she remained until 1905. She apparently had been ill when she left Vassar and became less well as the years went by.

In December 1905 Gentry reported to the Michigan alumni office that her address was Stilesville, and in December 1909 and October 1910 she reported an Indianapolis address. In 1910 she also gave her occupation as unsalaried nursing. A later note indicated that she was doing charity work in Stilesville. From 1911 to 1914 she travelled in the United States and Europe and arrived back in the United States from Europe in February 1914. In August 1916 she reported that she was living in Indianapolis. Ruth Gentry died at age fifty-five in Indianapolis in October 1917, having had breast cancer. She was buried in the family plot in Stilesville.

GIBBENS, Gladys. January 21, 1893–September 24, 1983.

H. Sophie Newcomb College (BA 1914), Tulane University (MA 1916), University of Chicago (PhD 1920). PhD diss.: A comparison of different line-geometric representations for functions of a complex variable (directed by Ernest Julius Wilczynski).

Gladys Elizabeth Corson Gibbens was born in New Orleans, Louisiana, the eldest of three children of Belle Frances (Arbour) and William Joseph Gibbens. Her mother attended a private academy before her marriage, and her father attended Louisiana State University and became an engineer who worked in a variety of positions in sugar mills. Gladys Gibbens had a younger sister and a younger brother.

Gladys E. C. Gibbens attended private kindergarten and public grade school, and was a boarding student at Ursuline Academy, a private, Roman Catholic school for girls in New Orleans. After graduating from Ursuline Academy in 1910, she entered H. Sophie Newcomb College, a degree-granting, coordinate college for women in Tulane University in New Orleans. She was a member of Kappa Alpha Theta social sorority.

Gibbens graduated in 1914, and from 1914 to 1917 she held a teaching fellowship in mathematics at Newcomb College while she was studying at Tulane University. She received her master's degree in 1916, one of five master's degrees granted by Tulane at that June commencement. In 1917 Gibbens moved to Chicago where she studied for nine quarters at the University of Chicago during the next three years. In 1918–20 she held a fellowship in the mathematics department.

After receiving her PhD from the University of Chicago in 1920, Gibbens took a position as an instructor at the University of Minnesota. She remained there throughout her career: as instructor 1920–25, assistant professor 1925–47, associate professor 1947–58, and associate professor emeritus after her retirement in 1958. Her slightly younger colleague **Elizabeth Carlson** had a similar employment history there. The two also shared an office for many years and co-taught a course in the summers of 1935 and 1936.

Gibbens taught elementary and advanced undergraduate courses in algebra, geometry, and analysis, and several dealing with the mathematics of finance. She was usually course chairman for at least one course every quarter. She was long associated with Raymond W. Brink, department chairman after 1940, in a course in the field of general education. She generally taught in one summer session each year; during the summers of 1943 and 1944 she taught in the War Training Program. She also served as acting chairman of the department for several weeks each summer from the mid-1940s until near the time of her retirement.

While at the University of Minnesota, Gibbens regularly attended meetings of the Minnesota Section of the MAA. At her first meeting in June 1921 she was elected to the executive committee of the section. The following year she served as acting secretary and was elected to the executive committee two more times during the 1920s.

Gibbens returned to New Orleans after her retirement in 1958. In the late 1930s she indicated that her hobbies were gardening and music and that her favorite recreations were golf and boating. In 1982 she wrote that she had worked as a volunteer teacher in a hospital for crippled children. Gibbens died at age ninety at Tulane Medical Center in New Orleans after a long illness.

GLASGOW, Josephine (Burns). July 22, 1887–January 22, 1969.
UNIVERSITY OF ILLINOIS (BA 1909, MA 1911, PHD 1913). PhD diss.: [Burns, J. E.] The abstract definitions of the groups of degree 8 (directed by George Abram Miller).

Josephine Elizabeth Burns was born in Greenville, Illinois, the daughter of Ida Jane (Carey) and James Clinton Burns. Her father graduated from Monmouth College, had a master's degree, and was later a professor of history at the Western Illinois State Normal School (now Western Illinois University) at Macomb. Her parents, who were college classmates, were married in 1878, and Josephine Burns was the third of four children.

Josephine Burns prepared in the Monmouth and Macomb high schools and attended Western Illinois State Normal School 1904–06, before it granted degrees. As an undergraduate at the University of Illinois she was a member of the Illiola Literary Society for women and the education honor society Kappa Delta Pi. Burns graduated in 1909, having earned Final Honors.

Josephine Burns remained at Illinois as a graduate student and held a mathematics scholarship in 1910–11. After earning her master's degree in 1911, she held a fellowship in mathematics at the University of Wisconsin 1911–12 and at the University of Illinois 1912–13. She received her PhD from Illinois in 1913 with a dissertation directed by G. A. Miller. Her work in group theory for her master's thesis and her doctoral dissertation resulted in a historical paper in the *Monthly* and a paper in the *American Journal of Mathematics*. Her critical reading of the page proofs of Miller's contribution to Miller, Blichfeldt, and Dickson's 1916 *Theory and Applications of Finite Groups* was acknowledged in the preface of that book.

Burns remained at Illinois as instructor 1913–15, 1918–20, first semester 1924–25, and 1925–26. On June 15, 1916, in Macomb, she married Robert Douglass Glasgow. Glasgow, born in Tennessee, Illinois, in 1879, had earned a BA in 1908 and a PhD in 1913 in entomology at the University of Illinois. He also remained at Illinois after obtaining his doctorate, as instructor 1913–24 and as associate 1924–28.

While the Glasgows were still at Illinois, Josephine B. Glasgow and **Mary Gertrude Haseman**, who was instructor there 1920–28, prepared a mimeographed freshman mathematics text. This work is likely to have formed the basis for the paper they read in 1926 to the Illinois Section of the MAA.

In 1928 Robert D. Glasgow took a position as state entomologist for New York, so they moved to Albany where they remained. Also in 1928 Josephine Burns Glasgow served as national secretary of Sigma Delta Epsilon and represented that society in the council of the AAAS. While in Albany she was active in the AAUW and in 1932 represented the state AAUW when the state Women's Joint Legislative Forum was organized. She became secretary-treasurer of the state AAUW in 1940, vice president of the North Atlantic region in 1947, and was appointed to the national board of directors as vice president in 1952. In 1951 the Albany branch named a fellowship in her honor.

The Glasgows had no children. Robert Glasgow retired in 1949, having reached the mandatory retirement age of seventy. He died in Albany in 1964. Josephine B. Glasgow died in or near Albany at age eighty-one in January 1969.

GOUGH, Sister Mary de Lellis. February 15, 1892–April 7, 1983.
CATHOLIC UNIVERSITY OF AMERICA (CATHOLIC SISTERS COLLEGE) (BA 1920, MA 1923), CATHOLIC UNIVERSITY OF AMERICA (PHD 1931). PhD diss.: On the condition for the existence of triangles

in-and-circumscribed to certain types of the rational quartic curve and having a common side (directed by Aubrey Edward Landry).

Margaret Gough was born in Kilmore, County Wexford, Ireland, the daughter of Ellen (Dunne) and Walter Gough. She attended convent schools in Wexford for her elementary and intermediate education. She arrived in Galveston, Texas, on a ship from Liverpool in 1909, as one of a group of novices from Ireland. She immediately entered the Congregation of the Sisters of Charity of the Incarnate Word in San Antonio, Texas, and took the religious name Mary de Lellis. Her profession of vows was in 1911. She continued her studies at Incarnate Word College, at that time a women's college and now the coeducational University of the Incarnate Word, in San Antonio.

Early in her teaching career Sister Mary de Lellis taught in Catholic elementary schools in Texas and in Missouri. She also taught mathematics in high school in St. Mary's Academy in Amarillo, Texas. She earned her bachelor's degree from the Catholic University of America in Washington, D.C., in 1920. She then began her college teaching of mathematics at Incarnate Word College and remained there for more than twenty years except for years when she was doing graduate work in mathematics. She first taught there until the second semester of 1921–22, when she was a graduate student at the University of Oklahoma. She then returned to Catholic University in 1922–23 and received her master's degree in 1923 before returning to Incarnate Word College until 1928. She studied at the University of Texas in the summer 1927.

Sister Mary de Lellis had a leave from Incarnate Word beginning in 1928 for further graduate study at Catholic University where she was in residence until 1931. She was one of four women religious who received PhD's in mathematics from Catholic in 1931. After she received her PhD with minors in education and physics, Sister Mary de Lellis returned to Incarnate Word College to teach mathematics until 1943. In 1943–44 she was a mathematics teacher at Incarnate Word Academy in St. Louis, Missouri. After she became ill in 1944, she changed from teaching to working in the finance office at St. Joseph's Hospital in Fort Worth, Texas, where she was chief accountant and treasurer for the next twenty years.

In 1964 Sister Mary de Lellis moved to a home for retired sisters in San Antonio. She remained until her death there at age ninety-one in 1983.

GRANT, Anna M. C. June 22, 1903–February 23, 1984.

Dalhousie University (BA 1925), Bryn Mawr College (PhD 1937). PhD diss.: Asymptotic transitivity on surfaces of variable negative curvature (directed by Gustav Arnold Hedlund).

Anna Margaret Catherine Grant was the youngest of five children of Jessie (Gordon) and Angus MacGillivray Grant, both of Nova Scotia, Canada; she was born in Springville, Pictou County, Nova Scotia. Both parents had attended country school, and her father was a foreman in weaving mills in Nova Scotia and in Providence, Rhode Island. Her two sisters received bachelor's degrees from Dalhousie University; her eldest sister graduated in 1911, with great distinction in mathematics, and became the first woman actuary in the department of finance with the Canadian federal government. Her other sister became an accountant, and her brothers were an electrical engineer and a businessman.

Anna M. C. Grant attended public schools in Nova Scotia and in New Brunswick, Canada, before entering Dalhousie University. She graduated in 1925 with great distinction in mathematics, English, and philosophy. After her graduation she taught in private schools for several years. They included St. Hilda's School, Calgary, Alberta, 1926–27; Kemper Hall, Kenosha, Wisconsin, 1927–29; and Miss Fine's School, Princeton, New Jersey, 1929–31.

Grant was a part-time graduate student in mathematics and physics at Bryn Mawr College 1931–34. She was also a full-time teacher at Miss Wright's School in Bryn Mawr starting in 1932 and continuing through 1934–35. She then returned to Bryn Mawr College as a full-time graduate student; she was a graduate scholar in mathematics 1935–36 and a graduate fellow in mathematics 1936–37. She also participated in a cooperative plan for graduate study with Swarthmore College and the University of Pennsylvania. She received her PhD in 1937, the same year as **Annita Tuller**, G. A. Hedlund's other student at Bryn Mawr. At some point, she took, but not for credit, mathematics courses at the University of Chicago.

All of Grant's postdoctoral positions were in the New York City area, most often on Long Island, where she lived much of the time after she was a student at Bryn Mawr. It appears that her first position began in 1938 when she taught mathematics from March to June at the Kent Place School in Summit, New Jersey. She was a mathematics teacher at the Brearley School in Manhattan 1939–42. From 1943 to 1945 she was a member of the technical staff for Bell Telephone Laboratories in New York, and from 1946 to 1949 she was a project engineer for Project Sparrow for Sperry Gyroscope Company in Lake Success, New York.

Grant returned to the Brearley School as mathematics teacher and head of the department 1949–51 and 1953–55. During this period she also took both undergraduate and graduate courses in electrical engineering at Brooklyn Polytechnic Institute (now Polytechnic Institute of New York) in order to compete with men in industry, where she had felt discrimination because of her sex.

From 1956 to 1959 Grant was principal engineer with the Dynamics Department at Republic Aviation Corporation in Farmingdale, New York. From 1959 to 1961 she was senior engineer with Fairchild Stratos Corporation in Wyandanch, New York, and from 1962 to 1967 she was an engineer working in dynamic analysis for the Airborne Instruments Laboratory in Deer Park, New York. When she retired she was a senior research engineer doing trouble-shooting on guided missiles.

In addition to her work, Grant was interested in all types of sewing, gardening, cooking, baking, and writing. She was also knowledgeable about eighteenth-century porcelains. In 1970 she indicated that she was a Democrat and a Protestant. She made her home in Medford, New Jersey, after her retirement. She died there in 1984 at age seventy-nine.

GRAUSTEIN, Mary F. (Curtis). April 12, 1884–July 18, 1972.

WELLESLEY COLLEGE (BA 1906), RADCLIFFE COLLEGE (MA 1915, PHD 1917). PhD diss.: [Curtis, M. F.] Curves invariant under point transformations of special type.

Mary Florence Curtis was born and grew up on a farm in Westminster, Massachusetts, the eldest of five children of Jennie Esther (Lucas) and Frank Abbott Curtis. She attended nearby Fitchburg High School 1899–1902.

Curtis entered Wellesley College in 1902, was an honors scholar 1904–06, and received her bachelor's degree in 1906. She then taught near her family's home: German, algebra, and geometry at Leominster High School 1906–08 and German and natural science at Cushing Academy in Ashburnham 1908–10. She studied botany and pedagogy at Cornell University in summer 1909 and mathematics and natural science at Universität Leipzig from May 1910 until August 1911.

Upon her return from Germany, Curtis joined the Wellesley College mathematics faculty. She was instructor 1911–14 and began her graduate studies in mathematics at Radcliffe College in September 1913. At Radcliffe she held a Mary E. Horton fellowship from Wellesley 1914–15. She earned her master's degree in 1915 and her doctorate in 1917. It is likely that she worked with Julian L. Coolidge, one of two faculty reviewers of her dissertation in differential geometry. Curtis was instructor and acting head 1917–18 at the College for Women, Western Reserve University, in Cleveland, Ohio, replacing a faculty member who was on leave. She returned to Wellesley, where she was instructor 1918–20 and assistant professor 1920–21. She was in Leipzig summer 1920. Curtis published five research papers; the first in 1918 was probably based on one of her two 1916 minor theses; the last in 1922 was based on her 1917 dissertation.

On June 10, 1921, Mary F. Curtis married William Caspar Graustein in Wellesley. Graustein, born in 1888 in Cambridge, Massachusetts, had earned both his BA and his MA from Harvard and his PhD in mathematics from Bonn in 1913. He was an instructor at Harvard 1913–14 and taught at Rice Institute in Houston 1914–18. In 1919, after serving with the Ordnance Department of the US Army, Graustein returned to Harvard. After her marriage, Mary C. Graustein took a leave of absence, returned to Wellesley in 1923 as assistant professor, and taught there until 1928. She was also a tutor at Radcliffe 1926–41, except for 1928–29 and 1937–38 when the Grausteins were in Europe. In addition to carrying out her Radcliffe tutorial duties, Mary Graustein assisted her husband, which included proof reading his book manuscript and correcting papers. The Grausteins had no children and spent every other summer in the Dolomites until World War II intervened.

William C. Graustein was professor of mathematics and assistant dean at Harvard when he was killed in an automobile accident in January 1941. Mary Graustein resumed teaching as instructor at Hunter College in the summer session 1941 and then held both substitute and part-time positions at Connecticut College and Abbott Academy in October and November 1941 and at Hunter College for the rest of 1941–42. She was assistant professor at Oberlin College 1942–44 but had a leave 1943–44 to care for her mother who was ill. She began teaching wartime navigation and calculus at Tufts College (now University) in 1944 and stayed on to teach veterans after the war. Graustein was promoted to associate professor at Tufts in 1950 and remained until her retirement in 1955. From 1946 until 1954, she was also faculty resident of a women's dormitory at Tufts.

After her retirement Graustein lived at her family home in Westminster. She continued to travel and to go to Cambridge for the symphony concert series. She described herself in the early 1940s as a Congregationalist and a Republican. Mary Curtis Graustein died at eighty-eight in Gardner, Massachusetts, in 1972 and was buried in Mount Auburn Cemetery in Cambridge.

GRAY, Alta (Odoms). October 1, 1910–December 30, 2001.

University of Cincinnati (BA 1931, MA 1932, PhD 1936). PhD diss.: [Odoms, A. H.] On the summability of double Fourier series (directed by Charles Napoleon Moore).

Alta Harriet Odoms was born in Fredonia, Indiana, the third of four children of Ethel (Nelson) and John Bert Odoms. In 1910 her father was a farmer in Crawford County, Ohio; in the 1920s and 1930s he was a plasterer, and the family was living in Clermont County, east of Cincinnati, Ohio.

After Alta Odoms graduated from high school in Milford, Ohio, all of her college education was completed at the nearby University of Cincinnati. Immediately after she received her BA in 1931, Odoms continued her graduate work at the University of Cincinnati and received her master's degree in 1932 and her PhD in 1936.

From 1935 to 1937 Odoms was in New York working as an editorial assistant for the American Mathematical Society. While in New York she was selected to sing with the New York City Oratorio Society. She resigned from the AMS job after two years and looked for a teaching position. In a letter to one of the authors in August 1992, Odoms wrote that Duke offered her part-time teaching in the Woman's College if she would work for the *Duke Mathematical Journal.* The teaching part of the job just lasted for a year after which she worked as secretary for the *Journal.*

On December 10, 1938, Alta Odoms married William Frederick Gray. Gray, born in Charleston, South Carolina, in about 1913, and Odoms met in Durham, North Carolina. Interested in a career in government and foreign service, William Gray entered the Fletcher School in Medford, Massachusetts, in 1941. This move caused Alta Gray to resign from her position at the *Duke Mathematical Journal*, but she found work as secretary of the radiation laboratory at MIT. She described herself as "useful as jack-of-all-trades" in the 1992 letter.

The Grays went to Washington, D.C., in 1942, where William Gray worked for the Board of Economic Warfare. Alta Gray "spent the next ten years producing and enjoying three children," according to her 1992 letter. In 1952 William Gray joined the State Department Foreign Service and served in countries in Europe and Latin America. Between assignments and after his retirement in 1964, the Grays lived in the Washington, D.C., area.

Alta Gray was interested in politics, history, and nineteenth-century literature. She wrote in 1992, "I have had a wonderful life.... Forgive me for having packed away all those mathematics books long years ago."

William Gray died in Charlottesville, Virginia, in 1986. Alta Gray continued to live in Charlottesville and was a long-time resident of a retirement community. In December 2001, during a holiday visit to a daughter in the Mechanicsville, Virginia, area, she became ill and died, at age ninety-one, at a local hospital.

GRAY, Marion C. March 26, 1902–September 16, 1979.

University of Edinburgh (MA 1922), Bryn Mawr College (PhD 1926). PhD diss.: A boundary value problem of ordinary self-adjoint differential equations with singularities (directed by **Anna Pell Wheeler**).

Marion Cameron Gray was born in Ayr, Scotland, the daughter of Marion (Cameron) and James Gray. She attended Ayr Grammar School (1907–13) and Ayr Academy (1913–19) before matriculating at the University of Edinburgh in 1919.

In 1922 Gray received her master's degree, the first degree given in universities in Scotland at that time, with first class honors in mathematics and natural philosophy. She remained at Edinburgh another two years as a postgraduate student in mathematics.

Gray came to the United States in 1924 to study at Bryn Mawr College. At Bryn Mawr she held a British graduate scholarship and a Carnegie research scholarship. She received her PhD in 1926, having written a dissertation in analysis and with allied subject physics. After receiving her doctorate, Gray returned to Great Britain, where she was a university assistant in natural philosophy at the the University of Edinburgh for one year and an assistant in mathematics at the Imperial College of Science in London for three years.

In 1930 Gray returned to the United States and was hired as an assistant engineer in the development and research department of American Telephone and Telegraph Company. She published two papers relevant to her work in the journal *Physics*. During this period she also discovered what has become known as the "Gray graph." The graph was rediscovered by I. Z. Bouwer and described in his 1968 paper "An edge but not vertex transitive cubic graph" in the *Canadian Mathematical Bulletin* (1:533–35). In a 1969 letter to Gray that Bouwer shared with the authors, he noted that her discovery of the graph was made "at a time when graph theory was almost nonexistent."

In 1934 Gray joined the technical staff of Bell Telephone Laboratories, where she remained for more than thirty years, first in New York City and later in Murray Hill, New Jersey. While there, she published several more articles and contributed to the field of mathematical physics in other ways as well. In addition to writing book reviews in journals she wrote 258 reviews in the first fourteen volumes of *Mathematical Reviews*. One hundred forty of her reviews were classified as "optics, electromagnetic theory, circuits" while another 52 were classified as "special functions." In 1954 Gray served on an ad hoc committee formed when the National Science Foundation asked the National Bureau of Standards to prepare a handbook of mathematical tables. The outcome of that project was the *Handbook of Mathematical Functions with Formulas, Graphs, and Mathematical Tables* published in 1964.

Gray became a naturalized United States citizen in 1937. After her retirement in 1967, Marion C. Gray returned to Edinburgh, where she died at age seventy-seven in 1979.

GREENFIELD, Bella (Manel). October 13, 1915–.
HUNTER COLLEGE (BA 1935), COLUMBIA UNIVERSITY (MA 1936), NEW YORK UNIVERSITY (PHD 1939). PhD diss.: [Manel, B.] Conformal mapping of multiply connected domains on the basis of Plateau's problem (directed by Richard Courant).

Bella Manel was born in New York City, the only child of Tillie and Jacob Manel. Both her parents were born in Poland, and they were married there before immigrating to the United States in 1914. In 1930 the family was living in the Bronx, and her father worked for a dress manufacturer.

Manel graduated from Morris High School in the Bronx in 1931 and entered Hunter College in the fall. She was elected to Phi Beta Kappa her junior year and graduated summa cum laude in 1935. Manel then studied at Columbia University and received her master's degree in 1936 with a thesis directed by Joseph Fels Ritt. Manel then began her studies at New York University and served as receptionist and secretary for Richard Courant until she received a Blumenthal fellowship in 1937. She held the fellowship for two years and received her PhD in 1939 with a physics minor.

On March 6, 1938, Bella Manel married Max Shiffman (1914–2000), a 1938 doctoral student of Courant. Shiffman was born in New York City and had earned a bachelor's degree and a master's degree from City College of New York. He was an instructor at City College in the evening division 1938–39 and then on the regular faculty until 1942. During this period Bella Shiffman, using the name Bella Manel, received her PhD and published two papers. She also gave birth to their first son. In 1942 Max Shiffman became a research mathematician mainly working for the Applied Mathematics Panel of the Office of Scientific Research and Development at NYU. The Shiffmans' second son was born in 1945. Max Shiffman was appointed associate professor at NYU in 1946 and was hired as full professor at Stanford University in 1948. According to a 2003 obituary, he suffered two schizophrenic breakdowns, the first in 1951 and the second in 1956.

In 1954 Bella Shiffman joined the Ramo-Wooldridge Corporation (now TRW Corp.) in Hawthorne, California, south of Los Angeles, where she first supervised a group that performed calculations on desk calculators and later worked in the machine language of digital computers. She and Max Shiffman were divorced in February 1957. On June 8, 1958, she married Emanuel Ernest Kotkin, a widower. Kotkin, who was born in 1897 in London, England, had immigrated to the United States in his late teens and was the president of a manufacturing plant at the time of their marriage. Bella Kotkin worked for the RAND Corporation in Santa Monica, California, 1959–65, first as mathematician and later as consultant. She worked closely with Richard Bellman, and they, with others, wrote papers and technical reports on mathematical and computer models for medical research.

Bella Kotkin left the RAND Corporation in 1965 and then volunteered in the field of human relations and spent time with her ailing husband and mother. She and Emanuel Kotkin later moved to northern California where he died in May 1981. During 1980–81 Bella Kotkin was professor of mathematics and acting chairman of the department at the College of Notre Dame in Belmont, north of Palo Alto; she returned to Los Angeles in 1981. Kotkin taught part time in the mathematics department at the University of California, Los Angeles, 1982–84. On September 17, 1984, she married Moses A. Greenfield, a medical physicist whose wife had died two years earlier. He was born in 1915 in New York City, earned a bachelor's degree from the City College of New York, and a master's degree and a PhD from NYU; he retired from the radiological sciences department of the medical school at UCLA in 1982. Since her own retirement, Bella Greenfield has devoted herself to her family and to music as a classical pianist.

In 1995 Moses Greenfield established the Bella Manel Prize for outstanding graduate work by a woman or other member of an under-represented group at the Courant Institute at New York University. In 2008 Bella and Moses Greenfield were living in the Los Angeles area.

GRENNAN, Elizabeth (Bennett). October 9, 1880–October 15, 1972.

Ohio University (BS 1903), University of Illinois (MA 1908, PhD 1910). PhD diss.: [Bennett, E. R.] Primitive groups with a determination of the primitive groups of degree 20 (directed by George Abram Miller).

Elizabeth Ruth Bennett, named Lizzie Ruth at birth, was the daughter of Emmeline (Loux) and Daniel Tinsman Bennett. She was born in Shawnee, in Monroe County, in east central Pennsylvania, the second of seven children. Her father was described as a miller in 1880 and 1900 and as proprietor of a livery stable in 1910. The other children in the family were four daughters and two sons.

Elizabeth Bennett graduated from the State Normal School in nearby East Stroudsburg, Pennsylvania, in 1898. Her occupation is listed as teacher (unemployed four months) in 1900. She attended Ohio University in Athens where she received her BA in 1903. She held a scholarship in mathematics at the University of Illinois 1907–08 and a fellowship 1908–10. She received a master's degree in 1908 and her PhD in 1910. She published two papers in the *Monthly* before receiving her doctorate, which was the second granted by the department and the first to a woman.

In 1910 Bennett went to the University of Nebraska in Lincoln as an instructor. During her first four years there she published two papers that appeared in the *American Journal of Mathematics* and made four presentations at AMS meetings.

Elizabeth Bennett and John Grennan, a faculty member in mechanical engineering at Nebraska, married in Lincoln on June 12, 1913. John Grennan was born in Michigan in 1881. In 1910 he was working as a moulder in a foundry in Ann Arbor, Michigan. His postsecondary education included a summer session at the University of Michigan in 1911, after which he was appointed instructor in forge and foundry work at the University of Nebraska. He continued to attend summer sessions at Michigan in 1912 and 1913 and at Nebraska in 1915.

The Grennans remained on the faculty at Nebraska until 1916, when John Grennan took a position as foreman of the forge shop at Michigan Agricultural College (now Michigan State University) in East Lansing. After two years they both moved to the University of Illinois. Elizabeth Grennan was appointed instructor of mathematics in September 1918, apparently on a term by term basis. John Grennan was appointed instructor in foundry practice and management in the College of Engineering in July 1918. He was then appointed assistant superintendent of the foundry in July 1919 and was reappointed in about March 1920. Both Grennans left Illinois in 1920, and John Grennan went to Camp Grant, an army training facility near Rockford, Illinois, as an instructor of foundry.

In 1922 John Grennan became an instructor of foundry practice in the metal processing department at the University of Michigan and retired in June 1947 as instructor emeritus of foundry practice. Elizabeth and John Grennan had no children and remained in Ann Arbor after his retirement. A John Grennan scholarship for studies related to foundry practice or technology in the materials science and engineering department was established as the result of a $20,000 bequest to the University of Michigan by Elizabeth B. Grennan.

John Grennan died in November 1964; Elizabeth B. Grennan died in Ann Arbor in October 1972, shortly after her ninety-second birthday. She also made a bequest to the University of Illinois that funds an Elizabeth R. Bennett scholarship in mathematics.

GRIFFIN, Harriet. April 6, 1903–January 13, 1991.

Hunter College (BA 1925), Columbia University (MA 1929), New York University (PhD 1939). PhD diss.: The abelian quasi-group (directed by Donald Alexander Flanders).

Harriet Madeline Griffin was born in Brooklyn, New York, the first of two daughters of Mary Madeline (Gully) and James Harry Griffin. Her mother had an elementary school education and her father received a high school education; he was a salesman for a furniture store in 1910 and a buyer for a furniture house in 1920. Her sister became a high school mathematics teacher after graduating from Hunter the same year as Harriet. Harriet Griffin recalled in 1985 that her father liked mathematics and encouraged her involvement in the subject.

Harriet Griffin attended Hunter College and won the Thomas Hunter Prize in Mathematics when she graduated in 1925. She was one of the charter members of Pi Mu Epsilon when it was established at Hunter in 1925 and served as the chapter's first vice-director; in the late 1950s she was one of the councilors-general at the national level. Griffin remained at Hunter as a tutor 1926–29 and as an instructor 1929–30. She earned her master's degree from Columbia University in 1929, having written a master's essay, "Modern Geometry in Three Dimensions."

In 1930 Griffin moved to a position at Brooklyn College where she remained most of the rest of her career. While there she earned her PhD in 1939 at New York University with a dissertation in abstract algebra. Griffin's first position at Brooklyn College was instructor in the women's division, but starting in 1932 the faculty was no longer divided between the men's and women's divisions. Griffin remained an instructor until 1940; she was assistant professor 1940–50, associate professor 1950–56, professor 1956–66, and professor emeritus after her retirement in 1966. She is listed on the faculty of the graduate division starting in 1946–47. She used her own course notes, "The concepts of the theory of numbers," printed in 1947 and revised in 1949, for number theory and "Systems of abstract algebra," printed in 1962, for abstract algebra. In 1954 she published a textbook in number theory that appeared in both hardback and paperback editions. From 1956 to 1964 she was advisor for a number of master's theses in number theory. After Griffin's retirement from Brooklyn College she taught two years, 1966–68, at Molloy College in Rockville Centre, New York.

During World War II Griffin taught mathematics and did vocational counseling at the United Service Organization in New York. She was a Roman Catholic and was a member of the Albertus Magnus Guild, an organization for Catholics in science. Harriet Griffin moved to Lakewood, New Jersey, after her retirement and died there in 1991 at age eighty-seven.

GRIFFITHS, Lois W. June 27, 1899–November 9, 1981.

University of Washington (BS 1921, MS 1923), University of Chicago (PhD 1927). PhD diss.: Certain quaternary quadratic forms and diophantine equations by generalized quaternion algebras (directed by Leonard Eugene Dickson).

Lois Wilfred Griffiths was born in Chagrin Falls, Ohio, the second of two children of Lena (Jones) and Frederick William Griffiths. Her mother graduated from the elementary course at the Kansas State Normal School in Emporia (now Emporia State University) in 1894 and taught school at various times. F. W. Griffiths was born in Wales, immigrated to the United States in 1880 and received a BA degree from Oberlin College in 1893. He earned a BD degree from Oberlin Theological Seminary in 1896 and was ordained a Congregational minister before marrying Lena Jones later that year. F. W. Griffiths was a minister in Michigan 1896–97; in Ohio, where Lois and her brother were born, 1898–99; and in Jennings, Oklahoma Territory, 1899–1900. He then worked for the *Minneapolis Journal* before the family moved to Seattle in 1904. He held various positions as pastor, as school administrator, and in the business community in and around Seattle.

Lois Griffiths received her elementary and secondary education in the public schools in Washington State before entering the University of Washington. While she was a student she also sometimes served as assistant to the comptroller of the university. She received her bachelor's degree in 1921 and her master's degree two years later.

Griffiths entered the University of Chicago in October 1925 and attended for seven quarters; she completed her dissertation under the direction of L. E. Dickson and received her PhD in 1927. Griffiths was immediately hired by Northwestern University in Evanston, Illinois, for what was to be her only teaching position. She was instructor 1927–30, assistant professor 1930–38, associate professor 1938–64, and emeritus associate professor after her retirement in 1964. Her promotion to associate professor came after she had spent part of her 1936–37 sabbatical in Cambridge, England.

Griffiths published a dozen research articles over a twenty-one year period. These articles appeared in the *American Journal of Mathematics*, the *Annals of Mathematics*, and the *Bulletin* of the AMS; most were concerned with polygonal numbers. In 1947 Griffiths published a textbook, *Introduction to the Theory of Equations*, that was an expansion of typewritten notes she had copyrighted earlier. In the late 1940s she produced two additional sets of notes, "Outline of the Theory of Groups" and "Matrices and Linear Dependence," but did not publish them. At various times in the 1930s and 1940s, she served as a referee for the *Bulletin* of the AMS and the *Monthly*. She also maintained a close relationship with the University of Chicago and attended seminars and meetings of the senior mathematics club there.

In 1952 Griffiths was given an honorary life membership in the Northwestern University Alumni Association. Her papers at Northwestern University include teaching files, material related to her publications, and mathematical correspondence, particularly in the 1940s with E. T. Bell, with whom she had studied at the University of Washington.

In the mid-1940s, Griffiths' mother came to live with her and remained until her mother's death in 1956. In about 1940, Griffiths indicated that her favorite recreation was walking. Acquaintances told one of the authors that she loved to garden and to cook and liked classical music and kept notes on what she listened to on the radio.

Lois Griffiths remained in Evanston after her retirement and died in 1981 in nearby Skokie; she was eighty-two years old at the time of her death. She bequeathed her house to the University of Chicago and had provided that her body be used for scientific study after her death. Her papers are deposited in the Northwestern University archives.

GUGGENBUHL, Laura. November 18, 1901–March 8, 1985.
HUNTER COLLEGE (BA 1922), BRYN MAWR COLLEGE (MA 1924, PHD 1927). PhD diss.: An integral equation with an associated integral condition (directed by **Anna Pell Wheeler**).

Laura Guggenbuhl (often Guggenbühl until the late 1930s) was the younger of two surviving children (of four born) of Emma Marie (Wildhaber) and Fritz Guggenbühl, both natives of Switzerland. In June 1900 the parents and infant son were living in Saratoga Springs, New York, where her father was a butcher in a hotel. Laura Guggenbuhl was born the next year in New York City; in 1910 the family of four was living in Manhattan, and her father was a baker in his own bakery. By 1920 her mother was widowed, and the two children and their mother, who was working as a clerk in a bakery, were still living in Manhattan. Laura's brother became a physician.

Laura Guggenbuhl attended public schools in New York City before entering Hunter College in 1918. While there she was a member of the mathematics club and played basketball. She attended Columbia University in summer 1920 and New York University in 1921. She was a teacher in training at Erasmus Hall High School in Brooklyn for six months before receiving her bachelor's degree with a major in mathematics from Hunter in June 1922. She then began her graduate studies in mathematics at Columbia that summer. Guggenbuhl was an instructor at Hunter for the year 1922–23 before continuing her graduate studies.

From 1923 to 1926 Guggenbuhl studied at Bryn Mawr College, first as a graduate scholar in mathematics 1923–24 before receiving her master's degree in 1924. She remained at Bryn Mawr as a resident fellow 1924–26. She continued to teach at Hunter during the summers and returned to Hunter as instructor in 1926. The following year she received her PhD in mathematics and education from Bryn Mawr.

Guggenbuhl remained on the faculty at Hunter College until her retirement in 1972. She was instructor 1926–32, assistant professor 1932–59, and associate professor 1959–72. Her research interests were largely in the history of mathematics. She had articles and reviews in a number of journals and wrote more than sixty reviews for *Mathematical Reviews*.

Guggenbuhl was a member of many professional societies and was a frequent participant in meetings. She attended the International Congress of Mathematicians in Zurich in 1932 and all six ICMs that were held 1950 through 1970; she served as official delegate from Hunter College to all but one of them. She made many other trips to Europe as well, including for meetings in Caen, France, in July 1955 and in Prague, Czechoslovakia, in September 1961.

In 1939 Guggenbuhl reported her hobbies as travel and photography and her favorite recreations as motoring, swimming, bridge, football, and basketball. She traveled extensively after her retirement and in 1976 participated in a study tour to China. She was a member of the Metropolitan Museum of Art and the Metropolitan Opera Guild. She was active in Hunter College alumnae activities and held a number of offices including treasurer and member of the board of directors of the Scholarship and Welfare Fund. She also engaged in fundraising for Bryn Mawr and the New York City Branch of AAUW. She was a Protestant. The last several years of her life she lived in White Plains, New York, as did her brother and his wife.

After her brother died, Laura Guggenbuhl and his widow set out on a three-month round-the-world cruise. She died March 8, 1985, aboard the Queen Elizabeth II as it was leaving Hong Kong.

GURNEY, Margaret. October 28, 1908–March 19, 2002.

Swarthmore College (BA 1930), Brown University (MA 1931, PhD 1934). PhD diss.: Some general existence theorems for partial differential equations of hyperbolic type (directed by Jacob David Tamarkin).

Margaret Gurney was the eldest of three children of Anna Elizabeth (Pickett) and Dayton Alvin Gurney. Her parents both received bachelor's degrees from the State Agricultural College in Michigan (now Michigan State University); her father also received a master's degree from there. Margaret Gurney was born in Washington, D.C., as were her younger brother and sister. Her father was an engineer in the US War Department and later became chief engineer in the Ordnance Office.

Margaret Gurney attended public schools and graduated from Central High School in Washington, D.C., in 1926. She held a White open scholarship for women at Swarthmore College 1926–30 and graduated with highest honors in mathematics, physics, and astronomy in 1930. Gurney studied at Brown University, where she was a University junior fellow 1930–31, earned her master's degree in 1931, and continued her graduate studies as a Joshua Lippincott fellow from Swarthmore College 1931–32. Two articles by Gurney appeared in 1932 dealing with convergence and summability in series.

In 1932–33 Gurney studied at the university in Göttingen, as a Miss Abbott's School Alumnae fellow from Pembroke College, Brown University. In 1933 she returned to Brown and completed her work for the PhD in 1934. In 1935–38 Gurney taught in girls' preparatory schools: one year at Ogontz School in Ogontz, Pennsylvania, just northwest of Philadelphia, and then at Wykeham Rise School in Washington, Connecticut.

In 1938 Gurney began her government career as a statistician. She first worked for the US Bureau of the Budget in Washington, D.C., as statistical consultant 1938–40 and as economist 1940–44. In 1944 she moved to the US Bureau of the Census, where she was a mathematical statistician until her retirement in about 1973. Much of her work at the census bureau involved planning and implementing sample surveys in demographic and economic fields. As part of her job Gurney programmed on the first UNIVAC I.

In addition to her work at the Bureau of the Census, Gurney taught a course in sampling theory as a visiting lecturer at Stanford University in 1952 and in the early 1950s was the Washington correspondent for the *American Statistician*, the news publication of the ASA. She also worked many years as a consultant. She served as a consultant in

Puerto Rico for the Bureau of Labor Statistics, US Department of Labor, six times in the period 1961–71. During the period 1962–76, she was a consultant for the US Agency for International Development and traveled to Thailand, the Dominican Republic, and Mexico in the 1960s; Brazil and Vietnam in the 1970s; and Kenya, Uganda, and Ethiopia in 1968 and 1970. Much of her consulting work dealt with aspects of agricultural census, sampling, and the computation of reliability.

Gurney received many professional honors. In February 1966 she was awarded a US Department of Commerce Silver Medal for her work in the theory and application of survey sampling methods. In 1968 she was elected a fellow of the American Statistical Association. In the late 1970s Gurney moved to Quilcene, Washington, on the Olympic Peninsula, near her sister. She died in Quilcene in 2002 at ninety-three. Her obituary noted that she was an Episcopalian with hobbies including reading mystery novels, gardening, weaving, and needlework.

HAGEN, Beatrice L. July 4, 1899–July 22, 1987.

University of Kansas (BA 1920), University of Chicago (MA 1926, PhD 1930). PhD diss.: Quintuples of three-dimensional varieties in a four-dimensional linear space (directed by Ernest Preston Lane).

Beatrice Liberty Hagen was the youngest child of Lena (Sessler) and Louis Hagen. She was born in Barton County, near Ellinwood, in central Kansas, where her father was a farmer. In the 1900 census it was reported that there were four children living of five born; the second child was the only son.

Beatrice Hagen received her elementary and secondary education in public schools and graduated from Ellinwood High School in 1916. She then attended the University of Kansas, where **Wealthy Babcock** was one year ahead of her. **Florence Black** was an instructor and graduate student during Hagen's final two years at Kansas. After receiving her bachelor's degree in 1920, she taught in Kansas high schools; she was in Burlington 1920–22 and in Dodge City 1923–24.

In 1924 Hagen entered the University of Chicago, where she studied for the next two years and received her master's degree in 1926, having written a thesis with E. P. Lane. She then taught at Ozark Wesleyan College in Carthage, Missouri, in 1926–27. She also taught at Nebraska State Normal College (now Chadron State College) in Chadron 1928–29. After further study at Chicago and completing a dissertation in projective differential geometry, she received her PhD in 1930.

Hagen's first position after her degree was at Fort Hays Kansas State College (now Fort Hays State University) 1930–31. In 1931 she took a position as instructor at Pennsylvania State College (now University), where she remained until her retirement in 1959. She was promoted to assistant professor in 1940, to associate professor in 1945, and to professor in 1956, three years before her retirement. In 1940 Hagen and several colleagues had notes for an introduction to statistics course printed; they revised these notes many times during the 1940s.

Hagen was appointed a temporary associate professor at Kansas State University in Manhattan after her retirement from Penn State and taught there 1960–62. In 1964 she moved to Great Bend, Kansas, where she made a home with her two sisters. Her eldest sister had helped run the family farm, and the younger had been a French teacher who moved back to Great Bend after she retired. Both Beatrice and the younger of her sisters did tutoring after retiring. Beatrice Hagen was a member of the Sunshine Club in nearby Ellinwood, and was a Presbyterian. In 1940 she described herself as a Republican.

In a 1978 letter to one of the authors, Beatrice Hagen reported that "women were quite frankly advised not to enter the higher fields of mathematics, physics and chemistry because there was difficulty in obtaining satisfactory positions." She also noted that "World War II and committee work filled my time, plus the feeling that what I did in the line of research was not very important made most of my 'playthings' go unpublished.

Only a few people do important research." Not all of her research went unpublished; in 1941 an article with **Ruth B. Rasmusen**, a fellow student of E. P. Lane, appeared in the *Bulletin* of the AMS. Beatrice Hagen died in Ellinwood, Kansas, a few days after her eighty-eighth birthday; she was the last of her immediate family.

HALLER, Mary E. June 15, 1901–June 16, 1970.
University of Washington (BA 1924, MS 1931, PhD 1934). PhD diss.: Self-projective rational octavics (directed by Roy Martin Winger).

Mary Elizabeth Haller was born in Cumberland, Maryland, the second of at least four daughters of Winifred (Buckey) and William Henry Haller. Her father worked for the railroad, as a clerk in 1900 and as a telegraph operator in 1910.

Haller attended public schools in Cumberland, Maryland, and graduated from Allegany County High School in 1916. The next two years she was enrolled in the Frostburg State Normal School (now Frostburg State University) in Frostburg, Maryland, from which she graduated in 1918. She taught in elementary schools for the next five years. She was an instructor in the public schools in Maryland 1918–20 and in Renton, Washington, 1920–23. While teaching in Renton, near Seattle, she attended the University of Washington and in 1924 earned her bachelor's degree in education with a major in mathematics and a minor in economics. The next five years, 1924–29, Haller taught in the high school in Renton and, after successfully teaching and demonstrating professional growth, earned the Washington State Life Diploma.

In 1929 Haller returned to the University of Washington, where she was a teaching fellow in the mathematics department 1929–31 while she worked toward her 1931 master's degree with a thesis in projective geometry and a minor in physics. While continuing her graduate work at the University of Washington she was an associate in mathematics 1931–34 and a Loretta Denny fellow 1933–34. She received her PhD in 1934 with a minor in physics and with a doctoral dissertation in algebraic geometry directed by R. M. Winger, who also directed her master's thesis.

In 1934–35 Haller was professor and dean of women at Gooding College in Idaho. It appears that she studied at the University of Chicago in the summer of 1935 and then returned to the University of Washington where she was instructor 1935–41, assistant professor 1941–49, associate professor 1949–65, and associate professor emeritus after her retirement in December 1965. In 1999 Haller was among fourteen faculty members whose teaching was recognized in the University of Washington alumni magazine. Two tributes and a letter to the editor indicate that she was known to be an extremely demanding teacher but that she was concerned about her students and had an extremely positive effect on those who were brave enough to remain in her classes.

Haller's continued interest in physics was demonstrated the summer before her promotion to associate professor when she attended a three-day symposium on electro-magnetic theory at MIT. She was a Methodist and a member of the Wesley Club and Kappa Phi, a club for Christian university women. Mary Haller remained in Seattle after her retirement and died there in 1970, the day after her sixty-ninth birthday.

HARSHBARGER, Frances. August 16, 1902–February 11, 1987.
Grinnell College (BA 1923), West Virginia University (MA 1925), University of Illinois (PhD 1930). PhD diss.: The geometric configuration defined by a special algebraic relation of genus four (directed by Arthur Byron Coble).

Frances Harshbarger was born in Quimby, Iowa, the middle child of Annie W. and Charles C. Harshbarger, a bank cashier. Two years separated her from her older brother and from her younger sister.

After graduating from Grinnell College in Iowa in 1923, Harshbarger was at West Virginia University for the next two years, where she taught half time and did graduate work in mathematics. She received her master's degree in 1925 and spent the next two years in Keyser, West Virginia, as head of the mathematics department of Potomac State School, a junior college that was established as a preparatory school for West Virginia University in 1901. It was also known as Potomac State College, which later became its official name; it is now a division of the university.

In 1927 Harshbarger went to the University of Illinois, where she was an assistant until 1929 and then a fellow for the year 1929–30. She minored in physics. After receiving her PhD in 1930, Harshbarger went to Istanbul, Turkey, where she was a professor in the American College for Girls (also known as Constantinople Woman's College) 1930–34. While in Turkey a published version of her dissertation appeared in the *Transactions* of the AMS. Upon Harshbarger's return to the United States, she taught at the high school associated with the University of Chicago.

In the fall of 1935 Harshbarger was appointed instructor at Kent State University in Ohio. She was assistant professor 1936–42, associate professor 1942–46, and professor 1946–72. She was designated emeritus professor upon her retirement in 1972. After her death at age eighty-four in 1987 in Cuyahoga Falls, Ohio, Frances Harshbarger was buried in Woodbine, Iowa.

HASEMAN, Mary Gertrude. March 6, 1889–April 9, 1979.

Indiana University (BA 1910), Bryn Mawr College (PhD 1917). PhD diss.: On knots with a census of the amphicheirals with twelve crossings (directed by Charlotte Angas Scott).

Mary Gertrude Haseman was born in or near Linton, Indiana, the seventh of nine children of Elizabeth Christine (Schultze) and John Dieterich (also Diedrich, Diederich) Haseman. All nine children from this farm family in central Indiana were college educated, and five of them earned doctorates. Gertrude Haseman's six older brothers all received bachelor's degrees from Indiana University; five also earned master's degrees from Indiana. Four of the six earned PhD's: one from the University of Pennsylvania in physics in 1907; Charles, the third eldest, from Göttingen in mathematics in 1907; one from Cornell in entomology in 1910; and one from Columbia in zoology in 1911. Her younger brother and sister both earned bachelor's degrees and did postgraduate work.

Gertrude Haseman attended high school in Linton, where her oldest brother was superintendent of schools. She entered Indiana University in 1907 and graduated cum laude in mathematics in 1910. Haseman first took a position as professor of mathematics at Vincennes University, then a two-year college in southwestern Indiana, for the year 1910–11. From 1911 to 1915 Haseman was a graduate student at Bryn Mawr College: as a graduate scholar in mathematics 1911–12 and as a resident fellow 1913–15. She worked as a tutor in the Bryn Mawr tutoring school from 1912 to 1915.

During 1915–17 Haseman taught at the Roland Park Country School in Baltimore, Maryland. She was also completing her graduate work at this time. In 1915–16 she was enrolled as a graduate student at the Johns Hopkins University, where she heard lectures of Frank Morley. She completed her PhD dissertation at Bryn Mawr and passed her final examination in May 1916. Although her name appears on the list of American doctorates conferred in 1915–16 in the *Bulletin* of the AMS, the degree was awarded by Bryn Mawr in 1917. It appears that she continued to hear lectures at Johns Hopkins during 1916–17, while she was still teaching in Baltimore. The results of Haseman's dissertation have appeared in books and papers in knot theory.

In 1917–18 Haseman taught at the Harcum School in Bryn Mawr and during 1918–19 at the high school in Linton, Indiana. She joined the University of Illinois faculty as instructor in 1920. While there she studied during two summers in the early 1920s at the University of Chicago. While she was at Illinois, Haseman and a colleague, **Josephine Burns Glasgow**, spoke at an MAA meeting about preparing students to take calculus who had minimal mathematics in high school. They also produced a mimeographed mathematics text. Also while at Illinois, Haseman served as second vice president of Sigma Delta Epsilon, the graduate women's scientific fraternity.

Haseman resigned her position at Illinois on October 28, 1927, and left on January 31, 1928. February 1 was the opening day for the new Junior College of Connecticut (now the University of Bridgeport), and Haseman had been hired as its first professor of mathematics. She was there briefly before serving as professor of mathematics and advisor of women at Hartwick College in Oneonta, New York, during the academic year 1928–29. She left after that year, with evidence suggesting a dispute with the college administration.

After leaving Hartwick, Gertrude Haseman spent some time with a brother who was then head of the department of entomology at the University of Missouri in Columbia. A nephew reported that she held a teaching job in the south before returning in the summer of 1936 to live in her former home in Linton, where she remained until about the last year of her life. Some of the time one of her brothers was there; later she lived alone and was cared for by two nephews.

In an interview with one of the authors in 1986, a nephew said Haseman played piano and harp and was very musical. He said she was interested in lawn work and that she gardened, canned, and was a good cook. Never in the hospital before her final illness, she spent about the last year of her life in a nursing home before her death at age ninety in Linton.

HAYNES, Nola (Anderson). January 9, 1897–December 21, 1996.
University of Missouri (BS 1922, MA 1926, PhD 1929). PhD diss.: [Anderson, N. L.] An extension of Maschke's symbolism (directed by Louis Ingold).

Nola Lee Anderson was the second of five children of Mary Estella (Swan) and Edward Lee Anderson, a farmer; both parents had an elementary school education. She and her siblings were born in or near Bucklin in Linn County in north-central Missouri. Her younger sisters both received bachelor's degrees from the University of Missouri; one became a teacher and homemaker, the other a librarian.

Nola Anderson received her early education in a one-room schoolhouse and graduated from the high school in Bucklin in 1915. She received a certificate and taught at a rural school for four years before enrolling at the University of Missouri in the fall of 1919. Anderson graduated three years later with a BS in education. She taught mathematics 1922–24 in the high school in St. Charles, Missouri; in 1924–25 she had charge of the mathematics department in Central College for Women, a junior college in Lexington, Missouri. She then returned to the University of Missouri to study mathematics and astronomy and received her MA after one year. She continued her graduate studies as a university scholar in mathematics 1926–27, as a Gregory scholar 1927–28, and as an instructor 1928–29. Anderson remained at the University of Missouri as an instructor for one more year after receiving her doctorate with a minor in astronomy in 1929.

In 1930 Anderson joined the faculty at H. Sophie Newcomb College, Tulane University, in New Orleans as associate professor and acting chair of the department. She was associate professor 1930–36 and professor 1936–38; she was also department chair during her years at Newcomb. In her first years there, Anderson supervised a master's thesis in geometry. She became active in the Louisiana-Mississippi Section of the MAA and served as secretary 1931–32 and vice chairman for Louisiana 1937–38.

Nola Anderson left Newcomb College at the end of the 1937–38 school year and married Eli Stuart Haynes on July 9, 1938. Haynes, who had served on her dissertation committee, was born in Missouri in 1880; he received his bachelor's and master's degrees from the University of Missouri and his PhD in astronomy from the University of California. He returned to the University of Missouri in 1923 as professor of astronomy and director of the Laws Observatory. After nearly twenty-six years of marriage and with two grown sons, Haynes was widowed in 1934.

After her marriage Nola Anderson Haynes moved to Columbia, Missouri, where she was subject to anti-nepotism policies at the university. However, in 1943–44 she was a special instructor and from 1946 to 1951 she served as acting associate professor. In 1951, the year after her husband retired, Nola Haynes became the first woman at the University of Missouri to hold the title of associate professor of mathematics. In 1952 she was elected to a one-year term as secretary-treasurer of the Missouri Section of the MAA. She served again as secretary-treasurer of the section 1960–63 and chairman 1963–64. E. S. Haynes died in 1956, and Nola Haynes continued teaching until her retirement as emeritus associate professor in 1967, at age seventy. After her retirement she was a professor at Randolph-Macon Woman's College (now Randolph College) 1967–68.

Nola Haynes was a member of Pi Lambda Theta, an education honor society. She was also a member of the National Society of Colonial Dames of America and was listed on their Honor Roll in 1980. Locally, she served as president of the Boone County League of Women Voters and was a member of the Readers' Club of Columbia. She was also a member of the First Christian Church in Columbia, where she was an honorary deaconess. She established the Eli Stuart Haynes and Nola Anderson Haynes Scholarship Fund at Missouri and in 1995 was awarded the first College of Arts and Science Silver Chalk Award for contributions in teaching. Nola Haynes died in Brookfield, Missouri, in December 1966, less than three weeks before her hundredth birthday.

HAZLETT, Olive C. October 27, 1890–March 8, 1974.

Radcliffe College (BA 1912), University of Chicago (MS 1913, PhD 1915). PhD diss.: On the classification and invariantive characterization of nilpotent algebras (directed by Leonard Eugene Dickson).

Olive Clio Hazlett was born in Cincinnati, Ohio, the only child of Olive Leonora (Binkley) and Robert Hazlett. In 1890 her father was a postal clerk for the Railway Mail Service in Cincinnati. In 1898 her mother graduated from the Laura Memorial Woman's Medical College in Cincinnati and was licensed to practice in Ohio and in Massachusetts. The following year Olive Clio Hazlett moved with her mother to the Boston area. In 1900 her mother was a physician at the Reformatory Prison for Women in Sherborn, Massachusetts, while her father, a postal clerk, was enumerated with his family in Zanesville, Ohio. Robert Hazlett was deceased by March 1905.

Olive C. Hazlett attended public schools in Massachusetts, including Maldin High School 1904–05. She and her mother traveled in France and England from September 1905 to August 1906, and then she attended Dorchester High School in Boston 1906–09. She entered Radcliffe College in 1909 and graduated magna cum laude in 1912.

Hazlett did graduate work in mathematics and mathematical astronomy at the University of Chicago 1912–15. She held a graduate scholarship from Chicago the first year and received her master's degree in 1913 with a thesis directed by L. E. Dickson. The next two years she held a graduate fellowship from Chicago, and during her final year she also held a Boston alumnae fellowship awarded by the Boston branch of the Association of Collegiate Alumnae. She received her PhD, magna cum laude, in 1915. In 1915–16 Hazlett was an Alice Freeman Palmer memorial fellow of Wellesley College for postdoctoral study at Harvard University. By 1916 she had published three papers on linear associative algebras and was later to become the most prolific of the women working in pure mathematics among those who earned PhD's in mathematics before 1940. Her later work concerned modular invariants and covariants.

In February 1916 Hazlett was offered a two-year appointment as an associate in mathematics at Bryn Mawr College. She was not reappointed in 1918, but was hired by Mount Holyoke College in South Hadley, Massachusetts, where she was assistant professor 1918–24 and associate professor 1924–25. Hazlett's years at Mount Holyoke were some of her most productive research years and included the start of her long tenure as an associate editor of the *Transactions* of the AMS. However, in 1925 she took a demotion in rank to join the faculty at the University of Illinois where she would have a better library and contact with other mathematicians doing research. She was promoted to associate professor in 1929 and was associate professor emeritus after her formal retirement from Illinois in

1959. While at Illinois she served a three-year term on the council of the AMS 1926–28, and in 1927 she became the second woman "starred" for mathematics in *American Men of Science*; the first was Charlotte Scott.

In January 1928 Hazlett requested a leave from the University of Illinois for study in Europe and received a Guggenheim fellowship for 1928–29. She spent the summer in the Italian Alps with her mother and then went to Bologna where she presented a paper at the International Congress of Mathematicians. She spent most of the year in Rome except for a few weeks in Zurich. Hazlett applied for and received a one-year renewal of her Guggenheim; she received a second leave of absence from Illinois under the condition that she return to the university and with a commitment for a promotion to associate professor. During her second year in Europe she also studied in Göttingen.

Hazlett returned to Illinois, but in December 1936 her department chairman wrote to the university president that she had suffered a nervous breakdown. She was granted a leave of absence for illness through August 1937. She was in Rogers Memorial Sanitarium in Oconomowoc, Wisconsin, for several months before spending some time in Chicago. In spring 1937, at her request, the leave was extended for another year, which Hazlett spent mainly in Colorado, first in Denver and then in Estes Park. Also in 1937 Hazlett described her interests as "rock climbing especially above timberline and when the technique is pressure climbing; landscape photography, especially working across the light or into the light; Oriental rugs and other" (Owens questionnaire). She returned to work at Illinois in September 1938. However, in November 1944 she was committed by court order to the Neuropsychiatric Institute of the University of Illinois in Chicago. At some point she was given a conditional parole but was committed by the Champaign County Court in March 1945 to the Kankakee State Hospital in Kankakee, Illinois. She was on a temporary disability leave with pay until May 1945, after which she was on permanent disability leave until her retirement in the late 1950s.

In 1953, with the help of a Kankakee physician and attorney, Hazlett petitioned successfully for her own release from the state hospital. She was discharged after which she moved to her cabin, Timeless Lodge, in Peterborough, New Hampshire. After Hazlett moved to New Hampshire, she was befriended by several local Discalced Carmelite brothers and was deeply involved as a lay woman in the order. After living in Peterborough for nearly two decades, Hazlett spent some months in a Keene, New Hampshire, nursing home before her death there at eighty-three in 1974.

HEDBERG, Marguerite (Zeigel). August 27, 1907–August 27, 2002.
DELTA STATE TEACHERS COLLEGE (BS 1928), UNIVERSITY OF MISSOURI (MA 1929, PHD 1932).
PhD diss.: [Zeigel, M. L.] Some invariant properties of a two-dimensional surface in hyperspace (directed by Louis Ingold).

Marguerite Lenore Zeigel was born in Kirksville, Missouri, the second of four children of Elizabeth (Neef) and William Henry Zeigel. Her father received a bachelor's degree from Missouri Valley College and a master's degree from the University of Missouri. From 1917 to 1925 he was a teacher, dean, and head of the mathematics department at the State Teachers College in Kirksville, Missouri. In 1924 he earned a PhD in education from George Peabody College for Teachers and the following year became dean and head of the education department at Delta State Teachers College (now Delta State University) in Cleveland, Mississippi.

Marguerite Zeigel received her elementary and secondary education in Kirksville, Missouri. She then attended Delta State Teachers College from 1925 until she received her bachelor's degree in education in 1928. The next three years she was a student at the University of Missouri. She was a scholar 1928–29, received her master's degree in 1929, and was a fellow 1929–31.

Zeigel returned to Delta State as an assistant professor of mathematics and physics for the year 1931–32. She received her PhD from Missouri in 1932 after completing her dissertation in geometry with a minor in physics. She then took a position as assistant professor at a women's college, Lander College (now the coeducational Lander University), in Greenwood, South Carolina. The following year she was promoted to professor.

On August 7, 1936, at the end of her fourth year at Lander, Zeigel married Ernest Albert Hedberg (1903–1961). Ernest A. Hedberg was born in Huntington, Missouri. He received his BS from Northeast Missouri State Teachers College in 1928, studied at Stanford University in 1928, and was a graduate student in mathematics at the University of Missouri during 1929–30, Marguerite Zeigel's second year there. He was an instructor at Wentworth Military Academy 1930–32, received his PhD from Missouri in 1935, and was a professor at Alabama State Teachers College in Livingston 1935–36.

The year following their marriage, 1936–37, Ernest Hedberg was professor at Lander, and Marguerite Z. Hedberg was not on the faculty. The following year he was professor and acting head of the department at the University of South Dakota. In 1938 Ernest and Marguerite Hedberg were both hired at Baylor University in Waco, Texas. On her 1982 Smithsonian questionnaire Marguerite Hedberg described herself as a teacher in 1938–40 and Ernest Hedberg as an associate professor 1938–42. They had no children.

During 1942–43 Marguerite Hedberg returned again to Delta State, this time as professor and acting head of the department. That year Ernest Hedberg was an instructor of mathematics, physics, and meteorology at the US Navy Preflight School on the campus of the University of Georgia in Athens. During 1943–44 Marguerite Hedberg was an assistant professor at the University of Georgia while her husband was teaching in a US Army Specialized Training Program. In 1944 she worked as a civilian for the US Office of Scientific Research and Development while he was a technical writer in California for a National Defense Research Committee (NDRC) Project. Ernest Hedberg then served as an associate professor of electrical communications in the radar school at MIT 1944–46.

In 1946, after the war, they both were recruited to the University of South Carolina in Columbia by Wyman L. Williams, the head of the mathematics department. Marguerite Hedberg was an adjunct professor 1946–49 and in 1949 became an associate professor; she retired as associate professor emeritus in 1976. Ernest Hedberg became a professor in 1950; he died in 1961.

Marguerite Hedberg was living in Columbia, South Carolina, at the time of her death in 2002 on her ninety-fifth birthday. After her death, Hedberg's estate left $1,000,000 for the establishment of the Wyman Loren Williams and Ernest Albert and Marguerite Zeigel Hedberg Chair of Mathematics at the University of South Carolina.

HENNEL, Cora B. January 21, 1886–June 26, 1947.

Indiana University (BA 1907, MA 1908, PhD 1912). PhD diss.: Certain transformations and invariants connected with difference equations and other functional equations (directed by Robert Daniel Carmichael).

Cora Barbara Hennel was born in Evansville, Indiana, the second of three daughters of Anna M. (Thuman) and Joseph Hennel. In early years her father was a teacher, served for three years on the Union side in the Civil War, and later was in business. Cora Hennel's sisters were Cecilia (b. 1883) and Edith (b. 1891). The family lived on a farm near Evansville. Cora Hennel and her older sister went to grade school together and both graduated from Evansville (now Central) High School in 1901. They then taught in country grade schools to save money for college and entered Indiana University in Bloomington together in the fall of 1903. The following year they convinced their parents and sister to move to Bloomington.

All three daughters received bachelor's and master's degrees from Indiana; Cora and Cecilia received BA's in 1907 and MA's in 1908 (Cora in mathematics and Cecilia in English), and Edith received a BA in 1911 and an MA in 1912 (in botany) and stayed an additional year as a teaching fellow. Cora Hennel was a member of Mortar Board; all three were elected to Phi Beta Kappa. Both of Hennel's sisters were on the faculty at Indiana before their marriages; her older sister returned to the English department in 1931 after homesteading with her husband in Wyoming and remained on the faculty until her retirement in 1953.

Cora Hennel continued her studies and began teaching at Indiana University immediately after receiving her bachelor's degree. She was a teaching fellow 1907–09, earned her master's degree in 1908, and was appointed instructor of mathematics in August 1909. She continued her work toward the doctorate, and wrote her dissertation during the year 1911–12. Her PhD in June 1912 was the first in mathematics and the first to a woman awarded by Indiana University. Hennel continued to hold a faculty position at Indiana until her death. She remained an instructor until 1916, was assistant professor 1916–23, associate professor 1923–36, and became professor in 1936. In about 1907 Hennel was instrumental in helping to start the Euclidean Circle, the department's mathematics club.

Hennel was extremely active on the campus and served, for example, as president of the Indiana University chapters of AAUP, AAUW, the YWCA Board, the Corda Fratres Association of Cosmopolitan Clubs (dedicated to fostering understanding between foreign and American students), and the Women's Faculty Club. She regularly participated in local, state, and national meetings of mathematical and scientific organizations. She attended the organizational meeting of the MAA at Ohio State University in December 1915 and was vice chairman 1939–40 and chairman 1940–41 of the Indiana Section. She was also chairman of the mathematics division of the Indiana Academy of Science for 1940.

Her obituary in the *Proceedings of the Indiana Academy of Science* records that Hennel was "well known throughout the State, having addressed numerous Indiana University alumni groups, and various educational, professional and civic organizations" (57 (1947): 4). It notes further that "Dr. Hennel's chief contributions were in her excellent teaching and her influence on many students, both native and foreign." Hennel's main hobby was writing poetry, and some of her poems appeared in various publications including *The Columbia Anthology of Verse*. She was affiliated with the First Methodist Church in Bloomington.

Cora Hennel died in Bloomington in 1947 at age sixty-one of colon cancer. In 1995 the Indiana mathematics department named its faculty-student lounge the Cora B. Hennel Room; the department also annually awards Cora B. Hennel memorial scholarships to outstanding undergraduate majors.

HENRIQUES, Anna (Stafford). August 20, 1905–November 28, 2004.
Western College for Women (BA 1926), University of Chicago (MS 1931, PhD 1933).
PhD diss.: [Stafford, A. A.] Knotted varieties (directed by **Mayme Irwin Logsdon**).

Anna Adelaide Stafford was born in Chicago, Illinois, the eldest of three sons, one of whom died in infancy, and three daughters of Caroline (Fleuchaus) and Richard W. Stafford. At the time of the 1910 census the family was living in Chicago where her father was manager of a pickle factory. After her parents died in 1919, the five surviving Stafford children lived with her father's sister and her family, first in St. Louis, Missouri, and later in Chicago.

Anna Stafford attended public schools in Chicago; Marshfield, Wisconsin; and Barnum, Minnesota; she graduated in 1922 from Frank Louis Soldan High School in St. Louis. She attended Western College for Women in Oxford, Ohio, with a four-year college scholarship and an AAUW scholarship from the St. Louis AAUW. While there her majors were mathematics and Greek and her minor was French. She was in the classical club, the French club, and the mathematics club; she participated in theater; and she was on the baseball, hockey, basketball, and swimming teams.

Following her graduation from Western College in 1926, Stafford taught mathematics and science for two years at St. John Baptist School, an Episcopal school for girls, in New York and later in Mendham, New Jersey. She taught at the Darlington Seminary in West Chester, Pennsylvania, 1928–29, and at St. John Baptist School in Mendham 1929–31. She attended graduate school at the University of Chicago five summers between 1927 and 1933. From 1931 to 1933 she completed two years residency at Chicago; she earned her master's degree in December 1931 and her PhD in August 1933 as a student of Mayme I. Logsdon. Her dissertation was in topology, but since there were no topology specialists at Chicago, she and Logsdon worked through the material together, learning Italian to do so.

From 1933 to 1935 Stafford was a member of the Institute for Advanced Study at Princeton, where she pursued her interest in topology. In the same period Stafford was on the faculty at St. John Baptist School, where she taught in the mornings so she could attend lectures at the Institute in the afternoons. From 1935 to 1937 she was an instructor at the University of Nebraska and attended the International Congress of Mathematicians at Oslo in July 1936.

In 1937 Stafford joined the faculty at the University of Utah in Salt Lake City where she remained for nearly twenty years. In 1942 Anna Stafford married Douglas E. Henriques. He was born in Nevada in 1910 and was an administrative law judge for the US Department of the Interior. He had a twelve-year-old son by a previous marriage, and they adopted a daughter, who was a member of the Navajo Nation.

At the University of Utah, Anna Stafford Henriques was instructor 1937–41, assistant professor 1941–47, and associate professor 1947–56. In 1946 she revived the mathematics club, which had become moribund some years earlier. She had a sabbatical leave during the spring quarter 1949 during which she taught one class while preparing a textbook for that course.

In 1956 Douglas Henriques's work took him from Utah to New Mexico, and so the family moved. In 1957 Anna Henriques became a lecturer at St. Michael's College (the College of Santa Fe since 1966). Two years later she also became a lecturer at the University of New Mexico and remained in both positions until 1962 when she became professor at St. Michael's. She retired in 1971 as professor emeritus. Shortly thereafter she and her husband were living in the Falls Church area of Virginia. Once retired Douglas and Anna Henriques made trips to every continent.

Anna Henriques held many positions in the Utah and New Mexico branches of AAUW and served as president, program director, and bulletin editor of the Utah Council of Teachers of Mathematics. During her lifetime her interests included mountain climbing, hiking, bowling, opera, theater, and travel. She was a generous contributor to the MAA and the IAS.

Douglas Henriques died in November 1987. Anna Henriques later moved to a retirement community in Bailey's Crossroads, Virginia, and died there in 2004 at ninety-nine.

HIGHTOWER, Ruby U. June 17, 1880–May 5, 1959.

Shorter Female College (BA 1896), University of Georgia (MA 1919), University of Missouri (PhD 1927). PhD diss.: On the classification of the elements of a ring (directed by Gustav Eric Wahlin).

Ruby Usher Hightower was born near Covington, Georgia, the youngest of four surviving daughters of Amarinth (Sims) and James Richard Hightower. Her father was in the Confederate army during the Civil War. In 1880 her father was a farmer, and in 1900 and 1910 he was described as a contractor.

Ruby Hightower attended high school in Jackson, Georgia, before completing work for her bachelor's degree in the academic year 1895–96 at Shorter Female College (Shorter College after 1923) in Rome, Georgia. In the twenty-three-year period between her graduation and the resumption of her teaching career at Shorter in 1919, she taught and studied at a number of places. She taught at: the grade school, at least in 1908–09, and high school in Dublin, Georgia; Central College (now closed) of Conway, Arkansas; Alabama Normal College (now the University of West Alabama); Cox College (now closed), a women's college in College Park, Georgia, at least 1915–16; Anderson College (now University), then a women's college in South Carolina, 1916–17; and Southwest Baptist College and Hardin

College in Missouri. She studied at the University of Chicago in 1899, at the University of London in 1913, and received her master's degree from the University of Georgia in 1919, the only one listed as Master of Arts for that June commencement. She was a charter member of the MAA.

After receiving her master's degree, Hightower returned to Shorter and remained there, except for a leave of absence in 1924–25, until her retirement in 1947. The college annual of 1920 notes that she was elected fellow in mathematics at the University of Missouri for 1919–20. Apparently she delayed her entrance to the graduate school there until the summer of 1921. She was resident for five summer sessions and was granted a leave to study at Missouri 1924–25. She was an honorary fellow in mathematics for the year and then returned to Shorter as head of the department. She received her PhD with minors in physics and astronomy in 1927. Hightower was the first Shorter graduate to receive a PhD degree.

During Hightower's years at Shorter College she was generally the only permanent member of the mathematics department. Her interests included mathematical astronomy and economics. After her retirement in 1947 she spent time with relatives in Quitman and Atlanta, but maintained her apartment on the college campus. She taught part time at the college 1949–50, 1954–55, and 1956–57, and was emeritus after 1952. Hightower had been living at the college apartments in Rome until a few weeks before her death in Quitman, Georgia, in 1959.

HILL, Agnes (Baxter). March 18, 1870–March 9, 1917.

DALHOUSIE UNIVERSITY (BA 1891, MA 1892), CORNELL UNIVERSITY (PHD 1895). PhD diss.: [Baxter, A. S.] On Abelian integrals: A resume of Neumann's "Abelsche Integrale," with comments and applications (directed by James Edward Oliver).

Agnes Sime Baxter Hill was born in Halifax, Nova Scotia, Canada, the eldest of three children of Janet (Methven) and Robert Baxter, originally of Scotland. Her father was for many years manager of the gas works in Halifax.

Baxter graduated from Dalhousie University in Halifax in 1891, the first woman at Dalhousie to take first class honors in mathematics and mathematical physics; she also received the Sir William Young Gold Medal for her academic accomplishments. She remained at Dalhousie as a graduate student and received her master's degree in 1892.

Agnes Baxter entered Cornell for graduate study in 1892. Her final year, 1894–95, she held an Erastus Brooks fellowship before receiving her PhD in 1895 with major subject pure mathematics, first minor mathematical physics, and second minor physics. The year after receiving her doctorate, she remained at Cornell editing the works of her advisor, James E. Oliver, who had died in March 1895.

On August 20, 1896, Agnes Baxter married Albert Ross Hill (1869–1943) in Halifax. A. Ross Hill, also from Nova Scotia, received a BA from Dalhousie in 1892, studied a year in Europe, and received a PhD from Cornell in 1895 with a dissertation in philosophy. At the time of their marriage A. Ross Hill was professor of psychology and education at the State Normal School in Oshkosh, Wisconsin. The following year he joined the philosophy faculty at the University of Nebraska. While they were in Lincoln, Nebraska, they had two daughters, born in December 1897 and in March 1903. In 1903 A. Ross Hill joined the educational psychology faculty at the University of Missouri as professor and dean of the school of education. During 1907–08 he was professor, director of the school of education, and dean of the college of arts and sciences at Cornell before returning the following year to Missouri as president of the University of Missouri.

Agnes Hill died shortly before her forty-seventh birthday in Columbia, Missouri, of pneumonia following operations during a lengthy illness. After her death, A. Ross Hill presented Dalhousie with a gift of books "to perpetuate the memory of one of its loyal graduates, who gave her life to assist in [his own] educational work instead of making an independent record for herself" (Dalhousie University press release, "Dalhousie University

Honors Alumna with Room Dedication," March 11, 1988). On March 15, 1988, the Agnes Baxter Reading Room, the library for the Department of Mathematics, Statistics, and Computing Science at Dalhousie, was officially opened and dedicated.

HILL, Sister Mary Laetitia. December 22, 1898–April 10, 1992.

Our Lady of the Lake College (BA 1922), Catholic University of America (Catholic Sisters College) (MA 1926), Catholic University of America (PhD 1935). PhD diss.: The number and reality of quadrilaterals in-and-circumscribed to a rational unicuspidal quartic with real tangents from the cusp (directed by Aubrey Edward Landry).

Maria Anna Hill was born in Koerth, Texas, the first of two surviving daughters of Mary Bradley (Watson) and Thomas James Hill. Her father had been married previously and had three sons with his first wife. After his first wife died, he attended medical school in Louisville, Kentucky, after which he moved to Koerth and remarried. He then practiced medicine and was in the stock business. In 1912 the family moved to Yoakum, Texas. Maria Hill's mother died in 1922, and her father remarried in 1923.

When she was four, Maria Hill attended public school for a year. She then attended the Sacred Heart School in Hallettsville, near Koerth, and in 1913–14 the St. Joseph School in Yoakum, where she was taught by the Sisters of Divine Providence. She entered the convent of the Congregation of Divine Providence (CDP) in San Antonio in 1914, at age fifteen, and made her first profession in 1916.

During the years 1916–22, Sister Mary Laetitia Hill taught and completed the work for her bachelor's degree. She first taught at St. Mary's School in San Antonio; while there she taught high school, fifth grade, and kindergarten in the years 1916–21; in second semester 1920–21, she taught in Our Lady of the Lake high school. She taught at the Holy Family School in Tulsa, Oklahoma, from 1921 to 1925. In the meantime, she received her bachelor's degree in 1922 from Our Lady of the Lake College (now University). In 1925–26, she studied at the Catholic University of America and received her master's degree in 1926.

Sister Mary Laetitia Hill taught at Our Lady of the Lake College for forty-one years: as instructor 1926–32 and as professor 1935–70. The first six years she taught both mathematics and physics. In 1932–35 she was at Catholic University studying for her doctorate in mathematics; she received her PhD in 1935 with first minor physics and second minor mechanical engineering.

For nearly twenty-five years, starting in summer 1935, Sister Laetitia taught mathematics or physics or both during the school year and in most summer sessions. She directed an NSF-sponsored summer institute for high school teachers at Our Lady of the Lake College in 1959. She then received an NSF grant to participate in a 1962 summer institute for college teachers of mathematics at Bowdoin College in Maine.

After Sister Laetitia retired from Our Lady of the Lake College in 1969, she worked in the college accounting office as a keypuncher until 1972. In 1972 she retired to the convent. In July 1974 she came out of retirement to go to the Congregation of Divine Providence house in Queretaro, Mexico, for one semester before returning to San Antonio. In August 1975 she joined the CDP community in Tuba City, Arizona, and remained for three years.

Sister Laetitia was a serious bird watcher, who was known to abandon class briefly, telescope in hand, if she spotted a particularly interesting bird out the window. She taught herself Spanish; shorthand; and to play the organ, flute, and violin; and was reported to have memorized all of Shakespeare's sonnets. Sister Laetitia finally retired in 1979. She died at age ninety-three in San Antonio in 1992.

HIRSCHFELDER, Elizabeth (Stafford). April 25, 1902–September 29, 2002.
Brown University (Women's College) (PhB 1923, MS 1924), University of Wisconsin (PhD 1930). PhD diss.: [Stafford, E. T.] Matrices conjugate to a given matrix with respect to its minimum equation (directed by Mark Hoyt Ingraham).

Elizabeth Thatcher Stafford, the eldest of three children, was born in Providence, Rhode Island, to Evangeline K. (Flagg) and Arthur Ervin Stafford, a banker. She had two younger brothers, the younger of whom graduated from Brown University in 1927 and became an accountant in Manhattan.

Elizabeth Stafford attended Women's College of Brown University while living at home. She graduated in 1923, remained an additional year, and earned her master's degree in 1924. Stafford was hired as an instructor in pure mathematics at the University of Texas in Austin for one year. She then moved to the newly opened Texas Technological College (now Texas Tech University) in Lubbock as an adjunct professor 1925–28. During the summer of 1927 Stafford studied with Mark H. Ingraham at Brown. Ingraham urged Stafford to apply to Wisconsin for a fellowship so she could do graduate work with him in his new position there. She took a leave from her position at Texas Technological College, studied at the University of Chicago in summer 1928, and entered the University of Wisconsin with a fellowship that fall. She held a fellowship again in 1929–30 and received her degree in 1930. That same year a paper Stafford coauthored with H. S. Vandiver appeared in the *Proceedings of the National Academy of Sciences.*

Stafford returned to Lubbock for the year 1930–31. On June 6, 1931, in Madison, she married Ivan S. Sokolnikoff (1901–1976), who had immigrated to the United States from Russia in 1921, did his undergraduate work at the University of Idaho, and received his PhD from Wisconsin in 1931. After their marriage Ivan Sokolnikoff progressed from instructor to professor at Wisconsin, while Elizabeth Sokolnikoff was an instructor there 1931–32, had no position 1932–33, and was hired primarily as a lecturer on an ad hoc basis for the next several years. Anti-nepotism practices and the Great Depression prevented her from being offered regular positions. During 1934–41 Elizabeth and Ivan Sokolnikoff coauthored five significant mathematical papers in analysis and a classic text *Higher Mathematics for Engineers and Physicists.*

Elizabeth Sokolnikoff wanted to volunteer for military duty during World War II but was told that she should teach mathematics instead. She spent the war years in Madison teaching army and navy groups, geologists, and engineers, while I. S. Sokolnikoff was in New York and Washington working with the National Defense Research Council. In 1946 he accepted a position at the University of California at Los Angeles, and Elizabeth Sokolnikoff remained in Wisconsin. They were divorced the following year, and she was then given a three-year appointment as assistant professor.

On March 7, 1953, Elizabeth S. Sokolnikoff married Joseph Oakland Hirschfelder (1911–1990), a native of Baltimore who had received a PhD from Princeton in physics and chemistry. He joined the chemistry department at Wisconsin in 1937 and remained until his retirement in 1981 except during World War II when he did military research. He was elected to the National Academy of Sciences in 1953 and was awarded the National Medal of Science in 1976. Joseph O. Hirschfelder acknowledged the help of his wife in papers he published in the mid-1950s.

Elizabeth "Betty" Hirschfelder was on leave 1953–54 and then resigned her assistant professorship so she could accompany her husband when he traveled. In the mid-1970s the Hirschfelders began to split their time between Madison, Wisconsin, and Santa Barbara, California, where he was an adjunct professor at the university. After Joseph Hirschfelder's death in 1990, Betty Hirschfelder helped endow the Broida-Hirschfelder graduate fellowship in the sciences at the University of California at Santa Barbara (UCSB). She also provided funding for the Joseph O. Hirschfelder prize in theoretical chemistry at the University of Wisconsin and was a leading donor for an addition to the chemistry building there.

Elizabeth S. Hirschfelder spent her winters in Santa Barbara and died in Madison, Wisconsin, at age one hundred in September 2002. She left significant bequests to the Broida-Hirschfelder endowment at UCSB and to the University of Wisconsin for the establishment of the Elizabeth Hirschfelder Fund for Graduate Women in Mathematics, Chemistry and Physics.

HOPKINS, Margarete C. (Wolf). November 3, 1911–April 3, 1998.
University of Wisconsin (BS 1932, MA 1933, PhD 1935). PhD diss.: [Wolf, M. C.] Symmetric functions of matrices (directed by Mark Hoyt Ingraham).

Margarete Caroline Wolf was born in Milwaukee, Wisconsin, the second of two daughters of Caroline (Kupperian) and John Theodore Wolf. Her mother was born in Germany and immigrated to the United States in 1892; her father was born in Wisconsin. Both were formally educated through elementary school. Her father was a truck gardener. Margarete Wolf's older sister, **Louise A. Wolf**, was born in Milwaukee in October 1898. In 1997 Margarete Wolf Hopkins said that she and her sister got their interest in mathematics and in education from their mother.

After attending the Jefferson School, a grade school in Greenfield, Wisconsin, and Bayview High School in Milwaukee 1924–28, Margarete Wolf entered the Milwaukee Extension Division of the University of Wisconsin for the year 1928–29 before going to Madison for the remainder of her undergraduate and graduate work. She held a Fanny P. Lewis scholarship 1930–32 and was president of the mathematics club her senior year. She held scholarships 1932–34 and was a research assistant 1934–35 before receiving her PhD in 1935 as a student of Mark H. Ingraham. Margarete's sister, Louise, thirteen years her senior, also earned her doctorate in 1935 with Ingraham serving as her advisor. Louise Wolf immediately took a position at the University of Wisconsin's Milwaukee Extension Center (now University of Wisconsin–Milwaukee), where she remained until her retirement shortly before her death in 1962.

Margarete Wolf continued at Wisconsin as research assistant 1935–36, part-time instructor 1936–38, and also research associate 1937–38. After receiving her doctorate, she continued to work with Mark Ingraham and published four papers, including her dissertation and two joint with Ingraham, by 1938. She was supported by grants for the work on three of these papers. She also gave several talks, including two joint with Ingraham and one joint with her sister. Results from Margarete Wolf's dissertation, published in 1936 in the *Duke Mathematics Journal*, were generalized thirty years later and continued to be referenced into the early 2000s.

In 1938 Margarete Wolf moved to Wayne University (now Wayne State University) in Detroit, where she was an instructor until 1941. On August 31, 1941, she married Edward John Hopkins, who was born in 1908 in New York City. He received his BS in electrical engineering from Wisconsin, worked for RCA 1936–37, and spent the rest of his career, 1938–71, working as an electronics engineer with the US Navy in Brooklyn, where they made their home.

Margarete Wolf Hopkins taught in the evening and extension sessions of Hunter College during the 1942–43 and 1943–44 academic years. Her first child, a son, was born in February 1945 in New York. Her daughter was born in January 1949, also in New York. Her son earned a doctorate from the University of Wisconsin in meteorology; her daughter earned a master's degree, also in meteorology from Wisconsin, and had several years of doctoral research there.

Margarete Hopkins returned to her mathematics career in 1958, when she joined the faculty at St. Joseph's College for Women in Brooklyn. She was assistant professor 1958–64, associate professor 1965–68, and professor and chairman of the mathematics department 1969–78. In 1970 the college changed its name to St. Joseph's College, New York; it became coeducational in 1971.

Edward Hopkins, Margarete's husband, died in October 1985. Margarete Wolf Hopkins later moved to Madison and lived with her daughter's family. She died in April 1998 at age eighty-six.

HOPPER, Grace (Murray). December 9, 1906–January 1, 1992.
VASSAR COLLEGE (BA 1928), YALE UNIVERSITY (MA 1930, PHD 1934). PhD diss.: New types of irreducibility criteria (directed by Oystein Ore).

Grace Brewster Murray was the daughter of Mary Campbell (Van Horne) and Walter Fletcher Murray. Her mother had attended a private girls' school, and her father graduated from Yale University and became an insurance broker. She and her two younger siblings were born in New York City but spent their summers at a family house on Lake Wentworth in Wolfeboro, New Hampshire, where she met her future husband in 1923.

Grace Murray first attended private schools in New York City. She was at the Graham School 1911–13 and the Schoonmaker School 1913–23, and in 1924 graduated from Hartridge School, an all-girls boarding school in Plainfield, New Jersey. She then enrolled at Vassar College in Poughkeepsie, New York, and graduated in 1928 with a degree in mathematics and physics, having attended beginning courses in all the sciences offered, as well as business and economics courses. She attended Yale University on a Vassar College fellowship 1928–29 and on a Sterling scholarship 1929–30, and received her MA in 1930.

Grace Murray and Vincent Foster Hopper (1906–1976) married on June 15, 1930, in New York City. Vincent Hopper had graduated from Princeton and received his MA there in 1928. That year he began his long association with New York University as an instructor of English in the School of Commerce, Accounts, and Finance (now the Leonard N. Stern School of Business). Grace Hopper studied at Yale 1930–31, again on a Sterling scholarship. In 1931 she returned to Vassar as an assistant in mathematics. In 1934 she received her PhD from Yale and was promoted to instructor at Vassar. In 1938 Vincent Hopper received his PhD in English and comparative literature from Columbia University.

In 1939 Grace Hopper was promoted to assistant professor, and she and Vincent Hopper built a home in Poughkeepsie. During 1941–42, Grace Hopper studied at NYU's Courant Institute on a Vassar faculty fellowship. After the bombing of Pearl Harbor in December 1941, she took a leave of absence from Vassar and and asked for a waiver to join the WAVES (Women Accepted for Volunteer Emergency Service), a branch of the US Naval Reserve for which she was over-age. During the summer of 1943 she taught an accelerated wartime calculus course as an assistant professor at Barnard College, and in December 1943 she joined the US Naval Reserve. In May and June 1944 she was an apprentice seaman and midshipman at the United States Naval Reserve Midshipman's School for Women in Northampton, Massachusetts. She graduated and was commissioned a lieutenant (jg). By this time she had been promoted to associate professor at Vassar, and she and Vincent Hopper had separated; they were divorced in 1945 and had no children.

Grace Hopper was assigned to work at Harvard University writing code for the Mark I computer, formally known as the Automatic Sequence Controlled Calculator, and, later, for the Mark II, the first multiprocessor. Working under the direction of Howard Aiken, a commander in the naval reserve, she compiled and edited notes for a manual for the Mark I that appeared in 1946. In June 1946 she was promoted to lieutenant and two months later she was released from active duty. She remained in the naval reserve and stayed at the Harvard Computation Laboratory as a research fellow in engineering science and applied physics. In 1949 Hopper left Harvard to become senior mathematician at the Eckert-Mauchly Computer Corporation in Philadelphia to write code for the Binary Automatic Computer.

Hopper remained with the company and its successors (Remington Rand and later Sperry Rand) until her military leave 1967–71 and formal retirement in 1971. She worked on the UNIVAC (Universal Automatic Computer) and in 1952 developed its first compiler. Starting in 1955 Hopper developed the first English-language data processing language,

FLOW-MATIC. She was one of the leaders in the movement to develop a standardized business language for computers. The result was COBOL, Common Business Oriented Language, which was greatly influenced by her work on FLOW-MATIC and for which she was a technical advisor to the executive committee overseeing the development of the language. Hopper held various positions, the last of which was staff scientist, systems programming. During this period Hopper also taught at the University of Pennsylvania's Moore School of Electrical Engineering, starting as a visiting lecturer in 1959 and ending as a visiting associate professor in 1963. In 1971 she was made professorial lecturer in management science at George Washington University in Washington, D.C., and held that position until 1978. In 1973 the Moore School made her an adjunct professor of engineering.

During her years in the naval reserve, Hopper was promoted to lieutenant commander in 1952 and to commander in 1957. She was involuntarily retired at the end of 1966. However, in August 1967 she was returned to active duty so she could standardize COBOL for the navy. She was to have served only six months but did not retire again until 1986. Until September 1968 she served in the office of the Special Assistant to the Secretary of the Navy as director of the navy programming languages group and then, until September 1976, she was assigned to the office of the Chief of Naval Operations as the head of the programming languages section. She was then with the Naval Data Automation Command as head of the training and technology directorate and special advisor to the commander. She was promoted to captain in 1973 and to commodore in 1983; two years later the rank was raised to rear admiral (lower half). She was a senior consultant for Digital Equipment Corporation after her final, again involuntary, retirement from the naval reserve in 1986. Hopper's retirement ceremony took place on August 14, 1986, aboard the USS *Constitution*, the navy's oldest commissioned warship. The *Constitution* was requested by Hopper, who at seventy-nine was the oldest officer on duty in all of the armed services. During the retirement ceremony she received the Navy Distinguished Service Medal, the highest award given by the Department of Defense.

Hopper prepared numerous papers and reports during her career, and her collected papers from 1944 to 1965 are in the Archives Center, National Museum of American History, Smithsonian Institution. She was a member of well over three dozen professional, scientific, honorary, military, historical, and genealogical organizations. Among the many honors and awards bestowed on her were nearly fifty honorary doctorates and seven military medals. Among other awards were: election to the National Academy of Engineering; becoming a Distinguished Fellow of the British Computer Society, the first American and the first woman so honored; and receiving the National Medal of Technology.

The Grace Murray Hopper Service Center of the Navy Regional Data Automation Center in San Diego, California, was dedicated in 1987, and in 1996 a guided missile destroyer was christened the USS *Hopper*. Hopper was seen by many on a segment of *60 Minutes* in 1983, the David Letterman show in 1986, and as grand marshal of the Orange Bowl Parade in 1987. The Grace Murray Hopper Award was established by the UNIVAC division of Sperry Rand in 1971 and is now supported by UNISYS. Since 1994 the Grace Hopper Celebration of Women in Computing Conference has been held. Grace Hopper died in Arlington, Virginia, on New Year's Day 1992 and was buried at Arlington National Cemetery with full military honors.

HOWE, Anna M. October 24, 1883–August 8, 1976.

Wells College (BA 1908), Cornell University (MA 1911, PhD 1917). PhD diss.: The classification of plane involutions of order three (directed by Virgil Snyder).

Anna Mayme Howe was born in Jordan, New York, the second of two children of Eleanor Caldwell (Reed) and Lewis B. Howe. In 1900 the family lived in Auburn, in upper New York State, and her father was a city missionary. He later was a farmer in Jordan, New York.

Anna Howe graduated from the high school in Auburn, New York, and entered nearby Wells College in 1904, shortly before she turned twenty-one. She graduated in 1908 and taught mathematics the next two years.

Howe studied at Cornell 1910–11 and received her master's degree in 1911 with a major in mathematics and a minor in education. She then became head of the mathematics department at Fairmont Seminary, a school for girls in Washington, D.C., that is now closed. She indicated later that she was teaching both high school and college work while there. She remained at Fairmont until 1915, when she returned to Cornell with a graduate scholarship. She wrote her dissertation in algebraic geometry, had a minor in mathematical analysis, and took courses in physics. Howe received her doctorate in September 1917 and returned to Fairmont to teach for an additional year. During 1918–19 Howe was overseer of technical shipments for W. R. Grace & Company in New York. In the fall of 1919 she became a teacher of mathematics at Dana Hall, a girls preparatory school near Wellesley College.

In 1920 Howe accepted the offer of an instructorship at H. Sophie Newcomb College, Tulane University, in New Orleans. She was on the faculty at Newcomb until 1930, as instructor 1920–21 and as assistant professor 1921–30. She attended the International Mathematical Congress in Toronto in 1924 and afterwards reported on bookstores at various women's colleges in the East before Newcomb established its own bookstore. In March 1927 Howe had to return home to New York State because of her father's illness but returned with her mother during the summer of 1928. On May 3, 1930, Howe submitted her resignation to the dean of Newcomb College; she was replaced by **Marie Weiss**.

In 1930 Howe and her widowed mother moved back to Jordan, New York, and Howe sought employment in the area. While her work history at this time is somewhat unclear, it appears that Howe found a position about forty miles away at the new Cazenovia Central School in the early 1930s. She indicated in 1937 that she was head of the department at Cazenovia Central School and that she had been supervisor of mathematics in Cazenovia, New York, since 1930. Later she reported that she was head of the department in schools in New York from 1931 to 1948. In 1943 she became professor and head of the department at Cazenovia Junior College (now Cazenovia College) and remained there until 1957.

Howe was involved in a number of professional and non-professional activities. She became a charter member of the MAA at the time of its founding in 1915 and attended many national and sectional meetings during the following fifty years. In the late 1930s she reported that she was president of the local county mathematics club, and that she was organizing a three-year course in mathematics for non-college-bound students. At that time she was also a member of a literary society, and she described her interests as traveling, music, and gardening.

Anna Howe moved back to Jordan in 1957 when she left Cazenovia. She died in 1976 at age ninety-two in Auburn, New York.

HSIA, Shu Ting (Liu). September 25, 1903–August 16, 1980.

UNIVERSITY OF MICHIGAN (BS 1926, MS 1927, PHD 1930). PhD diss. (1929): [Liu, S. T.] Theory of periodic orbits for asteroids of integral types (directed by Louis Allen Hopkins).

Shu Ting Liu was born in Beijing, China, the eldest of several children of Shien-Ying Chen and Te-Yuang Liu. In 1925 her father was described as a railroad engineer.

Apparently her precollege, and possibly some university education, was obtained in Beijing before she came to the United States. Shu Ting Liu entered the University of Michigan in 1923 with a Barbour Scholarship for Oriental Women and remained through August 1929. She earned a bachelor's degree in 1926, a master's degree in 1927, and finished her work for the PhD in August 1929. Her doctorate was conferred in March 1930.

Liu was a substitute instructor for an ill faculty member for two months in 1929 at Colorado College in Colorado Springs. In December 1929, Liu married Pin Fang (also

"Pinfang") Hsia (1902–1970) in Colorado Springs. Hsia, born in Anhui Province in eastern China, attended Tsinghua University in Beijing before coming to the United States as a Chung Hua scholar from Beijing and graduating from Colorado College and then from the Harvard Graduate School of Business Administration in about 1929.

In April 1930 the Hsias were living in Detroit, Michigan, where he was working with stocks and bonds. That summer they returned to China. In November 1930, Shu Ting Liu Hsia reported to the Michigan alumni office that she was a teacher in Shanghai College. Their two children were born in Shanghai, a son in November 1931 and a daughter in June 1937. Soon thereafter the family escaped to Hong Kong, presumably at the time of the Battle of Shanghai, which began in August 1937.

In August 1939, the Hsia family arrived in the United States, where they remained until 1946. During that period Pinfang Hsia was manager of the Bank of China in New York. In 1944, Shu Ting Hsia reported that, in addition to caring for her children, she was also engaged in teaching. She was listed as a temporary, part-time instructor at Hunter College for the academic year 1944–45.

In 1946 Pinfang Hsia was transferred to the branch in London, where he was resident manager of the Bank of China and was responsible for major expansion into Europe. Shu Ting Hsia and their daughter returned to New York in 1950 after the government changed in China. Their son was, at the time, a student at Oxford University. Pinfang Hsia stayed on in London until the legal question of bank ownership was resolved and then returned to New York, where he worked for Magnus Mabee & Reynard, an oil firm, until his retirement. After Shu Ting Hsia returned to New York, she did some statistical work for the city of New York. Her application for a social security account number in 1957 indicates that she was then working for the New York City Youth Board. Pinfang Hsia died in New York in December 1970, a year after becoming a US citizen.

In an e-mail to one of the authors in June 1998, Y. Edward Hsia wrote about his mother: "My mother ... enjoyed working in mathematics, and I believe she enjoyed teaching, but her primary commitment must have been to establish a home for her family, which prevented her from pursuing a more formal career in mathematics." He continued, "My mother was a rare example in her time of someone with a fairly classical Chinese upbringing – as a woman – who took the exceptional course of venturing to [the] USA for higher education, and was able to win a scholarship and earn a doctorate. She remained, however, a little uncomfortable with American mores and idiom."

After her retirement Shu Ting Liu Hsia moved to New Haven, Connecticut. Having been in ill health for some time, she died at age seventy-six in West Haven, Connecticut, in 1980. In 1981 a gift of $8000, a share of which was to augment the Barbour scholarship fund, was made to the University of Michigan from her estate.

HUGHES, Olive Margaret. December 13, 1899–December 29, 1936.

UNIVERSITY OF SASKATCHEWAN (BA 1925, MA 1926), BRYN MAWR COLLEGE (PHD 1934). PhD diss.: A certain mixed linear equation (directed by **Anna Pell Wheeler**).

Olive Margaret Hughes was the daughter of Martha (Rogers) and Daniel Hughes, both originally from farm families in Wales. By 1886 they had a dairy shop in London and lived there when their eleven children were born. Olive was the ninth in the family. The family migration to Canada began in 1905 when the eldest son arrived in western Saskatchewan, near Maidstone. The father and another son followed in 1908, the mother and youngest daughter in 1909, and finally four more daughters, including Olive, in 1912. The Hughes family farmed in western Saskatchewan after their emigration from England.

Olive Hughes began her education in London. After her arrival in Canada at age twelve she attended the Dee Valley School, a one-room school house some distance from their farm. In 1914 her mother rented a small house in North Battleford, about fifty miles away, so the four youngest girls could attend school there. Olive attended grades nine through eleven before teaching in the fall of 1917 at the Dee Valley School with a permit from Regina, the provincial capital.

Hughes finished grade twelve in North Battleford and apparently attended the normal school in Saskatoon and then taught. She entered the University of Saskatchewan, also in Saskatoon, in the fall of 1921 at age twenty-one and specialized in mathematics. She completed the standard three-year course in 1925, having taken off the year 1922–23. She was an assistant in the department her last year and was granted a BA with high honors in mathematics in 1925. In the summer of 1925 she studied mathematics at the University of Chicago and then returned to the University of Saskatchewan as a graduate student and assistant in the mathematics department. She received her master's degree in 1926.

Hughes remained at the university the next two years as an instructor and served as the faculty advisor to the mathematical society at least in 1927–28. She studied mathematics at the University of Chicago again during the summer of 1927 and left her instructorship at the University of Saskatchewan at the end of the academic year 1927–28 to accept a fellowship at Bryn Mawr College. She studied at Bryn Mawr 1928–31, the first two years as a resident fellow. Her dissertation was completed in 1931 and her PhD, with a minor in physics, was granted in 1934.

According to the Bryn Mawr College 1934 commencement program, Hughes was principal of the Islay School District in Alberta, Canada, 1931–34. Records supplied by Allen Ronaghan, a Canadian historian associated with the museum of the school in Islay, indicate that she was there 1932–34, and that she taught grades seven through ten in one room of the two-room school. It is unclear what Hughes did after receiving her PhD in 1934, although Bryn Mawr records show that she was residing in Chicago, Illinois, in 1935, possibly with a sister who was then working as a commercial artist there.

An extensive search for information about the date, place, and cause of her death has resulted in the following, written by a younger sister: "Her untimely demise in 1936 was due to a bus-train collision" (*Chain of Memories*, Maidstone Rural History Book Committee, North Battleford, Saskatchewan, 1984, 83). In addition, the grave marker at the cemetery in Maidstone, Saskatchewan, reads:

In Loving Memory
Olive Hughes, Ph.D.
Born Dec. 13, 1899
Died Dec. 29, 1936

HULL, Mary Shore (Walker). June 1, 1882–September 18, 1952.

UNIVERSITY OF MISSOURI (BA 1903, MA 1904), YALE UNIVERSITY (PHD 1909). PhD diss.: [Walker, M. S.] A generalized definition of an improper multiple integral (directed by James Pelham Pierpont).

Mary Shore Walker was born in Wentzville, Missouri, the daughter of Harriett F. (Shore) and Charles Joseph Walker. At the time of the 1900 census she was the eldest of the five living children of six born. By 1910 her father, an attorney, was widowed and had moved to Columbia, Missouri.

Mary Shore Walker went to Wentzville high school, just west of St. Louis, after which she entered Arkansas Industrial University (now University of Arkansas) in 1889. After one year at Arkansas, she transferred to the University of Missouri and was there from 1900 until 1907. She received her bachelor's degree in 1903 and her master's degree in 1904. From 1904 to 1907 she continued part-time graduate study and taught as an assistant in mathematics.

In 1907 Walker was appointed instructor at Missouri and was also granted a leave of absence to continue her studies. She was at the Yale graduate school for the academic years 1907–08 and 1908–09. After receiving her PhD in 1909, Walker returned to the University of Missouri, where she was instructor of mathematics for two years.

On June 14, 1911, Mary Shore Walker and Albert Wallace Hull, a physicist, were married in Columbia, Missouri. Albert Hull was born in Connecticut in 1880 and earned his BA in 1905 and his PhD in 1909 from Yale University. He was a member of the physics department at Worcester Polytechnic Institute from 1909 until 1914, when he joined the research staff of the General Electric Company as a physicist at its research laboratory in Schenectady, New York.

The Hulls remained in Schenectady where, from 1928 until his retirement in 1950, Albert Hull was assistant director of the laboratory. Among his many honors was election to the National Academy of Sciences. They had two children, a son born in 1917 and a daughter born in 1919. Their son earned a PhD in physics from MIT in 1943.

Mary S. W. Hull was active in a number of community organizations. She served on the board of trustees of Brown School in Schenectady and was a member of the First Presbyterian Church and the Daughters of the American Revolution. Mary Walker Hull died in Schenectady in 1952 when she was seventy. Albert W. Hull died in Schenectady in 1966 at age eighty-five.

HUMPHREYS, M. Gweneth. October 22, 1911–October 6, 2006.

University of British Columbia (BA 1932), Smith College (MA 1933), University of Chicago (PhD 1935). PhD diss.: On the Waring problem with polynomial summands (directed by Leonard Eugene Dickson).

Mabel Gweneth Humphreys was born in South Vancouver, British Columbia, the only child of Mabel Jane (Thomas) and Richard Humphreys. Her mother was born in London, England, and was a dressmaker, florist, and housewife. Her father was born in Pwllheli in Northwest Wales and was a machinist. She graduated from North Vancouver High School in 1928.

Humphreys attended the University of British Columbia, where she held scholarships all four years before graduating with honors in mathematics and earning the Governor General's Gold Medal in 1932. In 1981 she described how she came to study and live in the United States. "When I graduated in '32 the appropriation for the university had been cut so much that they stopped giving graduate work in mathematics that year. I was supposed to have an assistantship but it 'melted away.' So, [Professor F. S. Nolan] helped me apply for scholarships, fellowships in the United States and in Canada. And I received one at Smith College, a fellowship, and went there for my master's degree" (Smithsonian meeting tapes). After receiving her master's degree from Smith in 1933, Humphreys applied to the University of Chicago where she was awarded a fellowship that she held for the next two years. She earned her PhD in 1935, having written her dissertation in number theory.

Humphreys remained in the United States and became a naturalized citizen in 1941. Her first position was instructor of mathematics and physics, 1935–36, at Mt. St. Scholastica College in Atchison, Kansas. In 1936 Humphreys became an instructor at H. Sophie Newcomb Memorial College in New Orleans, Louisiana and was promoted to assistant professor five years later. In the summers of 1944 and 1946, she was assistant professor at Barnard College and Tulane University, respectively.

In 1949 Humphreys left Newcomb to become associate professor at Randolph-Macon Woman's College in Lynchburg, Virginia, where she was to remain for the rest of her career. At the end of her first year there she was named Larew Professor and head of the department. In 1973 she was also named Dana Professor; she remained as head until 1979 and retired the following year as the Gillie A. Larew and Charles A. Dana Professor Emeritus. She was on sabbatical leave 1955–56 and 1962–63. During her first leave she held a faculty fellowship from the Fund for the Advancement of Education, studied at the University of British Columbia, and examined undergraduate programs at several colleges and universities during short visits. She was a National Science Foundation faculty fellow and a visiting professor at the University of British Columbia while on her second sabbatical leave.

For most summers from 1959 to 1970 Humphreys taught in NSF summer institutes for high school teachers held at Randolph-Macon Woman's College. In the late 1960s she worked as a consultant for the Educational Testing Service; in the mid-1970s she was a consultant for the American Council on Education.

Humphreys was active in the MAA at both the sectional and national levels. She was twice vice chairman of the Maryland-District of Columbia-Virginia Section of the MAA in the 1950s. At the national level, she served on the Committee on Mathematical Personnel and Education in 1955, on the Board of Governors 1962–65 as governor of her section, and on the Committee on the Undergraduate Program in Mathematics (CUPM) 1965–67. She was an MAA consultant and visiting lecturer 1973–75.

From the 1950s through the 1970s Humphreys gave many presentations across Virginia and the rest of the country. In the 1960s she often spoke at conferences about the CUPM recommendations for the mathematics curriculum, while in the 1970s she spoke at schools about mathematics in music and art. She produced two sets of lecture notes. One, "Linear algebra and analysis," was joint with H. F. Davis and comprised notes for a course given at the University of British Columbia in 1956–57; another, "Linear algebra and geometry," appeared in many versions, the latest in 1974. While at Randolph-Macon Woman's College, Humphreys held many offices, including president, in the local chapters of AAUP and Sigma Xi.

In 1981 Humphreys listed her hobbies as gardening and reading. She was active in the Natural Bridge Appalachian Trail Club and served as a member of the council in 1954, 1955, 1970, and 1971. She continued to live in Lynchburg after her retirement in 1980 and died there at ninety-four in 2006.

HUNT, Mildred. May 8, 1888–December 14, 1975.

DENISON UNIVERSITY (SHEPARDSON COLLEGE) (BA 1909), UNIVERSITY OF CHICAGO (MA 1916, PHD 1924). PhD diss.: The arithmetics of certain linear algebras (directed by Leonard Eugene Dickson).

Mildred Hunt was born in Fairport, New York, the second of three children of Helen (Metcalf) and Horace Holmes Hunt. Her father attended Rochester Theological Seminary; he died when Mildred Hunt was nine years old. In 1900 Helen M. Hunt was matron of Shepardson College for Women in Granville, Ohio. The college became a part of Denison University that year.

Mildred Hunt prepared for college in the preparatory department of Denison University before attending the university as a student in Shepardson College (now fully merged with Denison). After her graduation in 1909, she taught in Woodland College in Jonesboro, Arkansas, and at high schools in Brownsville, Ohio, and Sistersville, West Virginia. She was an instructor of Latin and Greek at Hillsdale College in Hillsdale, Michigan, 1912–15. During the academic year 1915–16, Hunt studied at the University of Chicago; her master's thesis was directed by Dickson, and the work for her master's degree was completed in August 1916.

After receiving her master's degree, Hunt was an instructor of mathematics at Parsons College (now closed) in Fairfield, Iowa, 1916–17. She was professor of mathematics at Bessie Tift College (later Tift College, now closed), then a women's college in Forsyth, Georgia, 1918–22. She resumed her graduate studies at Chicago, wrote her dissertation under Dickson, and received her PhD in 1924.

Hunt's thirty-year career at Illinois Wesleyan University, in Bloomington, Illinois, began in the fall of 1924. She joined the two-person department as assistant professor and was promoted to professor two years later, when the department's previous professor left. She was the senior member of the department, which consisted of two people until a few years before her retirement.

Hunt served on many university committees and sometimes as freshman advisor at Illinois Wesleyan. She was secretary of faculty in the College of Liberal Arts after 1931 and assumed a major administrative position when she became registrar of the university in 1943. She remained in both of these latter positions and continued as professor of mathematics until she retired in 1954 as professor emeritus. Until she became registrar Hunt regularly attended meetings of the Illinois Section of the MAA and served as vice chairman 1939–40 and chairman 1940–41.

After her retirement Mildred Hunt moved to Lewisburg, Pennsylvania, to live with her family. Her death in 1975 at age eighty-seven in St. Petersburg, Florida, came after a lengthy illness.

HUSTON, Antoinette (Killen). November 23, 1904–July 13, 1993.

UNIVERSITY OF CHICAGO (BS 1926, MS 1930, PHD 1934). PhD diss.: [Killen, A. M.] The integral bases of all quartic fields with a group of order eight (directed by Abraham Adrian Albert).

Antoinette Marie Killen was born in Chicago, Illinois, the daughter of Anna (Harrold) and Mark Henry Killen, a salesman and later treasurer of a lumber company. Antoinette Killen received her elementary and secondary education in the Chicago public schools and in 1923 graduated from Nicholas Senn High School, where a teacher inspired her to study mathematics.

Killen entered the University of Chicago in the autumn of 1923. In 1978 she wrote to one of the authors that "all I ever wanted to be was a high school math teacher. But after graduating from U of C in 1926 no such job opened up so I settled for becoming librarian in charge of the math-physics-astronomy library." She later served as assistant secretary to G. A. Bliss, chairman of the mathematics department starting in 1928.

Killen began taking graduate courses in October 1927 and received her MS in 1930 and her PhD in 1934 as Adrian Albert's first doctoral student. Having just received her doctorate in June, on July 14, 1934, Antoinette Killen married Ralph Ernest Huston, a fellow graduate student. Ralph Huston was born in Indiana on September 16, 1902, and graduated from the University of Chicago in 1923. He was at Merton College, Oxford, as a Rhodes scholar for three years, studied at Grenoble for a year, and taught modern languages in Tennessee for two years before entering the University of Chicago for graduate study in mathematics in the summer of 1929. He earned his PhD in 1932, having written his dissertation in number theory under the direction of L. E. Dickson.

In 1934 Antoinette and Ralph Huston moved to Troy, New York, where Ralph Huston began a thirty-three year association with Renssalaer Polytechnic Institute, first as instructor and finally as professor emeritus. The Hustons had four sons; the first was born in 1935, the second in 1937, and the two younger boys within the next few years. Antoinette Huston joined the RPI mathematics department in 1955 and remained an assistant professor until 1967, the year Ralph Huston retired, when she was promoted to associate professor. Two years later, on October 8, 1969, Ralph Huston died. Antoinette Huston retired as associate professor emeritus in 1970. In 1973 Antoinette K. Huston and her sons established an annual departmental prize in Ralph Huston's honor for a graduate student who demonstrated unusual promise and ability as a teacher.

Antoinette Huston was active in the League of Women Voters and was an avid duplicate bridge tournament player. She moved to Santa Fe, New Mexico, in the 1970s and died there in 1993 at age eighty-eight. She was survived by her four sons, four grandchildren, and one great-grandchild. She was buried in Huntington, Indiana, her husband's birthplace.

INFELD, Helen (Schlauch). July 20, 1907–July 6, 1993.

NEW YORK UNIVERSITY (BA 1928), CORNELL UNIVERSITY (MA 1929, PHD 1933). PhD diss.: [Adams, H. S.] On the normal rational n-ic (directed by Virgil Snyder).

Helen Mary Schlauch was born in the Bronx, in New York City, the third child of Margaret (Brosnahan) and William Storb Schlauch. Her father earned a bachelor's degree from the University of Pennsylvania and a master's degree from Columbia University. He taught mathematics at the High School of Commerce in New York City and after 1929 was on the faculty of the School of Commerce, Accounts and Finance of New York University.

Helen M. Schlauch graduated from high school in Hasbrouck Heights, New Jersey, in 1924. She then attended Washington Square College of New York University on a scholarship and graduated in 1928 with a major in mathematics and minors in English and psychology. Schlauch next studied at Cornell University and received a master's degree in 1929. For the next several years, she served on the faculty at Hunter College in New York City and continued her graduate work at Columbia and Cornell. She was at Hunter as tutor 1929–31 and instructor 1931–32, while also taking a course yearly at Columbia during these years.

In May 1932 Helen Schlauch married Leonard Palmer Adams, who had received his master's degree from Cornell in 1930 and was studying for a PhD in labor economics. In June, Helen Schlauch Adams returned to Cornell as a scholar and took her PhD examination in July 1933 with major geometry, first minor algebra and second minor electricity. L. P. Adams was at Cornell through 1934 and received his doctorate in February 1935. They were divorced in 1936; there were no children.

Helen Adams remained on the faculty of Hunter as an instructor 1931–41. At a 1938 meeting of the AMS she met the Polish-born and -educated theoretical physicist, Leopold Infeld (1898–1968), who had received his PhD in 1921 from the University of Krakow. He had been at the Institute for Advanced Study in Princeton, had just published *The Evolution of Physics* with Albert Einstein, and was about to go to Canada to teach at the University of Toronto as a member of the applied mathematics faculty. During 1938–39 he often traveled from Toronto to New York to see Helen Adams. They were married in New Jersey on April 12, 1939, and had two children, a son and a daughter, born in 1940 and 1943, respectively, in Toronto.

While on the applied mathematics faculty, and later the mathematics faculty, of the University of Toronto, Leopold Infeld actively campaigned against nuclear weapons. After spending the summer of 1949 in Warsaw he requested a leave of absence to spend 1950–51 in his native Poland lecturing at the University of Warsaw. The leave was not granted, possibly because of unsubstantiated allegations of his link to an atomic spy ring. Nonetheless, in May 1950 he left for Poland; in August he resigned his professorship at Toronto. In Warsaw Leopold Infeld became the director of the Theoretical Physics Institute and served in this position until his death in 1968.

Before the family emigrated from Canada to Poland, Helen Infeld served in various positions in the Ajax division of the University of Toronto. The Ajax campus was created from a massive munitions plant after the war to help accommodate the returning veterans and was in existence for three and a half years. She was both assistant and reader the first six months of 1946 followed by assistant 1946–47 and instructor 1947–49, after which the Ajax campus was closed and she was unemployed. Helen Infeld sailed for Poland in July 1950, her first trip outside of North America.

Her son reported that Helen Infeld "faced the challenge of bringing up a family in a country that was completely new to her" and that "devotion to her family ... led to diminishing interest in mathematics (other than through her husband and son)" (Authors' questionnaire). Leopold Infeld had a heart condition and in 1959 suffered a stroke. Their son credits his father's survival to his mother's care.

In 1965–66 Helen and Leopold Infeld spent five months in the United States, living in Dallas, Texas, while Leopold Infeld was a visiting professor at the Southwest Center for Advanced Studies (now part of the University of Texas at Dallas). They then returned to Poland. After Leopold Infeld's death in 1968, and until 1982, Helen Infeld served as editor of *Poland (A Monthly)*, the English-language version of a periodical devoted to cultural events in Poland.

Helen Infeld received Poland's Gold Cross of Merit in 1954 and in 1970 the Chevalier Cross, Polonia Restituta, one of the highest distinctions awarded by the Polish government. She died in Warsaw in 1993 shortly before her eighty-sixth birthday.

JACKSON, Rosa L. May 7, 1883–January 15, 1967.

WESTERN COLLEGE FOR WOMEN (BA 1904), UNIVERSITY OF CHICAGO (MS 1922, PHD 1928). PhD diss.: The boundary value problem of the second variation for parametric problems in the calculus of variations (directed by Gilbert Ames Bliss).

Rosa Lea Jackson was the youngest of four children, a son and three daughters, of Mary Thom (Palmer) and James Knox Polk Jackson. Her father, a farmer, was widowed by 1900. Rosa Lea Jackson attended public elementary and secondary schools in Ripley, Tennessee, where she was born.

Jackson entered The Western, a College and Seminary for Women, in Oxford, Ohio, in 1900. After her graduation in 1904 from the renamed Western College for Women, she was at Athens Female College (now Athens State University) in Alabama until 1910, as head of the mathematics department 1904–07 and as dean and professor of mathematics 1907–10.

In 1910–11 and summer 1911 she attended the University of Chicago as a graduate student in mathematics. The following academic year she taught history and was dean at Central College for Women in Lexington, Missouri, after which she taught at Ripley High School in her home town in Tennessee 1912–14 and was an acting adjunct professor at Randolph-Macon Woman's College 1914–16. She returned to Athens College as professor of mathematics and dean from 1916 to 1921. She was also head of the department when she joined the MAA in 1920. Some of the time she was at Athens College, one of her sisters was a professor of history and the registrar there.

In 1921–22 Rosa L. Jackson again attended the University of Chicago full time, where she wrote her thesis in the calculus of variations and received her master's degree in 1922. She served as instructor at Northwestern University from 1922 until 1926 and attended the University of Chicago in the summers of 1925 and 1926. She continued at Chicago during the academic year 1926–27 as a fellow, and in summer 1927. In 1927–28 she was instructor in mathematics at Leland Stanford Junior University. She also finished her dissertation in calculus of variations and received her PhD from the University of Chicago in August 1928.

After receiving her doctorate, Jackson was an instructor at Hunter College from 1928 to 1930, following which she was instructor in the women's division of the newly opened Brooklyn College 1930–31. In 1931 she joined the faculty at Alabama College, State College for Women, in Montevallo, Alabama, where she remained as professor and head of the mathematics department until she retired as professor emeritus in 1959, when she was nearly seventy-six. The school had become coeducational and changed its name to Alabama College in 1956; it is now the University of Montevallo.

Eight years after her retirement, Rosa Lea Jackson died in Selma, Alabama. She was buried in Maplewood Cemetery in her hometown of Ripley, Tennessee.

JOHNSON, Roberta F. January 22, 1902–October 12, 1988.

WILSON COLLEGE (BA 1925), CORNELL UNIVERSITY (MA 1931, PHD 1933). PhD diss.: Involutions of order 2 associated with surfaces of genera $p_a = p_g = 0$, $P_2 = 1$, $P_3 = 0$ (directed by Virgil Snyder).

Roberta Frances Johnson was born in Philadelphia, Pennsylvania, the daughter of Mary Wallace (Abdill) and Jesse B. Johnson. Johnson later described her mother as a homemaker and her father as a mechanic.

Roberta Frances Johnson attended Frankford High School in Philadelphia before enrolling at Wilson College, a women's college in Chambersburg, Pennsylvania. After graduating from Wilson in 1925 with departmental honors in mathematics and a minor in history, she remained in Chambersburg three years, 1925–28, to teach mathematics at the Chambersburg high school. She taught mathematics and history for the next two years in the high school in Newfoundland, Pennsylvania.

Johnson attended summer sessions at Cornell University during 1929 and 1930. She enrolled at Cornell in the fall of 1930 and remained for three years. She held a fellowship from Wilson College all three years and an Erastus Brooks fellowship her second year. In February 1931 she received her master's degree, and in 1933 she received her doctorate with a dissertation in algebraic geometry and with first minor analysis and second minor philosophy.

Johnson had expected to remain at Cornell 1933–34 as a "resident doctor" in order to continue her research. However, the head of the mathematics department at Wilson became ill, and Johnson was to substitute for at least a month beginning in September. When the head took a leave of absence, Johnson remained for the academic year 1933–34 on a temporary appointment. Johnson taught at Wilson for twenty-five years. She was an instructor 1934–35, assistant professor 1935–44, and associate professor after 1944. During this time she taught all of the undergraduate courses and honors courses. She was acting head of the department second semester 1944–45 and became department head in 1946 when the previous chairman retired.

During the spring of 1957 Johnson decided that she needed to leave Wilson College and began to seek a new position. The following spring she received a promotion to full professor and a tenure contract at Wilson. Nonetheless, in 1958 Johnson moved to Fort Collins, Colorado, as an associate professor at Colorado State University. While there she directed the theses of about eight master's students. Johnson remained at Colorado State until her retirement as associate professor emeritus in 1967. The following spring the University of Colorado announced her appointment as an associate professor, and she taught at the University of Colorado at Denver for three years after her retirement.

Roberta Johnson died of bone cancer in 1988 at age eighty-six at her home in Fort Collins. During her illness she was cared for by friends and by a local hospice organization, to which she bequeathed her house. Her remaining assets were given to the Fort Collins public library.

KANARIK, Rosella (Kanarik). February 7, 1909–.

University of Pittsburgh (BA 1930, MA 1931, PhD 1934). PhD diss.: Fundamental regions in S_4 for the Hessian group (directed by Montgomery Morton Culver).

Rosella Kanarik was born in Bartfa, Hungary, the elder of two children of Sarah (Schondorf) and Albert Kanarik, both of Bartfa. Her mother had an elementary school education; her father had some high school and immigrated to the United States early in the twentieth century. He apparently returned to Hungary, where her parents were married in 1907. In 1912 her father returned to the United States where he then was in business. Rosella Kanarik and her mother arrived in New York about a year later. Her younger brother was born in Pittsburgh in 1926.

Rosella Kanarik attended public high schools, first Wadleigh High School in New York City 1923–25 and then the Fifth Avenue High School in Pittsburgh, from which she graduated with highest honor in 1926. Kanarik attended the University of Pittsburgh and received her BA degree as a student in the School of Education with high honor in 1930. She had a graduate scholarship at Pittsburgh 1930–34, was an assistant in mathematics 1931–33, and taught at Schenley Evening High School at least from 1931 until 1934. She received her master's degree in 1931 and her PhD with distinction in 1934.

Kanarik wrote in 1985, "I graduated during the depression. It was almost impossible for anyone, let alone a woman, to find a position in industry, college, or university. I was lucky to get into a high school to teach mathematics" (Smithsonian questionnaire). It appears that she taught some at the university, although not in a regular position, and in a high school in Pittsburgh 1932–36.

On July 25, 1936, Rosella Kanarik and Emery Kanarik were married. Emery Kanarik was born in 1909 in Bardejov Spa, Czechoslovakia, attended the College of the City of New York 1926–28, and earned a Bachelor of Architecture degree in 1932 after studying at the Columbia University School of Architecture. Later he was president of the architectural firm Emery Kanarik and Associates, organized in 1952 in California.

The Kanariks had two children, a son born in 1937 and a daughter born in 1940, both in Los Angeles. Rosella Kanarik taught in high schools in Los Angeles 1939–46. In 1946, soon after the end of World War II when there was a need for college mathematics teachers, she was hired as a lecturer at the University of Southern California. She remained in that position until 1952 and while there taught courses at all levels, including a graduate course. She also taught at a high school in Los Angeles 1951–53, and in 1953 was hired as the first woman member of the mathematics department at Los Angeles City College. She spent the rest of her career there, as instructor 1953–61, associate professor 1961–67, professor 1967–74, and professor emeritus after 1974. She also served as counselor 1956–62. She retired reluctantly when she reached the mandatory retirement age of sixty-five.

Since her retirement Kanarik has been a member of various retired teacher organizations and has done volunteer work tutoring high school and college students in mathematics. She has also been a member of a number of women's organizations related to education, architecture, and Zionism. Kanarik has enjoyed traveling, theatre, reading, playing bridge, and cooking. Emery Kanarik died in 1992, and Rosella Kanarik has continued to live in Los Angeles and to maintain her membership in the MAA. The Los Angeles City College awards the Rosella Kanarik scholarship each spring to a qualifying mathematics student.

KARL, Sister Mary Cordia. November 16, 1893–August 30, 1984.
HUNTER COLLEGE (BA 1916), JOHNS HOPKINS UNIVERSITY (MA 1927, PHD 1931). PhD diss.: The projective theory of orthopoles (directed by Oscar Zariski).

Elizabeth E. Mary Karl was born in New York City, the eldest of four children of Mary Anna (Klarmann) and Edward Philip Karl. Her father was an office clerk. In 1900 the family was living in the Bronx.

Elizabeth Karl attended Immaculate Conception School for eight years and then Normal College High School (later Hunter College High School) 1908–12. After her high school graduation she attended Hunter College, from which she graduated second in her class in January 1916 with a major in mathematics and minors in astronomy and physics. It was later reported that she hoped to become a nun after graduation but that her father objected strongly. Karl taught in New York City schools until 1918. She taught algebra and English at William Cullen Bryant High School in Long Island City, Queens, February 1916–February 1917; mathematics at Blessed Sacrament Academy in Manhattan February 1916–June 1917; and third grade at Public School 46 in the Bronx 1917–18.

In 1918 Elizabeth Karl entered the Order of the School Sisters of Notre Dame, whose motherhouse is in Baltimore, and took the name Mary Cordia. For most of the time from 1918 until her retirement in 1965, Sister Mary Cordia Karl was associated with the College of Notre Dame of Maryland, in Baltimore, a Roman Catholic women's college. She taught mathematics and education at the college 1918–20; mathematics, science, Latin, and history at Notre Dame High School in Baltimore 1920–22; and mathematics and physics at the College of Notre Dame after September 1922.

During most of the period from 1923 until 1931 Sister Mary Cordia was engaged in graduate work: in summer 1923 she studied at Fordham University; in fall 1923 she entered the Johns Hopkins University, where she studied, usually part time, most of the next eight years. She took courses in physics, chemistry, and mathematics. In June of 1927 she received a master's degree, based on her course work and on a master's essay on finite differences. She resumed her graduate work at Johns Hopkins in 1929 and finished her work for the PhD in 1931.

Sister Cordia was head of the mathematics department at the College of Notre Dame from 1922 until 1965 and was the religious leader of the nuns at the college from 1959 to 1965. It appears that in 1964–65 she was not teaching but served as coordinator of the science building planning project. After her retirement in September 1965 she moved to the Notre Dame Preparatory School in Towson, Maryland, where she taught and served as financial officer and then as administrative assistant until her second retirement in 1982. Later she lived in Baltimore at Villa Assumpta and the Maria Health Care Center for retired and ill School Sisters of Notre Dame.

Sister Mary Cordia Karl died at St. Joseph Hospital in Baltimore at age ninety in August 1984. In 1994 the first of the Sister Cordia Karl Mathematics Awards was conferred by the faculty of the mathematics department on an outstanding mathematics student. In 2004 an award was made by the National Science Foundation to fund a four-year program, The Cordia Karl Scholars Program, to support computer science, mathematics, and engineering students at the College of Notre Dame of Maryland.

KELLEY, Sister Mary Gervase. September 8, 1888–October 22, 1926.

Catholic University of America (Catholic Sisters College) (BA 1914, MA 1915, PhD 1917). PhD diss.: On the cardioids fulfilling certain assigned conditions (directed by Aubrey Edward Landry).

Helen Agnes Kelley was born in the Roxbury neighborhood of Boston, Massachusetts, the second of four surviving children of Mary (Callahan) and John P. Kelley, a plumber. Some records report the last name as Kelly instead of Kelley. Helen Kelley received her elementary education in St. Patrick's parochial school and graduated from St. Patrick's high school in 1905 in Roxbury. She entered the community of the Sisters of Charity of St. Vincent de Paul, Halifax, Nova Scotia, in 1906, took her first vows in 1908, and her final vows in 1915; she used the religious name Sister Mary Gervase.

Community records were destroyed in a fire in 1951, so much of the information about Sister Mary Gervase is based on material that was reconstructed after the fire. According to these records, Sister Mary Gervase taught in schools run by her religious community; she earned her "B" License in 1908 and her "A" License in 1912. According to her dissertation vita, "in 1910 she began work with the University of London, from which institution she received the Matriculation and the Intermediate Arts certificates."

It appears that Sister Mary Gervase was assigned to St. Mary's in Halifax 1908–13, to St. Patrick's in Roxbury, Massachusetts, 1913–17, and to Mount St. Vincent in Rockingham, Nova Scotia, 1917–23. In all of these assignments she was either a teacher or a student. From 1908 to 1913, she taught in Halifax public schools. From 1913 to 1917, including summer sessions, she was a student in residence at the Catholic Sisters College of the Catholic University of America, where she earned a BA in 1914, an MA in 1915, and a PhD in 1917. Her PhD was the first in mathematics earned by a woman religious in the United States. It was also one of the first two PhD's in mathematics awarded by Catholic University.

Sister Mary Gervase was a teacher at Mount St. Vincent Academy in Rockingham, Nova Scotia, from 1917 until 1923, when she became a patient at a tuberculosis sanatorium in Stellarton, Nova Scotia. Sister Mary Gervase remained there until her death at age thirty-eight in 1926.

KENDALL, Claribel. January 23, 1889–April 17, 1965.

University of Colorado (BA 1912, BEd 1912, MA 1914), University of Chicago (PhD 1921). PhD diss.: Congruences determined by a given surface (directed by Ernest Julius Wilczynski).

Claribel Kendall was born in Denver, Colorado, the daughter of Emma Gano (Reily) and Charles Martin Kendall, a lawyer who had graduated from Yale University before moving to Colorado. In 1900 Claribel Kendall had one living sister, Florence (1890–1971);

another sibling had died earlier. Emma Kendall, a widow, and her two daughters moved from Denver to Boulder in about 1910.

Claribel Kendall received both her elementary and secondary education in the Denver public schools. She entered the University of Colorado in Boulder in 1907 and earned both a bachelor of arts and a bachelor of education degree in 1912. It appears that she taught in public school at least a year before her graduation, most likely in 1908–09. During 1911–13, her senior year and her first year as a graduate student, she was an assistant in mathematics at Colorado. She became an instructor in 1913, one year before earning her master's degree in mathematics She then did further graduate work during summers at Colorado and during the summers of 1915, 1918, 1920, and 1921 at the University of Chicago. She also took a leave to study at Chicago during the academic year 1920–21. During her first two summers at Chicago she had scholarships; during her extended period of residence she had a fellowship. She received her PhD from Chicago in September 1921.

Claribel Kendall remained at the University of Colorado for her entire career except for 1920–21, the year she had a leave to study at Chicago, and the summer of 1925, when she taught at the University of California. She was promoted from instructor to assistant professor in 1922, to associate professor in 1928, and to professor in 1943. She served as acting chairman of the department 1954–55. Kendall directed the master's theses of nine or ten students, the last one in 1955, two years before she retired as professor emeritus. Among these advisees were at least six women. In addition to her work in the mathematics department, Kendall was active on various committees in the university.

Kendall was a member of Kappa Delta Pi, an education honorary society. She was active in the MAA through its Rocky Mountain Section and was president of the section in 1930. She presented papers at sectional meetings in the period 1922–42. She was secretary of the Alpha of Colorado chapter of Phi Beta Kappa 1922–64, and a scholarship in her honor was established by the chapter. She was also secretary, as well as treasurer and president, of the Colorado Mountain Club. She had an interest in bird watching, and at least twice in the 1920s she helped compile the Boulder Christmas bird census for the Audubon journal *Bird-Lore*. She held various offices in the First Church of Christ, Scientist in Boulder.

In 1956 Claribel Kendall was joined in her home in Boulder by her sister, Florence, who had retired as a high school mathematics teacher. The sisters remained in their home in Boulder after Claribel Kendall's retirement. Claribel Kendall died in 1965 at age seventy-six in Boulder. Both her local Phi Beta Kappa chapter and her local church were remembered generously in her will. Florence Kendall was eighty when she died in 1971, and contributions in her memory were requested to be sent to the Claribel Kendall Memorial Fund, in care of Phi Beta Kappa at the University of Colorado.

KETCHUM, Gertrude (Stith). August 4, 1903–September 27, 1958.

UNIVERSITY OF GEORGIA (BA 1924, MA 1928), UNIVERSITY OF ILLINOIS (PHD 1934). PhD diss.: On certain generalizations of the Cauchy-Taylor expansion theory (directed by Robert Daniel Carmichael).

Gertrude Stith was born near Nunez, Georgia, the eldest of five surviving children of Louella (Jones) and John R. Stith, a farmer. Her father had a stepdaughter, a son, and two daughters from a previous marriage.

A school superintendent who boarded with the family was an inspiration to Gertrude Stith, who was the first of her family to go to college. Immediately after graduating from the University of Georgia in 1924, Stith began teaching in high schools in Georgia. She taught in Hawkinsville 1924–25 and in Athens 1925–28. In 1928 she also completed her work for her master's degree at the University of Georgia.

From 1928 to 1930 Stith was a student assistant in the mathematics department at Brown University. She borrowed money from a local bank to finance her studies there; nevertheless, she was unable to continue at Brown for financial reasons. However, she was able to continue her graduate education after obtaining an assistantship at the University of Illinois for the year 1930–31.

After a year at Illinois, on July 6, 1931, Gertrude Stith married Pierce Waddell Ketchum, a member of the mathematics faculty. Ketchum, born September 5, 1903, in Salt Lake City, had earned his bachelor's degree in 1922 from the University of Utah and his master's degree in 1923 and doctorate in 1926 from the University of Illinois. Except for periods away as a member of the Institute of Advanced Study in 1937–38, as a fellow and visiting lecturer at Brown 1942–43, and as head of the analysis section of a gunnery school with the US Air Force 1944–45, he spent his career in the mathematics department at Illinois: as instructor 1926–28, associate 1928–37, assistant professor 1937–42, associate professor 1942–47, professor 1947–69, and professor emeritus after his retirement in 1969.

Because of anti-nepotism rules, Gertrude Stith Ketchum could not continue as a teaching assistant after her marriage. Although she was not allowed to teach, she was allowed to continue as a graduate student, which she did for the year 1931–32. Their first child, a son, was born in June 1933, after which Gertrude Ketchum resumed her studies in 1933–34, finished her dissertation in analysis, and received her PhD in 1934 with a minor in English. The following April 1935, their daughter was born.

Gertrude Ketchum had two publications in mathematics, one based on her dissertation and one that was coauthored with her husband. She continued her involvement in mathematics by proposing and solving problems that appeared in the *American Mathematical Monthly* and the *National Mathematics Magazine*. In a 1983 interview her husband said that she was hurt by not being able to teach when they first married but that some years later she was able to teach part time.

Gertrude Ketchum had always been active and athletic with a particular interest in riding. In high school she had organized a girls' basketball team. At the University of Georgia she had been captain of the rifle team and had ridden in university horse shows. Later she used her expertise in horsemanship as an instructor for five years at riding camps for girls.

In 1952 Gertrude Ketchum became ill and died at home in Urbana six years later at age fifty-five. In 1963 the daughter of Gertrude and Pierce Ketchum earned her PhD in mathematics from the University of California, Berkeley. P. W. Ketchum died in 1993.

KING, Eula (Weeks). September 13, 1882–June 30, 1967.

UNIVERSITY OF MISSOURI (BA 1908, BS 1908, MA 1909, PHD 1915). PhD diss.: [Weeks, E. A.] A symmetrical generalization of the theory of functions (directed by Earle Raymond Hedrick).

Eula Adeline Weeks was born near Louisville, Georgia, the third of five children of Luella Vienna (Tarver) and Caleb Garvin Weeks. Her father was successively a bookkeeper, a bank clerk, and later a county court clerk.

In 1900 and 1910 her family lived in Bates County, Missouri, near the Kansas border. Weeks was a teacher in the high school in Rich Hill, Missouri, from 1901 to 1907. In 1908, at age twenty-five, she received both a BA and a BS in education, and was valedictorian of the class of 1908 in the College of Arts and Science at the University of Missouri. The next year she was a university fellow in mathematics and in 1909 received a master's degree from Missouri.

Eula Weeks then went to Bryn Mawr College, where she was a fellow in mathematics 1909–10 and a fellow by courtesy 1910–11. Her occupation is listed as teacher in a preparatory school in the 1910 census, as well. After three years at Bryn Mawr, Weeks returned to Missouri, where she held a university fellowship 1914–15 and received her PhD in 1915.

In 1916 Eula A. Weeks was a teacher at Grover Cleveland High School in St. Louis. By 1919 she had been appointed to the National Committee on Mathematical Requirements as one of three secondary school representatives; the committee issued reports in 1920 and 1923 that were widely distributed. Weeks reported on the work of the committee at several meetings of the MAA, an organization she had joined in 1916 as a charter member.

Weeks continued teaching at Cleveland High School and was active in organizations for mathematics teachers. In 1920 she was elected secretary of the mathematics section of the Missouri Society of Science and Mathematics Teachers, and she served as vice president of the NCTM 1922–23 and on the NCTM board of directors 1923–26.

Eula Weeks married Harry Lane King (1885–1966) on June 15, 1924, in St. Louis. Harry King was a high school teacher in St. Louis and taught in the industrial arts department at Grover Cleveland High School at least in the mid-1940s. Eula Weeks King stopped teaching after she married, although she maintained her membership in the MAA until about 1940. The Kings had two children, a son born in 1925 and a daughter born in 1927. Their son became a civil engineer.

After her husband's death in 1966, King maintained the same residence in St. Louis she had had since at least 1930. She went to her son's home in Inglewood, California, about a month before her death there at age eighty-four in June 1967.

KLOYDA, Sister M. Thomas à Kempis. May 15, 1896–March 19, 1977.
College of Saint Teresa (BA 1920), University of Michigan (MA 1926, PhD 1936). PhD diss.: Linear and quadratic equations, 1550–1660 (directed by Louis Charles Karpinski).

Sister Mary Thomas à Kempis Kloyda was born Sophia Kloyda in Manly, Iowa, the first of three children of Mary (Kurash), a native of Iowa, and John L. Kloyda. Her father was born in Prague, Bohemia, and immigrated to the United States in 1882. By 1900 the family was living in Jackson, Minnesota, just north of the Iowa border, where her father was a bartender. Later they moved to Minneapolis, and her father worked as a machinist and her mother as a sewing machine operator. Sophia Kloyda had an older brother and a younger sister.

Sophia Kloyda earned her diploma from Columbus High School in Austin, Minnesota, in 1914. She taught second and third grades the following year. In the fall of 1915 she entered the College of Saint Teresa (now closed), a private college for women sponsored by the Sisters of Saint Frances in Winona, Minnesota. She was in residence there until 1918 and entered the Novitiate of the Sisters of Saint Francis in Rochester, Minnesota, in January 1918. She was again in residence at the college during 1919–20 and received her bachelor's degree in June 1920 with a major in mathematics and minors in botany and Latin. She was also an instructor at St. Claire Academy 1917–20.

After her graduation from college, Sister M. Thomas à Kempis taught for four years in high schools, in Minnesota 1920–22 and in Ohio 1922–24. In the summer of 1922 she began her graduate work at the University of Michigan and completed all of her work for the master's degree in November 1926 with courses taken every summer except that of 1925. She was principal of Saint Augustine High School in Minnesota during 1924–26.

In 1926 Sister Thomas à Kempis began teaching physics at the College of Saint Teresa, where she had done her undergraduate work. She continued her graduate studies at Michigan during the summers of 1928 and 1929, the academic years 1930–31 and 1934–35, and again in the summer of 1935 for work on her dissertation, which she subsequently completed in the history of mathematics under the direction of Louis C. Karpinski. She received her doctorate in February 1936, a year after **Sister M. Leontius Schulte**, also a student of Karpinski at Michigan and her colleague at the College of Saint Teresa.

Sister M. Thomas à Kempis continued her work at the College of Saint Teresa. She was head of the physics department for several years before serving as head of the mathematics department 1939–49. Her work at the college was interrupted for five years when she was asked to serve as principal of Saint Mary High School in Sleepy Eye, Minnesota, 1949–50 and of Saint Joseph High School in Ironton, Ohio, 1950–53; she was instructor of mathematics and science at Cotter Senior High School in Winona, Minnesota, 1953–54.

In 1954 Sister M. Thomas à Kempis returned to the College of Saint Teresa, where she was again head of the department of mathematics 1954–69 and professor of mathematics from 1969 until her retirement from full-time service in 1972. During her years at the College of Saint Teresa she was active professionally, especially with the Minnesota Section of the MAA, for which she was section chairman in 1934–35 and a member of the executive committee 1935–36, 1939–40, and 1962–63. She taught in NSF institutes for elementary school teachers during the summers of 1963 and 1964 and was a consultant at Florida Agricultural and Mechanical University during the summer of 1965. She was a member of the Minnesota Academy of Science. She published several historical articles; most were about individual women mathematicians. She also contributed a dozen entries on mathematicians to the 1967 *New Catholic Encyclopedia*.

Sister Mary Thomas à Kempis Kloyda worked part time at the college until September 1976, when she was eighty, at which time she was forced by failing eyesight and general health considerations to move to the Assisi Heights Motherhouse in Rochester, Minnesota. She died the following spring in Rochester.

KOHLMETZ, Dorothy Bothwell. January 25, 1906–December 18, 1941.
Oberlin College (BA 1931), Ohio State University (MA 1933, PhD 1937). PhD diss.: Certain problems of a special character in convex functions (directed by Tibor Radó).

Dorothy Bothwell Kohlmetz was the younger of two daughters of Alice Gray (Bothwell) and George William Kohlmetz. Her mother, of Albany, New York, was in the Wellesley College class of 1890. Her father's family was from Germany and immigrated to the United States when he was an infant. Most evidence indicates that Dorothy Kohlmetz was born in Albany, New York, although Cleveland, Ohio, also appears as her place of birth. Her father was a purchasing agent for the Cleveland Twist Drill Company and was also author of bibliographical notes for a collection of editions of a book of local interest. Her older sister graduated from the College for Women of Western Reserve University and became a junior high school teacher.

Kohlmetz attended public secondary school in Cleveland except for her last year. In 1924–25 she attended the Laurel School, a preparatory school also in Cleveland, from which she graduated in June 1925. She then studied at the Cleveland Institute of Music 1925–26.

Kohlmetz entered Oberlin in the fall of 1926 as a student in the Conservatory of Music but later changed her classification and graduated from the College of Arts and Sciences with a BA in December 1931. She entered Ohio State University in the winter quarter of 1932 and did course work there until spring quarter 1935, receiving her master's degree in 1933. She held a university scholarship in spring 1935 and, despite suffering from rheumatic heart disease, received her doctorate in December 1937.

Kohlmetz had been living at her family's home in Cleveland Heights, Ohio, before her death at the Cleveland Clinic Hospital at age thirty-five in 1941. She had been ill for more than six years and died a few days after major surgery.

KRAMER, Edna E. May 11, 1902–July 9, 1984.
Hunter College (BA 1922), Columbia University (MA 1925, PhD 1930). PhD diss.: [Part I] Polygenic functions of the dual variable w; [Part II] The Laguerre group (directed by Edward Kasner).

Edna Ernestine Kramer was born in New York, New York, the eldest of three children of Sabine (Elowitch) and Joseph Kramer. Her parents were born in Austria-Hungary and emigrated from there as children in the 1880s. They had high school educations, and her father was a salesman of men's clothing. Her younger sister graduated from Hunter College and was later a French teacher; her brother had a BA and an MA from Columbia and was a chemistry teacher.

Kramer attended Wadleigh High School in Manhattan and graduated in 1918 at age sixteen. She then attended Hunter College and graduated first in her class in 1922. In 1922–23 she taught at DeWitt Clinton High School in the Bronx, and in 1923–29 she taught at Wadleigh High School, while also doing graduate work at Columbia University. She earned her master's degree in 1925 and her PhD in 1930 with a minor in physics.

Kramer obtained a position at Montclair State Teachers College (now Montclair State University) in New Jersey in 1929. She remained there five years, as instructor 1929–32 and as assistant professor 1932–34. In 1934 she moved to Thomas Jefferson High School in Brooklyn as teacher and chairman of the mathematics department. The following year, on July 2, 1935, Kramer married Benedict Taxier Lassar, teacher of French and guidance counselor at Abraham Lincoln High School in Brooklyn. Lassar, born in New York City in 1906, had received a BA from CCNY, a JD from Columbia, and a Certificat d'Etudes Francaises from the University of Grenoble. He later received a master's degree from Columbia and a PhD in clinical psychology from New York University. He remained at Abraham Lincoln High School until 1962. He was also self-employed and, from 1964, was a staff psychologist at a psychotherapy center.

Edna Kramer-Lassar remained at Thomas Jefferson High School from 1934 until 1956. During this period she did further study, held several other professional positions, and was engaged in historical research and writing. She studied at the Courant Institute, New York University, 1939–40 and at the University of Chicago in 1941. She was instructor in the graduate division at Brooklyn College 1935–38; mathematics consultant in statistics for the US Office of Scientific Research and Development, the Division of War Research, Columbia University 1943–45; instructor at Polytechnic Institute of Brooklyn (now Polytechnic University) 1948–53; and instructor in the graduate division at New York University 1949–50. In 1942, during its second year, Kramer served as the vice chairman of the Metropolitan New York Section of the MAA. Her first book in history of mathematics, *The Main Stream of Mathematics*, was published by Oxford University Press in 1951. It subsequently was translated into Italian, Dutch, and Japanese.

In 1956 Kramer-Lassar left Thomas Jefferson High School but continued her teaching at Brooklyn Polytech as adjunct professor from 1953 until 1971, when she retired with emeritus rank. She was a consultant on a special project supported by the NCTM for the University of Oregon in 1963. From 1965 to 1969 she did further study at the Courant Institute, and in 1970 her book *The Nature and Growth of Modern Mathematics* first appeared. It was chosen as a Science Book of the Month Club selection. In the early 1970s, she wrote almost all of the biographies of women mathematicians that appeared in the *Dictionary of Scientific Biography*. In 1972 she was a consultant to a committee looking to place women mathematicians in Ivy League universities. In 1973 she was lecturer at Nanyang University in Singapore and was honored by election to the Hall of Fame at Hunter College. Kramer's interests included music and travel. Edna Kramer-Lassar died in Manhattan at age eighty-two in 1984.

LADD-FRANKLIN, Christine. December 1, 1847–March 5, 1930.

VASSAR COLLEGE (BA 1869), JOHNS HOPKINS UNIVERSITY (PHD COMPLETED 1882, AWARDED 1926). PhD diss.: [Ladd, C.] On the algebra of logic (directed by Charles Sanders Peirce).

Christine Ladd was born in Windsor, Connecticut, the eldest of the three children of Augusta F. (Niles) and Eliphalet Ladd. In 1850 the family lived in New York, and in 1853 they moved to her maternal grandmother's home in the Poquonock section of Windsor. Many years later Ladd described her father as having been both a merchant and a gentleman farmer. Ladd's mother died when Ladd was twelve, and soon thereafter she began keeping a journal, which is one of two diaries in the collections of Vassar College. In 1862 her father remarried and had two more children.

From 1860 to 1863, Ladd attended schools in Portsmouth, New Hampshire, while living with her paternal grandmother. She spent the next two years at the Wesleyan Academy, a coeducational college preparatory school in Wilbraham, Massachusetts, and graduated in 1865. During 1865–66 Ladd was at home in Poquonock with her infant half-sister. In September 1866 Ladd enrolled at Vassar, with financial support from her mother's older sister, Juliet Niles. After one year Ladd left Vassar because of lack of funds and taught in Utica, New York, the fall semester 1867. She returned to Vassar 1868–69 and graduated in 1869. Ladd then taught at a girls' school in Hollidaysburg, Pennsylvania. In May 1871 she took a position at a seminary in Washington, Pennsylvania. Through the end of July 1872 she studied with George C. Vose, a professor of mathematics at the local Washington and Jefferson College. Although there is no record of Ladd's having attended classes at Harvard, she wrote to her aunt in November 1872 that she was enrolled as a student there and was attending lectures of W. E. Byerly and J. M. Peirce. She was then living with relatives in the Boston area and remained there several years during which she taught at Chelsea High School and continued her informal work in mathematics.

By spring 1875 Ladd was teaching in Union Springs, New York, and in spring 1876 she also studied biology and mathematics at nearby Cornell University. Ladd remained in Union Springs through at least May 1878, when she wrote to J. J. Sylvester asking if she, as a woman, would be refused permission to attend his lectures at Johns Hopkins University. Although Sylvester strongly backed her admission to the university, the executive committee of the board of trustees only consented to her attending his lectures without charging her tuition. Ladd then spent the next four years studying at Johns Hopkins and traveling in Europe. Although she was not officially enrolled, at times she received the stipend of a fellow. She published substantial articles in early volumes of the *American Journal of Mathematics*. What should have been her dissertation for an 1882 PhD was published in *Studies in Logic by Members of the Johns Hopkins University*; the degree was not awarded because she was a woman.

On August 24, 1882, Ladd married Fabian Franklin (1853–1939), a member of the Johns Hopkins mathematics faculty. Franklin was born in Hungary, immigrated to the United States when he was four, received his bachelor's degree from Columbian College (now George Washington University), worked as a civil engineer and surveyor, and earned a PhD in mathematics at Johns Hopkins in 1880. In June 1883, their infant son died shortly after his birth. The following summer she gave birth to a daughter. While living in Baltimore Christine Ladd-Franklin sometimes taught private students; she also continued her mathematical research and began a career in the area of physiological optics with an 1887 paper that was a mathematical investigation of binocular vision. That same year Vassar College granted her an LLD, the only honorary degree it has ever conferred.

In 1891 Fabian Franklin took a sabbatical leave in Germany, where Ladd-Franklin worked in G. E. Müller's laboratory in Göttingen and in Hermann von Helmholtz's laboratory in Berlin. In 1892 Ladd-Franklin began publishing on the theory of color vision. In 1894 she returned to Berlin without her family to work for several months in Arthur König's laboratory. It was on the basis of her contributions to the theory of color vision that in 1906 Christine Ladd-Franklin was starred for psychology in the first edition of *American Men of Science*, and that she was offered an honorary doctorate by Johns Hopkins to be conferred during the university's fiftieth anniversary celebration in 1926. Rather than accept the honorary degree, she asked for, and received, the PhD in mathematics she had earned forty-four years earlier.

Ladd-Franklin also kept up her interest in logic. In 1904 she became the only woman on the Johns Hopkins faculty, teaching there as a part-time lecturer in logic and psychology. She left in 1909 when she moved to New York with her family. In 1895 Fabian Franklin had resigned his position at Johns Hopkins to become editor of the *Baltimore News*, and in 1909 he became associate editor of the *New York Evening Post*. In 1919 he started his own literary and political weekly publication, *The Review*. While in New York, Christine Ladd-Franklin taught as an unpaid part-time lecturer in logic and psychology at Columbia University.

Ladd-Franklin frequently wrote popular articles and letters to newspapers and news magazines that were mainly concerned with women's rights and women's higher education. In 1893 she gave **Mary Frances Winston (Newson)** $500 to help finance her first year of study in Göttingen. In 1897 she was involved in the formation of the Baltimore Association for the Promotion of the University Education of Women, which awarded fellowships and tried unsuccessfully to get Johns Hopkins to open its graduate school to women. At various times Ladd-Franklin served on the Association of Collegiate Alumnae (ACA) committee on fellowships and was instrumental in the establishment of the privately-funded Sarah Berliner fellowship.

Christine Ladd-Franklin's scholarly bibliography includes over one hundred articles and reviews in logic and color theory. Her correspondence was quite extensive and is preserved in the Rare Book and Manuscript Library of Columbia University. In 1929 a volume containing her collected works on color vision was published; it was reprinted in 1973 in the series Classics in Psychology. Her work in logic was examined in a 1999 article, "The Syllogism's Final Solution," in the *Bulletin of Symbolic Logic* (5:451–69). Christine Ladd-Franklin died at age eighty-two in New York City in 1930.

LANDERS, Mary (Kenny). February 5, 1905–November 18, 1990.

Brown University (Women's College) (BA 1926), Brown University (MA 1927), University of Chicago (PhD 1939). PhD diss.: The Hamilton-Jacobi theory for the problems of Bolza and Mayer (directed by Gilbert Ames Bliss and Magnus Rudolph Hestenes).

Mary Virginia Kenny was born in Fall River, Massachusetts, the eldest of six children of Katherine (Connell) and Bernard Francis Kenny. Her mother was a native of Massachusetts, and her father was born in England before immigrating to the United States in 1893. In the 1920 and 1930 US census her father's occupation was listed as letter carrier.

Kenny attended public schools in Fall River before entering Women's College in Brown University in 1922. After receiving her bachelor's degree in 1926, she stayed at Brown as a fellow in the mathematics department and received her master's degree in June 1927. Kenny was an instructor on a temporary assignment in the mathematics department at Hunter College the following year; the temporary nature of her appointment was dropped in 1928. She continued her graduate work in mathematics by taking one class at Columbia University each semester during the years 1928–30 while teaching at Hunter. Beginning in spring 1930, and continuing through spring 1932, Kenny taught in the evening and extension sessions at Hunter.

On July 30, 1932, Mary Kenny married Aubrey Wilfred Landers, Jr., a fellow mathematics student and an assistant at Brown 1926–27. Landers was born in New Hampshire in 1906, received a BA from Acadia University in Nova Scotia, and continued as assistant at Brown until 1929, when he received his MS. He was an instructor at Hunter College 1929–30 while taking classes at Columbia. Landers moved to Brooklyn College as instructor the next year, but in spring 1931 he joined Kenny teaching in the evening and extension sessions at Hunter. After their marriage Mary and Aubrey Landers continued in their positions at Hunter and Brooklyn College, respectively, and in summer 1933 they resumed their graduate work, this time at the University of Chicago. They were in residence 1933–34 and all summers 1933–39 except for 1936. They both worked in the calculus of variations and received their doctorates in December 1939.

Mary Landers continued at Hunter College as instructor 1928–47, assistant professor 1947–58, associate professor 1958–63, professor 1964–75, and professor emeritus after 1975. Except when he was on military leave during World War II, Aubrey Landers remained at Brooklyn College until he retired as professor emeritus in 1974. As an officer in the US Naval Reserve, he was on active duty in Washington, D.C., 1942–46 doing cryptographic work. Mary Landers joined him in Washington in January 1944, two days before their first son was born. On leave from Hunter College, Mary Landers remained in Washington and in September 1945 gave birth to a daughter. The family moved back to New York in May 1946, and, by adjusting their teaching schedules and later hiring a housekeeper, they both were able to resume full-time teaching. Their third child, a son, was born in January 1949.

Aubrey and Mary Landers were both engaged in early efforts to improve conditions for faculty in the public colleges of New York City. From 1959 to 1972, Mary Landers was secretary of the Legislative Conference, an organization that represented the professional staff of the public colleges in New York City and sought to influence legislation affecting the faculty and administrators. The conference merged with the United Federation of College Teachers in 1972 to become the Professional Staff Congress and the bargaining agent for the professional staff at City University of New York (CUNY). Mary Landers was co-secretary of the new organization and was co-chairman of the Hunter College unit. She also was trustee of the City University Welfare Board.

Aubrey Landers died in May 1986. Mary Landers died in Providence, Rhode Island, four and a half years later at the age of eighty-five.

LAREW, Gillie A. July 28, 1882–January 2, 1977.

RANDOLPH-MACON WOMAN'S COLLEGE (BA 1903), UNIVERSITY OF CHICAGO (MS 1911, PHD 1916). PhD diss.: Necessary conditions in the problem of Mayer in the calculus of variations (directed by Gilbert Ames Bliss).

Gillie Aldah Larew was born near Newbern, in Pulaski County, Virginia, the daughter of Gillie Augusta (Glendy) and Isaac Hall Larew. She had four older siblings, only one of whom was living at the time of her birth, and a younger sister. Her father was a farmer and a lawyer. Her mother died in 1887, and her father remarried in about 1890 and had five more children.

Gillie Larew received her primary and secondary education from private tutors before entering Randolph-Macon Woman's College (now the coeducational Randolph College) in Lynchburg, Virginia, in September 1899. Immediately after Larew's graduation in 1903 she became an instructor in mathematics there. She attended courses in the mathematics department at the University of Chicago in the summers of 1906 and 1909 and in the spring and summer quarters of 1911. She wrote her master's thesis in mechanics to complete the work for her master's degree in 1911.

Larew continued at Randolph-Macon Woman's College (R-MWC), except for leaves, for the remainder of her career. Having been made adjunct professor at R-MWC in 1909, she returned to the University of Chicago for seven quarters during the years 1914–16. She was a fellow in 1915–16, wrote her dissertation in the calculus of variations, and received her PhD in 1916. She returned to R-MWC and was promoted to associate professor in 1918 and to professor in 1921. She was acting head of the department for a year in the early 1920s. She spent the year 1929–30 in Munich studying mainly with Carathéodory.

In 1936 Larew became head of the department at R-MWC and remained so until 1950. She had become dean of the college the year before and became dean emeritus and professor emeritus upon her retirement in 1953 at age seventy. After her retirement she served on the college steering committee and helped with the alumnae development program. Shortly before her retirement, she presented her personal philosophy on the Edward R. Murrow radio program *This I Believe*. She also wrote about her thoughts on nonconformity for a 1954 article in the *AAUW Journal*.

In the 1920s Larew was secretary of the mathematics section of the Virginia Educational Conference. She was on the executive committee of the Maryland-District of Columbia-Virginia Section of the MAA in the early 1930s and was chairman 1937–38; she served as regional governor 1945–47. Larew was on the research committee of the Virginia Academy of Science in the early 1940s and was elected to the council in 1945. She was also active in the AAUW and was president of the local branch in 1922 and president of the state division 1938–40.

In 1948 the Randolph-Macon Woman's College alumnae association endowed the Gillie A. Larew Chair of Mathematics; in 1953 she was awarded the honorary degree of Doctor of Humane Letters by the college; and in 1968 the Gillie Aldah Larew Distinguished Teaching Award was established. Gillie Larew died at her home in Lynchburg, Virginia, in 1977. She was ninety-four. She was buried in the Larew Cemetery near her birthplace in Pulaski County, Virginia.

LEHR, Marguerite. October 22, 1898–December 14, 1987.
GOUCHER COLLEGE (BA 1919), BRYN MAWR COLLEGE (PHD 1925). PhD diss.: The plane quintic with five cusps (directed by Charlotte Angas Scott).

Anna Marguerite Marie Lehr was born in Baltimore, Maryland, the eldest of five children of Margaret (Kreuder) and George Lehr. Her mother was born in Maryland, and her father was born in Germany, immigrated to the United States in 1890, and became a grocer. Marguerite Lehr received her primary and secondary education in public schools in Baltimore and graduated from Western High School.

Lehr entered Goucher College in 1915 and graduated in 1919. She then went to Bryn Mawr College as an assistant to Charlotte Scott and spent her first two years there as a reader and graduate student, half time in each position. Although Lehr was awarded the M. Carey Thomas European fellowship for 1921–22, she postponed using it and remained at Bryn Mawr as a resident fellow in mathematics that year and as a scholar in mathematics 1922–23. She used the Thomas European fellowship and an AAUW European fellowship in 1923–24 to work on her dissertation and study in Rome with the algebraic geometers there. Soon after she returned from Italy in July 1924, she joined the Bryn Mawr faculty. She received her PhD in 1925 with a physics minor and was Scott's last doctoral student.

At Bryn Mawr, Lehr was instructor 1924–29, associate 1929–35, assistant professor 1935–37, associate professor 1937–55, and professor 1955 until her retirement as professor emeritus in 1967. She was an honorary fellow at the Johns Hopkins University 1931–32, her first sabbatical year, and published two more articles in algebraic geometry, one of which was coauthored by Virgil Snyder of Cornell University. During World War II Lehr taught mathematics in an engineering science and management war training program at Bryn Mawr and in a V-12 program at Swarthmore College. In 1950, during her second sabbatical, she visited the Institut Poincaré in Paris.

In 1952–53, Lehr presented a fifteen-week course, "Invitation to Mathematics," on the television show *University of the Air* in Philadelphia. This led to several articles and talks on presenting mathematics on television, and to her membership on the MAA committee on films for classroom instruction. She was a curriculum consultant for the state of Pennsylvania in 1954, a visiting fellow at Princeton University 1956–57, and chair of the Philadelphia Section of the MAA 1958–59. During 1958–59 she was also an MAA visiting lecturer and presented talks at colleges in seven states from New Hampshire to Minnesota. From 1957 until 1966 she served on a regional award committee of the Woodrow Wilson Fellowship Foundation. She was a member of the School Mathematics Study Group (SMSG) at Yale University, was an MAA representative to the NSF, and was a member of the examining committee on the new type of mathematics for the College Entrance Examination Board. In 1954 Lehr received a Goucher alumnae achievement citation, and upon her retirement from Bryn Mawr in 1967 she received the Lindback Foundation teaching award.

Among Lehr's interests were poetry, upholstering, and gardening in Maine, where she spent many summers. In 1968, the year after she retired, Lehr was living in Salisbury, Maryland, where her youngest sister had been living until her death in 1967. Lehr also spent time in Manset on Mount Desert Island in Maine, a place to which she was almost as devoted as she was to mathematics. In November 1972, she moved from Salisbury back to Bryn Mawr, Pennsylvania, where she died in 1987 at the age of eighty-nine. In 1988 gifts from former students and friends allowed the establishment of the Marguerite Lehr Scholarship Fund for undergraduates with need who have done excellent work in mathematics.

LESTER, Caroline A. April 6, 1902–December 29, 1996.

Cornell University (BA 1924, MA 1928), University of Wisconsin (PhD 1937). PhD diss.: A determination of the automorphisms of certain algebraic fields (directed by Cyrus Colton MacDuffee).

Caroline Avery Lester was born in Seneca Falls, New York, the daughter of Elizabeth (Campbell) and Frederick William Lester. Her mother had attended a school for young ladies in New York City; and her father, a surgeon, had received his MD from Columbia University in 1894. Her older siblings were a sister who was a bacteriologist with a BA from William Smith College and a brother who was a graduate of the US Naval Academy who retired as a rear admiral and then taught engineering; her younger sister earned a BA from Cornell University.

Lester attended the First Ward School and the Mynderse Academy, public schools in Seneca Falls, from kindergarten until her high school graduation in 1920. She majored in mathematics at Cornell University, which she attended on a free tuition scholarship from New York State. She graduated in 1924, having been elected to Pi Lambda Theta, a national honor and professional association in education, and to Phi Kappa Phi, a national honor society for all disciplines. After her graduation she taught for three years in high schools in New York State. During the summer of 1925 she studied education at Harvard University. In 1927 she returned to Cornell and at the end of the academic year received her master's degree in mathematics.

In 1929 Lester went to the New York College for Teachers (now the State University of New York at Albany) as an instructor and remained there except for subsequent leaves. She continued her graduate work in mathematics in the summers of 1933 and 1934 at the University of Chicago and studied at Ohio State University in the summer of 1935. At Ohio State she worked under C. C. MacDuffee and then followed him to the University of Wisconsin. After two years at Wisconsin, Lester completed her work for the PhD in 1937 with a major in algebra and a minor in analysis. She returned to Albany and was promoted to assistant professor by 1939 and to professor in 1950; she retired as emeritus professor in 1967. While at SUNY Albany she produced two manuscripts, on abstract algebra and on the theory of matrices.

Lester was on leave from her position at Albany when she was a lieutenant in the United States Coast Guard (Women's Reserve). Commissioned in February 1943, she attended the US Coast Guard Academy for a three-week course and then was stationed at the Naval Communications Annex in Washington, D.C., from March 1943 through September 1945. As a specialist in cryptanalysis, she decoded messages.

In May 1940 Lester was one of forty-three MAA members to petition for the formation of a section for upstate New York. She served as vice chairman of the Upper New York State Section, now called the Seaway Section, 1957–58 and as chairman 1958–59. From 1948 to 1951 she was an associate editor of the *American Mathematical Monthly.*

In the early 1980s Lester moved to Indianapolis, Indiana, where her younger sister was located. Lester was a member of the DAR, the National Society of Colonial Dames of America, and the Mayflower Society. She was a Republican and a member of a Presbyterian church, and was especially interested in travel and bridge. Her travels included several trips to Europe and Central America. Caroline Avery Lester was ninety-four when she died in Indianapolis in 1996. She is buried in Seneca Falls, New York.

LeSTOURGEON, Elizabeth. June 1, 1880–February 6, 1971.

Georgetown College (BA 1909), University of Chicago (MA 1913, PhD 1917). PhD diss.: Minima of functions of lines (directed by Gilbert Ames Bliss).

Flora Elizabeth LeStourgeon (also "Le Stourgeon") was born in Farmville, Virginia, the daughter of Elizabeth Mary (Vinyard) and Frederick George LeStourgeon, who farmed and operated canning companies in and near Farmville. She was the third of seven children.

Elizabeth LeStourgeon received her primary and secondary schooling in the State Female Normal School at Farmville (now Longwood University) and remained a year in postgraduate study. She graduated from the regular course with a full diploma in 1897 and remained 1897–98 as an irregular student. She had teaching positions for the next several years including in the public school in Bridgeton, New Jersey, from 1898 to 1901; in the Waynesboro, Virginia, public school starting in 1901; and at St. Katharine's School in Bolivar, Tennessee, from at least 1906.

LeStourgeon studied several summers at the University of Virginia, but, as a woman, could not receive course credit. She then entered Georgetown College in Kentucky in September 1908 at age twenty-eight. After one academic year there she received her bachelor's degree. She took additional course work in 1909–10 and also taught in the preparatory department 1908–10.

LeStourgeon was a professor of mathematics at St. Mary's College, then an Episcopal college for women in Dallas, Texas, 1910–12. She studied at the University of Chicago during the summer quarters of 1911 and 1912 and for three quarters in 1912–13 as a holder of a scholarship in mathematics. She received her master's degree in June 1913. LeStourgeon spent the next two years, 1913–15, teaching mathematics at Beaver College in Pennsylvania, then a women's college (now coeducational Arcadia University), before returning to the University of Chicago with a fellowship 1916–17 and completing her doctoral work, which made an important contribution to the formulation of some basic definitions in the calculus of variations.

After receiving her doctorate in 1917, LeStourgeon taught at the Liggett School for Girls in Detroit. She was an instructor at Mount Holyoke College for the year 1918–19 and an assistant professor at Carleton College in Minnesota the following year. In 1920 she went to the University of Kentucky, where she was assistant professor 1920–26 and associate professor 1926–46. She had leaves of absence during the years 1927–28 and 1944–45.

While at the University of Kentucky she often spoke to and served as an officer of the White Mathematics Club and the Pi Mu Epsilon chapter. She was a member of the Kentucky Academy of Science. In about 1940 LeStourgeon was active in a number of organizations including AAUW and AAUP and described herself as an Episcopalian and a Democrat.

In 1948 and 1950 LeStourgeon listed her address in the AMS membership list as Delray Beach, Florida. By 1952 she was living in Washington, D.C., but apparently continued to spend winters in Florida. By the early 1960s she had moved to Bridgeton, New Jersey, where several relatives lived. Flora Elizabeth LeStourgeon had been a member of AMS fifty-nine years at the time of her death in Bridgeton at age ninety in 1971.

LEWIS, Florence P. September 24, 1877–March 10, 1964.

University of Texas (BA 1897, MA 1898), Radcliffe College (MA 1906), Johns Hopkins University (PhD 1913). PhD diss.: A geometrical application of the theory of the binary quintic (directed by Frank Morley).

Florence Parthenia Lewis was born in Fort Scott, Kansas, the third of seven children of Monimia (Chase) and Walter Felix Lewis. She attended high school in Austin, Texas, and entered the University of Texas as a sophomore with advanced standing in 1894. When she entered she was in the group leading to a bachelor of literature; her last two years she was in the program leading to the bachelor of arts degree. Lewis graduated from the university in 1897.

Lewis remained at Texas 1897–98 as a graduate student with major subjects mathematics, philosophy, and pedagogy. Her master's degree in 1898 was awarded in philosophy. Lewis spent the next year as a fellow in philosophy at Bryn Mawr College. In 1899–1900 she was a traveling fellow from Bryn Mawr; she studied a half year at the Sorbonne and, during the summer quarter, at Zürich. Her work that year included lectures and reading on the philosophy of Spinoza, study of Aristotle (in Greek), a Kant seminar, and some lectures on experimental psychology.

After Lewis's return from Europe, it appears that she taught for a year in Mississippi. She then returned to the University of Texas where she remained for the next four years, as a graduate student in mathematics 1901–03 and as a tutor in mathematics 1902–05. During 1905–06 she earned her master's degree in mathematics at Radcliffe College after which she returned to Texas for a final year as a tutor in pure mathematics. As tutor in mathematics, Lewis taught regular mathematics courses.

In 1907 the trustees of the Johns Hopkins University voted to open graduate courses to women. In September of that year Lewis applied for admission to the graduate program and began her graduate studies at Johns Hopkins with mathematics as her major and the history of philosophy and psychology as subordinate subjects. She studied full time during 1907–08 and joined **Clara Latimer Bacon** on the faculty at Woman's College of Baltimore (Goucher College after 1910) as instructor in the fall of 1908. During her first year teaching at Woman's College, Lewis took two courses at Johns Hopkins. She remained at Women's College, and in January 1911 she made application at Johns Hopkins for her PhD. The committee on instruction in the department of philosophy, psychology, and education at Johns Hopkins accepted the records of advanced work that she had previously done at Texas, Byrn Mawr, the Sorbonne, and Zürich. Lewis was again a full-time student during 1911–12 and returned to teaching at Goucher in 1912. She received her PhD, the second granted to a woman in mathematics by Johns Hopkins, in 1913 with a dissertation in algebraic geometry.

Except for brief appointments of instructors, Bacon and Lewis were the mathematics faculty at Goucher until they were joined in 1925 by **Marion M. Torrey**. While Bacon and Lewis were on the Goucher mathematics faculty, nine women graduated who later received PhD's in mathematics, six from Johns Hopkins.

Apart from two leaves of absence, Lewis spent her entire career at Goucher. She was promoted from instructor to assistant professor in 1912, to associate professor in 1914, to professor in 1922, and to professor emeritus upon her retirement in 1947. Lewis was well respected in the mathematical community and served on the council of the AMS 1921–23, the first woman since Charlotte A. Scott last served at the turn of the century. Lewis was chairman of the Goucher mathematics department from 1931 to 1943. In addition to teaching mathematics, she was responsible for the founding of Goucher's astronomy program within the mathematics department. In 1954, seven years after she retired, the six-inch refracting Florence P. Lewis telescope was installed at Goucher.

Lewis was a charter member of the MAA, and, for various periods from the early 1920s through the early 1940s, she was on the executive committee of the Maryland-District of Columbia-Virginia Section of the MAA. She also served on the executive committee and the council of the AAUP.

Lewis remained in Baltimore, except for trips abroad, after her retirement. She was able to live in her apartment until illness required that she be in a nursing home for several weeks before her death in Baltimore at eighty-six in 1964.

LITTLE, Dorothy (Manning) Smiley. October 6, 1909–October 18, 1988.

STANFORD UNIVERSITY (BA 1933, MA 1934, PHD 1937). PhD diss.: [Manning, D.] On simply transitive groups with transitive Abelian subgroups of the same degree (directed by William Albert Manning).

Dorothy Manning was the eldest of five surviving children of Esther Mae (Crandall) and William Albert Manning. Esther Crandall earned a bachelor's degree and a master's degree in Greek from Stanford University and was a PhD student in psychology there at the time of her marriage. W. A. Manning earned a bachelor's degree from Willamette University in Oregon and a master's degree and a PhD in mathematics from Stanford. He spent his career at Stanford except for a postdoctoral year at the Sorbonne and a one-year visiting position at the University of Illinois. It was during the latter year that Dorothy Manning was born in Champaign. The other children were three daughters, twin sons who died as infants, and a son.

Dorothy Manning attended public schools in Palo Alto, California, and graduated in 1928 from Palo Alto Union High School. She entered Stanford in 1928 and received her BA with great distinction as a chemistry major in April 1933. The following year she completed the work for her MA in mathematics, with a thesis written under the direction of her father. In June 1935 her dissertation, also directed by her father, was approved. During 1935–36 and 1936–37 Dorothy Manning was an Abraham Rosenberg research fellow at Stanford, after which her PhD was granted with a minor in chemistry.

Dorothy Manning's four siblings all attended Stanford. Two sisters studied mathematics and earned BA's and MA's, each with a thesis directed by her father. One died two years later, and the other earned a PhD in mathematics at Stanford in 1941. Her youngest sister studied chemistry, and her brother earned a BA, MA, and PhD, all in electrical engineering, and became a professor at Stanford.

Dorothy Manning studied at the University of Chicago 1937–38 as a holder of a National Research Fellowship; the following year she was a member of the Institute for Advanced Study in Princeton. In 1939 she became an instructor at Wells College in Aurora, New York. She resigned after two years, and on August 20, 1941, Dorothy Manning and Malcolm Finlay Smiley (1915–1982) were married. Smiley received his PhD in mathematics from the University of Chicago in 1937, was a member of the Institute for Advanced Study in 1937–38, and taught at Lehigh University in Bethlehem, Pennsylvania, 1938–42. There were no children of the marriage.

Dorothy Manning Smiley had no formal employment after her marriage. During World War II, M. F. Smiley was in the US Naval Reserve and taught at the Post Graduate School of the US Naval Academy. He returned to Lehigh briefly after the war and then was at Northwestern University 1946–48, at the University of Iowa 1948–60, at the University of California at Riverside 1960–67, and at SUNY Albany from 1967 until his retirement in about 1981.

More than twenty years after her marriage Dorothy M. Smiley coauthored two mathematical papers with her husband, one of which also included a third author. In addition, M. F. Smiley acknowledged her assistance in articles he published in 1957 and 1968. Dorothy Smiley was quick to become involved in local activities after each of their moves; she also liked to paint.

After 1967 the Smileys lived in the Albany area, and it was there that Malcolm F. Smiley died in 1982. After his death Dorothy Manning Smiley moved from her home to an apartment. Later she met Jess Edward Little, a retired chemist, and they were married on March 26, 1988. Less than seven months after the marriage, on October 18, Dorothy Manning Smiley Little died suddenly in their home in Slingerland, New York. She was seventy-nine.

LITZINGER, Marie. May 14, 1899–April 7, 1952.

Bryn Mawr College (BA 1920, MA 1922), University of Chicago (PhD 1934). PhD diss.: A basis for residual polynomials in n variables (directed by Leonard Eugene Dickson).

Marie Paula Litzinger was born in Bedford, Pennsylvania, the eldest of five children of Katherine (O'Connell) and Rush C. Litzinger. Her father worked for the Pennsylvania Railroad as clerk, bookkeeper, and accountant. When the railroad moved its offices out of Bedford, he resigned his position and opened an insurance and real estate office there. The other children were three daughters and one son.

Marie Litzinger attended elementary and secondary school in Bedford and graduated from high school in 1916. She then attended Bryn Mawr College 1916–23, where she received both her bachelor's and master's degrees. During her final three years there as an undergraduate, 1917–20, she held several scholarships. In 1920 Marie Litzinger was awarded her BA in Latin and mathematics, magna cum laude. She was awarded the Bryn Mawr European fellowship and the Shippen foreign scholarship 1920–21, but did not study abroad until 1923. She also taught in the Devon Manor School in nearby Devon, Pennsylvania, during her two years as a master's student 1920–22.

Litzinger was awarded a mathematics resident fellowship for additional graduate study at Bryn Mawr in 1922–23, the year after she received her master's degree. During the year 1923–24 she went to Europe and attended courses at the University of Rome. The following year she taught at the Greenwich Academy in Greenwich, Connecticut.

Litzinger began her association with Mount Holyoke College in 1925. She was instructor 1925–28 and assistant professor 1928–37. During the years 1928 through 1934 she was in residence for four quarters at the University of Chicago, one in autumn 1929 on a leave of absence from Mount Holyoke; she received her doctorate in 1934.

In 1937 Marie Litzinger was promoted to associate professor and was made chairman of the department at Mount Holyoke. She remained department chairman and was promoted to professor in 1942. She held a chair in mathematics from the John Stewart Kennedy Foundation after 1948. While at Mount Holyoke, Litzinger notably strengthened the mathematics department. She was active in the Connecticut Valley Colloquium, the Mount Holyoke chapter of the AAUP, and the Connecticut Valley section of the Association of Teachers of Mathematics in New England, where she served as secretary 1936–38, vice president 1939–40, president elect 1940–41, and then president. In 1942 she and B. H. Brown of Dartmouth prepared the questions for the fifth Putnam Competition. Litzinger joined the AMS shortly after she went to Mount Holyoke and almost immediately started attending meetings of the society. From 1934 through 1949, with the exception of one year, she attended at least one, and usually several, meetings a year; most were in New York City.

Litzinger traveled in France and Italy in the summer of 1951. She was forced by illness to take a leave of absence from the college in November that year. At that time she returned to Bedford, Pennsylvania, where one of her sisters was a high school teacher. Marie Litzinger died several months later at age fifty-two. In a tribute after her death in the *Mount Holyoke Alumnae Quarterly*, a colleague recalled Litzinger's delicate humor, imaginative outlook, time for undergraduates, devotion to her family, and loyalty (36 (Aug 1952): 43).

LOGSDON, Mayme (Irwin). February 1, 1881–July 4, 1967.
UNIVERSITY OF CHICAGO (BS 1912, MA 1914, PHD 1921). PhD diss.: Equivalence and reduction of pairs of Hermitian forms (directed by Leonard Eugene Dickson).

Mayme Irwin was born in Elizabethtown, Kentucky, the daughter of Nan Belle (Farmer) and James David Irwin. She was the second of five children from her father's first marriage; there were three children from her father's second marriage. Her father, a constable in Elizabethtown in 1880, became an attorney in 1883. Mayme Irwin attended public primary and secondary schools and graduated from the Hardin Collegiate Institute, all in Elizabethtown, Kentucky. The Institute was primarily a preparatory school with a business course and a course for teachers, in which Irwin probably was enrolled.

Irwin was a high school teacher and principal 1900–11. On August 1, 1900, she married Augustus H. Logsdon (1859–1909), a businessman of nearby Munfordville. He was a widower with two young children; there were no children from his marriage with Mayme Irwin. Logsdon returned to school in 1911 at the University of Chicago, where she completed the work for her bachelor's degree in August of 1912. She began her graduate study there in 1912 and after a year took a position as the mathematics instructor and dean of women at Hastings College in Nebraska. After her first year there, she returned to Chicago for the summer and completed the courses and thesis for a master's degree. She remained at Hastings College as professor of mathematics and dean of women for three more years, 1914–17.

Logsdon was an instructor at Northwestern University 1917–19 before resuming her graduate work at Chicago, as a fellow in her first year and as an associate in her second. After completing her doctorate at age forty in 1921 she remained at the University of Chicago as instructor 1921–25, assistant professor 1925–30, and associate professor 1930–46; she retired as associate professor emeritus in 1946. In addition to her faculty duties, she served as a dean in the College of Science 1923–27 and was head of Kelly Hall, a graduate dormitory, for many years.

Logsdon's interests shifted to algebraic geometry early in her career at Chicago. During 1925–26 she studied in Rome on a foreign fellowship granted by the International Education Board. Shortly before Logsdon returned from Italy she gave a lecture in Rome based on her 1925 paper in the *Transactions* of the AMS. As noted in his autobiography, *Apprentissage d'un mathematicien*, Andre Weil attended the lecture, and it was from the offprints of this paper that he learned of Mordell's work on elliptic curves. In a private correspondence with one of the authors, Robert P. Langlands of the Institute for Advanced Study wrote, "She in fact put him on the trail of something important in his career.... [S]he recognized the interest and importance of the theorem of Mordell, and simply having drawn Weil's attention to it earns her a small place in the history of mathematics." Langlands also wrote of her, "Given her age at the time she finished her thesis, her subsequent development suggests a spirit and cast of mind that were intellectually open and adventurous."

Logsdon regularly gave advanced courses in algebraic geometry and directed the PhD dissertations of four students, including **Anna A. Stafford (Henriques)**, between 1933 and 1938. She was also interested in undergraduate education and wrote two textbooks: the two volume *Elementary Mathematical Analysis*, which appeared in 1932 and 1933 and is what might now be called a precalculus textbook, and the 1935 *A Mathematician Explains*, which treats topics prerequisite to calculus and briefly introduces both differential and integral calculus. She was an MAA governor-at-large 1940–42. She also was active in Sigma Delta Epsilon, a fraternal organization for graduate women in science, and was national president in 1939.

After her retirement from the University of Chicago in 1946 Logsdon moved to Florida, where she taught mathematics at the University of Miami until her second retirement at age eighty in 1961. She was an avid bridge player who traveled around the country to earn masterpoints. Her many travels included trips to the Caribbean and most of Europe. Logsdon died at the age of eighty-six in 1967 in Coral Gables, Florida.

MADDISON, Isabel. April 13, 1869–October 22, 1950.

University of London (BSc 1893), Bryn Mawr College (PhD 1896), Trinity College (Dublin) (BA 1905). PhD diss.: On singular solutions of differential equations of the first order in two variables and the geometrical properties of certain invariants and covariants of their complete primitives (directed by Charlotte Angas Scott).

Ada Isabel Maddison was born in Whitehaven, a seaport town in northwestern England. She was the daughter of Mary Jane (Anderson), born in Ireland, and John Maddison, a civil servant who was born in England. British census records indicate that Isabel Maddison was probably the youngest of four children.

In 1881 the family was living in Cardiff, Wales, and Isabel Maddison attended Miss Tallies School there. In June 1885 she passed the matriculation examination at the University of London, which was only an examining body and not a teaching institution at that time. She then entered the University College of South Wales and Monmouthshire in Cardiff and passed the University of London intermediate science examination in 1887. She remained at University College, Cardiff, until 1889, four years before it began awarding degrees as part of the University of Wales.

In 1889 Maddison went to Girton College in Cambridge on a scholarship given by the Clothworkers' Guild. With another first-year student, Grace Chisholm, she obtained permission to sit in on Arthur Cayley's lectures at Cambridge. Maddison and Chisholm took all their examinations at the same time including Part I of the Tripos in spring 1892 on which both earned first class honours. They both then unofficially sat for the Oxford Final Honours School in mathematics, the first women to do so.

In 1892 Isabel Maddison came to Bryn Mawr College as a graduate student where she studied mathematics and practical physics. In 1893 the University of London conferred a BSc with honours on her. During 1893–94, her second year at Bryn Mawr, she was a fellow in mathematics, and in April 1894 was awarded Bryn Mawr's first Mary E. Garrett European fellowship, which she used to study in Göttingen 1894–95.

Maddison was assistant secretary to M. Carey Thomas, the president of Bryn Mawr 1895–96 and received her PhD in 1896. Also during 1895–96 Maddison prepared a translation of an 1895 address by Felix Klein to the Royal Academy of Sciences of Göttingen, which appeared in the *Bulletin* of the AMS. From 1896 to 1904 Maddison served as secretary to President Thomas and as reader in mathematics. In 1904 her administrative position changed to assistant to the president, while in 1906 her departmental position changed to associate in mathematics. She remained in her dual departmental, administrative position until 1910 when she became recording dean and assistant to the president, positions she held until she retired in 1926. For at least a short period of time, Maddison lived with Charlotte Scott and Scott's cousin.

During her years as secretary to the president, Maddison compiled information for handbooks of universities open to women and for a statistical study of women college graduates. Soon after receiving her doctorate she published a number of book reviews in the *Bulletin* of the AMS as well as a short note on the history of map-coloring problems. Based on the work she had done at the University of Cambridge, Maddison received a BA degree from Trinity College, Dublin, in 1905.

Isabel Maddison lived in Bryn Mawr most of the time she was associated with the college. She maintained her British identification as a member of the London Mathematical Society and of the Daughters of the British Empire, and made a trip to England nearly every summer. Maddison spent her last years near Bryn Mawr in Wayne, Pennsylvania, and died there at age eighty-one in 1950.

MANGOLD, Sister Marie Cecilia. December 5, 1872–February 9, 1934.

TRINITY COLLEGE (WASHINGTON, D.C.) (BS 1910, MS 1914), CATHOLIC UNIVERSITY OF AMERICA (PHD 1929). PhD diss.: The loci described by the vertices of singly infinite systems of triangles circumscribed about a fixed conic (directed by Aubrey Edward Landry).

Josephine Margaret Mangold was born in Cincinnati, Ohio, the daughter of Mary Anna (Hemann), a native of Ohio, and Matthew Mangold, born in Bavaria, Germany. Eight of their children survived of ten born. Mangold's father was a dealer in imported wines, brandies, and liquor who died some time before 1900.

Josephine Mangold attended parochial school and later the Academy of the Sisters of Notre Dame de Namur in Cincinnati. In March 1898 she entered the novitiate of the Sisters of Notre Dame de Namur in that city and professed in August 1900. After Sister Marie Cecilia Mangold taught in the high school at Notre Dame Academy in Cincinnati, she was sent to Washington, D.C., in January 1901, to Trinity College (now called Trinity Washington University). There she was to assist in the teaching of German and to study mathematics with Sister Blandina of the original Trinity faculty.

Trinity College first offered instruction at the postsecondary level in 1900, and Sister Marie Cecilia began to teach mathematics there in 1902. The bachelor's degree was first conferred at Trinity in 1904; that year Sister Marie Cecilia succeeded Sister Blandina and then directed the mathematics department until her death in 1934. While teaching at Trinity College, she also took courses there. She completed the work for her BS degree in 1910 with major subjects mathematics and physics. She took no mathematics courses at Trinity, having been given credit for elementary mathematics courses from Notre Dame normal courses and having done all of her advanced work in both mathematics and physics privately, most likely under the guidance of faculty from the nearby Catholic University of America. She continued her private study and earned an MS from Trinity in 1914 with a mathematics major and chemistry minor.

Sister Marie Cecilia studied mathematics, history of science, and educational measurement at Catholic University. She wrote two papers in educational measurement, which appeared in 1927 and 1929. After writing a dissertation in algebraic geometry, she received her PhD from Catholic in 1929 with a minor in education. She was the second woman religious in this country to earn a PhD in mathematics; the first was **Sister Mary Gervase (Kelley)** in 1917.

During her more than thirty years on the faculty at Trinity College, Sister Marie Cecilia was either the only member, or one of two members, of the mathematics department. Among the many courses she introduced were some for teachers and students of physics and chemistry. The Pascal Circle, the Trinity College mathematics club, was organized in 1916, with Sister Marie Cecilia the faculty advisor and honorary president. In a memorial article after her death it was reported that "she was a lover of the classics, and a firm believer in the adage, 'The best mathematician is a student in every field'" ("In Memoriam," *Trinity Times*, March 1, 1934).

Sister Marie Cecilia was in ill health for a number of years before her death at the college in 1934 at age sixty-one. Among the pallbearers at her funeral were Aubrey E. Landry and Otto J. Ramler of the mathematics department at Catholic University. The Sr. Marie Cecilia Memorial Prize was awarded at least once, in 1942, to a senior mathematics major.

MARIA, May (Hickey). December 16, 1904–June 8, 2001.

RICE INSTITUTE (BA 1926, MA 1927, PHD 1929). PhD diss.: [Hickey, D. M.] A three dimensional treatment of groups of linear transformations (directed by Lester Randolph Ford).

Deborah May Hickey was born in Lumberton, Mississippi, the fifth of seven children of Edna May (Adams) and Charles Robert Hickey. In the early years of her parents' marriage, her father farmed in Alabama, where the first four children were born. The family was in Mississippi when Deborah May and a sister were born, and by 1910 was in Dallas, Texas, where May Hickey's father was working as a laborer. The youngest daughter was born in Texas.

When May Hickey was in the seventh grade, the family moved to Houston so the children could attend Rice Institute (now Rice University), which then provided tuition-free education to white residents of Houston. While there her father, who had not attended high school, worked for an oil refinery, an automobile assembly plant, and then the railroad. Her mother, who had attended high school and worked in dressmaking and millinery, took a course in stenography and became a secretary. All seven children eventually attended Rice, earning altogether six bachelor's degrees, one master's degree, and one PhD from that university, and two master's degrees from other universities.

May Hickey graduated as valedictorian from Houston Heights High School before entering Rice Institute in 1922. She was named a Hohenthal scholar after her first year and received the Graham Baker student award in 1924. She graduated from Rice in 1926 with honors in mathematics. She continued to study, as a fellow, the next three years in the graduate school there. Hickey earned her master's degree with honors in mathematics and physics in 1927. When she was awarded her doctorate in 1929, she became the first woman to earn a PhD in any field at Rice.

In 1929–30 May Hickey studied in Munich with an Alice Freeman Palmer fellowship of the AAUW. When she returned from Munich she took a position as professor of mathematics at Delta State Teachers College in Cleveland, Mississippi. Hickey remained as professor and chairman in the one- or two-person department at Delta State until 1938. She served as secretary of the Louisiana-Mississippi Section of the MAA 1932–33.

On June 2, 1938, May Hickey married Alfred Joseph Maria. Maria, born in Norfolk, Virginia, in 1896, graduated with a BS in chemical engineering from the Massachusetts Institute of Technology in 1922. He received a fellowship to Rice that fall and received a master's degree in 1923 and a PhD in 1925 in mathematics from Rice Institute. May Hickey was a student of his in both her freshman and junior years at Rice. Early in his career A. J. Maria taught at Rice, the University of Illinois, and Duke; he held an NSF fellowship abroad; he was a research fellow at Princeton; and he spent a year at the Institute for Advanced Study.

After their marriage May and Alfred J. Maria spent some time in Houston and then in Norfolk, Virginia, with his parents. In the summer of 1939, May Maria taught at the State Teachers College in Radford, Virginia. In 1939 they moved to New York City where both eventually became affiliated with Brooklyn College.

May Hickey Maria was an instructor in the evening session 1939–41 and in the summer session 1940–45 at Brooklyn College. She was a substitute in the day session 1941–42 when someone left for war-related work. She was an instructor at Queens College 1942–43. Her regular appointment at Brooklyn began after World War II, when she was an instructor 1946–55 and received tenure in 1948. She was an assistant professor 1955–62 and an associate professor 1962–75 and taught in the division of graduate studies from the mid-1950s. In summer 1959 she taught in an NSF institute for high school teachers at Arizona State University. Albert J. Maria first obtained a temporary appointment as tutor at Brooklyn College but then taught in the division of graduate studies. He was an associate professor when he died of a brain tumor in June 1964.

After her retirement in 1975, May Maria continued to live a few blocks from Brooklyn College in the apartment where she had lived since 1941. In 1993 she moved to a retirement home in Austin, Texas. She died in Austin in 2001 at age ninety-six.

MARTIN, Emilie Norton. December 30, 1869–February 8, 1936.

BRYN MAWR COLLEGE (BA 1894, PHD 1901). PhD diss.: Determination of the non-primitive substitution groups of degree fifteen and of the primitive substitution groups of degree eighteen (directed by James Harkness).

Emilie Norton Martin was born in Elizabeth, New Jersey, the eldest of three surviving children of Mary Holmes (Ford) and Robert Wilkie Martin. Her father was a surgeon, and her brother, the youngest child, became a physician and professor of medicine in Philadelphia.

Martin was prepared for college at Mrs. E. L. Head's School in Germantown, Philadelphia, and by private study. She entered Bryn Mawr College in 1890 and remained a student affiliated with the college for most of the decade. Her major areas of study as an undergraduate were Latin and mathematics, and she earned her bachelor's degree in 1894.

Martin was a graduate student in mathematics and physics during the first semester 1894–95 and was a teacher of Latin in the Bryn Mawr School in Baltimore during the second semester of that year. She resumed her studies as a fellow in mathematics 1895–96 and continued as a graduate student 1896–97. Martin and **Virginia Ragsdale**, another Bryn Mawr student, took courses with Felix Klein and David Hilbert in Göttingen during the year 1897–98; Martin was there as a holder of the Mary E. Garrett European fellowship.

Martin returned to Bryn Mawr College as a fellow by courtesy in mathematics for the year 1898–99. Although her name and dissertation title appear in the 1899 commencement program and on the 1899 list in *Science* of doctorates conferred, her PhD was dated 1901, the year her dissertation was published. The published version was mentioned in G. A. Miller's reports on progress in finite group theory that were published in the *Bulletin* of the AMS in 1902 and 1907. Martin later endowed a mathematical book fund at Bryn Mawr to express her appreciation for the guidance of Charlotte Scott, the head of the department when Martin was a student.

After leaving Bryn Mawr in 1899, Emilie Martin taught mathematics for a year in the Misses Kirk's School, Rosemont, Pennsylvania, and was a collaborator for the *Revue Semestrielle des Publications Mathématiques* 1899–1902. Martin did postgraduate work at Bryn Mawr in 1901–02 and was a private tutor of mathematics and Latin 1893–1903.

Martin took a position as instructor at Mount Holyoke College in 1903. She was an instructor there 1903–05 but was on leave of absence for the second semester of 1904–05. During this period she compiled the general index for the years 1891–1904 of the *Bulletin* of the AMS. Martin returned to Bryn Mawr for postdoctoral study 1906–07 after which she returned to Mount Holyoke College as instructor. Martin remained at that rank until 1911 when she was promoted to associate professor. She was promoted to professor in 1925 and became professor emeritus after her resignation in September 1935. She was also department chairman from 1927 through 1935.

Martin was a Presbyterian and maintained an interest in religious issues. She was a resident faculty member of Pearsons Hall, where she lived during the greater part of her career at Mount Holyoke. She also was secretary of the Mount Holyoke branch of the AAUP. For many years she spent summers with her sister in Montreat, North Carolina. In 1934 Emilie Martin was diagnosed with cancer. She resigned her position in September 1935 and died in February 1936 at age sixty-six in her apartment on the Mount Holyoke campus.

MAUCH, Margaret E. June 1, 1897–November 16, 1987.

HURON COLLEGE (BS 1919), UNIVERSITY OF CHICAGO (MS 1923, PHD 1938). PhD diss.: Extensions of Waring's theorem on seventh powers (directed by Leonard Eugene Dickson).

Margaret Evelyn Mauch was born in De Smet, South Dakota, the first of two daughters of Rose (Brekhus), who was born in Norway, and Henry Mauch, who was born in Wisconsin of German parents. Mauch grew up in the east central area of South Dakota near De Smet. At the time of the 1900 census the family was living in Lake Preston and her father's occupation was listed as harness maker. By 1910 Henry Mauch, now a banker, had been remarried for three years, and the family was living in Hetland. In much of the second decade of the century, Henry Mauch was secretary-treasurer of a local telephone company. By 1930 he was bank president. There were two sons from the second marriage.

Margaret Mauch attended elementary and secondary public schools in South Dakota. She studied at Huron College (now closed) in Huron, about thirty miles from De Smet. After receiving her bachelor's degree in 1919, she taught for a year at the Winner, South Dakota, high school in the south central part of the state and then was principal of the Edgerton high school in Minnesota for a year. The next two years, 1921–23, Mauch attended the University of Chicago; she received her master's degree with a thesis in theoretical mechanics in 1923.

Mauch taught in the high school in Jacksonville, Florida, 1924–25, and was an instructor and assistant professor at Randolph-Macon Woman's College in Lynchburg, Virginia, 1925–29. She returned to the University of Chicago to take graduate courses during the summers of 1926, 1928, and 1929. Following the summer of 1929 she remained at Chicago for the academic year 1929–30. She again took summer classes at Chicago during 1931 and 1938 and received her PhD in 1938 having written a dissertation in number theory under the direction of L. E. Dickson. Mauch had obtained some of her results considerably earlier; in a 1933 article in the *Bulletin* of the AMS on recent progress on Waring's theorem, Dickson mentions her work in a section he calls "Remarkable Empirical Generalizations of Waring's Theorem" (39:71).

Mauch was head of the mathematics department at the high school in Brookings, South Dakota, 1934–42 and then was an instructor and assistant professor at Carleton College in Northfield, Minnesota, 1942–44. She spent 1944–45 as an instructor at Michigan State College (now Michigan State University). In 1945 Mauch became assistant professor at the University of Akron, where she was to stay the remainder of her career. She was promoted to associate professor in 1950 and to professor in 1962. She retired in June 1963 but continued teaching at the University of Akron and was given the rank emeritus professor in 1966.

In 1966 Margaret Mauch and her sister moved back to Huron, where they both had attended college and where one of their younger brothers was an attorney and officer of the local utility company. Margaret Mauch died in Huron at age ninety in 1987.

MAYER, Joanna Isabel. March 6, 1904–February 28, 1991.

Dominican College of San Rafael (BA 1927), Marquette University (MA 1928, PhD 1931). PhD diss.: Projective description of plane quartic curves (directed by Harvey Pierson Pettit).

Joanna Isabel Mayer was born in Pettis County, near Sedalia, Missouri, the third of five children of Anna (Poinsignon) and Edward John Mayer. In 1910 the family was living in Dresden Township, near Sedalia, and Edward Mayer was a farmer; in 1920 the family was living in Portland, Oregon, where he was a farmer and a laborer; in 1930 they were living in Salina, Kansas, and Joanna Mayer was listed as a parochial school teacher.

Letters written during 1984–85, when Joanna Mayer was about eighty years old, describe her family's many moves and the schools she attended before she entered college. Although there is also sketchy material about places she lived and jobs she held, it has been impossible to confirm all of the details. She first attended the Sacred Heart school in Sedalia, Missouri. In about 1917 the family moved to Hollywood, California, where she attended Cathedral High School. After the "great flu" of 1918 they moved to Phoenix, Arizona, where her father and a brother sold cars. They moved back to Los Angeles, from there to San Jose, and then to Portland, Oregon, for a year. She was in the first academic class at St. Mary's Academy in Portland during the school year 1919–20, when she was fifteen years old. The family then moved to Seattle, then briefly to Spokane, and then to Kansas City, Missouri, where she went to Loretto High School, a Catholic girls' school. They moved to Florida and ended up in Nashville, Tennessee, where she went to St. Bernard's High School. From there they moved to Kansas City and then to Salt Lake City, and back to San Jose, in the San Francisco Bay area, where, with some interruptions, she and some members of her family were to spend much of the remainder of their lives.

Joanna Mayer graduated from Notre Dame High School in San Jose before entering Dominican College of San Rafael (now Dominican University of California). Apparently the family first lived in Oakland and then in San Rafael while she attended college. In May 1927 Mayer received her BA degree from Dominican College with a major in mathematics and a minor in philosophy. Mayer then entered Marquette University graduate school and remained enrolled until 1931; she received the MA degree in June 1928. It appears that during the academic year 1929–30 she was head of the department at Marymount College in Salina, Kansas, where she and her parents were living. In June 1931 she received her PhD, the first awarded by Marquette in mathematics. Her dissertation showed the construction of some of the curves enumerated in **Ruth Gentry**'s dissertation.

Joanna Mayer's name appears on the November 1931 MAA membership list with an address in Sedalia, Missouri. By some time in 1932 she was a teacher in San Jose, California. In 1937–38 she was an instructor at Seton Hill College (now Seton Hill University), a Catholic school in Greensburg, Pennsylvania, and at that time a women's college. In a March 15, 1938, article, "Please! Dr. Mayer is Math Professor, Not a Freshman!," in the *Setonian*, the Seton Hill student newspaper, she is described as "brimming over with enthusiasm for Saturday morning hikes and moonlight toboggan slides." The article concludes, "Enthusiastic, peppy and eager as she is, she cannot imagine what Seton Hill does without the Pacific Ocean. 'Don't you really miss surf riding?' was her closing remark to the reporter." Material from Seton Hill indicates that she was there to replace a faculty member on leave for the year; that she had come from San Francisco College for Women, a Catholic college in California that is now closed; and that she had previously taught at

Saint Mary-of-the-Woods College and at Clarke College. Mayer then taught 1939–41 at Xavier University in New Orleans (now Xavier University of Louisiana), the only Catholic institution among the historically black colleges and universities.

In 1956 Mayer wrote to the Marquette graduate school dean that she left teaching in 1942 and had supported herself since by buying and selling stocks. Listings in the AMS and city directories indicate that after she returned to California following her father's death in 1941, she spent most of the rest of the 1940s in Santa Clara, California, working at least some of the time as a stenographer. Mayer wrote in 1985 that during World War II she served in Washington, D.C., that in 1950 she did secretarial work for the Guided Missiles Committee, at the Pentagon, and that she worked for over thirteen years at the military personnel records center in St. Louis; we have not been able to confirm this information.

At various times during her life, Mayer was unable to work because of illness. She was a devout Roman Catholic who lived much of the latter part of her life in San Jose, California. She died in San Jose in 1991, a week before her eighty-seventh birthday.

MAZUR, Miriam F. (Becker). March 30, 1909–March 5, 2000.

Hunter College (BA 1930, MA 1932), Yale University (PhD 1934). PhD diss.: [Becker, M. F.] On relative fields (directed by Oystein Ore).

Miriam Freda Becker was born in New York, New York, the eldest of three children of Lena (Silverman) and Joseph David Becker. Her mother emigrated from Russia and her father from Austria. They lived in Manhattan at least through 1920 before moving to the Bronx some time before 1930. Joseph Becker was a life insurance salesman.

Miriam Becker graduated from Hunter College in 1930 with a mathematics major and a physics minor. She remained at Hunter as a graduate student in mathematics, finished her master's thesis in October 1931, and received her master's degree in 1932. She then entered Yale University, with a university fellowship for mathematics, for further graduate work. While at Yale she was awarded a Susan Rhoda Cutler fellowship in mathematics in 1932 and was elected a member of Sigma Xi in 1933. She received her PhD in 1934.

Becker was appointed tutor at Hunter College in 1934 and taught there three years, including the summer of 1935. She was terminated from Hunter College at the end of August 1937 with no charges preferred against her. Hers was part of a larger case brought by the Teachers Union against the Board of Higher Education regarding the extension to instructors in municipal colleges of regulations governing tenure for teachers in the public school system. Although a December 1937 ruling directed her reinstatement, in March 1938 the Court of Appeals reversed that decision. After leaving Hunter College, Miriam Becker spent the year 1937–38 at the Institute for Advanced Study in Princeton. Becker then returned to New York City and taught at George Washington High School, a public high school in Manhattan, 1938–41.

On July 7, 1940, Miriam Becker married Abraham Mazur (1911–2000), a biochemist who had received his PhD from Columbia in 1938. He went from fellow to professor and chair in the chemistry department at City College from 1932, when he received his bachelor's degree there, until his retirement in about 1975. Abraham Mazur was also on the faculty 1941–75 at the Cornell University Medical College, in New York City, and was at one time vice president for research at the New York Blood Center. The Mazurs had two children, a daughter and a son. Miriam Mazur returned to teaching in 1954 and spent the next eleven years at Brooklyn Technical High School. While at Brooklyn Tech she directed student teachers and taught advanced placement courses.

Miriam Mazur was hired by City College (CUNY) in 1964. She was assistant professor from September 1964 until she was promoted to associate professor effective January 1, 1972. She remained in that rank until her retirement in September 1977. While an assistant professor she was course supervisor for several remedial and introductory courses; from about 1970 she was course supervisor for elective courses. The Abraham and Miriam Mazur Award for Outstanding Achievement in Mathematics at CCNY was awarded in

2004. Miriam Mazur had been living in New York when she died in 2000 shortly before her ninety-first birthday. Her husband died four months later.

McCAIN, Gertrude I. July 15, 1879–April 4, 1955.

INDIANA UNIVERSITY (BA 1908, MA 1911, PHD 1918). PhD diss.: Series of linear iterated fractional functions – character of the functions.

Gertrude Iona McCain was the daughter of Alice A. (Neff) and Charles Edwin McCain. In 1880 her parents, her two-year-old brother, and she were living with her maternal grandparents in Three Rivers, Michigan, where she was born. At that time, her father was a grocer. By 1900 her family was living in Delphi, Indiana, and her father was a manufacturer of lime. In 1910 he was described as a foreman in the lime business; in 1920 her widowed father was bailiff at the county court.

McCain graduated from the Delphi high school in 1896, at sixteen. During the next four years she attended Indiana State Normal School for a year and a half and was a teacher in the high school in Delphi from 1897 to 1899. During most of the following decade, she worked, some as a librarian, and did her undergraduate work at Indiana University.

In the fall quarter of 1900, McCain entered Indiana and remained through the spring term 1902. She then interrupted her studies for five years, some of this time as librarian for the Delphi Public Library. She returned to the university for the academic year 1907–08 and graduated in 1908 shortly before she turned twenty-nine.

During the next ten years, McCain taught and did graduate work in mathematics. She was principal of the Friends' High School in Salem, Indiana, 1908–09, and began her graduate studies at Indiana in summer 1909. She continued in the fall term 1909–10, and, after a break, resumed her graduate work in summer 1910. She was a teaching fellow during the academic year 1910–11 and earned one of four master's degrees in mathematics granted by Indiana in 1911. The following year McCain was a fellow in mathematics at Bryn Mawr College.

In at least 1912 McCain was again teaching at the high school in Delphi, giving instruction in both mathematics and English, although she was also doing graduate work at Indiana in the winter and spring terms of 1913–14. She continued with a fellowship at Indiana the academic year 1914–15. In 1915 McCain took a position as professor of mathematics at Oxford College for Women (later merged with Miami University) in Oxford, Ohio, where she remained for six years. During this period, in 1918, at nearly thirty-nine, she completed the work for her PhD at Indiana with a minor in philosophy.

McCain left Oxford College in 1921, taught at Hunter College in New York City that summer, and in September 1921 became professor at Westminster College in New Wilmington, Pennsylvania, where she remained until June 1928. She started teaching at the State Teachers College (now Radford University) in East Radford, Virginia, in the summer quarter 1929 and remained as professor until the end of the academic year 1930–31. In 1931–32 she taught mathematics as a professor at Marymount College in Salina, Kansas. According to MAA membership lists, she was living in her hometown of Delphi in 1933–34 and in East Radford, Virginia, in 1935–36, in both cases without a professional affiliation. She was professor of mathematics and physics at Brenau College, a private women's college in Gainsville, Georgia, in the academic year 1937–38. While there she also served as the "head of the house" for a sorority on campus. There is no indication of any professional activity after this time although she did not stop seeking academic employment.

From December 1915, when McCain attended the organizational meeting of the MAA at Ohio State University, until the late 1930s, she maintained her affiliation with the MAA and submitted problem solutions to the *Monthly* over a period of about twenty years. At some point McCain returned to Delphi, Indiana.

Having been in ill health for several years, Gertrude McCain entered a nursing home near Delphi in Carroll County, Indiana, shortly before her death there at age seventy-five in 1955.

McCOY, Dorothy. August 9, 1903–November 21, 2001.

BAYLOR UNIVERSITY (BA 1925), UNIVERSITY OF IOWA (MS 1927, PHD 1929). PhD diss.: The complete existential theory of eight fundamental properties of topological spaces (directed by Edward Wilson Chittenden).

Dorothy McCoy was born in Waukomis in Oklahoma Territory, the elder of two children of Susan Ellen (Holmes) and Stephen Franklin McCoy. Her father was a farmer who homesteaded in Oklahoma. After he died in 1906, her mother and the two young children moved to Chesapeake, Missouri, where her mother had family. There Dorothy and her brother, Neal, attended a one-room elementary school before the family moved to nearby Marionville where the children attended high school. After Neal skipped the seventh grade, the two of them did almost all of their further schooling together.

After Dorothy and Neal McCoy graduated from high school in 1921, the family moved to Waco, Texas, where they both attended Baylor University. After they graduated with honors in 1925, both taught in high school for a year in Texas; Dorothy was in Port Arthur. Together they then began their graduate work in mathematics at the University of Iowa in Iowa City in July 1926. They earned their master's degrees a year later and continued their work for the doctorate. Both Dorothy and Neal received their doctorates in 1929 with dissertations written under the direction of E. W. Chittenden; Dorothy McCoy's first minor was education and her second minor was psychology. While at Iowa Dorothy held a scholarship for one year, a fellowship for the year 1927–28, and a graduate assistantship during 1928–29.

In 1929 Dorothy McCoy went to Belhaven College in Jackson, Mississippi, as professor and head of the mathematics department. She taught as a visitor at Baylor in summer 1930 and attended classes at the University of Chicago as a visiting PhD in summer 1931. She also studied during a summer at Columbia University. McCoy was on the editorial board of the predecessors of *Mathematics Magazine* 1933–45 and was active in the Louisiana-Mississippi Section of the MAA; she was secretary 1932–37, chairman 1937–38, and vice chairman for Mississippi 1938–39.

McCoy moved to Wayland Baptist College (now University) in Plainview, Texas, in 1949. She went as professor of mathematics and chairman of the division of physical and biological sciences, a position she kept until 1972. She was a Fulbright professor at Baghdad University in Iraq in 1953–54. She attended institutes for college teachers and worked summers at Patrick Air Force Base in Florida and Aberdeen Proving Ground in Maryland. She also taught summers at Northwestern State Teachers College (now Northwestern Oklahoma State University), the University of Hawaii, Baylor, and the University of New Mexico. She served on the council of the AAUP in the early 1960s. In 1975 McCoy retired as distinguished professor emeritus of mathematics. She then taught children of missionaries in Malawi in southeastern Africa for six months, and a year later she taught at a similar school in Indonesia for three months. These experiences caused her to found a mission interest group in Plainview, with which she remained involved for the rest of her life.

Honors McCoy received include the establishment of the Dorothy McCoy lecture series in 1980, the Wayland Alumni Association plaque for distinguished service to students in 1982, and the Roy McClung service plaque for her concern for foreign students at Wayland Baptist. In 1999 she was the first to receive the Distinguished Lifetime Service Award from the Association of Former Students. In 2001 she was inducted into the Division of Mathematics and Science hall of honor at Wayland Baptist, and an honors dormitory there was dedicated in her name. A generous gift from her estate to Wayland Baptist was used to establish the Dorothy McCoy scholarship fund for mathematics.

Dorothy McCoy was active in the First Baptist Church in Plainview. Her hobbies included travel, photography, and in later years, bird watching. In addition to her trips to Iraq and Africa, she traveled to Central and South America and to India. She maintained her home in Plainview and often visited her brother, Neal, in Northampton, Massachusetts, where he taught at Smith College. Neal McCoy died in January 2001, and in September 2001 Dorothy McCoy moved to an assisted living facility in Plainview. She died there at age ninety-eight less than two months later.

McDONALD, Emma (Whiton). August 3, 1886–July 7, 1948.
COLORADO COLLEGE (BA 1909), UNIVERSITY OF CALIFORNIA (MA 1916, PHD 1932). PhD diss.: Magic cubes which are uniform step cubes (directed by Derrick Norman Lehmer).

Emma Kirtland Whiton was born in Brooklyn, New York, the younger daughter of Ella E. (Lewis) and Erastus T. Whiton. In 1900 the family was living in Pueblo, Colorado, and two children were living of three born. Her father was an undertaker.

Emma Whiton graduated from Centennial High School in Pueblo, Colorado, in 1904. It appears that she taught for a year there before entering Colorado College in Colorado Springs, where she earned her bachelor's degree cum laude in 1909. While at Colorado College, her activities included Hypatia (a literary and social club), the Dramatic Society, and the YWCA. After her graduation, she returned to Centennial High School, where she taught history 1909–11 and then mathematics until 1916, with the possible exception of 1914–15 when she is not listed in the school yearbook. Whiton was a graduate student at the University of California living in a hotel in Berkeley in fall 1916, and her master's thesis, written under the direction of J. H. McDonald, was completed and approved by December 1916. She was in residence at the university 1916–17 and the first semester 1917–18, when she was a PhD candidate in mathematics.

Whiton was professor of mathematics at the University of Redlands in southern California from January 1918 until 1923, when she joined the faculty as associate professor at Mills College. She was also head resident of Mills Hall in January 1925 when she resigned from the college following her December 19, 1924, marriage to her master's thesis advisor, John Hector McDonald (1874–1953). J. H. McDonald was born in Toronto, Ontario, Canada, earned his PhD at the University of Chicago in 1900, and joined the faculty at California in January 1902.

By about 1929 Emma Whiton McDonald had resumed her graduate studies at the University of California, and it appears that she and J. H. McDonald were living separately by 1930. She took her final exam for the PhD in May 1932 and received the degree that year with a dissertation in number theory and a minor in astronomy. By 1933 she was using the name Kirtland McDonald, and at some point she and J. H. McDonald were divorced. In 1944 he married Sophia Levy, a member of the mathematics faculty at Berkeley who had a PhD in astronomy.

The last entries in MAA directories for Emma Kirtland McDonald appear under Mrs. Kirtland McDonald in 1933–34 at a hotel in Los Angeles and in 1935–36 with her address unknown. She lived in San Marino and in South Pasadena and worked for some years as a real estate broker in southern California. Emma Kirtland McDonald was a member of the Daughters of the American Revolution. She had been living in South Pasadena, California, for seven years before her death there from cancer at age sixty-one in 1948.

McFARLAND, Dora. April 18, 1895–November 26, 1987.
MONMOUTH COLLEGE (BA 1916), UNIVERSITY OF OKLAHOMA (MA 1921), UNIVERSITY OF CHICAGO (PHD 1936). PhD diss.: Division algebras defined by non-Abelian groups (directed by Leonard Eugene Dickson).

Dora McFarland was born near Aledo, Illinois, the first of two children of Nell (Blayney) and Edmund Curry McFarland. In 1900 this farm household also included a son, as well as Edmund McFarland's father and a boarder.

Dora McFarland attended grade school and high school in Aledo, a small town in the northern part of the state near the Mississippi River. She then attended nearby Monmouth College. After her graduation from college in 1916, she taught in high schools for a number of years: in Gravity, Iowa, 1916–17; Aledo, Illinois, 1917–19; and Sterling, Illinois, 1919–20.

In the summers of 1919 and 1920 McFarland took classes at the University of Oklahoma before completing her work for her master's degree in 1921. She joined the faculty there as instructor in 1921. Except for the years she was in residence at Chicago for her doctoral work and for leaves, McFarland remained at Oklahoma the rest of her career. She was instructor 1921–27, assistant professor 1927–39, associate professor 1939–48, professor 1948–65, and emeritus professor after her retirement in 1965 at the age of seventy.

McFarland attended the University of Chicago for four quarters beginning in fall 1924 while on leave from Oklahoma. While on a sabbatical in 1928–29, she returned to Chicago on a fellowship for four more quarters. She also took courses in spring and summer 1936 and received her doctorate in 1936.

During the 1930s and early 1940s McFarland often served as an officer or sponsor of the Oklahoma chapter of Pi Mu Epsilon, and in 1945–46 she served as chair of the Oklahoma Section of the MAA. In the mid-1950s she contributed to a series of tape recordings made by members of the faculty and sold under the name "Tapes for Teaching." McFarland also ran several summer institutes for school mathematics and science teachers. Much of her teaching at the University of Oklahoma was in mathematics education. In 1966 she and Eunice M. Lewis, a colleague in the education department, published a comprehensive textbook that was compared favorably with the National Council of Teachers of Mathematics' 29th Yearbook, *Topics in Mathematics for Elementary School Teachers*, in a review in *Choice*, where it was "highly recommended" (4 (1967): 712).

McFarland was an elder in the Presbyterian church; was a member of the national board of the YWCA; and was the Oklahoma founder, president, and international first vice president, of Delta Kappa Gamma, an honorary society for women in education. She described herself as a Democrat in the early 1960s.

In 1949 McFarland was a recipient of the Matrix Table award given by Theta Sigma Phi (now Women in Communications, Inc.), and in 1953 she was given the University of Oklahoma Foundation teaching award. In 1960 she received the Delta Kappa Gamma International achievement award. Gamma State, the Oklahoma state organization of the Delta Gamma Kappa Society International, now offers a Dora McFarland scholarship for graduate study to a member of the society who is a resident of Oklahoma. McFarland took her last leave in the fall of 1959 to take a trip to India and Southeast Asia. Dora McFarland was living in Norman, Oklahoma, at the time of her death at age ninety-two in 1987.

McKEE, Ruth (Stauffer). July 16, 1910–January 9, 1993.

Swarthmore College (BA 1931), Bryn Mawr College (MA 1933, PhD 1935). PhD diss.: [Stauffer, R.] The construction of a normal basis in a separable normal extension field (directed by Emmy Noether).

Ruth Caroline Stauffer was the second of three children of Hannah (Henry) and Charles Christian Stauffer, a physician. Her father received degrees from Franklin and Marshall College and the Medical School of the University of Pennsylvania. Her mother earned a teaching certificate and was an elementary school teacher in a one-room schoolhouse before her marriage; after her marriage she worked as the first assistant in her husband's office. All three children were born in Harrisburg, Pennsylvania, and received their undergraduate degrees from Swarthmore College. Ruth Stauffer's older sister received an MA from Columbia, studied at Harvard, and taught mathematics in a high school and in a community college. Her brother received a PhD in physical chemistry from Harvard, and was a college and university chemistry professor.

Ruth Stauffer received her elementary and secondary education in the public schools of Harrisburg. She graduated from the all-girls Central High School at age fifteen in 1926 and then did an additional year at the new coeducational William Penn High School before enrolling at Swarthmore College. She earned her undergraduate degree at Swarthmore in 1931 with a major in mathematics and with honors in mathematics, physics, and chemistry.

Arnold Dresden, her teacher and advisor there, urged Stauffer to go to Bryn Mawr College and helped her obtain the mathematics graduate scholarship for her first year, 1931–32. She received her master's degree in 1933 and continued her work at Bryn Mawr the following year as a scholar of the Society of Pennsylvania Women in New York. Stauffer, who was Emmy Noether's only doctoral student in the United States, was one of four women, the only one without a doctorate at the time, who studied with Noether at Bryn Mawr in 1934–35. Noether died suddenly in April 1935, a great blow to Stauffer. Her doctoral examination was given by Richard Brauer, then at the Institute for Advanced Study. After receiving her degree in June 1935 Stauffer spent 1935–36 teaching mathematics at the Bryn Mawr School in Baltimore and doing postdoctoral work in algebra with Oscar Zariski at the Johns Hopkins University. She taught mathematics at Miss Fine's School in Princeton, New Jersey, 1936–37.

During her last year at Bryn Mawr College, Ruth Stauffer met George W. McKee, then a law student at the University of Pennsylvania. They married in 1937. George McKee was born in 1908, graduated from Princeton in 1931, and received his law degree in 1934. He had his own law practice in Harrisburg from 1936 until 1990, although he was semi-retired from about 1980.

During 1938–39, while living in Harrisburg, Ruth McKee taught algebra as an instructor at Bryn Mawr until a pregnancy prevented her from commuting to Bryn Mawr. Their first daughter was born in October 1939 and their second in December 1941. Their elder daughter received a bachelor's degree from Bryn Mawr and a master's degree in English from Yale. The younger received her bachelor's degree from Swarthmore and did graduate work in psychology at Temple University and the University of Pennsylvania.

Ruth McKee did not work outside the home until 1953, when she began her employment with the Joint State Government Commission in Harrisburg. As an analyst she did mathematical statistics for this non-partisan research agency for the Pennsylvania state legislature until her retirement in 1980. She planned and directed analyses and reports in areas such as health, education, and benefits.

Ruth McKee's non-mathematical interests included story telling, painting, weaving, bird watching, and teaching Sunday school. She was involved with parent education groups through various local civic clubs in Harrisburg. In the mid-1960s the McKees moved to Ruth McKee's parents' former summer home just northwest of Harrisburg.

In 1980 George and Ruth McKee moved to Pennswood Village, a retirement community in Newtown, Pennsylvania. While there they were both active in volunteer work. She resumed her interest in art, developing her abilities in sculpture, and became active in the Society of Friends. Ruth McKee died at Pennswood Village at age eighty-two in 1993. George McKee died there just over two years later.

McMILLAN, Audrey (Wishard). July 7, 1914–January 10, 2008.

Northwestern University (BS 1935), Radcliffe College (MA 1936, PhD 1938). PhD diss.: [Wishard, A.] Functions of bounded type (directed by Lars Valerian Ahlfors).

Elizabeth Audrey Wishard was born in Jubbulpore (now Jabalpur), India, the only child of Lena (Linn) and Glenn Porter Wishard. Her parents both received bachelor's degrees from Northwestern University. Her father was YMCA secretary, foreign service, and served in Jubbulpore 1910–15; Bangalore, India, 1919–22; Colombo, Ceylon (now Sri Lanka), 1923–33; and Manila, Philippines, 1936–45. Between his two India postings he apparently attended the University of Chicago and received a master's degree. Audrey Wishard first attended the Kodai School (now the Kodaikanal International School), a boarding school founded primarily for the education of children of American missionaries, in southern India. In 1927 she and her mother moved to Pasadena, California, where she continued her schooling and attended Pasadena Junior College 1929–31. Her parents were divorced in 1932.

Audrey Wishard entered Northwestern University in 1931 and graduated in 1935. She studied at Radcliffe College 1935–38 and was a graduate assistant her last two years there. She earned a master's degree in 1936 and a doctorate in 1938, after which she was an instructor at Vassar College 1938–42.

In June 1942, Wishard moved to Princeton, New Jersey, where she soon met and, on September 2, married Brockway McMillan. Brockway McMillan was born in 1915 in Minneapolis, Minnesota. He received a PhD in mathematics in 1939 from MIT and was a Proctor fellow, Fine instructor, and research associate at Princeton University 1939–42. In 1942–43 Audrey McMillan was a member of the School of Mathematics and an assistant to Hermann Weyl at the Institute for Advanced Study in Princeton. Brockway McMillan was in the US Naval Reserve after completing officer training school. They were both in Dahlgren, Virginia, starting in early summer 1943, and Audrey McMillan worked for the Bureau of Ordnance as an associate in mathematics at the US Naval Proving Ground there. They remained in Dahlgren through 1945, when, at the end of December, he was ordered to Los Alamos in New Mexico.

In 1946 Brockway McMillan began a long association with Bell Telephone Laboratories in New Jersey: in Murray Hill 1946–61 and in Whippany 1965–79. His last position with Bell Labs was vice president for military systems. From 1961 to 1965 he was with the US Air Force in Washington, D.C., as assistant secretary for research and development 1961–63, and as under secretary 1963–65. He was elected to the National Academy of Engineering in 1969.

The McMillans had three children; their daughter was born in 1946 when they were still in Los Alamos, and their sons were born in 1947 and in 1952. Their daughter earned a PhD from Yale in Far Eastern languages and literature, and has worked primarily as a computer scientist in computational linguistics. Their older son earned a master's degree in architecture from the University of California, Berkeley, and an MBA from the Wharton School, University of Pennsylvania, and went into business for himself. Their younger son received a bachelor's degree from the University of California, Berkeley, and has worked mainly as a computer programmer and computer software consultant.

In the early 1960s Audrey McMillan turned her attention to elementary school teaching and consulting. When Brockway McMillan was working for the Air Force in 1961–65, Audrey McMillan was an instructor at Georgetown Day School in Washington, D.C. She was a consultant to the Summit, New Jersey, board of education 1966–76 and led seminars in local schools. She also taught a course in "new math" for elementary school teachers in summer 1967 at New York University and developed teaching materials for early primary students.

McMillan was a Democrat from 1936 and a Unitarian 1948–70. Her hobbies were painting and music. In 1980 the McMillans moved from Summit, New Jersey, to Sedgwick, Maine, where they had summered since 1967. Brockway and Audrey McMillan both served on the board of overseers for the Kneisel Hall Chamber Music Festival in nearby Blue Hill, Maine. After retiring, they traveled to Greece, France, Britain, and Alaska. Audrey McMillan died in Sedgwick, Maine, in January 2008, at age ninety-three.

MEARS, Florence M. May 18, 1896–December 3, 1995.

GOUCHER COLLEGE (BA 1917), CORNELL UNIVERSITY (MA 1924, PHD 1927). PhD diss.: Riesz summability for double series (directed by Wallie Abraham Hurwitz).

Florence Marie Mears was born in Baltimore, Maryland, the daughter of Florence Kate (Waidlich) and Frank G. Mears. Her mother attended college for one year, while her father had a high school education; they married in 1895. Her father was manager of a farm implements store. Her twin brother was her only sibling.

The Mears family lived in Baltimore while Florence Mears was growing up. She graduated as a mathematics major from Goucher College in 1917. She assisted in the physical and chemical laboratories at Eastern High School in Baltimore for five years, presumably just after her graduation from Goucher.

Mears began her formal graduate studies during the summer of 1922 at the Johns Hopkins University. The following summer she was at Cornell University, where she enrolled in a course in education and two mathematics courses. She remained at Cornell during the academic year 1923–24 and earned her master's degree in June 1924. Mears returned to Cornell in 1925 and received her doctorate in June 1927 with a minor in physics. She held a graduate scholarship in mathematics 1925–26 and a fellowship her last year. After receiving her PhD, Mears served for one year as head of the department at Alabama College, then a women's college and now the coeducational University of Montevallo. She was acting assistant professor at Pennsylvania State College the following year, 1928–29.

In 1929 Mears began her career at George Washington University in Washington, D.C., as an assistant professor. In 1930 she was living with her widowed mother in nearby Bethesda, Maryland, where she continued to reside. She was promoted to associate professor in 1936 and to professor in 1944. The last two of her six research papers appeared after her promotion to professor. Mears directed nine master's theses between 1949 and 1964 and two doctoral dissertations at George Washington. Her first doctoral student received his degree in 1958, when Mears was sixty-two, and her second two years later. She was also a member of the examining committee for doctoral dissertations in mathematics for Allahabad University in India in 1962.

In 1939 Mears was elected president of the George Washington University faculty club; she was the first woman elected to that office. The following year she was an officer of the local chapter of Phi Beta Kappa. She was also active in the Maryland-District of Columbia-Virginia Section of the MAA; she was secretary of the section 1949–50 and was on the executive committee 1951–52.

Mears retired from George Washington University as professor emeritus in June 1966. At that time the president noted her excellence in teaching and her contributions to research and to the department. After her retirement from George Washington, Mears was immediately hired by Howard University, also in Washington, D.C., where she taught for another ten years. After her second retirement, at age eighty, Mears remained in Bethesda, Maryland, where her hobbies included gardening, with a special interest in roses, and textile-based handicrafts. She was particularly noted as a weaver. Florence Mears died in Bethesda at age ninety-nine in 1995.

MERRILL, Helen A. March 30, 1864–May 2, 1949.

Wellesley College (BA 1886), Yale University (PhD 1903). PhD diss.: On solutions of differential equations which possess an oscillation theorem (directed by James Pelham Pierpont).

Helen Abbot Merrill was born in Orange, New Jersey, the fifth of seven children of Emily Dodge (Abbot) and George Merrill. Her father graduated from high school and was in the life insurance business, first in Newburyport, Massachusetts, and later in New Brunswick, New Jersey; from about 1860 he was a merchant. The three sons who survived to adulthood all graduated from theological seminary.

Merrill attended public schools in Newburyport and in New Brunswick. She took the classical course as an undergraduate at Wellesley, with her main subjects being mathematics, Greek, Latin, science, and history, and graduated in 1886. After her graduation, Helen Merrill taught at the Classical School for Girls in New York City 1886–89 and at Walnut Lane School in Germantown, Pennsylvania, 1891–93. As a volunteer, she taught mill girls for the Dutch Reformed Church in New Brunswick in the years 1889–91. She began her long tenure as a faculty member at Wellesley College in 1893. During her first year teaching Merrill also studied mathematics at Wellesley. She was an instructor during 1893–1901 except for the year 1896–97, when she was a graduate student at the University of Chicago. Helen Merrill was promoted to associate professor in 1901 but had a leave of absence for study at Göttingen and for travel in England and Italy during the year 1901–02 and for study at Yale during the academic year 1902–03. She received her PhD from Yale in 1903.

During the rest of her career at Wellesley, Merrill was associate professor 1901–15, professor 1915–31, Lewis Atterbury Stimson professor 1931–32, and professor emeritus 1932–49. She was chairman of the department from 1916 until her retirement in 1932. She had a leave of absence in 1914–15 for the Napier Conference in Edinburgh and for travel in Great Britain, Canada, and the United States; she also attended a summer session at the University of California in 1915. In 1922–23 she had another leave for work in English university libraries and for travel in France and Italy.

Merrill was a particularly active member of the MAA; she was an associate editor of the *American Mathematical Monthly* 1916–19, served on the executive council 1917–19, and was vice president in 1920, the year the executive council became the board of trustees. Her publications include a textbook, *A First Course in Higher Algebra*, coauthored with her colleague **Clara E. Smith**, published in 1917. Merrill also wrote mathematical poetry and songs, some of which appeared in the *Mathematics Teacher*. Her highly acclaimed book, *Mathematical Excursions*, often used as a source for mathematics club presentations, was published in 1933 and was reprinted by Dover Publications in 1958. In 1944, a dozen years after her retirement, she presented to the department "A History of The Department of Mathematics, Wellesley College from the Opening of the College in 1875."

Merrill was a member of the College Settlements Association, Consumers' League, National Child Labor Association, and American Society For Judicial Settlement of International Disputes. In the summer of 1917 she taught trigonometry in an aviation camp. She also worked with the College Entrance Examination Board. She was a member of the Religious Education Association and the Massachusetts Civic League. In 1942 she described herself as a Republican and a member of the Congregational church, formerly a Presbyterian. She was also interested in music and in model-making and other handicrafts. Helen Merrill died in 1949 at age eighty-five in Wellesley and was buried in Brooklyn, New York. A special AMS fund was established by her estate to further mathematical research.

MERRILL, Winifred (Edgerton). September 24, 1862–September 6, 1951.

WELLESLEY COLLEGE (BA 1883), COLUMBIA COLLEGE (PHD 1886). PhD diss.: [Edgerton, W.] Multiple integrals (1) Their geometrical interpretation in Cartesian geometry; in trilinears and triplanars; in tangentials; in quaternions; and in modern geometry. (2) Their analytical interpretation in the theory of equations, using determinants, invariants and covariants as instruments in the investigation (directed by John Howard Van Amringe).

Winifred Haring Edgerton was born in Ripon, Wisconsin, the daughter of Clara (Cooper) and Emmet Edgerton. In 1860 her father was a farmer in Ripon, but in the mid-1860s the family moved to New York City, where he was a real estate operator.

Winifred Edgerton received her precollege education from private tutors and graduated from Wellesley College in 1883. She then moved back to New York City and lived with her parents, who supported her desire to continue her study of mathematics and astronomy at Columbia College. Although the college, which offered graduate degrees in pure science until the formation of Columbia University in 1896, had not awarded degrees to women, Edgerton had the backing of Columbia's president, Frederick A. P. Barnard. During 1883 Edgerton did independent work in mathematical astronomy, and in January 1884 she applied to Columbia to pursue her studies in the observatory. While working in the observatory, Edgerton also taught at Mrs. Sylvanus Reed's Boarding and Day School for Young Ladies and attended classes at Columbia. In June 1886 Winifred Edgerton was awarded the PhD cum laude and, thus, became the first woman to receive a degree from

Columbia and the first American woman to receive a PhD in mathematics. Although she later reported writing two dissertations, one in mathematics and one in astronomy, and her degree is often listed as being in astronomy, all official records indicate that the degree was awarded on the basis of a dissertation in mathematics.

After receiving her PhD, Edgerton remained at Mrs. Reed's school for a year as vice-principal until her marriage on September 1, 1887, to Frederick James Hamilton Merrill. Frederick Merrill was born in 1861 in New York City, received a bachelor's degree from the Columbia School of Mines, and worked on the New Jersey Geological Survey. At the time of their marriage, F. J. H. Merrill was an instructor in paleontology and geology at Columbia; he received his PhD in 1890. The first of their four children, a daughter, was born in 1888 in New Jersey. From 1890 to 1904 Frederick Merrill held various positions with the state of New York in Albany, including director of the New York State Museum. The Merrills' sons were born in 1890 and 1901, and a daughter was born in 1897. In 1904 Frederick Merrill left Albany to become a mining geologist in New York City. In 1907 he moved to Arizona and in 1913 to California, where he died in 1916. Winifred Edgerton Merrill and her children remained in the New York City area.

While raising her family, Winifred Merrill taught at the Emma Willard School in Troy 1894–95, served as an alumna trustee of Wellesley 1898–1904, was president of the Wellesley Alumnae Association 1889–90, and served on the Association of Collegiate Alumnae committee on collegiate administration 1899–1904. Earlier, in 1888, Winifred Edgerton Merrill had been asked to serve on a small committee to form a separate woman's college to be part of Columbia. Because there were men on the committee and it met in an office in downtown Manhattan, Frederick Merrill disapproved. Although she resigned from the committee, her name appears on the request to Columbia's trustees asking for the establishment of a woman's college. The resulting school, Barnard College, was founded in 1889.

After her return to the New York City area in 1904, Winifred Edgerton Merrill resumed her teaching career as the head of the mathematics department at Highcliff Hall, a girls' school in Yonkers. In 1906 she founded the Oaksmere School for Girls, a non-sectarian boarding school in Westchester County, New York, which closed in 1928. The school was first in New Rochelle but soon moved to her estate in Mamaroneck on Long Island Sound. She opened a branch in Paris, Oaksmere Abroad, in 1912. She also helped finance the American team that participated in the first international track meet for women, which was held in Paris in 1922. On September 26, 1928, her appointment as "director of the Three Arts Wing of The Barbizon for students of music, drama, and art" was announced in an advertisement in the *New York Times*. She apparently had moved from Mamaroneck to the Barbizon Hotel in New York City in 1926 and was the librarian for that residence hotel for women.

Merrill served as president of the Diocesan Branch of the Woman's Auxiliary of Albany, of the New York Branch of the Intercollegiate Alumnae Association, and of the Barbizon Book and Pen Club. She was a member of several women's clubs including the Woman's Organization for National Prohibition. She held a life membership in the AAAS. In 1933, for the fiftieth anniversary of her graduation from Wellesley, her class together with the Woman's Graduate Club of Columbia University and the Zeta Chapter of Phi Delta Gamma presented Columbia with a portrait of Winifred Edgerton Merrill. The painting bears the inscription, "She opened the door," and was first hung in Philosophy Hall but later moved to Low Library. For the last two years of her life, Merrill lived with her elder son in Fairfield, Connecticut, and died in nearby Stratford in 1951, a few weeks before her eighty-ninth birthday.

METCALF, Ida M. August 26, 1856–October 24, 1952.

Boston University (PhB 1886), Cornell University (MS 1889, PhD 1893). PhD diss.: Geometric duality in space (directed by James Edward Oliver).

Ida Martha Metcalf was born in Texas, the daughter of Martha C. (Williams) and Charles A. Metcalf. In 1850 Charles and Martha Metcalf were living in Louisiana, where he was a bookkeeper. Ten years later, they and two daughters were living in New Orleans, where Charles was a clerk. In 1952 Pearson Hunt of the Harvard Business School sent the Cornell University alumni office a short account of Ida Metcalf's life in which he reported that she had come to Boston from Texas on a sailboat with her brother and mother soon after her father died. In 1870 Martha Metcalf, her son, and Ida were living in Needham, Massachusetts, with Martha's father and sisters. Hunt reported that while she was still in her teens, Ida Metcalf taught in small schools in New Hampshire and worked in the fields after school. In 1880 she was living in Framingham, Massachusetts, and was a student. It appears she was also a third assistant in the Hillside School in Jamaica Plain, Massachusetts, in the late 1870s and early 1880s while in her early to mid-twenties.

Metcalf attended Boston University as a special student 1883–85 and after a year as a regular student received her PhB with the class of 1886. She wrote a thesis, "The Origin and Development of Styles of Architecture." During 1888–89, Metcalf was a graduate student in mathematics at Cornell University and received her master's degree with a thesis, "The Theory of Illumination by Reflected and Refracted Light." In February 1891, she was teaching at the newly opened Bryn Mawr School in Baltimore.

Metcalf studied at Cornell 1892–93 and helped George W. Jones prepare exercises for his *Drill-Book in Algebra*. James E. Oliver was chair of Metcalf's doctoral committee, although she had taken a course in projective geometry, the area of her dissertation, with Jones. Her doctorate was awarded magna cum laude in 1893. Metcalf later made a gift to Cornell to assist women students there.

In September 1897 a list of appointments to the three public high schools in New York City (then Manhattan and the Bronx) showed Ida M. Metcalf of Boston having been appointed a second assistant for history at the Girls' High School. In the 1900 census report Metcalf was listed with her mother and brother in Wellesley, Massachusetts, and was described as a teacher of mathematics; her brother was a carpenter. Metcalf lived for a time in Newton Lower Falls, Massachusetts. She also worked for a banking house in New York as a securities analyst.

In April 1910, Metcalf was a teacher and living in Brooklyn. At about that time she passed a New York City civil service examination and left the teaching profession. In October 1910 Metcalf was appointed a monitor and in December 1910, at the age of fifty-three, she became a civil service examiner. Pearson Hunt reported that she was the first woman to take such an examination, and was appointed despite her sex because of her performance. In 1912 she became a statistician for the Department of Finance for New York City. She attempted to take the examination for statistician in the Education Department but was not permitted to do so when her education, teaching experience, and experience in the finance department were not deemed sufficient. She remained in the Department of Finance until her retirement at the end of 1921. She worked again as a civil service examiner on a per diem basis until 1939, when she was eighty-one years old.

In October 1915 Metcalf placed an advertisement in the *New York Times* offering a free home and education to a young girl. She then took as a ward an eight-year-old Austrian girl who became a commercial illustrator and author after studying at New York University, Columbia University, the Grand Central School of Art, and the Art Students League.

By 1931 Metcalf had moved to Jamestown, Rhode Island, where she had spent summers for many years. In 1948 she became ill and moved to a nursing home in Washington, Connecticut, before moving to a nursing home in Lexington, Massachusetts, where she lived just over a year before her death at age ninety-six.

MILLER, Bessie Irving. November 4, 1884–February 4, 1931.
WOMAN'S COLLEGE OF BALTIMORE (BA 1907), JOHNS HOPKINS UNIVERSITY (PHD 1914). PhD diss.: A new canonical form of the elliptic integral (directed by Arthur Byron Coble).

Bessie Irving Miller was born in Baltimore, Maryland, the only child of Bessie (Knotts) and Irving Miller. Her father was a surgeon, and in 1900 the household in Baltimore included four boarders, a cook, two house maids, and two trained nurses in addition to the three family members.

Miller attended the Girls Latin School, the preparatory department for the Woman's College of Baltimore (Goucher College after 1910), which she then attended and from which she graduated in 1907. In 1907–08 Miller was a graduate student in mathematics and physics at the University of Chicago with a fellowship sponsored by the Woman's College of Baltimore. In October 1908 she entered the Johns Hopkins University as a graduate student in mathematics, astronomy, and classical archaeology. This was just a year after **Clara Latimer Bacon**, her college mathematics instructor, entered Johns Hopkins when the trustees voted to allow women to be admitted to graduate courses without special permission.

In 1911 Miller interrupted her graduate studies to become an instructor of mathematics at Kemper Hall in Kenosha, Wisconsin. Kemper Hall was an Episcopal school for girls that had preparatory and collegiate departments when Miller was there. Two years later she returned to Johns Hopkins and finished her doctoral work as a university fellow during the year 1913–14. Miller was the first of ten Goucher graduates to obtain a PhD in mathematics before 1940; seven of these were granted by Johns Hopkins. Miller remained at Johns Hopkins the year after receiving her doctorate and was later described as a research worker that year.

In 1915 Miller became head of the mathematics and physics departments at Rockford College, then a women's college, in Illinois, where she was to remain for thirteen years. During her tenure at Rockford, Miller was usually the professor teaching the majority of the mathematics, physics, and astronomy courses. In addition to her regular teaching and administrative duties at Rockford, Miller remained intellectually active in broader areas. She again studied at the University of Chicago during the summer of 1920. During her years at Rockford, she greatly expanded and broadened the course offerings in mathematics and science. She introduced many courses at Rockford including an elective course that covered a broad range of topics in mathematics and science. Her 1924 book, *Romance in Science*, was based on lectures developed for this course.

While Miller was at Rockford, she was especially active in the MAA. She was a charter member of the MAA and was elected secretary of the Illinois Section in May 1924. The following three years she was secretary-treasurer of the section.

Miller had a leave of absence from Rockford College in the spring term 1927–28. She then took a position as instructor at the University of Illinois in 1928. At Illinois she directed the master's thesis of **Josephine Chanler** and possibly others. She also published another research paper in her field, the first since her earlier work based on her dissertation.

Miller played violin and attended movies, theater productions, and the symphony. Her eyesight, which had never been good, became increasingly bad during the 1920s, and she frequently used a scribe for her correspondence. Bessie Irving Miller died at age forty-six in February 1931 in Champaign, Illinois, as the result of a streptococcal infection. The Bessie Irving Miller scholarship is given to a female mathematics major at Rockford College each year.

MONTAGUE, Harriet F. June 9, 1905–March 19, 1997.
UNIVERSITY OF BUFFALO (BS 1927, MA 1929), CORNELL UNIVERSITY (PHD 1935). PhD diss.: Certain non-involutorial Cremona transformations of hyperspace (directed by Virgil Snyder).

Harriet Frances Montague, born in Buffalo, New York, was the younger of two children of Alice Barbara (Haffa) and Laurence Hibbard Montague. Her mother had a high school education and later was a cashier in retail stores and a homemaker; her father had an elementary and private musical education and was an organist, choirmaster, and composer.

Montague attended Lafayette High School in Buffalo before entering the University of Buffalo, where she spent many years as a student and all of her career as a faculty member. (The University of Buffalo, originally private, was incorporated into the state system in 1962 and has since been the State University of New York at Buffalo.) Montague received her undergraduate degree from the University of Buffalo magna cum laude in 1927 in the honors division. She stayed at the university as a graduate assistant in mathematics 1927–29 and received her master's degree in 1929. She was then appointed instructor in the department and continued some graduate work in mathematics while also teaching.

Montague studied informally with Virgil Snyder at Cornell University in summer 1934 and continued her formal course work in the academic year 1934–35. Montague was admitted to Cornell without financial support because she was unknown to any member of the department but was recommended for a tuition scholarship for spring 1935. Having completed her dissertation in geometry, Montague received her PhD in June 1935 with minors in algebra and logic.

After receiving her doctorate, Montague returned to the University of Buffalo and served there as instructor 1929–39, assistant professor 1939–44, associate professor 1944–47, and professor 1947–73. She was acting chairman of the mathematics department during the first semester 1945–46 and in the academic years 1962–65, and was director of undergraduate studies 1970–73. From 1967 she was also professor of education. She retired as professor emeritus in 1973.

Montague taught both undergraduate and graduate courses, primarily in algebra, geometry, and mathematics education and often served as faculty advisor to the mathematics club. She directed one mathematics PhD dissertation, awarded in 1962, and seven dissertations for EdD degrees, awarded between 1969 and 1972. Already in 1953 her concern with mathematics education was evident when she organized the Inter-School Math Society for gifted high school students in six Buffalo area schools. Montague was director of various NSF summer institutes for secondary teachers from 1957 until 1970. Her 1963 textbook for non-science students, *The Significance of Mathematics*, was coauthored with her colleague Mabel D. Montgomery.

Montague was active in mathematics and mathematics education organizations. She was particularly involved with the Upper New York State Section (now the Seaway Section) of the MAA. She was one of the original founders in 1940, served as vice chairman 1952–53, chairman 1953–54, and governor 1961–64. She was also a member of the council, and later historian (1974–79), of the Association of Mathematics Teachers of New York State.

Montague was also active in organizations outside of mathematics, especially in Buffalo chapters of the AAUP and the AAUW. She was a Presbyterian with an involvement in church activities at the local, regional, and national levels. She was a ruling elder of the United Presbyterian Church USA beginning in 1957 and was the first woman to serve as vice-moderator (1963) and moderator (1964) of the Presbytery of Western New York. She served on the boards of directors of the Auburn Theological Seminary in New York City and of the Buffalo Zoological Society, one of her many local community interests.

In 1972 the State University Alumni Association presented Montague with its Distinguished Alumni Award. The Harriet F. Montague Award, given to an undergraduate mathematics major, was established at the time of her retirement in 1973. After a long illness, Montague died at age ninety-one in her Buffalo home, which she shared with Mabel D. Montgomery, her only listed survivor.

MOODY, Ethel I. February 9, 1905–April 11, 1941.

WELLS COLLEGE (BA 1926), CORNELL UNIVERSITY (MA 1927, PHD 1930). PhD diss.: A Cremona group of order thirty-two of cubic transformations in three-dimensional space (directed by Virgil Snyder).

Ethel Isabel Moody was born in Rushville, New York, the second of two children of Alice Arminda (Stearns) and Edward L. Moody. The family lived in Gorham in Ontario County, New York, about six miles from Rushville. Her father was a farmer; in the late 1920s he was town supervisor. Her brother graduated from Cornell University in 1918, the year Ethel Moody entered Rushville High School. He became a teacher and was, after his retirement, the Gorham, New York, town historian.

After her high school graduation in 1922, Moody entered Wells College in Aurora, New York, about fifty-five miles from her home. She graduated with high honors in mathematics in 1926. As a result of her work on the honors examinations, Virgil Snyder of Cornell University, who was one of the professors conducting them, invited her to come to Cornell for graduate work. Moody attended Cornell on a scholarship, wrote a thesis under Snyder's direction in algebraic geometry, and received her master's degree in 1927.

Ethel Moody returned to Wells as an instructor for the academic year 1927–28, substituting for **Evelyn Carroll (Rusk)**, who was absent on leave for graduate study at Columbia University. Moody then returned to Cornell to complete her doctoral work. She was appointed an Erastus Brooks fellow upon her return, and at the end of the first year she was reappointed, the first time such a reappointment had occurred. Her 1930 dissertation in algebraic geometry was directed by Snyder.

In the fall of 1930 Moody took a temporary position at Sweet Briar College, a college for women in Virginia, and was an instructor there for three years. In 1933 Moody moved to Pennsylvania State College; she spent the last eight years of her life there as an instructor. Her publications from this period consist of reviews and a short note in the *Bulletin* of the AMS. Moody was active at the national level in Sigma Delta Epsilon, a fraternity for graduate women in science, and was treasurer in 1939 and 1940.

In the fall of 1939 Moody bought an automobile, a 1935 Chevrolet Deluxe Coupe, for $275. At age thirty-six, on April 11, 1941, Ethel Moody died as the result of a one-car accident near her hometown of Rushville, New York.

MOORE, Nina M. (Alderton). August 19, 1890–November 23, 1973.

MOUNT HOLYOKE COLLEGE (BA 1914), COLUMBIA UNIVERSITY (MA 1915), UNIVERSITY OF CALIFORNIA (PHD 1921). PhD diss.: [Alderton, N.] Involutory quartic transformations in space of four dimensions (directed by Derrick Norman Lehmer).

Nina May Alderton was born in Berkeley Springs, West Virginia, the second of two daughters of Arah Eleanor (Easter) and Joseph Wilton Alderton. In 1900 the family was living in Washington, D.C., where Joseph Alderton was a real estate agent; shortly thereafter, her father was living in Ohio. Her parents presumably divorced as her father remarried in about 1902 and had two more children. In April 1910, she and her mother were living in South Hadley, Massachusetts.

Nina Alderton entered Mount Holyoke College in South Hadley in 1910 and graduated in 1914. She did graduate work between February 1914 and August 1915 at Yale University and Columbia University and received her master's degree from Columbia in October 1915 with a thesis on multidimensional geometry.

Alderton was an instructor of mathematics at Mount Hermon School in Massachusetts 1915–18. In 1918, before the end of World War I, she became a laboratory assistant involved with X-ray work at the National Bureau of Standards (now National Institute of Standards and Technology) in Washington, D.C., and taught at night in a high school. In 1919 Alderton and her mother moved to Berkeley, California, and Alderton resumed her graduate studies at the University of California. She was an assistant in mathematics

1919–21 and received her PhD in 1921 after taking her final examination in April of that year.

In August 1921 Nina M. Alderton went to Mills College, a women's college in nearby Oakland, as instructor of mathematics and physics. She was assistant professor 1922–26 and associate professor 1926–34. In 1923 she became chairman of what was normally a two-person department. Alderton had a leave of absence during the academic year 1932–33 and taught at John Brown University in Siloam Springs, Arkansas. She returned to Mills but resigned effective January 1934.

Nina May Alderton married William Harrison Moore on July 16, 1934. Moore was born in 1878 in Oregon and was a real estate broker. Nina Moore later described herself as a Baptist and her professional activity as assistant in the real estate business in California during the period 1934 to 1939. Her husband died in 1939, and Nina Alderton Moore again sought academic employment. From 1940 to 1942 she was a teacher at the Drew School in San Francisco, and the next year she returned to John Brown University as a professor of mathematics and physics. Moore taught at West Liberty State College in West Virginia 1943–45.

In 1945 Moore returned to Washington, D.C., and the National Bureau of Standards, this time as a mathematician. According to a Yale alumni directory, she was still living there in 1968 and held, or had held, an appointive position in government service. Nina Alderton Moore moved to Vancouver, British Columbia, Canada, by 1970 and died there in 1973 at age eighty-three.

MORENUS, Eugenie M. February 21, 1881–October 15, 1966.

VASSAR COLLEGE (BA 1904, MA 1905), COLUMBIA UNIVERSITY (PHD 1922). PhD diss.: Geometric properties completely characterizing the set of all the curves of constant pressure in a field of force (directed by Edward Kasner).

Eugenie Marie Morenus was born in Cleveland, New York, the first of two children of Marie Euphemia (Van Blarcorn) and Eugene Morenus. In the 1900 census her father was described as manager of a glass works in Cleveland, New York, and later as a manufacturer of thermometers.

In 1900, when she was nineteen, Morenus was a boarder in East Washington, Pennsylvania, while attending school. After receiving her bachelor's degree from Vassar College in 1904, she remained there on a Coykendall scholarship for the year 1904–05, wrote a thesis in algebraic geometry, and received a master's degree in 1905. From January 1906 until June 1907 Morenus was a teacher of mathematics and Latin at the Watertown, New York, high school. She was a substitute in mathematics at Vassar 1907–08 and was a teacher of mathematics at Poughkeepsie High School 1908–09.

In 1909, at the age of twenty-eight, Morenus began her long association with Sweet Briar College, about a dozen miles from Lynchburg, Virginia. She was instructor in mathematics and Latin 1909–16, associate professor of mathematics 1916–18, and then professor of mathematics. She was on leave 1918–19 and returned as head of the department, a position in which she continued until her retirement in 1946.

After Morenus went to Sweet Briar she continued her studies at the University of Chicago in the summer of 1912 and at Göttingen in the summer of 1913. She studied at Columbia University in the summers of 1915 and 1916 and held an alumnae fellowship from Vassar for study at Columbia 1918–19. She studied at Columbia again during the summers of 1920 and 1921 and received her PhD in 1922. She spent the year 1927–28 as an AAUW Anna C. Brackett memorial fellow in Cambridge, England. In the summer of 1930 she did additional work at the University of Chicago. She later reported having studied at the University of California, having spent the winter of 1941–42 in Berkeley.

Morenus was active within Sweet Briar College, where for many years she was secretary of the faculty and was chairman of the committee on instruction. The Sweet Briar memorial tribute by two colleagues at the college also conveys a sense of her non-scholarly interests: "She sang in the choir, was considered an excellent photographer and furnished numerous pictures for the early yearbooks.... However, riding was her chief joy, and she and her horse, October, affectionately known as Toby, were familiar sights in the countryside as well as on college campus.... She planned and directed faculty plays which were an annual event in those days, starring in a number of them herself" (tribute by Gladys Boone and Bertha Wailes, November 1, 1967, Sweet Briar College Archives).

Morenus was a charter member of the MAA, was a member of the Virginia Academy of Science, and was active in the AAUW. For the latter she served as president of the Sweet Briar branch and as chairman of the education committee of the Virginia division at various times. During World War I, in response to calls for volunteers to help staff factories, Morenus worked for six weeks one summer at an overall factory in Lynchburg. She made trips abroad in addition to those to study in Göttingen and Cambridge. They included travel in India and at least two more trips to Europe, the last when she was seventy-four.

After her retirement from Sweet Briar in 1946, Morenus taught at Connecticut College for Women in 1947. For some time after that she spent her summers at her family home in Cleveland, New York, northeast of Syracuse, and her winters in Lynchburg. Sometime later she moved to Babson Park, Florida. Eugenie Morenus died in 1966 in Lake Wales, Florida, at age eighty-five. She was cremated, and her ashes were sent back to her home town of Cleveland, New York. In 1960 the Eugenie M. Morenus scholarship endowment was established by one of the first five graduates of Sweet Briar in the class of 1910.

MORRISON, Sister Charles Mary. July 19, 1895–January 15, 1953.

Fordham University (BA 1922, MA 1925), Catholic University of America (PhD 1931). PhD diss.: The triangles in-and-circumscribed to the biflecnodal rational quartic (directed by Aubrey Edward Landry).

Rose Mary Morrison was born in Hyde Park, Massachusetts, in 1895, before the community was annexed to Boston. She was one of five children of Mary Etta (Kennedy) and Charles F. Morrison. In 1900 and 1910 her father was a grocer; in 1920 he was a clerk for the city of Boston. Rose Mary Morrison received her early education in parochial and public schools of Boston and entered the Congregation of the Sisters of Charity of Nazareth, Kentucky, in September 1916, at age twenty-one.

After making her vows, Sister Charles Mary was a high school teacher at Presentation Academy, Louisville, 1918–21. She was in residence at Fordham University from June 1921 to August 1922, when she received her BA degree. She then taught Latin, Greek, mathematics, and German at Nazareth College, the Louisville women's college opened by the Sisters of Charity in 1920, where she was to spend most of the next twenty-eight years. In 1925 she received an MA from Fordham, after which she taught mathematics at Nazareth College.

From 1928 to 1931 Sister Charles Mary was in residence at the Catholic University of America. After receiving her PhD in 1931 with minors in philosophy and physics, she returned to Nazareth College, where she remained until August 1950. While at Nazareth College, Sister Charles Mary also served as head of the mathematics department, as registrar 1925–42, and as dean after 1942. After receiving her PhD and until she became dean, Sister Charles Mary attended almost every meeting of the Kentucky Section of the MAA. She presented three papers to the section, all in algebraic geometry. In 1934 she was president of the Kentucky Association of Collegiate Registrars.

In 1950 Sister Charles Mary moved to Massachusetts, where she joined the faculty of Archbishop Williams High School in Braintree and served as archdiocesan superintendent for Community High School. She died suddenly in Boston less than three years later as the result of a myocardial infarction; she was fifty-seven. Her remains were interred in the Nazareth Cemetery in Nazareth, Kentucky. In August 1970 a residence hall at Spalding

College, the coeducational successor to Nazareth College that is now Spalding University, was named Morrison Hall in her honor.

MULLIKIN, Anna M. March 7, 1893–August 24, 1975.

Goucher College (BA 1915), University of Pennsylvania (MA 1919, PhD 1922). PhD diss.: Certain theorems relating to plane connected point sets (directed by Robert Lee Moore [at the University of Texas from 1920]).

Anna Margaret Mullikin was born in Baltimore, Maryland, the youngest of four children of Sophia Ridgely (Battee) and William Lawrence Mullikin. William Mullikin worked as a leather hider in 1900 and as a leather dealer in 1910. The three Mullikin daughters graduated from Goucher College and became public high school teachers; the son earned a PhD in chemistry and worked in industry.

After graduating from Goucher College in 1915, Mullikin taught in private schools for three years. She was at Science Hill School in Kentucky 1915–17 and was an instructor at Mary Baldwin Seminary in Virginia 1917–18. She entered the University of Pennsylvania with a university scholarship in mathematics for 1918–19 and received her master's degree at the end of the academic year. The next year she continued her studies at the University of Pennsylvania, and at the same time she taught at the Stevens School in the Germantown neighborhood of Philadelphia.

At the University of Pennsylvania, Mullikin studied with R. L. Moore, who went to the University of Texas in fall 1920. He was clearly interested in having Mullikin at Texas and secured her an instructorship there for the year 1920–21. The following year she returned to study at the University of Pennsylvania and simultaneously taught at the Lane Country Day School. Even though Moore remained at Texas, he was advisor for Mullikin's dissertation culminating in her doctorate from Pennsylvania in 1922.

Mullikin's dissertation was an important work in topology, and her main result was often referred to as "Miss Mullikin's theorem" or the "Janiszewski-Mullikin theorem." Although Mullikin's result was obtained independently, Janiszewski had published the result some years earlier in Polish in a mathematical physics journal. Thomas Bartlow and David Zitarelli give biographical information and discuss the mathematics in great detail in "Who Was Miss Mullikin?" in the *American Mathematical Monthly* (forthcoming).

After receiving her PhD, Mullikin became a high school teacher. She taught at William Penn High School in Philadelphia the academic year 1922–23 and moved to Germantown High School, also in Philadelphia, in 1923. She was a teacher at Germantown High School and became head of the mathematics department in 1952. She remained there until her retirement in 1959. During the late 1950s and early 1960s she coauthored textbooks with Ethel L. Grove, who had been at Cuyahoga Heights High School in Cleveland, and Ewart L. Grove of the University of Alabama.

In 1954 Goucher College honored Mullikin with an alumnae achievement citation. Two years later Mullikin endowed the Julia Morgan Fund at the First United Methodist Church in Germantown. This fund honored her friend of about thirty years who had been a Methodist medical missionary in China and who had taught medicine in Philadelphia after her return to the United States. Anna Mullikin was residing in Philadelphia at the time of her death at age eighty-two in 1975.

NEE, Henrietta (Terry). December 30, 1904–June 5, 1981.

Shurtleff College (BS 1926), University of Illinois (MA 1929, PhD 1934). PhD diss.: [Terry, H. P.] Abelian subgroups of order p^m of the I-groups of the Abelian groups of order p^n type $1, 1, 1, \ldots$. (directed by Henry Roy Brahana).

Henrietta Pearl Terry was born in Virden, Illinois, the seventh of eight surviving children of nine born to Mary Etta (Kirwin) and Herman R. Terry, a farmer in central Illinois. By 1920 the family had moved to the Alton, Illinois, area, near the Mississippi River. Henrietta Terry graduated from Alton Community High School in 1922 and attended the local

Shurtleff College, a Baptist college that closed and was later absorbed by Southern Illinois University at Edwardsville.

After Terry received her bachelor's degree in 1926 with a major in mathematics, she taught until 1928 in the high school in Patterson, Illinois. She then did graduate work at the University of Illinois, received her master's degree in 1929, and taught for two years at the newly opened Muscatine Junior College (now Muscatine Community College) in Iowa.

In 1931 Terry returned to the University of Illinois and received her doctorate in 1934 with a minor in chemistry. She taught for one year at the American College for Girls in Turkey. Upon her return she was an instructor in mathematics at the University of Illinois High School, Urbana, where she remained until she took a leave for war service in 1942. She enlisted in the WAVES (Women Accepted for Volunteer Emergency Service), did her basic training at Smith College, trained in weather forecasting at the Massachusetts Institute of Technology, and served at the Norfolk Naval Air Station in Virginia, where she was an aerologist stationed in the weather bureau there. She accompanied flight crews and was also trained to fly. She became a lieutenant (jg) and later a full lieutenant before her discharge on December 1, 1945.

On December 22, 1945, Henrietta Terry married Raymond Moore Nee, an engineer. Raymond Nee was born in Portsmouth, Virginia, in 1903; he received a bachelor's degree from Lynchburg College, a master's degree from Virginia Polytechnic Institute, and a certificate in electrical engineering from Lowell Institute. He held various positions with American Cyanamid Company in Wayne, New Jersey, and worked as a private consultant after his retirement.

The Nees had a daughter born in 1948. She graduated as a mathematics major from Bryn Mawr College, studied mathematics for a year at Oxford University, and was a tax attorney in New York City after studying law at Harvard and at New York University. From 1951 the Nees lived in Upper Nyack, New York, except for the year 1955–56, when they lived in Argentina. Henrietta Terry Nee, known as Terry Nee, was active in a number of local community groups, including AAUW, the League of Women Voters, PTA, the Nyack Garden Club, and the Morning Music Club; she was a trustee of the Nyack Library. She played piano and enjoyed going to concerts and plays.

H. Terry Nee died at age seventy-six in Nyack in 1981 after a lengthy illness. The H. Terry Nee Book Fund at Bryn Mawr College was established as a memorial fund by her husband, who died in 1990, and her daughter, who died in 1999. There are two separate collections of her papers at the University of North Carolina at Greensboro. Those that relate to her military service in the WAVES are part of the Women Veterans Historical Collection; other personal and professional documents are in the Henrietta Pearl Terry Papers.

NELSON, Sara L. January 17, 1903–March 6, 1995.

GEORGIA STATE COLLEGE FOR WOMEN (BS 1926), CORNELL UNIVERSITY (MS 1930, PHD 1939).
PhD diss.: Cremona transformations belonging to a family of cubic curves (directed by Virgil Snyder [retired 1938] and Robert John Walker).

Sara Louise Nelson was born in Oglethorpe, Georgia, the daughter of Mattie (Trussell) and John Pendleton Nelson. It was a second marriage for her father, who was described as a dry goods merchant in 1900, but who later was a teacher and then county superintendent of schools. She had one older brother and at least two half-brothers.

Sara Nelson received her elementary and secondary education in the public schools of Oglethorpe, Georgia. She then attended Georgia State College for Women (now Georgia College and State University). Nelson was enrolled there continuously from the summer of 1922 until June 1926, when she received her bachelor's degree with a major in mathematics and a minor in chemistry. A year earlier she had received a collegiate normal diploma.

For three and a half years Nelson taught mathematics, chemistry, and history in two high schools in Georgia. She attended the summer session at Cornell University in 1929 and then resumed her studies there in February 1930. She continued in the summer of 1930 and received her master's degree in September, having studied mathematics and geology and physical geography.

In fall 1930 Nelson joined the faculty at her alma mater, the Georgia State College for Women, where she was to spend the rest of her career except for leaves for further graduate study. She was an instructor for two years before being promoted to assistant professor. Nelson again took courses at Cornell during the 1933 summer session and then returned for full-time study in 1937. She had requested and received some financial support from Georgia State College for Women. At Cornell she held a tuition scholarship 1937–38 and an Erastus Brooks fellowship in 1938–39. She received her doctorate in 1939 with a major in geometry, a first minor in algebra, and a second minor in analysis.

When Nelson returned to Georgia State College for Women in 1939, she was promoted to associate professor; in 1942 she was promoted to full professor and became department chair. Nelson was particularly interested in the training of teachers and attended an NSF summer workshop on that subject in 1963. She remained as department chair until her retirement in 1968, a year after the school became coeducational and was renamed Georgia College at Milledgeville. Nelson was also active in the school's alumni association and served as its president from 1943 until 1947 and again from 1978 until 1980. In 1987 the association honored her with the Georgia College Alumni Heritage Award.

Nelson was a member of the Georgia Academy of Science; Phi Kappa Phi, an honor society for excellence in scholarship; and Pi Gamma Mu, an honor society in the social sciences. For many years she taught Sunday school at the First United Methodist Church of Milledgeville. Sara Nelson died in Milledgeville at the age of ninety-two in 1995. The Sara L. Nelson Fund for Mathematics and Science Education was established at the college.

NEWSON, Mary (Winston). August 7, 1869–December 5, 1959.
UNIVERSITY OF WISCONSIN (BA 1889), GEORG-AUGUST-UNIVERSITÄT ZU GÖTTINGEN (PHD 1897).
PhD diss.: [Winston, M. F.] Ueber den Hermite'schen Fall der Lamé'schen Differentialgleichung (directed by Felix Klein).

Mary Frances Winston was born in Forreston, Illinois, the fourth of seven surviving children of Caroline Eliza (Mumford), who was born in Pennsylvania, and Thomas Winston, who was born in Wales and immigrated to the United States when he was two years old. Mary Winston's father was a physician, and her mother, who had taught French, art, and mathematics, prepared the children for college. The family lived in Forreston, in northern Illinois, until 1892 when they moved to Chicago. All of the seven children were college graduates, one earned a bachelor of law degree, two earned master's degrees, and two earned PhD's.

In 1884 Mary Frances Winston enrolled at the University of Wisconsin in Madison. She interrupted her studies to teach one year in a country school before returning to Wisconsin and graduating in 1889 with honors in mathematics. Winston taught mathematics at Wisconsin Female College in Fox Lake 1889–91. She studied at Bryn Mawr College with a graduate fellowship 1891–92 and the following year lived with her parents while attending the newly opened University of Chicago with an honorary fellowship that paid her tuition. Although she was encouraged to stay at Chicago, Winston wanted to study in Germany. In late August 1893 she met with Felix Klein of Göttingen who was in Chicago as the representative of German mathematics at the International Mathematical Congress held in connection with the World's Columbian Exposition. He encouraged her to come to Göttingen to study, although he could not promise her admission there. **Christine Ladd Franklin**, who had previously corresponded with Klein about the admission of women to Göttingen, heard of Winston's desire to study in Germany and sent her $500 to help finance her first year there.

At the end of September 1893 Mary Frances Winston sailed for Europe, and in Göttingen she met Grace Chisholm and Margaret Maltby, who had also come to Göttingen to try to enroll: Chisholm from England in mathematics and Maltby from the United States in physics. Winston's experiences in Europe are chronicled in letters to her family now in the Sophia Smith Collection at Smith College. From them we learn that the three women petitioned for admission, and by the end of October 1893 their petitions were accepted, but only as exceptions, and they were able to attend lectures as enrolled students at Göttingen.

Winston remained in Europe, studying at Göttingen and occasionally traveling. She was awarded the Association for Collegiate Alumnae (ACA) European fellowship for 1895–96, her final year at Göttingen. Her dissertation was completed by the middle of that year, and she passed her examination in July 1896. Winston received her PhD magna cum laude in 1897 after finally finding a publisher in Germany for her dissertation. Chisholm and Maltby had both received their PhD's in 1895, so Winston was the second woman, and the first American woman, to receive a doctorate in mathematics from Göttingen as an enrolled student. (Sofia Kovalevskaya had received a PhD from Göttingen in 1874 but had never attended classes there.)

Winston returned to the United States in 1896 and at the beginning of September found a teaching position at St. Joseph High School in St. Joseph, Missouri. In 1897 Winston was appointed professor of mathematics at Kansas State Agricultural College (now Kansas State University) in Manhattan. Winston stayed at Kansas State for three years before resigning in order to marry Henry Byron Newson, a mathematician at the University of Kansas, on July 21, 1900. H. B. Newson was born in 1860 in Ohio and was educated at Ohio Wesleyan University, at the Johns Hopkins University, in Heidelberg, and with Sophus Lie in Leipzig. Anti-nepotism regulations prevented Mary Winston Newson from obtaining a position at Kansas, but she was able to teach during the summers. Also during this period, with David Hilbert's permission, she translated his famous "Mathematical Problems" paper for the *Bulletin* of the AMS.

The Newsons had three children, daughters born in 1901 and 1903 and a son born in 1909. Henry Byron Newson, who was not yet fifty, died in February 1910. He had been unable to purchase life insurance because of a bad heart and had taught only nineteen and a half years of the twenty years required in order for his widow to receive payments from his retirement pension. Mary Newson still could not obtain a position at the university because her younger sister had just accepted a position in the English department. Newson remained in Lawrence for a few years after her husband's death.

In 1913 Newson became assistant professor at Washburn College in Topeka, Kansas, close enough so that she could come back to Lawrence to spend weekends with her children, who were at first living at her father's house in Lawrence. While at Washburn, Newson was active in the Kansas Association of Teachers of Mathematics and was its chairman in 1915. She was still assistant professor when she left Washburn at the end of the 1920–21 academic year to become head of the department of mathematics at Eureka College in Illinois. Newson taught courses in both mathematics and astronomy and shared the leadership of a combined mathematics and science division from 1928 until 1934 when she became chairman of that division. She was required to retire from full-time teaching in 1937 when the college instituted a mandatory retirement policy, but she remained chair of the division another two years and continued teaching part time until 1942.

At Eureka, Newson was chairman of the International Relations Round Table of the Eureka branch of the AAUW 1928–38. In 1940 she was honored by the Women's Centennial Congress as one of one hundred women who held positions that were not open to them one hundred years earlier. She was a member of the Woman's Christian Temperance Union, was raised a Unitarian and kept that affiliation as an adult, and in 1914 reported that she was a political independent. After her retirement Newson moved to Lake Dalecarlia in Lowell, Indiana, where she had previously spent vacations. In 1956 she moved to a nursing home in Poolesville, Maryland, near where her elder daughter lived. She died there at age ninety and was buried in Lawrence, Kansas. After Newson's death her children endowed a Mary Newson lecture series on international relations at Eureka College.

NEWTON, Abba V. February 19, 1908–May 5, 1996.
MOUNT HOLYOKE COLLEGE (BA 1929), UNIVERSITY OF CHICAGO (MA 1931, PHD 1933). PhD diss.: Consecutive covariant configurations at a point of a space curve (directed by Ernest Preston Lane).

Abba Verbeck Newton was the younger of two daughters of Sarah C. (Verbeck) and Samuel Smith Newton. She and her sister were both born in Ballston Spa, New York, where their father worked in the office of the American Hide and Leather Company.

Abba Newton attended Ballston Spa High School 1920–24 and St. Margaret's School in Waterbury, Connecticut, 1924–25 before entering Mount Holyoke College. While there she was in the class choir and was a Sarah Willison scholar before graduating magna cum laude in 1929 with major subjects mathematics and chemistry. She taught mathematics at Science Hill School, a girls' preparatory school in Shelbyville, Kentucky, the following year and studied mathematics education at Teachers College, Columbia, in the summer of 1930.

Newton attended the University of Chicago from the fall of 1930 through the summer of 1933, taking off only one quarter, summer 1932, during that time. She received her master's degree in 1931 with a thesis directed by **Mayme Logsdon**, held a scholarship during 1932–33, and received her doctorate in 1933.

Having received her PhD in the midst of the Depression, Newton was unemployed until April of 1934 when she began teaching in an instructor's absence at American International College in Springfield, Massachusetts. She remained there as instructor of mathematics and chemistry until 1938. Newton then joined the faculty at Hartwick College in Oneonta, New York, where she remained until 1943. She then taught two classes in the summer of 1943 at the University of Chicago and held a one-year appointment as assistant professor at Smith College 1943–44 as a substitute for a member of the faculty on leave for war work.

In 1944 Newton accepted an assistant professorship at Vassar College, where she was to remain for the next twenty-nine years. She was promoted to associate professor in 1950 and to professor in 1957. She chaired the department at Vassar 1950–51, 1953–58, and 1966–67, and retired as professor emeritus in 1973. While at Vassar she was a faculty fellow at the Institute Henri Poincaré at the Sorbonne in 1951 and at Duke University in 1966. She was an NSF science faculty fellow at the University of Michigan 1958–59 and a visiting research fellow at Princeton in 1971.

After her retirement, Abba Newton continued to live in Poughkeepsie, New York. She was a member of the Adirondack Mountain Club from the mid-1940s to at least the early 1980s. She served as an alumnae trustee of St. Margaret's School 1965–70 and was a member of the Dutchess County Council on World Affairs. Newton died at age eighty-eight in Poughkeepsie and was buried in the family plot in the Ballston Spa Cemetery.

NOBLE, Andrewa. March 28, 1908–November 21, 1993.
PACIFIC UNIVERSITY (BA 1929), UNIVERSITY OF CALIFORNIA (MA 1934, PHD 1935). PhD diss.: On the enumeration of uniform squares (directed by Derrick Norman Lehmer).

Andrewa Rebecca Noble was the second of four children of Grace (Marshall) and Emory James Noble. Her mother attended high school and later was a teacher and a housewife. Her father had an LLB degree from the University of Oregon and was an attorney. Andrewa Noble and her siblings were born in Oregon City and all attended college.

Andrewa Noble attended public elementary school 1913–21 and high school 1921–25 in Oregon City. At graduation she received an honor scholarship to Pacific University in Forest Grove, Oregon. Four years later she graduated magna cum laude from Pacific. While at Pacific, she was involved with the drama club, was a member of the Women's Athletic Association, and played soccer, basketball, and volleyball. She was active in the debate club and earned the Witham debate cup her junior year. In her senior year she was vice president of the student association and was editor of the college newspaper. She served as the permanent secretary of her graduating class of 1929 and later was active in the alumni council.

The year following her graduation Noble was living in Gladstone, Oregon. In 1930 she became a high school teacher in Tosten, Montana, and taught there until 1932. She was a graduate student at the University of California 1933–35. While in Berkeley she earned a master's degree in 1934 and a doctorate in 1935, having written a dissertation in number theory.

During 1935–36 Noble was a teacher in Colstrip, Montana, and the following year she was superintendent of schools there. She taught in a high school in the state of Washington 1937–38. In 1940 Noble became an instructor at San Francisco Junior College and stayed in that position until 1945, serving as coordinator of the remedial arithmetic program from 1942. She was appointed instructor at Montana State University (now University of Montana) in Missoula sometime before July 1946 and was an assistant professor there 1946–47.

In 1947 Noble returned to her alma mater, Pacific University, and was an associate professor for two years. It appears she then lived with her brother in Oakland, California, for some time before joining the faculty at Mills College in 1952. From 1952 until 1957 she was assistant professor of mathematics and physics and head of the department at Mills, a women's college in Oakland. From 1957 to 1959 she was assistant professor at Chico State College before returning to Pacific University, where she taught the last eight years of her career. She was associate professor 1959–61 and professor from 1961 until her retirement in 1965. She also was chairman of the department at Pacific. In the 1960s she was chairman of the chemistry, physics, and mathematics section of the Northwest Scientific Association.

Noble belonged to the Protestant Episcopal Church and maintained a life membership in the Townsend, Montana, Order of the Eastern Star. She was a Republican and served as precinct committeewoman in Forest Grove in the 1960s. She counted stamp collecting among her hobbies. After her retirement, Noble lived for a number of years in Auburn, California, and in Forest Grove, Oregon. She died in Placer County, in which Auburn, California, is located, at age eighty-five in 1993.

O'BRIEN, Katharine. April 10, 1901–April 10, 1998.
Bates College (BA 1922), Cornell University (MA 1924), Brown University (PhD 1939).
PhD diss.: Some problems in interpolation by characteristic functions of linear differential systems of the fourth order (directed by Jacob David Tamarkin).

Katharine Elizabeth O'Brien was born in Amesbury, Massachusetts, the younger of two daughters of Catherine (Higgins) and Martin William O'Brien, both of whom had emigrated from Ireland about a decade before their marriage in 1896. Early in the twentieth century Martin O'Brien was a carriage painter; by 1920 he was an auto painter in a repair shop. The family moved to Portland, Maine, when Katharine was an infant.

Katharine O'Brien was the class valedictorian when she graduated from Deering High School in Portland at age seventeen. She then entered Bates College in nearby Lewiston, Maine, and graduated from Bates in 1922 with double honors in mathematics and science and with a minor each in the language and philosophy groups.

O'Brien's interests throughout her life focused on three areas: mathematics, music, and poetry. The year after her college graduation, she played piano as an assistant in the department of hygiene and physical education at Smith College. She also took a graduate course in music and one in English that year. The following year, 1923–24, she did graduate work in mathematics at Cornell University and earned a master's degree in 1924 with a major in geometry and a minor in analysis, although she had requested a minor in English.

From January to June 1925 O'Brien taught at Jordan High School in Lewiston, Maine, and from 1925 until 1936 at the College of New Rochelle, then a Catholic college for women in New Rochelle, New York. At New Rochelle, O'Brien was instructor 1925–29 and professor and department head 1929–36. She was also a scholarship participant in a master class in piano with Sigismond Stojowski in New York City in 1934–35.

In 1936 O'Brien returned to graduate studies in mathematics at Brown University, where in 1937–38 she was a University scholar. She received her PhD in 1939 with a dissertation in analysis.

O'Brien joined the mathematics faculty of Deering High School in Portland, Maine, in 1940. The first five years there she was also director of the Girls' Glee Club. She became head of the mathematics department in 1945 and remained head until her retirement in 1971.

In 1962 O'Brien became a lecturer at Gorham State Teachers College (now the University of Southern Maine). She continued lecturing there until 1973. She also lectured in several NSF summer institutes for teachers at Brown during the 1960s. O'Brien's many professional and extracurricular activities included participating in the activities of the NCTM and its affiliate, the Association of Teachers of Mathematics in New England (ATMNE). She refereed for the NCTM journal *Mathematics Teacher* for more than thirty years and, during the summer of 1951, helped run an ATMNE-sponsored institute at Connecticut College for teachers of mathematics. She was also active in Mu Alpha Theta, the national high school and junior college mathematics club. She published poetry and light verse in magazines and newspapers, including the *Saturday Review*, *Christian Science Monitor*, *New York Times*, *Ladies Home Journal*, and *Scientific Monthly*. She was a member of the International Platform Association, dedicated to public speaking and performing, and the Poetry Society of America. She also composed music and published choral octavos using the words of Thomas Hardy.

O'Brien was recognized in a number of ways for her contributions to mathematics, poetry, and music. In 1960 the University of Maine awarded her an honorary Doctor of Science in Education degree, and in 1965 Bowdoin College gave her an honorary Doctor of Humane Letters degree. The Dr. Katharine E. O'Brien Mathematics Award to be given annually to a senior at Deering High School was established by the Deering class of 1964. In 1985 Westbrook College, which later merged with the University of New England, gave O'Brien the Deborah Morton Award, an annual award in Maine honoring women's achievement.

Katharine O'Brien died on her ninety-seventh birthday at a nursing home in Falmouth, Maine, after a lengthy illness. After her death, the University of Maine System received a gift of $400,000 from her estate in support of its library holdings. A bequest of a comparable size was made to Bowdoin College, a portion of which was earmarked for the library. Her papers are housed in the Maine Women Writers Collection, University of New England Libraries.

OFFERMANN, Jessie (Jacobs). October 1, 1890–July 7, 1954.

McPherson College (BA 1914), University of Kansas (MA 1916), University of Illinois (PhD 1919). PhD diss.: [Jacobs, J. M.] The trilinear binary form as a cubic surface (directed by Arthur Byron Coble).

Jessie Marie Jacobs was born in Wilmington, Delaware, the second of three daughters of Annie Amelia (Wright) and William C. Jacobs, a carriage maker, cabinet maker, and house carpenter. Her family moved to Kansas soon after she was born and was living in McPherson in central Kansas by 1900. She graduated from McPherson High School in 1907. In 1910 both Jessie Jacobs and her older sister were public school teachers. She was also a high school teacher in Kansas during the year 1911–12 and attended summer school, at least in the summer of 1913, before completing her undergraduate work at the local McPherson College in 1914.

Jacobs taught in high school in Kansas for a year after her graduation. She then went to the University of Kansas on a fellowship in 1915–16, one of the first two awarded to students in mathematics by the graduate school, and received her master's degree in 1916. Although awarded another fellowship by Kansas for 1916–17, she accepted a graduate assistantship at the University of Illinois for that year. She remained at Illinois with a fellowship from 1917 until she received her PhD in 1919.

Jacobs was associate professor of mathematics and physics at Rockford College in Illinois the next year. In 1920 she became an instructor in pure mathematics at the University of Texas. While in Austin she met Hermann Joseph Muller (1890–1967), a new associate professor of biology. Born and raised in New York City, he received his PhD from Columbia in 1915. Jacobs and Muller were married June 11, 1923, and she remained an instructor through the academic year 1923–24. She lost her position at Texas because she became pregnant, and, according to her husband's biographer, "her colleagues felt that a mother could not give full attention to classroom duties and remain a good mother" (Elof Carlson, *Genes, Radiation, and Society* (Ithaca, N.Y.: Cornell University Press, 1981), 133). Their son was born in Austin in November 1924. In 1925 Jessie Jacobs-Muller collaborated with her husband and received acknowledgement in one of his papers and joint authorship of another. In the early 1930s Hermann Muller was continuing his pioneering work on the effects of radium radiation on the genetics of fruit flies, and Jessie Muller helped him in the laboratory while also taking care of their son and the house. H. J. Muller was awarded a Nobel Prize in 1946 for work done during this period.

During the early 1930s H. J. Muller's socialist views became more public and tensions in his marriage increased. In autumn 1932 Muller left the United States to spend a year as a Guggenheim fellow in Berlin. In July 1933 his family joined him, and two months later they all moved to Leningrad, where Muller had become senior geneticist. In July 1934 Jessie and her son returned to Austin, where Jessie Muller obtained a job with the Works Progress Administration. The marriage was dissolved in the Soviet Union in January 1935, and Jessie Muller obtained a divorce in Texas that summer. Sometime between October 1935 and January 1936 Jessie Muller married Carlos Alberto Offermann.

Carlos Offermann (1904–1983) was born in Argentina and had come to the University of Texas in December 1930 to work in Muller's laboratory. He joined Hermann Muller in Leningrad 1933–34 and married Jessie Muller in Georgetown, Texas, on a visit to Austin the following year. Carlos Offermann then returned to his position in Muller's laboratory, which had moved to Moscow. A Texas judge refused to permit Jessie Offermann's son to leave Texas as long as his father was in the Soviet Union, so they remained in Austin, unable to join her new husband. In Austin Jessie Offermann supervised a WPA group that was writing a history of Travis County and made additional money by renting a room in their small apartment and by tutoring university students in mathematics. H. J. Muller left the Soviet Union in 1937 after the start of Stalin's political persecutions.

Carlos Offermann returned to Austin in 1938, and that fall the Offermanns moved to Chicago, where Carlos Offermann, with his wife's help, worked to complete the experimental work for a PhD at the University of Chicago. However, the following academic year she was diagnosed with tuberculosis. In summer 1940 they moved to the San Gabriel Valley in Los Angeles County, California, where they hoped she could rest and recover and where perhaps he could finish his dissertation.

Jessie Offermann remained at home until 1942, when her health deteriorated and she entered a local sanitarium. By 1946 she was in a facility in Duarte, California, where she died in 1954 at age sixty-three. Jessie Offermann's son, who earned a PhD in physics from the California Institute of Technology in 1951, wrote to one of the authors in 1997, "There is no doubt that my interest in, and satisfaction with, mathematics began with my mother who spent many hours talking to me about mathematics and teaching me many things while I was still very young. During my 40 years at Illinois in the mathematics department I have had over a dozen Ph.D. students of my own who may be thought of as her mathematical grandchildren."

OLSON, Emma J. December 7, 1892–November 14, 1981.

UNIVERSITY OF SOUTH DAKOTA (BA 1915), UNIVERSITY OF CHICAGO (MA 1924, PHD 1932). PhD diss.: Conjugate systems characterized by special properties of their ray congruences (directed by Ernest Preston Lane).

Emma Julia Olson was born in Fairview, South Dakota, the fourth of seven surviving children of Julia (Erickson) and Sven Olson, both natives of Norway. Her parents farmed in Lincoln County, South Dakota, near the Iowa border. She had four sisters and two brothers.

Olson received her high school and her college education at the University of South Dakota and earned her bachelor's degree in June 1915. From 1916 to 1919 she taught mathematics, physics, and chemistry at a high school in Milbank, South Dakota. From 1920 to 1923 she taught at the College of Industrial Arts (now Texas Woman's University) in Denton, Texas. In September 1923 she entered the University of Chicago, studied there for four consecutive quarters, and received her master's degree in August 1924 with a thesis written under the direction of E. P. Lane. During the second semester of 1924–25, she was acting head of the department at the River Falls Normal School (now University of Wisconsin-River Falls). From 1926 to 1930 she taught at Northwestern University in Evanston, Illinois.

Olson resumed her graduate work at the University of Chicago and completed the work for her PhD in 1932 with a dissertation in projective differential geometry, again directed by E. P. Lane. She was acting head of the department at Mount Saint Scholastica College in Atchison, Kansas, during the second semester of 1934–35 and was assistant professor at Concordia College in Moorhead, Minnesota, 1935–36. In January 1937 Olson joined the faculty at Kent State University in Kent, Ohio. She was instructor 1937–40, assistant professor 1940–45, associate professor 1945–51, professor 1951–63, and emeritus professor after her retirement in 1963.

Emma Olson traveled widely in Europe, Scandinavia, and "the Holy Land." After her retirement, she returned to South Dakota. She died in Canton, South Dakota, at age eighty-eight in 1981.

OWENS, Helen (Brewster). April 2, 1881–June 6, 1968.

UNIVERSITY OF KANSAS (BA 1900, MA 1901), CORNELL UNIVERSITY (PHD 1910). PhD diss.: Conjugate line congruences of the third order defined by a family of quadrics (directed by Virgil Snyder).

Helen Barten Brewster was born in Pleasanton, Linn County, Kansas, the only child of Clara (Linton) and Robert Edward Brewster. In 1880 her parents were living in Linn County, where her father, a Civil War veteran, was a farmer. It appears that her father died in 1892, and that later her mother was a school teacher.

Helen Brewster did her college preparation at the Pleasanton High School and entered the University of Kansas in Lawrence in September 1897, when she was sixteen. She earned her bachelor's degree in 1900 and her master's degree in 1901. The following year she was a teaching fellow at Kansas and published an article, possibly her master's thesis, in which she expressed her gratitude to Henry B. Newson, the husband of **Mary Winston Newson**. Brewster continued her graduate studies at the University of Kansas until June 1903 and taught in the high school in Lawrence from 1902 until 1904.

On June 22, 1904, in Pleasanton, Kansas, Helen Brewster married Frederick William Owens, who had also studied mathematics with H. B. Newson at the University of Kansas. F. W. Owens was born in Rockwell City, Kansas, in 1880. He received his bachelor's and master's degrees from Kansas in 1902 and was a teaching fellow there 1902–03. During

1903–05 he was a resident graduate student in mathematics and astronomy at the University of Chicago. Helen B. Owens studied at the University of Chicago in the summer and autumn quarters of 1904. Their first daughter was born in May 1905 and died in 1928 while a medical student. F. W. Owens taught at the academy of Northwestern University in Evanston, Illinois, 1905–07, and Helen Owens taught in high school in Evanston 1906–07. In 1907 F. W. Owens received his PhD from Chicago and went to Cornell University in Ithaca, New York, as an instructor. Their second daughter was born in February 1908; she was widowed shortly after graduating from medical school and raised a son with the help of her parents.

Helen Owens enrolled as a graduate student at Cornell in fall 1908 and received her PhD in 1910 with minors in physics and mathematical physics. She was a private teacher of mathematics in Ithaca starting in 1908 and taught at Wells College, a women's college about thirty miles from Ithaca, as an instructor second semester 1914–15 and as an acting assistant professor second semester 1916–17. Owens also campaigned for women's suffrage in Kansas, as a state organizer for the Kansas Equal Suffrage Association during the successful campaign of 1911–12, and in New York State, as a district chairman of the Empire State Campaign Committee 1913–15. She remained active in the New York state campaign until women won the right to vote there in 1917. Her interest in civic affairs led her to presidencies of the Political Study Club of Ithaca and the Tompkins County Equal Suffrage Association, as well as memberships in the Woman's Christian Temperance Union, the Civic Improvement League, and the Cornell Consumers' League. Owens was an instructor at Cornell University 1917–22 and was intermittently employed as a secondary school teacher and as a private tutor 1908–26. In 1922 she was the first national treasurer of Sigma Delta Epsilon, a fraternal organization for graduate women in science that was formed at Cornell in 1921.

In 1926 F. W. Owens assumed the headship of the mathematics department at Pennsylvania State College (now University) in State College. Helen Owens returned to Ithaca for private tutoring for three months each academic year until 1928. Although state antinepotism policies kept her from teaching at Penn State at that time, she would regularly get together with the women faculty, **Teresa Cohen**, **Beatrice Hagen**, and **Aline Huke Frink**, to read advanced mathematical textbooks. From 1935 to 1938 Helen Owens was an associate editor of the *American Mathematical Monthly* and edited the Mathematical Clubs section with her husband. During the late 1930s she corresponded with various institutions and organizations and sent questionnaires to those North American women she was able to identify as holding PhD's in mathematics, and at the summer mathematics meetings of 1937 in State College she organized a luncheon honoring women pioneers in mathematical research in America. The questionnaires she collected and the correspondence about that meeting are in her papers in the Schlesinger Library, Radcliffe Institute, Harvard University.

In 1940 Helen Owens was hired as an instructor by Pennsylvania State College and was promoted to assistant professor in 1945. In 1949 she retired as assistant professor emeritus and her husband retired as professor emeritus. F. W. Owens died in 1961 in State College. Three years before her death, Helen Owens moved to the home of her surviving daughter in Martinsburg, West Virginia. She died in Martinsburg in 1968 at age eighty-seven.

PEIRCE, Leona May. August 4, 1863–September 27, 1954.

SMITH COLLEGE (BA 1886, MA 1893), YALE UNIVERSITY (PHD 1899). PhD diss.: On chain-differentiants of a ternary quantic (directed by William Edward Story [Clark University]).

Leona May Peirce was born in Norway, Maine, the daughter of Mary Hobbs (Foster) and Levi Merriam Peirce (also "Pierce"). Her father received a bachelor's degree from Waterville College (now Colby College) in Maine in 1860 and was a teacher for nearly a dozen years before starting a music company specializing in the sale of pianos and organs in Springfield, Massachusetts. Leona Peirce had a younger brother who served as president of Kenyon College in Ohio from 1896 until 1937.

Leona Peirce attended Springfield High School and taught elementary subjects in a private school in Springfield for a year before entering Smith College in 1882. Her main subjects at Smith were rhetoric and history, with only the required mathematics courses. After her graduation in 1886, she taught mathematics, physics, and chemistry at the Springfield Collegiate Institute for three years. In the summer of 1887 she studied chemistry and French at Amherst College.

Peirce enrolled at Cornell 1889–90 as a candidate for an advanced degree studying mathematics, physics, and philosophy, and took courses in mathematics and history of philosophy. The following year she taught mathematics at the Mount Hermon School for boys in Mt. Hermon, Massachusetts, before returning to her studies. During 1891–92 she studied mathematics at Newnham College, Cambridge, and in 1892–93 she was at Smith College, where she completed work for a master's degree in 1893. She then returned to Cornell and again studied mathematics and philosophy. She was there for a year before leaving because of illness in her family.

Following the suggestion of the Cornell mathematics department, Peirce arranged to study privately with William E. Story of Clark University, a school in Worcester, Massachusetts, that did not yet formally admit women to its doctoral programs. This arrangement was made with the understanding that Peirce would return to Cornell for her final examinations and present a thesis written under Story's direction. After working with Story for three years, she submitted her thesis at Cornell in spring 1897 but did not receive a Cornell PhD owing to an apparently irresolvable disagreement concerning the requirements for the final examination in one of her minor subjects. Following Story's advice, Peirce did not return to Cornell but continued to work with him; she enrolled at Yale a year later. She received her PhD from Yale in 1899 after having taken a year of course work there.

Peirce did not return to teaching mathematics for almost thirty years. During most of the intervening years she was at her family home in Springfield where, upon the death of her father in 1908, she undertook the management, and was treasurer, of the L. M. Pierce Co., her father's music company. Peirce was also involved in numerous civic and women's organizations including the Springfield school board and the Springfield Civil Service Reform League. She was a delegate to the biennial convention of the General Federation of Women's Clubs in St. Louis in 1904. In 1914 she traveled for several months with her brother in Italy.

In 1928, when she was in her mid-sixties, Peirce began teaching mathematics and physics at the Barrington School for Girls in Great Barrington, Massachusetts. During the summers of 1930 through 1932 she studied mathematics at Harvard. She left the Barrington school in 1932 and for the next two years taught mathematics and English at the Leominster, Massachusetts, high school. Leona Peirce had retired by 1937. She died in 1954 at ninety-one in Springfield after a short illness.

PENCE, Sallie E. September 5, 1893–July 26, 1989.

University of Kentucky (BA 1914, MA 1928), University of Illinois (PhD 1937). PhD diss.: The configuration of the double points of cubics of a pencil (directed by Arnold Emch).

Sallie Elizabeth Pence was born in Lexington, Kentucky, the seventh of eight children of Belle (Kelly) and Merry Lewis Pence. Her father graduated in 1881 from what was then the Agricultural and Mechanical College of Kentucky and earned a master's degree in 1885. He taught for over fifty years and for nearly forty years was at what eventually became the University of Kentucky, first as professor of civil engineering and physics and then as head of the physics department that he established. Pence Hall on the campus is named for him. Four of his five daughters graduated from the University of Kentucky.

Sallie Pence attended public schools in Lexington for ten years and the academy of the State University of Kentucky for one year. She graduated in 1910 and entered the State University, renamed the University of Kentucky in 1916. She received her bachelor's degree in 1914, having majored in modern languages. While at the university she was a member of the women's basketball team, a campus drama production company, Mortar Board, and Alpha Gamma Delta, a social sorority.

From 1914 until 1927 Pence taught in high schools in Owensboro, Bellevue, Hazard, and Morganfield, Kentucky. In summer 1921 she attended George Peabody College for Teachers in Nashville, Tennessee, and began work in mathematics at the University of Kentucky in summer 1922. She was a graduate student in summers 1926 and 1927, was a fellow the following year, and received her master's degree in 1928. Pence was an instructor at Murray Normal School and Teachers College (now Murray State University) in Kentucky 1928–29. The following year she was hired as an instructor at the University of Kentucky. Pence continued to take mathematics courses there before attending Cornell University during the summers of 1932 and 1933. She took a leave of absence to attend the University of Illinois 1934–35 and continued there fall 1935 through summer 1936, as an assistant in mathematics 1935–36. She returned for the second semester of 1936–37 to complete the work for her PhD. Her major field was geometry with a minor in astronomy.

Pence was promoted to assistant professor at Kentucky in 1938, a year after receiving her doctorate. She was promoted to associate professor in 1947, was chairman of the division of physical science 1954–55, was promoted to professor in 1956, and retired in 1963. She was active in the Kentucky Section of the MAA for which she served as secretary 1947–49 and as chairman 1953–54. She was official MAA lecturer to Kentucky high schools in 1958–59 and was executive director of the MAA high school lecture program under the sponsorship of the National Science Foundation 1962–63. Pence directed a summer seminar for in-service high school mathematics teachers in 1957 and was executive director of the University of Kentucky high school honors examinations in mathematics 1959–63. Immediately after her retirement, Pence took a four-month trip in Europe with a friend. She then did some part-time teaching at the University of Kentucky and was professor at East Carolina College (now University) in Greenville, North Carolina, 1964–66. She was a visiting distinguished professor there 1966–67.

Pence was interested in golf, philately, photography, and travel, and was frequently invited to give slide presentations about her travels. She was a member of the the Lexington Business and Professional Women's Club, the Kentucky Archaeological Society, the Order of the Eastern Star, and the Daughters of the American Revolution, among others. She was also active in the Calvary Baptist Church and described herself as a Democrat. Sallie Pence died at the age of ninety-five in Lexington in 1989. The Sallie E. Pence Award was established by the University of Kentucky board of trustees and is given annually to outstanding mathematics students at the University of Kentucky who plan to pursue a career in teaching.

PEPPER, Echo D. June 3, 1897–January 5, 1979.

UNIVERSITY OF WASHINGTON (BS 1920, MS 1922), UNIVERSITY OF CHICAGO (PHD 1925). PhD diss.: Theory of algebras over a quasi-field (directed by Leonard Eugene Dickson).

Echo Delores Pepper was the second child of Josephine LaBena (Sanderson) and Albert Ephraim Pepper, natives of Ontario, Canada. She was born in Spokane, Washington, about a year after her parents and their infant daughter moved to the United States and before the family moved to Seattle. Both parents were trained as accountants, and for most of their time in the United States her father was self-employed and her mother helped in his businesses: first a plumbing company and a funeral home, and later hotels and apartment houses. Echo Pepper's brother was born in 1927.

When Echo Pepper was seven, she was ill and lost the use of her right hip, although an operation improved the situation so that she could walk without crutches by about the time she entered high school. She received all her education in Seattle but was taught at home for several years because of her hip problem. She also attended a convent school and then a public high school, from which she graduated in 1916. She entered the University of Washington, where she studied mathematics and physics for her bachelor's and master's degrees. She received her BS in 1920 and held a teaching fellowship in mathematics in 1920–22 before receiving her MS in 1922. In the summer of 1922 she studied at the University of California. In 1922–23 she served as an associate at the University of Washington, replacing a faculty member who was on leave.

During the years 1923–25 Pepper was a fellow at the University of Chicago. In 1925–26, the year after her PhD (awarded December 1925), she studied at Oxford University as a holder of a National Research Council Fellowship.

Upon Pepper's return from England she went to Bryn Mawr College as instructor for the years 1926–28. In 1928 Echo D. Pepper moved to the University of Illinois where she remained most of the rest of her career. She was instructor 1928–29 and associate (similar to senior instructor) 1929–45. During this period she was active in Sigma Delta Epsilon, a society for graduate women in science, and served as national treasurer in 1937 and 1938 and as first vice president in 1944 and 1945. Later, in 1955–57, she was the business manager of the *Pi Mu Epsilon Journal.*

Pepper requested and was denied a leave for 1945–46, but was promised reappointment at the end of the year. She spent the year in Seattle and taught mathematics and physics as assistant professor at Seattle College (now University). She returned to Illinois in 1946 as assistant professor and remained in that rank until 1960.

During the spring of 1952 Pepper had a corrective bone operation and was granted a leave for the fall semester of 1952–53. From about that time she was connected with the rehabilitation program at the University of Illinois. She was later awarded an honorary life membership in a rehabilitation service fraternity on the Illinois campus.

Pepper taught some summers at Southern Illinois University in Carbondale to instruct teachers how to teach the "new math." In the spring of 1960 Pepper was considering moving to Southern Illinois at a considerably higher salary than she was making at the University of Illinois. Mathematics department papers show that the Illinois department chairman's recommended promotion met with initial resistance from the dean after which several senior members of her department wrote in her behalf. She and her teaching were given high praise. It was also noted that her promotion was long overdue and that younger men in the department had been successful in their requests for promotion. She was promoted to associate professor in 1960 and remained in that rank until she was forced to retire from the University of Illinois in 1965 because of her age, sixty-eight. From 1965 to 1970, after retirement from Illinois, Pepper was at Notre Dame University in Nelson, British Columbia, as associate professor and department head.

According to her brother, Echo Pepper was about five feet tall and weighed no more than a hundred pounds. She was completely ambidextrous and could write with both hands simultaneously. She was a pacifist and an Independent but more Republican than Democrat. She was raised as a "high" Episcopalian but became a Roman Catholic when she was young. About 1940 she became an oblate of the Order of St. Benedict, a lay position.

After her second retirement, Pepper moved to the Washington, D.C., area to live in an adult community in Silver Spring, Maryland, near her brother. When she was past eighty, she moved to a Carmelite retirement home. She died in Hyattsville, Maryland, in 1979 at age eighty-one and was buried in Seattle.

PETERS, Ruth M. December 11, 1906–May 12, 1961.

BRYN MAWR COLLEGE (BA 1928, MA 1931), RADCLIFFE COLLEGE (PHD 1933). PhD diss.: Parallelism and equidistance in Riemannian geometry (directed by William Caspar Graustein).

Ruth Margaret Peters was born in Gettysburg, Pennsylvania, the second of three children of Julia (Clutz), a 1901 graduate of Goucher College, and Robert John Peters. Julia Clutz was the daughter of the president of Midland College in Aitchison, Kansas, where Robert J. Peters was a professor of English 1904–09. Ruth Peters had an older brother and a younger sister. In 1910 the family was living in Pittsburgh, where her father was a college professor. In the 1920 census her father's occupation was listed as educational secretary; in 1930 he was called an executive with the US government.

Peters attended grade schools in Gettysburg and prepared for college at the Seiler School, a private school in Harrisburg, Pennsylvania. As an undergraduate mathematics major at Bryn Mawr College, she held the Elizabeth Wilson White memorial scholarship her senior year. She graduated in 1928; the following year, 1928–29, Peters worked for the traffic engineering department of the Bell Telephone Company doing trunk analysis.

In 1929 she returned to Bryn Mawr for graduate work in mathematics and physics and studied there as a scholar of the Society of Pennsylvania Women in New York 1929–30 and as a fellow in mathematics 1930–31. Peters received her master's degree in 1931, a year after her sister received her bachelor's degree from Bryn Mawr and the same year that her sister received her master's degree in chemistry from Radcliffe College.

In 1931 Ruth Peters entered Radcliffe College to begin her doctoral work and received her PhD in 1933. Her first position after receiving her PhD was as a personnel assistant doing job analysis for the Pennsylvania Emergency Relief Board. The following year Peters went to Judson College, a small Baptist women's college in Marion, Alabama, where she was assistant professor 1934–35 and associate professor 1935–36. She then took a position as professor of mathematics and physics at Lake Erie College, also a women's college, in Painesville, Ohio. She was there 1936–43 and 1945–47. Peters reported later that she was a technical aide in airborne fire control, guided missiles, and rockets for the Office of Scientific Research and Development at the Massachusetts Institute of Technology 1943–45 and that she did research on the theory of errors in bombing.

Two years after returning to Lake Erie College, Peters went to St. Lawrence University in Canton, New York. She was associate professor 1947–55 and professor 1955–58. While there she served as department chairman for two years. She also held a National Science Foundation faculty fellowship at Harvard University the second semester 1957–58 and was president of the New York Lambda chapter of Phi Beta Kappa in the late 1950s. In the mid-1950s, Peters indicated that she was a member of the League of Women Voters and was interested in oil painting and gardening.

In 1958 Peters went to the University of New Hampshire as associate professor. She was there only two years before she had to take a medical leave in October 1960. Ruth Peters then moved to her sister's home in Belmont, Massachusetts. She died from cancer in a Boston hospital at age fifty-four in 1961 and was survived by her mother, sister, and brother. The Dr. Ruth Peters Memorial Award, established by her sister in 1964, is given each year to a University of New Hampshire student majoring in mathematics who has a deep interest in mathematics and shows signs of creativity.

PIXLEY, Emily (Chandler). August 19, 1904–August 28, 2000.

RANDOLPH-MACON WOMAN'S COLLEGE (BA 1926), UNIVERSITY OF CHICAGO (MS 1927, PHD 1931). PhD diss.: [Chandler, E. M.] Waring's theorem for fourth powers (directed by Leonard Eugene Dickson).

Emily McCoy Chandler was born in Knoxville, Tennessee, the daughter of Mayme (McCoy) and David Sanders Chandler. The 1920 census lists two other daughters, age ten and age two months. Her father was a district passenger agent with the railroad.

Chandler attended primary and secondary school in the Knoxville public school system and graduated from Knoxville High School in 1922. She then entered Randolph-Macon Woman's College (now Randolph College) in Lynchburg, Virginia, where she majored in psychology and mathematics and minored in English. While at Randolph-Macon, Chandler was active in student government, the YWCA, and the Inter-racial Committee. Immediately after her graduation, Chandler entered graduate school at the University of Chicago in the summer quarter of 1926 and received her master's degree in 1927. She then took a position as instructor and acting head of the mathematics department at Saint Xavier College for Women (now Saint Xavier University) in Chicago. She continued her graduate work part time at the University of Chicago and received the PhD in 1931.

On September 8, 1931, Emily Chandler married Henry Howes Pixley, a fellow mathematics graduate student at Chicago who also earned a master's degree in 1927 and a doctorate in 1931. Pixley was born in New York State in 1902 and received a BA and an MA from John B. Stetson University in Deland, Florida. Although inquires for a position in Detroit for Emily Chandler were made in the spring of 1931, she remained at Saint Xavier in Chicago, where she continued as professor and department chairman 1930–36, and he stayed at the College of the City of Detroit as instructor 1930–34. In 1933–34 Emily and Henry Pixley had leaves of absence and worked in areas having to do with labor productivity and mathematical economics for the National Recovery Administration.

In 1933 the College of the City of Detroit united with other public colleges in Detroit, and in 1934 became Wayne University (Wayne State University since 1956). Henry Pixley remained as assistant professor 1934–37 and associate professor from 1937 until his retirement in 1972. From 1945 until his return to the classroom in 1969, his responsibilities were largely administrative. In September 1936 Emily Pixley was hired by Wayne University as a special instructor with an hourly salary. She worked as a special instructor part time 1936–42 and full time 1942–47. Also, during her first years there, the Pixleys had three children: a daughter born in 1937 and sons born in 1939 and 1942.

In September 1947 the title of Emily Pixley's position changed from special instructor to regular substitute assistant professor with a teaching load of sixteen hours per week. In spring 1948 she learned that her services would be terminated in June 1949 to comply with the university anti-nepotism policy. She did not wait to be fired, but found employment at the University of Detroit (now University of Detroit Mercy). She was assistant professor 1948–51, associate professor 1951–56, and professor from 1956 until her retirement in 1973 as professor emeritus. Although, as reported by her son, her being a woman prevented her from becoming chair of the department, she was acting chair in the summers 1963–69 and in the academic year 1968–69.

In the late 1940s Emily Pixley indicated that her church preference was Congregational and that she had traveled in Alaska, Europe, and the United States. She also noted that she read French, German, and Italian. Henry Pixley died in Detroit at age eighty-three in 1985. Emily Pixley remained in the metropolitan Detroit area and died at age ninety-six in 2000 in Redford Township, Michigan.

PORTER, Goldie (Horton). September 4, 1887–May 11, 1972.

University of Texas (BA 1908, PhD 1916), Smith College (MA 1910). PhD diss.: [Horton, G. P.] Functions of limited variation and Lebesgue integrals (directed by Milton Brockett Porter).

Goldie Printis Horton was the youngest of three surviving daughters, of four children born, of Colistia (Polk) and John Thomas Horton. She was born in Athens, Texas. Her father was a physician, and in 1900 and 1910 the family lived in Quanah, Texas, just below the southeastern part of the Texas panhandle. Goldie Horton attended public schools in Quanah before entering the University of Texas in 1904. While at the university she was secretary-treasurer of her senior class and was a member of the YWCA. She graduated from Texas in 1908. Her next older sister also received a BA from Texas in 1908. Her eldest sister earned a BA in 1924 and an MA in 1927 from Texas and was a member of the history department until her retirement in 1959.

After receiving her bachelor's degree, Goldie Horton taught in the high school in Grandview, Texas, for a year before going to Smith College in Northampton, Massachusetts, with a fellowship in 1909. She received her master's degree in 1910 and returned to Texas, where she taught at the high school in Amarillo for the next two years, 1910–12. Horton then studied at Bryn Mawr College as a fellow 1912–13.

In 1913 Goldie Horton was hired as a tutor at Texas. She also continued her graduate work and in 1916 completed the work for her PhD with a dissertation in analysis written under the direction of M. B. Porter. This was the first doctoral degree in mathematics awarded at Texas. She was also the first woman to receive a PhD in any field at Texas. Goldie Horton remained a tutor until 1917 when she was promoted to instructor in pure mathematics; in 1926 she was promoted to adjunct professor and in 1935 to assistant professor. In the summer of 1921 she worked on the preparation of ballistic tables at the Aberdeen Proving Ground in Maryland.

In 1934 Horton and Milton Brockett Porter, her colleague and dissertation advisor of eighteen years earlier, had notes for a text on plane and solid analytic geometry reproduced. On December 26 of that year she and Porter were married. This was the third marriage for M. B. Porter, who had been twice widowed. Milton Brockett Porter was born on November 22, 1869, in Sherman, Texas. He received a BS in 1892 from the University of Texas, and an MA in 1895 and a PhD in 1897 from Harvard University. After teaching at the University of Texas and at Yale University, he returned in 1902 as professor in pure mathematics to the University of Texas, where he remained until his retirement in 1945. M. B. Porter died in Austin on May 27, 1960.

Goldie Porter, still an assistant professor, stopped teaching full time and went on modified service in 1958. She continued part-time teaching until her retirement in 1966, when she was granted the title professor emeritus. Goldie Horton Porter was active in various organizations in Texas, particularly before she married. In 1919 she was chair of the mathematics section of the Texas State Teachers' Association. She was a founder in 1920 of the University of Texas chapter of Alpha Phi, a social sorority, and in 1923 of the Austin chapter of AAUW, which she also served as treasurer, secretary, and president. In the late 1930s she described herself as a Democrat and indicated interests in gardening and housekeeping.

In 1930 Goldie Horton and her mother, then almost seventy years old, had a house built in Austin where the Porters lived after their marriage. Their house was considered a library and museum and is listed in the National Register of Historic Places as the Goldie Horton-Porter House. The Porters collected books, antique furniture, glass, porcelains, oriental rugs, and other *objets d'art*. They were also music lovers; he played flute and she piano.

Goldie Porter remained in Austin and died in 1972 at eighty-four. Among her bequests was one to the Texas governor's mansion of Meissen porcelain figurines; she also endowed the Goldie Horton Porter Book Fund at Bryn Mawr College.

PRICE, Irene. January 16, 1902–March 13, 1999.

INDIANA UNIVERSITY (BA 1926, MA 1927, PHD 1932). PhD diss.: On a certain type of polynomials (directed by Kenneth Powers Williams and Harold Thayer Davis).

Irene Price was born in Parr, Indiana, the second of five children of Ruth Elizabeth (Schroer) and William Edward Price; her parents farmed in northwestern Indiana. Irene Price attended high school in Rensselaer, Indiana, 1915–19 before teaching grade school in Rensselaer for the next four years. During the summer of 1921 she studied at Indiana State Normal School (now Indiana State University) in Terre Haute.

Price completed the work for her bachelor's degree at Indiana University in 1926 and stayed there the next three years for her graduate work in mathematics. She was a part-time instructor at Indiana 1926–29, was awarded a Clara Javen Goodbody scholarship in 1927, and was an instructor in the extension division 1927–29. She earned her master's degree in 1927. While at Indiana she also worked on a tables computation project. Price's doctorate was granted by Indiana University in 1932.

After her graduate studies, Price had three somewhat distinct careers. She was a college professor for about fifteen years, then worked as a statistician for the US Air Force for a dozen years, and finally was self-employed as a real estate agent for more than forty years. From 1929 to 1944 she was professor of mathematics at Oshkosh Teachers College (now University of Wisconsin Oshkosh). In addition to maintaining memberships in various professional associations, she was active in the Wisconsin Section of the MAA, and served on the program committee 1934–35 and 1939–40 and as chairman of the section 1941–42. She also belonged to the Oshkosh Education Association, for which she served as treasurer for several years after 1936, and the AAUW, for which she was president of the Oshkosh branch 1942–43.

In 1944 Price began her employment as a statistician for the Headquarters of the Air Materiel Command at Wright-Patterson Air Force Base in Ohio. While in Ohio she taught several evening courses in college mathematics in Dayton and at Miami University in nearby Oxford. Price remained in Ohio until 1949, when she moved to Alamogordo, New Mexico, where she directed mathematical research on guided missiles at the Holloman Air Force Base until 1953. She then took a position as a mathematician at White Sands Proving Ground where she did similar work and stayed until 1956.

In 1956 Irene Price became owner of Price Realty and after 1956 was engaged in a number of professional activities associated with this business. She held various offices in the Alamogordo Board of Realtors and was district director of the New Mexico Realtors Association 1962–64. In 1971 she taught a college course in real estate practice to retiring military at Holloman Air Force Base, and in 1978 she was named Alamogordo Realtor of the Year by the Alamogordo Board of Realtors. She retired from the real estate business in January 1998 at ninety-six.

Price was affiliated with the Christian Church and was a member of the John Birch Society, Business and Professional Women (treasurer 1957), and the Alamogordo Chamber of Commerce. She was also a trustee of the Betty Dare Foundation and a member of the advisory board of the Betty Dare Good Samaritan Center, a nursing home. Irene Price was ninety-seven when she died in Alamogordo, New Mexico, in 1999.

QUINN, Grace (Shover). December 20, 1906–February 4, 1998.

Ohio State University (BA 1926, BS 1926, MA 1927, PhD 1931). PhD diss.: [Shover, G.] On the class number and ideal multiplication in a rational linear associative algebra (directed by Cyrus Colton MacDuffee).

Carolyn Grace Shover was the second of two daughters of Margaret (Seeds) and Cyrus F. Shover. She was born in Grove City, Ohio, just outside Columbus. Her parents attended rural schools; her mother held a teacher's certificate, while her father was a carpenter and was in the lumber business. Her older sister became a high school teacher. The family was living in Columbus, Ohio, by 1900.

Grace Shover attended public schools in Columbus from 1912 to 1922, when she entered Ohio State University as a commuter. She graduated in 1926 with a BA with honors and a BS in education. Shover remained at Ohio State and earned a master's degree in 1927, after having been an assistant in psychology during 1926–27. Her master's thesis was on scoring methods in college reading tests. She continued her work at Ohio State with a university scholarship in mathematics 1927–29; she also studied at the University of Chicago during the summer of 1928. She received her PhD from Ohio State in 1931.

Shover was an instructor at Connecticut College for Women 1929–33 and worked the following year at the Federal Reserve Bank in Cleveland, Ohio. During 1934–35 Grace Shover was at Bryn Mawr College to study with Emmy Noether, who had come to Bryn Mawr from Germany the previous fall; she held an Emmy Noether fellowship, which paid for her room, board, and tuition. The next year Shover was a teacher at the Shipley School in Bryn Mawr, a position she had taken the previous winter with the intention of continuing her studies with Noether, who had died suddenly in April 1935. In the summer of 1936 Shover attended the International Mathematical Congress in Oslo.

Shover was an instructor at the New York College for Teachers (now State University of New York at Albany) 1936–37 and then at Carleton College in Minnesota 1937–42. On June 9, 1942, in Cleveland, Ohio, Grace Shover married Robert Byron Quinn, who had been an instructor of physics at Carleton the previous three years. Robert Quinn, born in 1907 in Indiana, received his bachelor's degree and master's degree from Indiana University and his doctorate in physics from the University of Chicago. After their marriage the Quinns moved to Washington, D.C., where from 1942 to 1948 Robert Quinn was a staff member at the US Naval Research Laboratory. Robert Quinn continued his professional work with the US Navy and retired in 1970 from the Naval Air Systems Command.

During World War II, Grace Quinn was first a cryptanalyst for the US Navy in 1942 and then a research assistant for the National Defense Research Committee 1943–45. She was also a part-time lecturer at George Washington University 1942–44. The Quinns' two daughters were born in 1946 and 1949. The elder daughter earned a bachelor's degree in mathematics and computer science from Iowa State University and a master's degree in computer science from Rutgers University; the younger daughter earned a bachelor's degree in English Literature from Ripon College and a certificate from the Morven Park International Equestrian Institute in Virginia.

Grace Quinn resumed her career in 1956 as an assistant professor at American University in Washington, D.C. She retired as professor in 1970, having been promoted to that rank in 1963. She continued as an adjunct professor for a year after her retirement. Quinn directed several master's theses while at American University.

In her 1981 Smithsonian questionnaire, Quinn described her religious affiliation as Congregational since 1928 and listed playing the piano among her interests. At that time she was treasurer of the American University Women's Association, was secretary of the Women's Guild at her church, was on the executive board of the Bethesda Garden Club, and was vice president of a social club.

In July 1991 the Quinns moved to Foxdale Village, a Quaker retirement community in State College, Pennsylvania. Less than two years later Robert B. Quinn was in an automobile accident, and after nearly seven months in a coma he died in October 1993. Grace Quinn died at ninety-one in State College in 1998.

RAGSDALE, Virginia. December 13, 1870–June 4, 1945.

Guilford College (BS 1892), Bryn Mawr College (BA 1896, PhD 1904). PhD diss.: On the arrangement of the real branches of plane algebraic curves (directed by Charlotte Angas Scott).

Virginia Ragsdale was born in Jamestown, North Carolina, the second of three surviving children of Emily Jane (Idol) and Joseph Sinclair Ragsdale. Her mother taught in county public schools before her marriage, and her father fought in the Civil War, owned and taught in a small private school, and then became a cotton manufacturer in Jamestown and a prominent businessman in the state.

Virginia Ragsdale first attended the private Flint Hill School in Jamestown. She also studied privately and attended both the high school in Jamestown and the Salem Female Academy. She entered the Salem Female Academy as a junior and graduated as valedictorian with an extra diploma in piano in 1887, three years before the academy was renamed Salem College. In 1891 Ragsdale entered the coeducational Guilford College in Greensboro, North Carolina, and graduated after one year.

Ragsdale received a fellowship from Bryn Mawr College and studied there for a second bachelor's degree that she received after four years in 1896. After two years she began taking mathematics courses open to graduate students and undergraduates who had completed the major. She was elected European fellow for the class of 1896 but remained at Bryn Mawr for a year as an assistant demonstrator in physics and a graduate student in mathematics. She used the fellowship to study with Felix Klein and David Hilbert at the university in Göttingen 1897–98.

When she returned from Germany, Ragsdale taught mathematics and science at the Bryn Mawr School in Baltimore. It was not until 1901, when she received a one-year fellowship awarded by the Baltimore Association for the Promotion of University Education of Women, that Ragsdale returned to her studies at Bryn Mawr College. The following year, 1902–03, she held the resident fellowship in mathematics and passed her examination for the PhD in September 1903. Ragsdale then returned to teaching, this time in New York City at Dr. Sachs' School for Girls. Her PhD was conferred at the 1904 commencement.

Ragsdale continued to teach at Dr. Sachs' School for Girls until 1905 and then was at her home in North Carolina the following year. She resumed teaching and was head of the department of mathematics at the Baldwin School in Bryn Mawr 1906–11. During this time she maintained her connections with Bryn Mawr College, and from 1908 to 1910 she assisted Charlotte A. Scott, her dissertation advisor, as a reader.

In 1911 Ragsdale moved back to North Carolina to take a position at the State Normal and Industrial College in Greensboro, renamed North Carolina College for Women in 1919 and now the University of North Carolina at Greensboro. During the years 1911 through 1926 she was instructor, associate professor, and then professor, and was head of the department from 1926 until 1928. She was on leave 1913–15 and spent most of the first year at a sanatorium with a lung lesion, later determined not to be tubercular. In 1928, when she was not yet fifty-eight, Ragsdale retired to be with her mother and help manage a family farm. After her mother's death in 1934, Ragsdale had a house built on the Guilford College campus and became part of the Guilford community. She was a member of the executive committee of the Guilford College Alumni Association and of the Society of Friends at New Garden Friends Meeting. She was an avid gardener, was a devoted and much-loved aunt to her nieces and nephews, and was interested in genealogy.

Virginia Ragsdale died in Greensboro at age seventy-four in 1945. Her house was bequeathed to Guilford College and now stands as Ragsdale House, the home of the president of the college. In 1950 the Dr. Virginia Ragsdale Residence Hall was built on the campus of the University of North Carolina at Greensboro.

What has become known as the Ragsdale Conjecture, the statement that the conclusion of a conditional statement in her dissertation always holds, was never explicitly made by Ragsdale. It was proved to be incorrect in 1979 by Oleg Viro (*Soviet Math. Dokl.* 22 (1980): 301–6), and articles have continued to appear that address the characteristics of counterexamples to the conjecture.

RAMBO, Susan M. April 3, 1883–January 7, 1977.

SMITH COLLEGE (BA 1905, MA 1913), UNIVERSITY OF MICHIGAN (PHD 1920). PhD diss.: The point at infinity as a regular point of certain linear difference equations of the second order (directed by Walter Burton Ford).

Susan Miller Rambo was born in Easton, Pennsylvania, the eldest of five children of Annie Roberta (Kortz) and George Green Rambo. Her father was a wholesale grocer in Easton after his marriage.

Susan Rambo attended high school in her hometown of Easton before entering Smith College in Northampton, Massachusetts. She graduated from Smith in 1905 and taught mathematics in the high school in Hoosick Falls, New York, near the Massachusetts border for two of the years between 1905 and 1908.

In 1908 Susan Rambo was hired as an assistant in mathematics at Smith College. Except for leaves of absence, Rambo remained in the department at Smith the remainder of her career: as assistant 1908–11, instructor 1911–18, assistant professor 1918–22, associate professor 1922–37, professor 1937–48, and professor emeritus after her retirement in 1948. During her early years on the faculty at Smith, Rambo also took graduate courses in mathematics and astronomy and received a master's degree from Smith in 1913. She also had attended lectures at Columbia University in 1907 and at the University of Chicago in 1910. In 1916, having taken a leave of absence for the year 1916–17, Rambo entered the University of Michigan. She was in residence at Michigan for two years and received her PhD in 1920, two years after returning to her position at Smith.

Rambo's many contributions at Smith include serving as department chairman 1934–40. She attended the International Congress of Mathematicians in Bologna in 1928. Ten years later she made a trip around the world during a one-semester leave of absence from Smith. In the Smith College "Twenty-fifth Reunion Class Book" in 1930 it was noted that Susan Rambo "drives a Ford, gardens, and keeps a charming home with a friend" (Smith College Archives). She shared a house with her Smith colleague and friend **Suzan Benedict** from 1918 until Benedict's death in 1942. Their home was typically the location for the department Christmas party. In 1945 Rambo relinquished her life tenure on the house, and it went to Smith to be sold and the proceeds to be used for scholarships.

Rambo died at ninety-three in Northampton, Massachusetts, in 1977 and was buried in Easton, Pennsylvania. The Susan M. Rambo Fund that provides support for summer student research in mathematics at Smith was established in her memory by a 1931 graduate of Smith.

RASMUSEN, Ruth B. March 29, 1899–October 27, 1974.

Concordia College (BA 1921), University of Chicago (MS 1926, PhD 1936). PhD diss.: Conjugate osculating quadrics associated with the lines of curvature (directed by Ernest Preston Lane).

Ruth Beatrice Rasmusen was born in Moorhead, Minnesota, where her father was a barber. She was the first child of Oline (Steen), born in Norway, and Robert Rasmusen, born in Minnesota. In 1910 there were three daughters living of five children who had been born.

Ruth Rasmusen attended the public schools of Moorhead and graduated from Moorhead High School in 1918. After three years at Concordia College in Moorhead, she graduated summa cum laude with a mathematics major and a Latin minor. For the next four years, 1921–25, Rasmusen taught in high schools in Minnesota and North Dakota. In the summer of 1925 she enrolled at the University of Chicago and studied there for five consecutive quarters before receiving her master's degree in September 1926.

Rasmusen was a teacher at the senior high school in Muskegon, Michigan, for three semesters in 1927–28, an instructor at North Dakota State College (now University) in Fargo 1928–29, and an instructor at South Dakota State College (now University) in Brookings 1929–33. She continued her graduate work at the University of Chicago, spending twelve quarters in residence between summer 1930 and summer 1936, after which she received her doctorate with a dissertation in differential geometry. It appears that Rasmusen was living in her home town of Moorhead for a while after completing her graduate work and was appointed instructor at the Platteville State Teachers College in Wisconsin (now University of Wisconsin-Platteville) some time before February 1938.

Rasmusen began her employment with the Chicago Board of Education at Woodrow Wilson Junior College (now one of the City Colleges of Chicago) in October 1938. She was transferred from Wilson to Chicago Teachers College (now Chicago State University) in September 1953. When rank was first established there in 1959, she was made a full professor, one of 18 of the faculty of 137 to be granted that rank. Rasmusen's compulsory retirement occurred in June 1964. She then taught at Kearney State College (now University of Nebraska at Kearney) 1964–65. In September 1965 the Chicago Teachers College became a state institution, Illinois Teachers College Chicago-South, with a more lenient retirement policy, and Rasmusen returned to work there until her second retirement in 1967.

Rasmusen continued some research in differential geometry during the late 1930s and early 1940s. Her published work includes a joint article in 1941 with **Beatrice Hagen** of Penn State and a text for use at Chicago Teachers College written with two of her colleagues and published in 1960. Rasmusen also was an associate editor of *Mathematics Magazine* 1964–68. In 1970 the Dr. Ruth B. Rasmusen Award was established at Chicago State; it is given annually to a mathematics major who shows outstanding promise as a scholar in the field of mathematics research. In 1972 she received an Alumni Achievement Award from the Concordia College Alumni Association.

For many years Rasmusen lived in Chicago with her sister Esther Rasmusen, who was a biology teacher at a Chicago high school. Esther Rasmusen died in April 1974; six months later Ruth Rasmusen died at age seventy-five in Minneapolis. She was buried in Moorhead.

RAYL, Adrienne S. September 25, 1898–November 27, 1989.

Tulane University (BA 1924, MA 1934), University of Chicago (PhD 1939). PhD diss.: Stability of permanent configurations in the problem of four bodies (directed by Walter Bartky).

Adrienne Sophie Rayl was born in New Orleans, Louisiana, the daughter of Sophie Catherine (Schick) and James John Rayl. Her father had an elementary school education, was a cigar classer in 1900, and later was a postal employee; her mother received a high school and normal school education and became an elementary school teacher. Adrienne Rayl was the eldest of eight children.

Adrienne Rayl attended public grade school and Girls High School in New Orleans, graduating from the latter in 1915 first in her class of fifty-one. She attended the New Orleans Normal School from 1915 to 1917 and graduated, first in her class, with a teaching diploma in 1917.

During the next twenty-two years Rayl taught in the New Orleans public school system and completed all of her undergraduate and graduate work. In 1917 she began teaching seventh and eighth grades at the Beauregard School and began working toward her bachelor's degree by taking courses for teachers and summer school classes at Tulane University. She received her BA in education in 1924 and continued teaching at the elementary school until 1929, when she was transferred to the Edward Douglas White High School to teach mathematics. While teaching at the high school, where she remained until 1939, she attended summer school and afternoon classes, including classes by **Nola Anderson (Haynes)** and by **Marie Weiss**, at Tulane and earned her MA in 1934. She immediately began further graduate work at the University of Chicago in the summer of 1934. She took classes there for five consecutive summers and the academic year 1936–37 and finished her dissertation in August 1939. Rayl's dissertation concerned the four-body problem. She had begun work in this area while at Tulane and had solved a problem related to the three-body problem when taking a celestial mechanics course there.

Adrienne S. Rayl spent the next three decades in Birmingham at the University of Alabama Extension Center (now the University of Alabama at Birmingham), which had opened three years before her arrival. She was instructor 1939–41, assistant professor 1941–47, associate professor 1947–52, and professor 1952–69. She retired in 1969 with the rank of professor emeritus. She was also chairman of the Division of Mathematics.

Rayl was a member of the Lutheran Church, Missouri Synod. In the early 1980s she moved back to New Orleans, where she was living at the time of her death in 1989 at ninety-one.

REAVIS, Mabel (Griffin). August 28, 1907–November 19, 1999.
Duke University (Woman's College) (BA 1928), Duke University (MA 1930, PhD 1933).
PhD diss.: [Griffin, M. J.] Invariants of Pfaffian systems (directed by Joseph Miller Thomas).

Mabel Jeanette Griffin was born in Durham, North Carolina, the youngest of four children of Naomi Frances (Burke) and Charles A. Griffin. Her mother attended the Pittsboro Academy in North Carolina; her father attended county schools and became a carpenter and home builder.

Mabel Griffin was one of the first undergraduates at Duke University's Woman's College, which was founded in 1924 as a coordinate college for women. She received her bachelor's degree magna cum laude in 1928, taught in high school for a year, and then returned to Duke as an assistant with a graduate scholarship. She remained an assistant through 1932 with a scholarship through 1931 and a fellowship 1931–32. She received her master's degree in 1930 and her PhD in 1933, having taught in high school 1932–33. After receiving her doctorate, Mabel Griffin taught one more year in high school before teaching 1934–36 as professor at Campbell Junior College (now Campbell University) in Buies Creek, North Carolina.

On August 3, 1936, Mabel Griffin married L. Boyd Reavis, who was born in 1907 in North Carolina and graduated from Campbell Junior College in 1930. He received further education from several schools, including Howard Payne College (now University) in Brownwood, Texas, from which he earned a Doctor of Divinity degree. For the first four years of their marriage, 1936–40, Mabel Reavis taught as assistant professor at Mary Hardin–Baylor College (then a women's division of Baylor College, now the coeducational University of Mary Hardin-Baylor) in Belton, Texas. She later served as a trustee of the college.

From 1939 to 1955 L. Boyd Reavis was pastor of four Baptist churches, the first two in Texas: in Plainview 1939–45 and in Paris 1945–49. Their three children were born in Texas. The elder son was born in 1940, graduated from Baylor University, and became an executive with an aerospace corporation; a daughter was born in 1943, earned both a bachelor's and a master's degree from Baylor, and became co-host of a radio talk show in Chicago; the younger son was born in 1946, graduated from Texas Christian University, and earned a PhD in education from North Texas State University before holding positions in academia.

In 1949 the Reavis family moved to High Point, North Carolina, and in 1950 Mabel Reavis resumed her career as professor and head of the mathematics department at High Point College. After three years they moved to Denton, Texas, and two years later to Fort Worth, where L. B. Reavis was director of development of the Southwestern Baptist Theological Seminary 1955–73. In 1955 Mabel Reavis began an eighteen-year period as a mathematics faculty member at Texas Christian University (TCU). She was assistant professor 1955–58, associate professor 1958–73, and retired as professor emeritus in 1973. During five summers in the 1960s she was director of the mathematics section of the NSF summer science institute at Texas Christian. She was vice president of the TCU chapter of Phi Beta Kappa and served on the TCU faculty senate.

Reavis included various church related activities among her interests. She was a member of various study clubs, lectured for local clubs and societies, and engaged in Bible teaching. In 1991 she and her husband established the L. B. and Mabel Reavis professorship and scholarship program at Campbell University; Reavis scholars are honored for leading their churches in growth and evangelism. L. Boyd Reavis died in March 1995. Mabel Reavis remained in Fort Worth until 1998, when she moved to Chicago to be close to her daughter. Mabel Griffin Reavis died in Chicago in November 1999 at age ninety-two.

REES, Mina S. August 2, 1902–October 25, 1997.

Hunter College (BA 1923), Columbia University (Teachers College) (MA 1925), University of Chicago (PhD 1931). PhD diss.: Division algebras associated with an equation whose group has four generators (directed by Leonard Eugene Dickson).

Mina Spiegel Rees was the youngest of two daughters and three sons of Alice Louise (Stackhouse) and Moses Rees. Her mother was born in England and immigrated to the United States in 1883, and her father was born in New York of German parents. Mina Rees was born in Cleveland, Ohio, but the family moved to New York when she was a baby, and her early schooling was in the New York City public schools.

After graduating from Hunter College High School as valedictorian of her class in 1919, Mina Rees attended Hunter College, graduated summa cum laude in 1923 as a mathematics major, and was awarded an "H" pin as one of the seniors who had performed exemplary service. After she graduated from Hunter College, Rees became an assistant teacher at Hunter High School and attended Columbia University as a full-time graduate student. She received her master's degree from Teachers College of Columbia University in 1925 and was hired as instructor of mathematics at Hunter College. In 1929 she took a leave of absence from Hunter to study with Leonard Eugene Dickson at the University of Chicago. Although she wrote her dissertation with Dickson, she worked mainly on her own since Dickson was no longer engaged in the field in which she was working.

Having had a fellowship her last year at Chicago, Rees received her PhD in 1931 and returned to Hunter College as instructor. She was assistant professor 1932–40 and became associate professor in 1940. In 1943 she took a leave of absence to work for the newly established Applied Mathematics Panel (AMP) of the National Defense Research Committee and served as a technical aide and executive assistant to Warren Weaver, chief of the AMP. Rees was with the AMP until 1946, when she went to Washington, D.C., to work for the Office of Naval Research (ONR). At ONR she was head of the mathematics branch 1946–49, director of the mathematical sciences division 1949–52, and deputy science director 1952–53. Rees remained officially on leave from Hunter until 1950.

The significance of her role during and just after the war was recognized in a resolution adopted by the council of the AMS at its annual meeting in December 1953. It reads in part: "Under [Mina Rees's] guidance, basic research in general, and especially in mathematics, received the most intelligent and wholehearted support. No greater wisdom and foresight could have been displayed and the whole postwar development of mathematical research in the United States owes an immeasurable debt to the pioneer work of the Office of Naval Research and to the alert, vigorous and farsighted policy conducted by Miss Rees" (*Bull. Amer. Math. Soc.* 60 (1954): 134).

Mina Rees returned to Hunter College as professor of mathematics and dean of the faculty in 1953 and became head of a new Office of Institutional Research there the following year. On June 24, 1955, she married Leopold Brahdy (1892–1977), a physician who was born in Vienna, Austria, and immigrated to the United States when he was six. Rees remained as professor and dean until 1961 when the City University of New York (CUNY) was founded and she became professor and dean of graduate studies. She was provost of the graduate division 1968–69 and was president of the Graduate School from 1969 until her retirement as president emeritus in 1972.

Starting in the 1950s, Rees served in a number of important positions in scientific organizations. Especially noteworthy were her positions on the National Science Board 1964–70 and within the AAAS, where she became the first woman president in 1971. She served on the original appointed executive council of the ACM 1947–48 and on the elected executive council 1948–50. For the AMS she was a trustee 1955–59. In the MAA she was vice president 1963–64, having earlier served as vice chairman 1955 and chairman 1956 of the New York Metropolitan Section. For SIAM she was councillor 1953–56 and on the board of directors of the SIAM Institute for Mathematics and Society. She also served as a member of the executive committee of the American Conference of Academic Deans 1960–62.

Among Rees's many honors are at least eighteen honorary degrees from US colleges and universities; the King's Medal for Service in the Cause of Freedom; the President's Certificate of Merit 1948; MAA's first award for service to mathematics 1962; Hunter College High School's first distinguished graduate 1965; AAUW achievement award 1965; a AAAS Symposium to Honor Mina Rees, January 1982; and the Public Welfare Medal, National Academy of Sciences 1983. The library at the Graduate School and University Center of CUNY was dedicated as the Mina Rees Library in 1985.

Rees had broad cultural interests that included painting, music, dance, and literature. She was a frequent traveler abroad and made trips to South America and Europe. She was a member of the Unitarian Church of All Souls in Manhattan for many years.

Mina Rees died in Manhattan a few months after her ninety-fifth birthday. She left $1.7 million to the CUNY Graduate Center to establish a fellowship and to endow a chair in mathematics. She is the subject of two doctoral dissertations.

REILLY, Sister Mary Henrietta. February 14, 1895–April 21, 1964.

St. Xavier College (BA 1923, MA 1925), Catholic University of America (PhD 1936).
PhD diss.: Self-symmetric quadrilaterals in-and-circumscribed to the plane rational quartic curve with a line of symmetry (directed by Aubrey Edward Landry).

Margaret Reilly was born in Chicago, Illinois, the daughter of Agnes (Finn) and Michael Reilly. In 1900 the family lived in Chicago, and there were three children, an older sister, and a younger brother. Her father's occupation was listed as sewer builder.

Reilly received her high school diploma from Our Lady of Mercy Academy in Cincinnati in June 1914 and entered the congregation of the Religious Sisters of Mercy in Cincinnati that same month. She took the religious name Mary Henrietta and was known as Sister Henrietta. She took her perpetual vows in August 1917.

Except for her years as a doctoral student at the Catholic University of America in Washington, D.C., all but one of Sister Mary Henrietta's assignments were in Cincinnati, where she taught second grade at St. Edward School 1914–16; grades five and six at St. Andrew School 1917–19; grades nine and ten, and music at Our Lady of Mercy Academy 1919–22; grade nine and English at Elder High School 1922–25; and grades seven through ten and twelve at Mother of Mercy Academy 1925–28. During this period she obtained her bachelor's degree in 1923 from St. Xavier College (Xavier University since 1930), a Jesuit college for men in Cincinnati, by attending special classes scheduled in late afternoons, Saturdays, and summers for the sisters from local and neighboring congregations. She obtained her state high school life certificate in December 1923 having also earned credits from what was called in congregation records Our Lady of Mercy Normal. She received a master's degree from St. Xavier College in August 1925 with a major in sociology and a minor in philosophy.

In the period 1928–36 Sister Henrietta taught mathematics and science, and she continued her studies. She taught at Teachers' College of the Athenaeum of Ohio 1928–31 and apparently earned some credits there as well. She then studied at Catholic University, where she was in residence 1931–32. She taught again at Teachers' College 1933–34 before returning to Catholic University where she was again in residence 1934–36. She wrote her dissertation in algebraic geometry and received her PhD in 1936.

Sister Henrietta then returned to Cincinnati to resume her college teaching of mathematics in the Athenaeum of Ohio. Sister Henrietta was head of the department at Teachers' College at least from 1939 until 1949 and remained there until 1953, when the Teachers' College was discontinued.

Sister Henrietta's teaching assignments were at high schools for the rest of her life. She taught at a high school in Piqua, Ohio, 1953–55; at Our Lady of Mercy High School in Cincinnati 1955–56, and at Mother of Mercy High School in the Westwood neighborhood of Cincinnati 1956–64. At the high school level she taught English, religion, mathematics, history, chemistry, and Latin. Sister Mary Henrietta Reilly died in Mercy Hospital, Hamilton, Ohio, in 1964 at age sixty-nine.

REKLIS, Virginia (Modesitt). June 15, 1910–November 29, 2000.

Mount Holyoke College (BA 1931), University of Illinois (MA 1932, PhD 1937). PhD diss.: [Modesitt, V.] Some singular properties of conformal transformations between Riemannian spaces (directed by Harry Levy).

Virginia Modesitt was born in Bloomington, Indiana, the only child of Floy (Underwood) and Raymond Lyons Modesitt. Both of her parents received bachelor's degrees from Indiana University in 1908; her father also earned a master's degree the following year. The 1910 census lists her father as a high school teacher in Bloomington. He became a teacher of mathematics at Eastern Illinois State Normal School in Charleston (later Eastern Illinois State Teachers College, now Eastern Illinois University).

Virginia Modesitt graduated from Eastern Illinois State Teachers College High School in 1927. Her father died suddenly in December of that year, and she remained in Charleston and attended the teachers college for a year before entering Mount Holyoke College in September 1928. She did honors work in mathematics and minored in astronomy. Modesitt attended Butler University in Indianapolis, Indiana, in the summer of 1930 and graduated from Mount Holyoke magna cum laude in 1931.

In 1931 Modesitt entered the University of Illinois for graduate work. She received her master's degree in 1932 and her doctorate, with a minor in astronomy, in 1937. While at Illinois she was a secretary in the astronomy department 1933–34 and was an assistant in the mathematics department 1934–37. In 1934 she published an article with a fellow graduate student in the *Astrophysical Journal*.

Modesitt took a temporary instructorship at Randolph-Macon Woman's College in Virginia to substitute for someone on leave of absence 1937–38. She studied at the University of Michigan in the summer of 1938 and returned to Randolph-Macon for one more year. In 1939 Modesitt moved to Chicago, where she was an instructor at Wright Junior College for the next four years. At Wright Modesitt worked with **Ruth Mason Ballard** on the development of survey courses for junior colleges, and both remained there until 1943, when Wright was taken over by the Navy as barracks and all the junior colleges in the city were combined into one.

On January 20, 1942, Virginia Modesitt married Ernest Peter Reklis, whom she had met while both were students at the University of Illinois. Ernest Reklis, born in Illinois in 1912, was an engineer who worked at Commonwealth Edison in Chicago after receiving his BS in engineering from Illinois. He was in the army from 1943 to 1945; they moved to the East, where both worked at the Aberdeen Proving Ground in Maryland. Virginia Reklis was first a "computer" and then a mathematician at the Ballistic Research Laboratories, Army Service Forces, 1944–47. After the war Ernest Reklis remained at the Ballistic Research Laboratories at the Aberdeen Proving Ground. Their son was born in September 1947, and Virginia Reklis did not seek employment while he was young. Later, however, she was an instructor of extension courses in Maryland: for the University of Maryland at the Edgewood Arsenal 1960–65 and for the University of Delaware at the Aberdeen Proving Ground 1962–66. Her son earned a PhD in physics and became a research scientist.

Virginia Reklis was a Presbyterian. She was a member of the Daughters of the American Revolution and the P.E.O. Sisterhood, a philanthropic educational organization. Beginning in 1967, she published several genealogies. Virginia Reklis's husband died at age sixty-two in 1975; her mother died at ninety-nine in 1984. In the middle 1980s Virginia Reklis lived in her mother's former home in Danville, Indiana. She later moved to California, closer to her son's family. Virginia Reklis had been living in an assisted living facility in Palo Alto, California, when she died in November 2000 at age ninety. She was buried in Danville.

ROE, Josephine (Robinson). May 5, 1858–April 29, 1946.
OBERLIN COLLEGE (BA 1894), DARTMOUTH COLLEGE (MA 1911), SYRACUSE UNIVERSITY (PHD 1918). PhD diss.: Interfunctional expressibility problems of symmetric functions *and* Interfunctional expressibility tables of symmetric functions (directed by Edward Drake Roe, Jr.).

Josephine Alberta Robinson was born in Meredith, New Hampshire, the daughter of Frances Eliza (Weld) and Joseph Wadleigh Robinson, a farmer. The marriage was the second for her father. She had two older sisters and four younger brothers. The youngest three all attended Dartmouth College; one earned a PhD in economics, one was a minister, and all had positions at some time with academic institutions.

Josephine Robinson did her secondary work at the New Hampton Literary Institution, a few miles west of Meredith. She entered in November 1873, did the English course, and graduated in 1880, having completed both the commercial and the regular courses. She taught for more than a decade in New Hampshire schools before entering college. Her early teaching positions included: teacher in New Hampshire public district schools for about fifteen months; principal of the high school in Laconia 1880–82; and teacher, 1882–89, and preceptress, 1889–90, New Hampton Literary Institution and Commercial College. Robinson entered Oberlin College at age thirty-two in 1890 and graduated four years later having done her major studies in mathematics.

After graduating from Oberlin, Robinson was at Kimball Union Academy in Meriden, New Hampshire, 1894–97, where her main duty was teaching Latin. She had also been engaged in postgraduate work at Oberlin in Latin and German education when she was "called" to Berea College in Kentucky in 1897. Upon her arrival at Berea she taught Latin, English literature, and mathematics; she was principal of the Ladies Department and acting professor of mathematics 1897–1901, dean of women 1901–07, and professor of mathematics 1901–11. During the summers of 1907 to 1910 she studied at Dartmouth College and received a master's degree in mathematics in June 1911.

On February 1, 1911, in Berea, Kentucky, Josephine Robinson married Edward Drake Roe, Jr., a widower and mathematics professor at Syracuse University. E. D. Roe was born in Elmira, New York, in 1859 and earned bachelor's degrees from Syracuse and Harvard and a master's degree from Harvard. He went to Oberlin College as associate professor of mathematics in 1892, when Josephine Robinson was in her third year there. He studied at Erlangen 1897–98 and received a PhD from there in 1898, the same year his first wife died. E. D. Roe was on the faculty at Syracuse University from 1900 until his death twenty-nine years later; he was the founder of the national honorary mathematics society Pi Mu Epsilon in 1914.

After her marriage, Josephine Robinson Roe left Berea College and moved to Syracuse, where she subsequently engaged in graduate studies in mathematics. In 1918, at age sixty, she received her PhD with a dissertation that began: "This investigation is a continuation of published work of Professor E.D. Roe, Jr., to whom I am greatly indebted for assistance and advice." She taught as assistant professor at Syracuse in 1920.

In response to an inquiry from **Helen Owens**, Roe wrote on her 1940 questionnaire, "Assisted my husband somewhat in his private Observatory in our residence. Interested in birds and wild flowers, housekeeping, support of Christian missions at home and abroad, and of temperance work. A lover of travels, a wide reader. Have been active in Daughters of the American Revolution. Especially interested in people, just people, everywhere. A firm believer in immortality" (Owens questionnaire). Josephine Robinson Roe died in Newton Center, Massachusetts, in April 1946, shortly before her eighty-eighth birthday.

ROSENBAUM, Louise (Johnson). January 21, 1908–January 16, 1980.
UNIVERSITY OF COLORADO (BA 1928, MA 1933, PHD 1939). PhD diss.: [Johnson, L. L.] On the diophantine equation $x(x+1)\cdots(x+n-1) = y^k$ (directed by Aubrey John Kempner).

Laura Louise Johnson, known as Louise, was born in Carrollton, Illinois, the youngest of five children of Ida Jane (Taylor) and William Foster Johnson. The family moved from

Illinois to a farm outside of Boulder, Colorado, when Louise Johnson was ten. She attended country school through grade nine and at age thirteen moved by herself into Boulder in order to attend a school with a stronger preparatory program.

After completing high school at sixteen, Johnson entered the University of Colorado, where she did all of her undergraduate and graduate work. She received her bachelor's degree in 1928 and taught in high schools in eastern Colorado 1928–30. She returned to the University of Colorado to begin her graduate studies in fall 1930 and worked as an assistant in mathematics at least two years. She was an instructor in the extension division of the university most of the rest of the time she was there and taught mathematics at a Civilian Conservation Corps (CCC) camp one year in the early 1930s. Johnson received her master's degree in 1933 after completing a thesis directed by Aubrey Kempner, who also directed her 1939 doctoral dissertation.

In fall 1939 Louise Johnson and Robert A. Rosenbaum were two of four fellows participating in a mathematics teaching seminar sponsored by the General Education Board at Reed College in Portland, Oregon. They both stayed on at Reed as instructors when the program ended in 1940 and were married on August 1, 1942. Rosenbaum, who was born in 1915 in Connecticut, had graduated from Yale in 1936; he was in the US Navy 1942–45 and served as an aviator in the Pacific. Louise Rosenbaum was instructor until 1943 and assistant professor after 1943; she reported an address in Hollywood, Florida, in the 1944 MAA membership list. R. A. Rosenbaum was assistant professor at Reed 1945–46, finished his PhD from Yale in 1947, returned to Reed as associate professor, and was promoted to professor in 1949. He was a visiting professor at Swarthmore College 1950–51. Louise and Robert Rosenbaum remained on the faculty at Reed until 1953, when Robert Rosenbaum became professor of mathematics at Wesleyan University in Middletown, Connecticut. He later was dean of science, provost, academic vice president and provost, and acting president. In 1985 he was named University Professor of Mathematics and the Sciences, emeritus. During 1958–59 he was an NSF fellow at the Mathematics Institute at Oxford.

The Rosenbaums had three sons born in 1943, 1946, and 1948. Louise Rosenbaum taught at least one course every semester when she was at Reed; she also directed undergraduate theses and served on major college committees. After the Rosenbaums moved to Connecticut, Louise Rosenbaum's regular teaching stopped. She had occasional visiting appointments at Trinity College in Hartford and at Connecticut College in New London, and was a visiting faculty member at Smith College in 1973–74 when Robert Rosenbaum was a visiting professor at the nearby University of Massachusetts. She also was a professor at Saint Joseph College in West Hartford, served on a committee of the School Mathematics Study Group (SMSG), and directed summer institutes for teachers in Connecticut and Oregon. During the 1950s and 1960s, she and her husband published, and revised several times, a pamphlet listing mathematics books recommended for high school libraries. She also wrote a short book on mathematical induction.

Chief among Louise Rosenbaum's interests was hiking. While a student at Colorado she helped to arrange and guide tours of students into the mountains by bus or backpacking and climbed many of the high peaks of Colorado. She was also an expert skier, who skied in Colorado, Oregon, and Europe. Her husband reported in a 1998 communication with one of the authors that she became a member of the Society of Friends, "finding its philosophy, values, and social concerns particularly appealing." Louise Rosenbaum died in Middletown, Connecticut, in 1980 shortly before her seventy-second birthday.

RUSK, Evelyn (Carroll). September 28, 1898–December 5, 1964.

Wells College (BA 1920, MA 1922), Cornell University (PhD 1932). PhD diss.: [Carroll, E. T.] Systems of involutorial birational transformations contained multiply in special linear line complexes (directed by Virgil Snyder).

Evelyn Teresa Carroll was the daughter of Teresa (Caraher) and William J. Carroll. In June 1900 the family was living in Rome, New York, where she was born. At that time her mother had had two children, although only Evelyn Carroll was living, and her father was owner of a knitting mill. In 1910 she had a six-year-old brother. Although a birth year of 1900 for Evelyn Carroll appears in later documents, the 1900 census and corroborating evidence point to the 1898 date above as correct.

Evelyn T. Carroll received her preparation for college at Miss Lamphier's Private School and Rome Free Academy and then studied at Wells College, a women's college in Aurora, New York. After receiving her bachelor's degree in 1920, Carroll became an instructor at Wells; she also continued her studies there and received a master's degree in 1922. She was promoted to assistant professor in 1924 and served as acting head of the department 1924–25. She studied at Columbia University during the summer of 1924 and while on leave in 1927–28. Also in 1928 she worked for the General Motors Export Company. She studied at Cornell University 1929–31 and became associate professor at Wells upon her return in 1931, a semester before receiving her doctorate from Cornell in February 1932. She wrote her dissertation in geometry and had minors in analysis and algebra. After publishing her dissertation and another paper, both in the *American Journal of Mathematics*, she collaborated with Snyder on three more publications that appeared in 1936, 1937, and 1939.

Evelyn Carroll married William Sener Rusk, an art historian at Wells, in Aurora on August 31, 1932. William S. Rusk was born in 1892 and received a PhD in 1933 from the Johns Hopkins University. He first joined the faculty at Wells College in 1921. He taught for three years, 1925–28, at Dartmouth College before returning to Wells as professor of fine arts; he remained at Wells for thirty years until his retirement in 1968.

Evelyn Carroll Rusk was promoted to professor in 1934. Several of her papers were published under the name Evelyn Carroll-Rusk, and she used that form of her name in some other professional contexts. She served as acting dean at Wells 1937–38 and became dean in 1938. While she was dean she was active in national and state associations for deans and was a member of the Cooperative Bureau for Teachers, where she was on the board 1938–51 and served as chairman of the board 1940–42. She remained as dean and professor until 1951, when she resigned the deanship and took a year's sabbatical to study mathematics curricula at colleges and universities in the United States. During this leave she held a fellowship from the Ford Foundation Fund for the Advancement of Education. After returning from her sabbatical year, Rusk returned to the mathematics department, and in 1954 she became chairman of the department.

Rusk was Roman Catholic and a Republican. In about 1940 she indicated that her favorite recreations were golf and bridge and that her hobby was dramatics; while at Wells she was very active in faculty theatre productions. She served as chairman of the mathematics department until her death at age sixty-six in Aurora in 1964. William Sener Rusk died in 1984. In 1970 the Evelyn Carroll Rusk '20 Theatre Series at Wells College was funded by a student from the class of 1934.

RUSSELL, Helen G. September 28, 1901–October 24, 1968.

Wellesley College (BA 1921), Columbia University (Teachers College) (MA 1924), Radcliffe College (PhD 1932). PhD diss.: On the convergence and overconvergence of sequences of polynomials of best simultaneous approximation to several functions analytic in distinct regions (directed by Joseph Leonard Walsh).

Helen Gertrude Russell was born in Gorham, Maine, the third of six children of Winifred Parker (Stone) and Walter Earle Russell. Her mother graduated from Maine Wesleyan Female College and taught at private schools before her marriage. Her father, a graduate of Wesleyan University, was a faculty member and principal of the Gorham Normal School, a predecessor of the University of Southern Maine. She had three brothers and two sisters, one of whom died in infancy. Her two older brothers both received PhD's: one in education from Yale and the other in history from American University. Her younger brother attended Wesleyan, and her sister attended Wellesley.

Helen Russell studied at Gorham schools before entering Wellesley College. After her college graduation in 1921, she taught mathematics and Latin in the high school in Mount Holly, New Jersey, 1921–23, and mathematics in the Horace Mann School for Girls in New York City 1923–27. She earned a master's degree from Teachers College, Columbia University, in 1924, having studied education and mathematics.

Russell returned to Massachusetts in 1927 to continue her graduate work at Radcliffe College. She was also an instructor at Wellesley 1928–29. She completed her work at Radcliffe in 1932. Her dissertation in analysis was directed by J. L. Walsh, with whom she collaborated for more than thirty years.

Russell again joined the mathematics department at Wellesley in 1932 and remained there for the rest of her career except for leaves. She was instructor 1932–36, assistant professor 1936–45, associate professor 1945–51, professor 1951–66, and emeritus professor after her retirement in 1966. She was also named chairman in 1952 and the Helen Day Gould professor of mathematics in 1955; she held that professorship until her retirement.

While at Wellesley, Russell continued her research in complex analysis and contributed research articles and reviews to journals. She took several leaves; the first, in 1942–43, was at the University of California. Her next three leaves were all at Harvard University: the first semester of 1949–50 and the academic years 1956–57 and 1963–64. During her last leave she was a research fellow in mathematics. After each of these leaves Russell published a paper jointly with J. L. Walsh. The last of these papers appeared in 1966, the year they both retired.

Russell was an active participant in college committee work and other extracurricular activities. Her professional activities included serving on the council of the Association of Teachers of Mathematics in New England and on a committee on summer institutes for mathematics teachers. She was also on a committee on a cooperative Harvard plan for the preparation of teachers.

Russell was a Radcliffe alumna member of Phi Beta Kappa and held offices in the Wellesley chapters of Phi Beta Kappa, Sigma Xi, and AAUP. She was a Republican and a Methodist. She was an enthusiastic traveler, whose trips included travel in Europe, in the western United States, and in the Orient in 1963–64 with her mother who was then ninety-five. Helen G. Russell died in 1968 at age sixty-seven in Portland, Maine. She was survived by her mother, her three brothers, and her sister.

SAGAL, Mary Helen (Sznyter). June 21, 1893–March 24, 1975.

University of California (BA 1915, MA 1916, PhD 1918). PhD diss.: [Sznyter, M. H.] The hypersurface of the second degree in four-dimensional space (directed by John Hector McDonald).

Mary Helen Sznyter was born in Woodhaven, New York, the only surviving child of three children born to Mary (Mogilska) and Valentine Sznyter. Her parents had married in their native Poland before immigrating to the United States, her father in 1890 and her mother in 1891. In 1900 the family was living in Withee, Wisconsin, where her father was a farmer. At some point the family moved to California.

Mary Helen Sznyter did all of her undergraduate and graduate work at the University of California in Berkeley, starting in 1911 as a student in social sciences. The following year, on November 2, 1912, she married John Boleslaw Sagal, a student in commerce at the university who was in residence there for the first half of 1912–13. John Sagal was born in about 1887 in Bialystok, Poland, apparently with the family name of Sawoinewsky according to his draft registration form for World War I. In 1917 he was a surveyor's helper and in 1930 he was a civil engineer in Alameda County, California.

Mary Helen Sznyter continued to use the name Sznyter as a student and on her mathematical publications, the last of which was a published version of her dissertation that appeared in 1924. By 1912–13 she had switched from social sciences to natural sciences, and for the year 1913–14 she held a Levi Strauss scholarship for the third congressional district. She received her bachelor's degree from the University of California in 1915. The following year she was a graduate student and a reader in mathematics. She completed the work for her master's degree in May 1916 with a master's thesis directed by J. H. McDonald. A paper based on her thesis appeared in the *American Mathematical Monthly* the following year. The Sagals' only child, a daughter, was born a month before her mother received her master's degree.

Sznyter continued her studies at California, with her major subject mathematics and her minor subject astronomy. She was again a reader in mathematics 1917–18, her final year at the University of California. According to the mathematics department at the University of California, Mary Helen Sznyter's dissertation advisor was J. H. McDonald who chaired the committee for her public final examination in May 1918. However, in her dissertation she expresses her gratitude to D. N. Lehmer "whose course on algebraic surfaces of three dimensional space served to arouse the interest in the study of such surfaces and has been a foundation for much that is included here" (*Univ. Calif. Publ. Math* 2 (1924): 2).

At the time of the census in 1930, John and Mary Sagal were living in Eden Township in Alameda County, California, with their thirteen-year-old daughter and Mary Sagal's widowed mother. No occupation was indicated for Mary Sagal. Their daughter received both a bachelor's degree and a master's degree from the University of California, Berkeley, married, and worked as a laboratory technician.

Mary Helen Sagal taught in the Chowchilla School District for about twenty-five years according to the report made by her daughter on the death certificate. She died at age eighty-one in 1975 in Castro Valley, California. Her husband had died in 1968.

SANDERSON, Mildred Leonora. May 12, 1889–October 15, 1914.
MOUNT HOLYOKE COLLEGE (BA 1910), UNIVERSITY OF CHICAGO (MS 1911, PHD 1913). PhD diss.: Formal modular invariants with application to binary modular covariants (directed by Leonard Eugene Dickson).

Mildred Leonora Sanderson was born in Waltham, Massachusetts, the daughter of Edna E. (Pratt) and Horace M. Sanderson. Her father was a florist and truck farmer. In the census of 1900, there were four children living of five born; she had two older sisters and a younger brother.

Sanderson attended North Grammar School and Waltham High School, public schools in Waltham. After graduating from high school as valedictorian in 1906, Sanderson was a student at Mount Holyoke College until her graduation in 1910. At Mount Holyoke she earned general honors at the end of her sophomore year and honors in mathematics when she graduated. The year following her college graduation Sanderson studied at the University of Chicago as the holder of the Bardwell Memorial Fellowship from Mount Holyoke. She wrote both her master's thesis and her doctoral dissertation under the direction of L. E. Dickson and received her master's degree in 1911 and her PhD in 1913 with minor work in astronomy.

In 1915 Dickson described the work she did under his direction noting that her master's thesis on generalizations in the theory of numbers and theory of linear groups "might well have served for her doctor's thesis; but she was quite willing to undertake a new investigation in a wholly different field" ("A Tribute to Mildred Lenora [*sic*] Sanderson," *American Mathematical Monthly* 22:264). About her doctoral dissertation he wrote, "Her main theorem has already been frequently quoted on account of its fundamental character. Her proof is a remarkable piece of mathematics." E. T. Bell wrote in 1938 that this theorem "has been rated by competent judges as one of the classics" in the field of modular invariants ("Fifty Years of Algebra in America, 1888–1938," 22).

Sanderson was an instructor at the University of Wisconsin during the first semester 1913–14 but left in February 1914 when she became ill with pulmonary tuberculosis. She died at age twenty-five in East Bridgewater, Massachusetts, the following October. A

Mildred L. Sanderson prize for excellence in mathematics was established in 1939 and is awarded annually by Mount Holyoke College.

SCHULTE, Sister M. Leontius. September 4, 1901–March 20, 2000.

College of Saint Teresa (BA 1923), University of Michigan (MA 1931, PhD 1935). PhD diss.: Additions in arithmetic, 1483–1700, to the sources of Cajori's "History of Mathematical Notations" and Tropfke's "Geschichte der Elementar-Mathematik" (directed by Louis Charles Karpinski).

Sister Mary Leontius Schulte was born Catherine Mary Schulte in Cleveland, Wisconsin, the youngest of seven surviving children of Mary (Wagner) and Joseph Schulte, both of whom had some grade school education and were farmers. Catherine Schulte attended primary school in Cleveland and graduated from high school in Manitowoc in 1919. That fall she entered the College of Saint Teresa in Winona, Minnesota, with the intention of studying home economics. While there she played hockey, tennis, and basketball, and became interested in mathematics. In 1923 she received her bachelor's degree, with a major in chemistry and minors in mathematics, French, and English. After her graduation Schulte taught mathematics at high schools in Minnesota; she was at Holy Trinity High School in Rollingstone 1923–24, Saint John High School in Rochester 1924–26, and Saint John High School in Caledonia 1927–28. She entered the Sisters of Saint Frances in Rochester on January 19, 1926, and made her profession of vows in 1927.

Sister Mary Leontius Schulte began her graduate work at the University of Michigan in the summer session of 1928 and became an instructor at the College of Saint Teresa that fall. She took graduate courses in mathematics and physics during the four summers of 1928–31 and received her master's degree in September 1931. Sister M. Leontius continued teaching except for the two academic years 1932–34, when she was in residence full time at the University of Michigan to finish the course work for her PhD. During these two years she lived at Mercy Hospital in Ann Arbor. She was enrolled during the summer of 1934, when her dissertation was accepted, and the degree was conferred in March 1935. **Sister M. Thomas à Kempis Kloyda**, her slightly older colleague at the college, had a similar educational and professional history, which included working with the same dissertation advisor.

Sister M. Leontius was instructor and associate professor at the College of Saint Teresa until 1948 when she was promoted to full professor. She retired as professor emeritus in 1975. During her years on the faculty she was particularly involved with issues concerning mathematics education and was a participant at NSF institutes and conferences at Ball State Teachers College 1959, Montclair State College 1961, Oklahoma State University 1961, Washington State University 1962, and Teachers College, Columbia University, 1969. In the summer of 1966 she attended a Cuisenaire workshop in Denver. Sister Leontius did a television series of nineteen telecasts, *Parents Ask About Arithmetic*, that were shown in Rochester in 1961–62, and fifteen kinescope recordings, *Teaching Mathematics K–6*, shown in St. Paul the next year. Among the mathematics and educational organizations in which Sister Leontius participated were the Minnesota Council of Teachers of Mathematics, where she was a district board member 1970–75, and the MAA, where she was on the executive committee of the Minnesota Section 1953–54 and 1955–56 and was chairman of the section 1954–55. As a faculty member she was a sponsor of the Legion of Mary, which ministered to the sick; she served as a pastoral minister and visited shut-ins.

Sister Leontius received an alumnae citation from the College of Saint Teresa in 1973. She remained there for several years after her retirement and did volunteer work in the college offices. After the college closed in 1987, Sister Leontius lived at the Motherhouse, Assisi Heights, Rochester, Minnesota. She died in 2000 at age ninety-eight in Rochester.

SEDGEWICK, Rose (Whelan). June 16, 1903–June 7, 2000.

Brown University (Women's College) (PhB 1925), Brown University (MA 1927, PhD 1929). PhD diss.: [Whelan, R. A.] Approximate solutions of certain general types of boundary value problems from the standpoint of integral equations (directed by Jacob David Tamarkin).

Rose Alice Whelan was born in Brockton, Massachusetts, the fifth of six children of Mary Theresa (Manchester) and Daniel Edward Whelan. Her mother was born in Liverpool, England, and immigrated to the United States in 1868. Daniel Whelan was born in Massachusetts, and his formal education stopped at two years of high school; he worked, mainly as a cutter, in a shoe factory. Rose Whelan had three sisters who all earned master's degrees, and two brothers, one of whom earned a college degree.

Rose Whelan received her primary and secondary education in the public schools of Brockton, Massachusetts. After her graduation from Brockton High School in 1921, Whelan entered Women's College in Brown University (later Pembroke College, which then merged with Brown). She received the PhB magna cum laude in 1925, having been elected to Phi Beta Kappa in her junior year. Whelan remained at Brown University during 1925–26 as a graduate student and part-time assistant in the mathematics department. She completed the work for her master's degree in September 1926 and received the degree in 1927, after having spent the academic year 1926–27 as a resident fellow at Bryn Mawr College. She was offered reappointment at Bryn Mawr for the year 1927–28 but returned to Brown with an Arnold fellowship.

Whelan also taught three summers at her former high school and, while studying for her doctorate and writing her dissertation, taught a section of freshmen at Women's College. During 1928–29 Whelan was an instructor at the University of Rochester in New York State. Her 1929 degree was the first PhD in mathematics granted by Brown. After receiving her doctorate, Whelan continued to teach at Rochester, as an instructor until 1932 and as an assistant professor 1932–34.

On December 25, 1932, Rose Whelan married Charles Hill Wallace Sedgewick, an instructor at the University of Connecticut at Storrs (then called Connecticut Agricultural College). C. H. W. Sedgewick was born in Nova Scotia in 1902. He completed his bachelor's degree at Dalhousie University in 1925, taught as a high school teacher, and in 1927 went to Brown, where he was an assistant, instructor, and then fellowship holder. He received his master's degree in mathematics in 1930 and left Brown in 1932 to take the position at Connecticut. After their marriage Rose and Charles Sedgewick returned to their teaching positions in Rochester and in Storrs. In October 1933 the first of their four children, a daughter, was born. In 1934 C. H. W. Sedgewick received his PhD in mathematics from Brown, he was promoted to assistant professor at Connecticut, and Rose Sedgewick gave up her position at Rochester. Their three other children were a daughter born in December 1935 and sons born in July 1940 and in December 1946. The elder daughter received an EdD and became a professor of nursing, the younger daughter majored in physics and became an electronic systems engineer, and both sons majored in mathematics; the youngest son also received a PhD and became a professor of computer science.

Rose Sedgewick was an instructor at Connecticut for some years beginning in 1943 and ending in 1956. She also taught at Hillyer College (now part of the University of Hartford) 1955–58. In 1958 C. H. W. Sedgewick left his professorship at Connecticut and moved to the US Census Bureau in Suitland, Maryland, where he remained until he retired. Rose W. Sedgewick was hired as an instructor at the University of Maryland in 1958 and was promoted to assistant professor in 1961; she retired at that rank in 1969. In 1971 they moved from Maryland to Dunedin, on the west coast of Florida.

The Sedgewicks played contract bridge and were both life masters in the American Contract Bridge League. Charles H. W. Sedgewick died in Dunedin in 1988 at the age of eighty-six. Rose W. Sedgewick moved to a continuing care facility in Dunedin and died there in June 2000, shortly before her ninety-seventh birthday. She was survived by her four children and ten grandchildren. In October 2001, the Rose Whelan Society was founded at Brown University as an organization for women graduate students and postdoctoral appointees in mathematics and applied mathematics.

SEELY, Caroline E. August 3, 1887–May 17, 1961.
BARNARD COLLEGE (BA 1911), COLUMBIA UNIVERSITY (MA 1912, PHD 1915). PhD diss.: Certain non-linear integral equations (directed by Edward Kasner).

Caroline Eustis Seely was born in Delhi, New York, the youngest of four children of Sarah Augusta (Wheeler) and Henry Bates Seely. Her father was a naval officer who had graduated from the US Naval Academy in 1857. Before her birth he served on ships in the Civil War, and after 1865 he served in various capacities as lieutenant commander and commander. From 1884 until 1891 her father was captain, and then commandant, of the League Island Navy Yard in Philadelphia. He retired in 1892 and died in 1901 when Caroline was thirteen.

In 1910 Caroline Seely and her widowed mother were living in New York City, where Caroline Seely was attending Barnard College, from which she graduated with honors in mathematics in 1911. She then did graduate work at Columbia University and was an assistant in mathematics to David Eugene Smith at Teachers College, Columbia, in 1911–13. She received her master's degree in 1912 with a thesis in analysis and continued her graduate work in mathematics at Columbia. Her doctoral dissertation was completed in 1914, and the degree was awarded by Columbia the following year.

In 1913 Seely began a twenty-two-year association with the American Mathematical Society when she became the first mathematician to be employed full time by the society. Her first position with the AMS was clerical and editorial assistant. In addition to Seely's duties involving work of the society as clerk to the secretary of the AMS, Frank Nelson Cole, she was engaged in editorial work for the *Bulletin* and the *Transactions* of the AMS. Seely continued her work with the society when R. G. D. Richardson of Brown replaced Cole as secretary in 1920. Seely was officially associate editor of the *Bulletin* 1925–34 and was cooperating editor of the *Transactions* 1924–36.

Seely lived in New York City, where she worked for the AMS and served as secretary to Cole at Columbia. During World War I, she worked with the Army Ordnance Department in Washington, D.C. She continued her mathematical research in analysis during and after the war. In 1918 she and David Eugene Smith compiled a *Union List of Mathematical Periodicals* for the Bureau of Education.

In November 1934, Seely purchased nearly seventeen acres of land in the village of Willseyville in Tioga County, New York, with the plan to retire the following spring. Although the leadership of the AMS tried very hard to convince her to stay, on April 1, 1935, a date her friends thought quite appropriate, Seely left her position at the AMS to establish a chicken farm. The next year, while she was getting her farm established, she also did editorial work part time for the *Transactions*. Seely continued to live in Willseyville most of the rest of her life. She died in 1961 at age seventy-three in nearby Ithaca, New York.

SHEA, Sister Ann Elizabeth. September 19, 1900–May 14, 1957.
UNIVERSITY OF KANSAS (BA 1927), UNIVERSITY OF WISCONSIN (MA 1931, PHD 1934). PhD diss.: Regular Cremona transformations in S_4 (directed by Theodore Lake Bennett).

Mary Gertrude Shea, known as Gertrude when she was young, was born and reared in St. Joseph, Missouri, the second of four children of Elizabeth (Raney) and Martin J. Shea. In 1900 her parents were living in St. Joseph with their two-year-old son as well as Elizabeth's mother and sister and Martin's sister. By 1910 all of the family's children were born. Her father was a sergeant in the city police department.

In the 1920 census Gertrude Shea's occupation was listed as stenographer for a railroad clearing house. She entered the Novitiate of the Sisters of Charity of Leavenworth in 1922, made first vows in 1924, and was thereafter known as Sister Ann Elizabeth Shea.

She graduated from the newly named Saint Mary College, a junior college for women in Leavenworth, Kansas, in 1923, the first year instruction at the postsecondary level was added. The school, which was chartered in 1860, was formerly St. Mary's Academy; in 2003 it became the University of Saint Mary. After completing her bachelor's degree in 1927 at the University of Kansas, Sister Ann Elizabeth taught mathematics until 1930 in Immaculata High School in Leavenworth. In 1930 she entered the University of Wisconsin to study for her master's degree, which she received in 1931.

From 1931 until her death in 1957, Sister Ann Elizabeth was associated with Saint Mary College. The college, from which she had graduated in 1923, had just added an upper division curriculum in 1930 and became a four-year institution in 1932. From 1931 to 1933 she was instructor of mathematics. At the same time, she continued her graduate studies at the University of Wisconsin, where she was in residence 1933–34 and completed her work, including a dissertation in algebraic geometry, in 1934.

After receiving her PhD, the first Sister of Charity of Leavenworth to earn a doctorate, Sister Ann Elizabeth was professor and chairman of the mathematics department and later registrar of the college. In her latter role, she was also a member of the Association of Collegiate Registrars and the Kansas Association of Collegiate Registrars, for which she served as secretary. In 1940 she described her positions as registrar and chairman of the division of natural sciences and mathematics. She later became a member of the board of control of the college. She also taught in the summer of 1937 at Mount Mary College and in the summer of 1942 at Nazareth College.

In 1954 Sister Ann Elizabeth Shea was appointed treasurer-general of the corporation of the Sisters of Charity of Leavenworth. Her 1957 death in Leavenworth at age fifty-six from a cerebral hemorrhage was sudden and unexpected.

SIMOND, Ruth G. March 7, 1904–September 15, 1958.

BOSTON UNIVERSITY (BA 1927, MA 1929), UNIVERSITY OF MICHIGAN (PHD 1938). PhD diss.: Relations between certain continuous transformations of sets (directed by William Leake Ayres).

Ruth Gertrude Simond was the youngest of four surviving children, of five born, of Grace M. (Fifield) and Walter A. Simond. She was born and grew up in Franklin, New Hampshire. According to census records, in 1900 her father was a knitter of hosiery (a common occupation in Franklin at that time); he was described in 1910 as foreman of a hosiery mill, in 1920 as a draftsman in a machine shop, and in 1930 as a machinist in a hosiery mill. Her mother taught vocal and instrumental music.

Simond entered Boston University in 1923 and graduated in 1927. She then entered the Boston University graduate school and received her master's degree in June 1929 with a thesis on using graphic calculation in physical chemistry. Simond classified mathematics books for the Harvard University library during 1929–30; was an instructor at Hampton Institute (now Hampton University), one of the historically black colleges and universities in Virginia, 1930–31; was a teacher at the Gorham (Maine) Normal School 1931–32; and returned to Hampton Institute as instructor 1932–33.

Ruth Simond began her graduate work at the University of Michigan in the summer of 1933 and was registered for courses every term until 1937. She took courses part time during the regular academic years 1934–35 through 1936–37, and her PhD was awarded in February 1938. Simond was in charge of the mathematics courses in the correspondence study department of the University of Michigan extension service 1936–38 and was among those who prepared or revised materials for supervised correspondence courses under the aegis of the Michigan Works Progress Administration (WPA).

After leaving Michigan in 1938, Simond returned to Hampton Institute, where she was associate professor the next four years. From August 1943 until March 1944, Simond taught in the Army Specialized Training Program at Heidelberg College in Tiffin, Ohio. She was a cryptanalyst for the Navy Department in Washington, D.C., during World War II, presumably during 1942–43 or 1944–45. Relatives indicate that she also worked as a "Grey Lady" or nurse's aide during the war. From September 1945 to June 1947 Simond was assistant professor at Berea College in Kentucky, where she had gone to teach mathematics to members of the Navy V-12 unit. The following year she taught at Morningside College in Sioux City, Iowa. In 1948 she was appointed assistant professor at the University of Vermont and State Agricultural College and remained in that position through the spring semester 1958.

Simond lived in Essex Junction, just outside Burlington, Vermont, and belonged to the Essex Junction Methodist Church. She died in Burlington of complications from diabetes at age fifty-four in September 1958. She was buried in Franklin, New Hampshire.

SINCLAIR, Mary E. September 27, 1878–June 3, 1955.

OBERLIN COLLEGE (BA 1900), UNIVERSITY OF CHICAGO (MA 1903, PHD 1908). PhD diss.: Concerning a compound discontinuous solution in the problem of the surface of revolution of minimum area (directed by Oskar Bolza).

Mary Emily Sinclair was the fourth of five children of Marietta S. (Fletcher) and John Elbridge Sinclair. Her parents met when, in 1869, they received the two new faculty appointments at the Worcester County Free Institute of Industrial Science (Worcester Polytechnic Institute after 1887) in Massachusetts, then a men's institution. Her mother, who had studied in France and Germany, taught English and modern languages 1869–72. Her father, recently widowed and the father of two young daughters, arrived from Dartmouth College, where he had been on the mathematics faculty. He remained at Worcester Polytechnic Institute (WPI) for nearly forty years. The John E. Sinclair chair of mathematics, the first endowed professorship at WPI, was established in 1915 with a gift from John Sinclair and his children.

Mary Emily Sinclair was born in Worcester, attended public schools there, and graduated in 1896 from Worcester Classical High School. She enrolled at Oberlin College in Ohio and graduated in 1900. Sinclair was an assistant teacher in Woodside Seminary, Hartford, Connecticut, 1900–01 and then studied at the University of Chicago. She received her master's degree in 1903 and taught in spring 1903 at Lake Erie College in Painesville, Ohio. She held a fellowship at Chicago 1903–04 and was an instructor at the University of Nebraska 1904–07. In 1907 Sinclair was hired as an instructor at Oberlin, where she remained for thirty-seven years. In 1908 she received her doctorate from the University of Chicago and was promoted to associate professor. Sinclair was at Columbia University and the Johns Hopkins University 1914–15, her first sabbatical year. During that period she adopted an infant daughter in 1914 and an infant son in 1915.

Sinclair resumed her research in the calculus of variations while she was a Julia C. G. Piatt fellow of the AAUW 1922–23, first at Cornell University and then at Chicago. She then presented four papers in late 1923. In 1925 she was promoted to professor and began a year's sabbatical in Rome and at the Sorbonne. That summer she visited in Freiburg with Oskar Bolza, her Chicago advisor who had returned to Germany in 1910. During 1927–28 she was on leave at the University of Miami, Coral Gables, Florida. It appears that her daughter was then at a school there; also in the 1930 census her son was listed as a cadet at a military academy in Coral Gables. In spring 1935 she was on leave at the Institute for Advanced Study in Princeton, New Jersey. She became head of the department at Oberlin in 1939 and was named Clark professor of mathematics in 1941. Except for a leave at Columbia second semester 1941–42, Sinclair remained in that position until she retired as professor emeritus in 1944. She taught mathematics to Navy V-12 students part time at Berea College, Kentucky, 1944–46 and then returned to Ohio.

Sinclair served on the auditing committee of the MAA 1917–19 and was appointed MAA librarian in 1918 and assistant librarian in 1922; she was a member of the MAA board of trustees 1936–38. Soon after she retired from Oberlin, a Mary Emily Sinclair prize was established there. In recognition of the support given her by the AAUW, Sinclair contributed generously to their fellowship fund. She also contributed to Oberlin to allow for the establishment of a loan fund for women to use for professional reasons. In May 1950 Sinclair was severely injured when she was assaulted on the outskirts of Oberlin, and in 1953 she moved to Belfast, Maine, where she made her home with her daughter-in-law. Mary Emily Sinclair died in Belfast at age seventy-six in 1955. Interment was in Worcester, Massachusetts.

SMITH, Clara E. May 20, 1865–May 12, 1943.

MOUNT HOLYOKE COLLEGE (BA 1902), YALE UNIVERSITY (PHD 1904). PhD diss.: Representation of an arbitrary function by means of Bessel's functions (directed by James Pelham Pierpont).

Clara Eliza Smith was born in Northford, Connecticut, the only child of Georgiana (Smith) and Edward Smith, a farmer and cabinet maker in Northford. She was educated with private tutors and entered Mount Holyoke Female Seminary (Mount Holyoke College since 1893) in 1882. She graduated in 1885, four years before the first bachelor's degree was awarded. Smith attended the Yale School of the Fine Arts 1886–87 and 1888–89 after which she was a teacher of drawing and assistant in mathematics at the Bloomsburg Literary Institute and State Normal School (now Bloomsburg State College) in Pennsylvania for eight years, 1889–97.

Smith entered the Yale University graduate school in 1901 to study mathematics. She was awarded a bachelor's degree from Mount Holyoke College in 1902 after passing an exam on one year's additional work in French and on the presentation of a certificate from Yale for courses in mathematics. She received her PhD in 1904 and remained at Yale as a reader in mathematics for two years, after which she was an acting instructor at Wellesley College 1906–07. The following year Smith was an instructor at the Western College for Women in Oxford, Ohio. In 1908 she returned to Wellesley as assistant and, except for leaves, spent the rest of her career there. She was an instructor 1909–14, associate professor 1914–24, professor 1924–34, (some years as the Helen Day Gould Professor), and emeritus professor after her retirement in 1934. She had three leaves of absence: in 1911–12, during which she traveled and studied one semester at the university in Göttingen; in 1918–19, when she was an exchange professor at Goucher College; and in 1926–27, when she made a trip around the world. Smith was department chair 1932–34.

While at Wellesley, Smith coauthored two textbooks with **Helen A. Merrill**; the first, printed in 1914, was to be used in a required course at Wellesley, and the second was published in 1917. Clara Smith served in a number of ways on the College Entrance Examination Board: as a reader starting in 1919, on a committee on revision, and on a board of examiners. She was permanent secretary of the AMS membership committee during the 1920s and was a trustee on the original board of the AMS at the time of its incorporation in 1927. Smith was a trustee of the MAA 1923–25 and was vice president in 1927. In a resolution voted on by the board of trustees at the annual meeting of the MAA, December 31, 1924–January 1, 1925, she was singled out for the efforts she made while she was in charge of a joint committee on membership of the AMS and the MAA. Her work for the AMS was also recognized in a resolution of the council in September 1925.

Smith and **Lennie Phoebe Copeland**, her long-time friend and colleague in the mathematics department, shared a home in Wellesley from the early 1920s. In 1935 they traveled together in Europe, the Middle East, and Africa. After having spent the summer at Smith's farm in Connecticut, they both attended the September 1937 AMS meeting in State College, Pennsylvania, where Smith was one of three guests of honor at a luncheon honoring women pioneers in mathematical research in America.

In May 1943, shortly before her seventy-eighth birthday, Clara Eliza Smith died in Wellesley. She is buried at the Northford Cemetery in Connecticut. She had been a member of the Congregational Church.

SPEER, Mary (Taylor). March 27, 1906–November 23, 1966.

University of Pittsburgh (BA 1926, MA 1928, PhD 1935). PhD diss.: [Taylor, M. M.] Reciprocals of certain curves and surfaces with respect to a space cubic curve (directed by Forest Almos Foraker).

Mary Margaret Taylor was the only daughter and third of four children of Hallie Blanche Virginia (Criss) and Albert Aaron Taylor, both of Washington County in the southwestern part of Pennsylvania. Taylor's mother attended common school and several terms at the state normal school and was a teacher in a one-room school and a housewife; her father attended common school and was a carpenter, mason, cabinet maker, and general contractor, having engaged in home study in addition to common school.

Mary Taylor was born in Midway, Pennsylvania, and attended the three-year high school there 1918–21. The following year she took the classical course in the Carnegie High School on the outskirts of Pittsburgh and graduated in 1922. She was awarded the four-year college scholarship, given annually on the basis of a competitive examination, from Washington County. She then attended the University of Pittsburgh where she carried majors in both mathematics and Latin and graduated with highest honor, the only recipient of a BA so honored in 1926.

Taylor held a graduate teaching assistantship at Pittsburgh from 1926 to 1930. She received her master's degree in 1928 and studied at the University of Chicago that summer. From 1930 to 1934 Taylor was instructor of mathematics at the University of Pittsburgh at Johnstown, and in the academic year 1934–35 she was back at the Pittsburgh campus as a graduate assistant to complete her PhD.

In 1935 Taylor married Eugene R. Speer (1903–1978), who also graduated from the University of Pittsburgh in 1926. He did graduate work there in mathematics and physics 1926–28 and graduated from the University of Pittsburgh law school in 1931. The Speers remained in Pittsburgh, where Mary Speer taught at her alma mater, and Eugene Speer practiced law. In 1936–37 Mary Speer is listed in the catalogue as a graduate assistant. From 1937–38 to 1942–43 she is listed as an instructor in mathematics.

The Speers had three children, daughters born in 1941 and 1947, and a son born in 1943. All three children did their undergraduate work in mathematics. The elder daughter did a year of graduate work in mathematics in England as a Fulbright scholar and became a computer software manager. Their son earned a PhD in mathematics, and their younger daughter received a PhD in computer science.

In 1986 Mary Speer's son reported that she continued her position at the university until sometime after the birth of their elder daughter and before his birth, and that his mother "recounted that she left the University of Pittsburgh, and mathematics teaching, because men junior to her were promoted to Associate Professor while she was not, and because she was told that this was because the men had families to support and needed the money" (Smithsonian questionnaire).

She was an involved Presbyterian throughout her life and was an active member and administrator in Girl Scouts of Allegheny County, Pennsylvania, from the early 1950s until her death. Mary Speer died of cancer at age sixty in Pittsburgh.

SPENCER, Vivian E. October 27, 1907–September 14, 1980.

Oberlin College (BA 1928, MA 1929), University of Pennsylvania (PhD 1936). PhD diss.: Persymmetric determinant and Jacobi matrix expressions for orthogonal Tchebycheff polynomials (directed by James Alexander Shohat).

Vivian Eberle Spencer was born in New Castle, Pennsylvania, the elder of two daughters of Ina M. (Eberle) and Andrew Berger Spencer. Her father was a traveling salesman throughout his life and was able to be with the family only infrequently; he sold furniture in 1910 and stock in 1920. Her sister, Domina Eberle, was born in September 1920 and earned a PhD in mathematics from MIT in 1942 under the direction of Dirk J. Struik; she has been on the faculty at the University of Connecticut in Storrs for most of her career.

Vivian Spencer was taught at home, mainly by her mother, before enrolling in Oberlin College in 1924. At that time the family moved to Oberlin, Ohio, where they remained for most of the next decade. Vivian Spencer majored in mathematics and English and received a bachelor's degree in 1928 and a master's degree in mathematics the following year. She was a graduate assistant in mathematics at the University of Pittsburgh during the academic years 1929–33 and studied at the University of Chicago during several of the summers. In fall 1933 Vivian Spencer entered the graduate school at the University of Pennsylvania; she was a Bennett fellow 1933–34 and a Moore fellow 1934–35. During fall 1935 she was a research assistant, and in June 1936 she received her PhD.

In January 1936 Spencer became a statistical research assistant for the National Research Project (NRP) of the Works Progress Administration (WPA) in Philadelphia; she also worked as an associate statistician for the US Bureau of Mines. She did research on fuel efficiency and labor production, and, in addition to authoring NRP reports, she contributed to NRP volumes on manufacturing industries and on petroleum and natural gas production. During 1936–37 she audited a graduate course in advanced economic statistics at the University of Pennsylvania.

In January 1940 Spencer moved to Washington, D.C., where she became a mineral economist and statistician for the US Department of Commerce. She worked for the US Bureau of the Census within the Department of Commerce, first as senior professional assistant for the mines and quarries division of the census bureau. She was statistician, senior statistician, senior economic analyst, and assistant to the chief of the minerals section before becoming chief of the minerals section in July 1948. She directed the statistical staff for the President's Materials Policy Commission 1950–52 and in 1964 became special assistant for raw materials and production indexes. In 1969, two years before she retired, Spencer moved from the Bureau of the Census to the US Department of the Interior's Bureau of Mines to become chief of the commodity staff. From 1957 to 1965 Spencer was a professorial lecturer in statistics at American University and from 1973 was an adjunct professor in the mathematics department of the University of Connecticut, where her sister taught. She was a consultant to the census bureau in 1972 and 1977–78.

In the mid-1930s Vivian Spencer reported to the University of Pennsylvania placement office that she played violin and piano and that she read English, Latin, German, French, and Italian. After her retirement from the government, Spencer maintained her home in Washington, D.C. She also had a home in Boston, where her sister lived. She was seventy-two years old when she died in 1980 in Boston. She was survived by her sister and was buried in New Castle, Pennsylvania.

SPERRY, Pauline. March 5, 1885–September 24, 1967.

SMITH COLLEGE (BA 1906, MA 1908), UNIVERSITY OF CHICAGO (MS 1914, PHD 1916). PhD diss.: Properties of a certain projectively defined two-parameter family of curves on a general surface (directed by Ernest Julius Wilczynski).

Pauline Sperry was born in Peabody, Massachusetts, the second of three children of Henrietta (Learoyd) and Willard Gardner Sperry. Her mother studied at Abbot Academy in Andover, Massachusetts, and served five years as a teacher and vice principal there. Her father graduated from Yale University, was master of a high school in Massachusetts, served for many years as pastor at Congregational churches in Massachusetts and New Hampshire, received a doctor of divinity degree, was president of Olivet College in Michigan 1893–1904, and then became a minister in York Beach, Maine. Her brother was in the first group of American Rhodes scholars to go to Oxford in 1904, earned master's degrees from Oxford and from Yale, became a clergyman and writer, and spent most of the latter years of his life as dean of the Harvard Divinity School. Her sister graduated from Smith College and became a writer.

Pauline Sperry studied at Olivet College from 1902 until her father left the presidency there in 1904. She then attended Smith College in Northampton, Massachusetts, and received her BA in 1906. She was a teacher of mathematics at Hamilton Institute in New York City 1906–07. The next year she was a fellow in music and mathematics at Smith, where she earned a master's degree in music in 1908 with a thesis on the modern English oratorio. Sperry remained at Smith as an assistant in mathematics 1908–11 and as instructor 1911–12. The following year she was a traveling fellow from Smith.

Sperry then studied mathematics at the University of Chicago; she received her master's degree in 1914, was a fellow 1915–16, and received her doctorate in 1916. She returned to Smith as assistant professor for the year 1916–17. During the 1917 summer session she taught at the University of California in Berkeley and joined the faculty there as an instructor that autumn.

Sperry remained at the University of California for the rest of her career. She was instructor 1917–23, assistant professor 1923–31, associate professor 1931–52, and associate professor emeritus from 1952. In 1931 Sperry published a bibliography of projective differential geometry and later supervised five doctoral dissertations. Sperry was well respected among the students at Berkeley and was known to mentor women there. She was also involved in the MAA and was vice chairman 1944–45 and then chairman 1945–46 of the Northern California Section.

Although Sperry is listed as an associate professor 1931–52, she did not serve as such 1950–52. In July 1950 she was one of thirty-one professors dismissed from the university because they would not sign a loyalty oath affirming that they were not members of the Communist Party. In October 1952 the California Supreme Court ordered that the nonsigners be reappointed, and Sperry was given the rank associate professor emeritus as of July 1952; further litigation brought her back pay for 1950–52 in 1956.

During her years at the University of California, Sperry had a close friend in Alice Tabor (1878–1959) of the German department, who had gone to the University of California a year before Sperry joined the faculty and who had also received a Chicago PhD in 1916. For many years, starting at least by 1920, Sperry lived with Tabor in Berkeley. By the early 1930s, they had built a home in Carmel, California, on the Monterey Peninsula, in which they spent many summers, and to which they moved permanently in 1950. After she retired to Carmel, Sperry was active in the Monterey Peninsula Friends' Meeting, the American Civil Liberties Union (ACLU), the American Friends Committee on Legislation, the Committee for a Sane Nuclear Policy (SANE), and the League of Women Voters. In the mid-1960s her chief interest was the Step-by-Step School for impoverished children in Haiti.

Pauline Sperry received the 1961 Olivet College distinguished alumni award. By early 1961 she had sold her home in Carmel and moved to a retirement community in Pacific Grove, California. She continued to travel and in the early 1960s made a three-month trip by freighter to England and New York. In 1967 Pauline Sperry died in Pacific Grove at the age of eighty-two.

STARK, Marion E. August 23, 1894–April 15, 1982.

Brown University (Women's College) (BA 1916), Brown University (MA 1917), University of Chicago (PhD 1926). PhD diss.: A self-adjoint boundary value problem associated with a problem of the calculus of variations (directed by Gilbert Ames Bliss).

Marion Elizabeth Stark, born in Norwich, Connecticut, was the daughter of Ella E. and Charles L. Stark. In 1900 and 1910 her father was a salesman, and in 1920 a buyer, for a dry goods store. Stark received her elementary education at the Pearl Street School and the Mt. Pleasant Street School, both in Norwich; her secondary education was at the Norwich Free Academy.

After graduating from Brown (Women's College) in 1916, Stark remained the next year as a fellow and received her master's degree in 1917. From 1917 until 1919 Stark was the professor of mathematics, and sole mathematics instructor, at Meredith College, a women's college in Raleigh, North Carolina.

Stark then became an instructor at Wellesley College where she was to remain, except for further graduate work, for the rest of her career. During her first year there, 1919–20, she took courses from **Helen A. Merrill** and **Mabel M. Young** while teaching part time. She studied at the University of Chicago during the summer quarters of 1923 and 1925 and in 1924–25, when she held a fellowship. She received her doctorate in 1926 with a dissertation in the calculus of variations. The year after receiving her PhD, Stark was promoted to assistant professor at Wellesley. She was promoted to associate professor in 1936 and to professor in 1945. She retired as Lewis Atterbury Stimson professor emeritus of mathematics in 1960.

In 1932 Stark attended the International Congress of Mathematicians in Zurich. She continued to study mathematics: at the University of Chicago during the fall quarter of 1938 and at Harvard 1939–40. Between 1940 and 1950 she published ten book reviews in the *National Mathematics Magazine* (renamed *Mathematics Magazine* in 1947). During this period, and continuing through 1960, Stark was on the editorial staff of this journal. Throughout her career she was involved with mathematics clubs. Stark participated in summer mathematics institutes for high school and college teachers from 1955 to 1958 at Middlebury College, Williams College, Dartmouth College, and Brandeis University. She continued writing reviews in the 1950s and early 1960s, first for *Scripta Mathematica* and then for the *American Mathematical Monthly.* She served on the council of the New England Association of Mathematics Teachers.

In at least the early 1930s, Marion Stark's widowed mother lived with her in Wellesley. Later Stark shared her home with Grace Ethel Arthur, the secretary to presidents of Wellesley for nearly four decades who retired in 1952. During World War II, Stark assisted several children from Europe and a child from Korea, at least some through the Foster Parents Plan and the Christian Children's Fund.

Stark was an enthusiastic traveler whose summer trips included, among many others, Europe in 1921 and Glacier National Park in 1956. After her retirement in 1960, Stark lived in Rockport, Massachusetts; in 1974 she moved to New Jersey. In addition to being an avid gardener, she wrote poetry, collected nautical art, and in her later years taped recordings of mathematical books for blind students. Marion Stark died at age eighty-seven at her home in Waterford, New Jersey, in 1982. Funeral services and burial were in Norwich, Connecticut.

STOKES, Ellen Clayton. September 27, 1900–May 23, 1974.

Brown University (Women's College) (BA 1923), Brown University (MA 1924), University of Chicago (PhD 1939). PhD diss.: Applications of the covariant derivative of Cartan in the calculus of variations (directed by Gilbert Ames Bliss).

Ellen Clayton Stokes, born in Hackensack, New Jersey, was the middle child of Louisa Cartwright (Stoney) and John Edgar Stokes. Louisa Stokes died in childbirth in 1906 when their younger son was born; John Edgar Stokes was a manufacturing representative with New England Yarn Company until after World War I. He later remarried, had another daughter, and owned an appliance store.

Ellen Stokes received most of her elementary and all of her secondary education at boarding schools in Beaver Dam, Wisconsin, where she first attended Hillcrest School and then Wayland Academy. The Stokes family was living in Providence, Rhode Island, when Ellen Stokes entered Women's College in Brown University in 1919. She earned her bachelor's degree in 1923 and her master's degree the following year.

In 1924 Stokes accepted a position as instructor in mathematics at Coker College for Women (now Coker College) in Hartsville, South Carolina. After two years at Coker, Stokes moved to the New York State College for Teachers in Albany (now the University at Albany, State University of New York), in part to be near family members who were living in the area. She spent the rest of her career there. In the summer of 1928, Stokes took three mathematics courses at Cornell University. Then, beginning in the summer of

1933, she studied for a total of nine quarters at the University of Chicago before receiving her doctorate there in December 1939 with a dissertation in the calculus of variations.

During the period of her studies at Chicago, Stokes remained an instructor in mathematics at the New York College for Teachers in Albany. In 1943 she became dean of women and stayed in that position until her retirement in 1965. A mathematics colleague at Albany was **Caroline A. Lester**, who joined the faculty in 1929 and lived in the same apartment building as Stokes in some of their earlier years in Albany. In about the early 1950s, Stokes and two other women on the faculty, one in psychology and one in biology, bought a house in Albany, painted it pink, and shared the house for many years.

Stokes excelled at handwork, especially weaving and knitting, played bridge, and made two or three trips abroad with friends. Ellen Stokes died in Albany in 1974 at age seventy-three. She was survived by her stepmother, who had been living with her, her half-sister, two nieces, and a nephew.

STOKES, Ruth W. October 12, 1890–August 27, 1968.

Winthrop Normal and Industrial College of South Carolina (BA 1911), Vanderbilt University (MA 1923), Duke University (PhD 1931). PhD diss.: A geometric theory of solution of linear inequalities (directed by Joseph Miller Thomas).

Ruth Wyckliffe Stokes was born in Greenville, South Carolina, the youngest of six children of Frances Emily (Fuller) and William Henry Stokes. In the 1880 census her father was listed as an MD and a farmer; in 1900 he was listed as a farmer.

Stokes graduated in 1911 from Winthrop Normal and Industrial College of South Carolina (now Winthrop University), a women's college in Rock Hill, South Carolina. She then had six teaching positions during the next dozen years. In 1911–12 she taught in grade school in Denmark, South Carolina; the next year she taught high school mathematics, English, and French in the same town; in 1913–16 she was principal of the Ebenezer Graded School in Rock Hill, South Carolina; the following year she was head of the mathematics department at Synodical College in Fulton, Missouri; in 1917–20 she taught high school mathematics and English in Spartanburg, South Carolina; and in 1920–22 she taught high school mathematics in Greenville, South Carolina. During this time, she also took courses at Columbia University, the University of Virginia, and the University of Chicago.

Stokes held a fellowship at Vanderbilt University 1922–23 and received her master's degree in 1923. For the next five years she was assistant professor at Winthrop College, the then current name of her alma mater. She also attended summer school at the University of Wisconsin 1926 through 1928. In 1928 Stokes enrolled at Duke University, where she was a graduate student and an assistant 1928–31. She received her PhD, the first awarded in mathematics by Duke, in 1931 and remained at Duke as an instructor 1931–32.

The next three years, 1932–35, Stokes was an associate professor at North Texas State Teachers College (now North Texas State University) in Denton. The following year she was professor and head of the department at Mitchell College in Statesville, North Carolina. She also taught in the summer of 1936 at the Asheville Normal and Teachers College in North Carolina. In 1936 Stokes returned to Winthrop College, where she remained for most of the next ten years as professor of astronomy and mathematics and as department head. While there she was active in a number of professional organizations and, in 1940, was a member of the Solar Eclipse Expedition to St. Augustine, Florida. In 1941 Stokes attended the summer session at Brown University for advanced instruction and research in mechanics. During World War II, she taught military cryptography and cryptanalysis.

In the 1940s Stokes had disputes with the administration at Winthrop, partly over issues of funding for the mathematics department. These resulted in her leaving in 1946, when she joined the faculty at Syracuse University as assistant professor. She held a dual position in mathematics and education in 1947–48 and was promoted to associate professor in 1953. Stokes was granted a year's leave of absence in 1956–57 to accept a visiting professorship at the American University in Beirut but returned home in December 1956 to be with a sister who was ill. Ruth Stokes retired from Syracuse as associate professor emeritus in June 1959 after which she was associate professor, 1959–60, at Longwood College (now Longwood University) in Farmville, Virginia.

Stokes was always professionally active. In the 1940s she was chairman of the Southeastern Section of the MAA, president of the mathematics section of the South Carolina Education Association, and a member of the board of directors of NCTM. Stokes was particularly interested in mathematical models; she gave a number of talks on this subject in the 1940s and exhibited a collection of models at the International Congress of Mathematicians at Cambridge, Massachusetts, in 1950. She was editor-in-chief of the *Pi Mu Epsilon Journal* when it first appeared in 1949. After her retirement, Stokes moved to her family home in Mountville, South Carolina. She died at age seventy-seven in the hospital in nearby Clinton in 1968.

SULLIVAN, Sister M. Helen. April 10, 1907–December 22, 1998.
St. Benedict's College (BA 1930), Catholic University of America (MA 1931, PhD 1934).
PhD diss.: The number and reality of the non-self-symmetric quadrilaterals in and circumscribed to the rational unicuspidal quartic with a line of symmetry (directed by Aubrey Edward Landry).

Monica Elizabeth Sullivan was born in Effingham, Kansas, the fourth of nine children of Mary E. (Majerus) and John Edward Sullivan. Her mother graduated from high school; her father attended Nebraska State Normal School (now Peru State College) and St. Benedict's College in Atchison, Kansas. He was a teacher before becoming a banker, land owner, and community leader in Kansas. All of the children had some college education.

Elizabeth Sullivan attended parochial schools in Kansas: in Effingham 1913–19, at St. Peter's parish grade school in Mercier 1919–20, and at Mount St. Scholastica Academy in Atchison 1921–25. She graduated as class valedictorian in 1925 and attended Mount St. Scholastica Junior College 1926–28 and St. Benedict's College 1928–30. Sullivan joined the Benedictine Sisters in 1925, professed into the Order of St. Benedict in 1930, and as Sister Mary Helen Sullivan continued her studies at the Catholic University of America, where she had been sent by her order to study physics. She received her master's degree in physics in 1931, taught at Mount St. Scholastica College that summer, and returned to Catholic where she earned her doctorate in mathematics with a minor in physics and mechanics in 1934.

Sister M. Helen Sullivan returned to Mount St. Scholastica College in 1934 as an instructor. She was on sick leave second semester 1934–35, all of 1935–36, and two more early periods. She was instructor 1934–37, assistant professor 1938–45, professor 1945–70, and professor emeritus after 1970, after which Mount St. Scholastica and St. Benedict's officially merged to become Benedictine College. Sister Helen was away from Mount St. Scholastica 1937–38, 1945–46, and 1954–57, and during some of that time was assigned to be principal at schools in Louisiana, Iowa, and Kansas. Sister Helen was chairman of the mathematics department 1934–54 and after 1957; she was chairman of the division of mathematics and natural sciences 1963–70. During her last few years at Mount St. Scholastica, Sullivan was the campus coordinator of the junior year abroad program for the Institute of European Studies.

Sister Helen taught or studied several summers at schools in the United States, Canada, and England. In 1948–49 she was a visiting professor at Loyola University (New Orleans), and during 1964–65 she was visiting professor at the University of Minnesota and, part time, at the College of St. Catherine in St. Paul, Minnesota. Among her many professional activities were several sponsored by NSF: an undergraduate research program at Mount St. Scholastica, visiting lectureships, institutes for college mathematics teachers, a writing team to produce college geometry materials, and review panels. Sister Helen contributed to the *Guidance Pamphlet* issued by the Post-War Commission of the MAA and was on the state committee appointed by the MAA to evaluate standards proposed by the Committee on the Undergraduate Program in Mathematics (CUPM). She was vice chairman of the Kansas Section of the MAA 1946–47 and chairman 1947–48. A scholarship established by the alumnae of the local chapter of the mathematics honor society Kappa Mu Epsilon was named the Sister Helen Sullivan scholarship in 1967. She was national historian of Kappa Mu Epsilon 1943–47 and assistant editor of its journal, *Pentagon*, 1943–47 and 1961–70; in 1991 she received the society's George R. Mach Distinguished Service Award.

After Sullivan left Mount St. Scholastica College in 1970, she was visiting professor under the International Cultural Exchange in Ireland at University College, Galway, and held various positions at the Institute of European Studies at the University of Vienna in Austria. She was director of development at Lillis High School in Kansas City 1972–73 and assistant director, Diocesan Office of Education in Billings, Montana, 1973–75. The following year she engaged in graduate study in theology at Gonzaga University in Seattle.

In 1976 Sister Helen moved to Berkeley, California, and was research assistant to the dean and faculty, Jesuit School of Theology, 1976–77; assistant to the president, Graduate Theological Union, 1977–78; assistant to the president of the School of Applied Theology 1978–79; and development research specialist for Catholic Charity Services in Oakland 1978–81. In the early 1980s, she was a member of spiritual direction and retreat teams at the School of Applied Theology in Berkeley. Sister Helen Sullivan returned to Atchison, Kansas, in 1985 and lived in a care facility at Mount St. Scholastica Monastery, where she died in 1998 at age ninety-one. Her publications are preserved in the Helen Sullivan Collection, Mathematics, National Museum of American History, Smithsonian Institution.

SULLIVAN, Mildred M. November 16, 1907–August 30, 1958.
RADCLIFFE COLLEGE (BA 1929, MA 1930, PHD 1932). PhD diss.: On the derivatives of Newtonian and logarithmic potentials near the acting masses (directed by Oliver Dimon Kellogg).

Mildred Marie Sullivan was born in Boston, Massachusetts, the youngest of two surviving children of three born to Alice T. (Sullivan) and John F. Sullivan, both born in Massachusetts of Irish parents. In 1900 her parents had been married six years, and there was one living child, a son, of two children who had been born. Her father was a letter carrier for the post office.

Sullivan did her undergraduate and graduate work at Radcliffe College and earned her doctorate at the age of twenty-four. Her minor thesis was "The theory of quadratic residues," and her 1932 PhD dissertation in potential theory was supervised by O. D. Kellogg. In Kellogg's November 1931 application for funds for a book project, he wrote that Sullivan "has unusual originality and power to carry through, with a minimum of guidance, on assigned programs. Her availability appears to present a rare opportunity for help on the above project" (Papers of William Fogg Osgood, HUG 4659.10, Harvard University Archives). Sullivan planned to work as Kellogg's research assistant the year after she received her doctorate in June. However, Kellogg died suddenly in July 1932. Sullivan was a research assistant in mathematics at Harvard under a Milton Fund grant 1932–33. During that year she presented their joint work at an AMS meeting and wrote a paper based on that work that was published in 1935 under her name and Kellogg's in the *Memoirs of the American Academy of Arts and Sciences.*

The year following her postdoctoral year at Harvard, Sullivan was a National Research Fellow working at Rice Institute in Houston and at the University of California in Berkeley. She was an instructor at the University of Houston 1934–35, during its first year as a four-year institution. This was during the Great Depression, and there is no indication that Sullivan was able to secure academic employment at the end of that year. In January 1937 Sullivan became an editorial assistant for the AMS in order to succeed **Alta Odoms (Gray)** who had resigned effective April 1, 1937.

In 1938 Sullivan became an instructor at Queens College in Flushing, New York. She was an instructor until 1946, when she was promoted to assistant professor. The year after her promotion Sullivan joined the MAA and from then on regularly attended meetings of the Metropolitan New York Section. Also starting when she became editorial assistant at the AMS in 1937, and continuing through the 1940s, Sullivan attended several meetings a year of the society in the New York City area. Mildred Sullivan died suddenly in August 1958 of heart failure at age fifty while visiting relatives in Boston.

SUTTON, Flora Dobler. June 7, 1890–June 23, 1976.

GOUCHER COLLEGE (BA 1912), JOHNS HOPKINS UNIVERSITY (PHD 1921). PhD diss.: On certain chains of theorems in reflexive geometry (directed by Frank Morley).

Flora Dobler Sutton, born in Baltimore, Maryland, was the third of seven children of Ann (or Annie) Elizabeth (Dobler) and John Robert Sutton. The family lived in Baltimore and her father was in the dry goods business.

Flora Dobler Sutton attended public schools in Baltimore and entered nearby Woman's College of Baltimore (Goucher College after 1910) immediately after her graduation from Western High School in 1908; she received her BA from Goucher in 1912. At that time the mathematics department was staffed by **Clara Latimer Bacon** and **Florence P. Lewis**, who earned PhD's from Johns Hopkins in 1911 and 1913, respectively.

In 1915, three years after graduating from Goucher, Sutton entered the Johns Hopkins University, where she took courses in mathematics, education, and statistics. At the time of her initial application for graduate study, she indicated that she did not expect to become a candidate for a degree. However, she did course work during the next six years. In addition to her mathematics and education courses, she took a course in statistical methods in the political economy department and then enrolled in January 1920 as a special student in the Johns Hopkins School of Hygiene and Public Health to take additional courses in statistics. Her PhD dissertation in algebraic geometry is dated 1920, although the degree was granted in June 1921.

During 1922–23 Sutton was an instructor of statistics replacing an associate who was on leave from the Department of Biometry and Vital Statistics in the School of Hygiene and Public Health at Johns Hopkins. In 1930 Sutton was a substitute teacher in high school and was living at her family home in Baltimore. She was still living there when she applied in February 1936 to take graduate courses in the Department of Political Economy at Johns Hopkins.

Sutton spent the major part of her career as a statistician for a stock brokerage firm in Baltimore that was known as Mackubin, Legg & Co. when she joined them, and is now Legg Mason, Inc. In 1959, several years after her retirement, Sutton took a trip with two of her sisters and a sister-in-law to visit a brother who was a missionary in Southeast Asia. In thirty-five days they traveled around the world. On the way to Tavoy, Burma, to see their brother, they visited Copenhagen, Rome, Istanbul, the Taj Mahal, and Rangoon; their trip home passed through Thailand, Singapore, Hong Kong, and Japan. Sutton had been retired for more than twenty years when she died in 1976 at age eighty-six in Baltimore.

TAPPAN, Helen. October 22, 1888–November 10, 1971.

WESTERN COLLEGE FOR WOMEN (BA 1909), CORNELL UNIVERSITY (MA 1912, PHD 1914). PhD diss.: Plane sextic curves invariant under birational transformations (directed by Virgil Snyder).

Anna Helen Tappan was born in Mount Pleasant, Iowa, the ninth of eleven children of Anna (Grand-Girard) and David Stanton Tappan. Her father earned a bachelor's degree and a master's degree from Miami University in Oxford, Ohio, and graduated from the Western Theological Seminary in Allegheny, Pennsylvania, in 1867. He was ordained as a Presbyterian minister and held various academic and pastoral positions before and after serving as president of Miami University 1899–1902. Tappan Hall there is named for him. Seven of Helen Tappan's siblings survived to adulthood: three older sisters, three older brothers, and a younger sister. Of these, one brother became a physician, another became a missionary in China, and her youngest sister earned a doctorate in education.

In 1905 Helen Tappan graduated from Everts High School in Circleville, Ohio, where her father was pastor of the Presbyterian Church. She then attended Western College for Women in Oxford, Ohio; she graduated with honors in 1909 and was an instructor there for the next two years. In 1911 Tappan enrolled at Cornell University where she was to remain for the next three years. She received her master's degree in 1912 with a major in pure mathematics and a minor in mathematical physics. Her master's thesis, "The relations between the theorems of Pappus and Desargues in the foundations of geometry," was directed by Virgil Snyder. She was a scholar 1912–13 and an Erastus Brooks fellow 1913–14 before receiving her PhD in 1914, again as a student of Snyder, with work in algebraic geometry and minors of mathematical analysis and philosophy.

After receiving her doctorate, Tappan joined the faculty at Iowa State College of Agriculture and Mechanic Arts (now Iowa State University) in Ames. She remained for eleven years: as instructor 1914–18, assistant professor 1918–22, and associate professor 1922–25. Others on the mathematics faculty at Iowa State during that period included **Julia T. Colpitts** and **Fay Farnum**.

In 1925 Tappan returned to Western College for Women as professor of mathematics and head of the department. She also served as dean of women from 1927 until 1941 and as academic dean from 1941 until 1944, when she resigned that position to return to full-time teaching. She was head of the department at the time of her retirement in 1954 as academic dean emeritus. In 1957–58 she served as acting head of the mathematics department.

Western College honored her in 1953 as one of the ten outstanding living alumnae and again in 1954 with an honorary Doctor of Humane Letters degree. In 1974 the college merged with Miami University, and the Western College Program was created within the university. In 1990 the Western program established the Anna Helen Tappan Center for Computer Assisted Learning, whose tutors are called Tappan Tutors. Tappan House, two floors of a dormitory at Iowa State University, is also named in her honor.

In about 1940 Tappan described herself as a Presbyterian and a Republican with interests in reading and dramatics. After her retirement Tappan moved to South Pasadena, California, where she had spent parts of summers with relatives. Helen Tappan died in 1971 at age eighty-three in Hamilton, Ohio, after a long illness. A memorial service was held at nearby Western College where the Helen Tappan Memorial Fund was established.

TAYLOR, Mildred E. July 25, 1898–November 3, 1978.

OXFORD COLLEGE FOR WOMEN (BA 1921), UNIVERSITY OF ILLINOIS (MA 1922, PHD 1931).
PhD diss.: A determination of the types of planar Cremona transformations with not more than 9 F-points (directed by Arthur Byron Coble).

Mildred Ellen Taylor, born in Virginia, Illinois, was the eldest of eight children of Emily Elizabeth (Treadway) and Angus Taylor, natives of Illinois. Her father was a farmer in the central part of the state.

Mildred Taylor graduated from the high school in Virginia, Illinois, in 1917 and entered Oxford College for Women in Ohio that fall. In 1918–19 and 1920–21 she was a special student at Miami University doing course work in physics. After her graduation from Oxford College in 1921, she received a scholarship from the University of Illinois for the year 1921–22. She received her master's degree in 1922 and taught the next year in the Johnston City Township high school in southern Illinois.

From 1923 until 1929 Taylor was an instructor at Knox College in Galesburg, Illinois. During this time she was working toward her PhD at the University of Illinois. She took a leave of absence from Knox 1929–30 to spend her only full year in residence at the University of Illinois. All of her other course work occurred during the summers 1924 to 1930. She received her doctorate in 1931 with a dissertation in algebraic geometry and with a minor in physics.

In 1930 Taylor was hired as professor of mathematics and department head at Mary Baldwin College, a college for women in Staunton, Virginia. In 1937 her title changed to professor of mathematics and astronomy. She remained at Mary Baldwin until her retirement in 1968. Throughout her career at Mary Baldwin, Taylor attended meetings of the AMS and the MAA. In 1938 she was elected to the executive committee of the Maryland-District of Columbia-Virginia Section of the MAA, and in 1953 she was elected vice chairman of the section. Taylor was also active in Pi Mu Epsilon, Sigma Delta Epsilon, and the AAUW. She attended a number of annual meetings of the NCTM in the 1950s and gave talks about teaching geometry at the college level in 1953 and 1954.

Mildred Taylor was living in Staunton, Virginia, at the time of her death at age eighty. She was buried in Urbana, Illinois.

THORNTON, Marian (Wilder). July 18, 1905–September 30, 1992.
UNIVERSITY OF MINNESOTA (BA 1927, MA 1930, PHD 1933). PhD diss.: [Wilder, M. A.] Some problems in closest approximation over a discrete set of points (directed by Dunham Jackson).

Marian Augusta Wilder, born in Shakopee, Minnesota, just southwest of Minneapolis, was the second of two surviving children of Minnie Florence (Buchanan) and George D. Wilder, a salesman. The family was living in St. Paul in 1910 and in Minneapolis in 1920 and 1930.

All of Marian Wilder's university work was done at the University of Minnesota, where she earned her bachelor's degree in 1927, her master's degree in 1930, and her doctorate in 1933. After she received her BA, Wilder was a high school teacher in Minnesota for the next two years, 1927–29. She returned to the university in 1929, where she was an assistant in mathematics 1929–31 and an assistant in biometry 1931–34. Wilder wrote three articles with the noted biometrician and botanist J. Arthur Harris shortly before his death in 1930. She received her master's degree in 1930 with a major in mathematics and a minor in education; her thesis dealt with the mathematical theory of statistics. She then completed her doctoral work with a major in mathematics and a minor in biometry. She remained at the university as a graduate fellow in the biometric laboratory 1933–34, the year after she received her PhD.

Wilder worked as a statistician at the Mayo Clinic 1934–35 and then returned to the University of Minnesota, where she was hired as a research statistician and instructor for 1935–36. However, in November 1935 she entered a sanitarium, most likely Glen Lake state tuberculosis sanitarium, and remained there about one year. She was a research statistician for the committee on educational research in the University of Minnesota's college of education and produced about a dozen technical reports during 1938 and 1939. From 1938 or 1939 until 1942, she was a statistician with the division of preventable diseases of the Minnesota State Department of Health. Her work in the late 1930s and early 1940s focused on the use of statistics in medical and psychological data.

On August 27, 1938, Marian Wilder married G. Parnell (Barney) Thornton (1907–1968). They remained in the Minneapolis–St. Paul area. In February 1954, when she applied for her social security number, she was a substitute at the College of St. Catherine in St. Paul. By 1966 she was on the graduate faculty in biometry in the School of Public Health of the College of Medical Sciences at the University of Minnesota and appears to have still been at the university in 1970–71. Marian Thornton was living in St. Louis Park, Minnesota, just west of Minneapolis, at the time of her death at eighty-five in September 1992.

THUENER, Sister M. Domitilla. October 25, 1880–September 29, 1977.
ST. XAVIER COLLEGE (BA 1920), CATHOLIC UNIVERSITY OF AMERICA (CATHOLIC SISTERS COLLEGE) (MA 1923), CATHOLIC UNIVERSITY OF AMERICA (PHD 1932). PhD diss.: On the number

and reality of the self-symmetric quadrilaterals in-and-circumscribed to the triangular-symmetric rational quartic (directed by Aubrey Edward Landry).

Eleanor Margaret Thuener was born in Allegheny, Pennsylvania, the daughter of Josephine and August Thuener. Her mother was born in Pennsylvania and her father in Germany. He immigrated to the United States in about 1870. In the 1900 census her father is described as a driver; in 1910 as a brewery driver. It appears that three children survived of seven who were born.

Eleanor Thuener received her elementary education in public and parochial schools in Allegheny. In the 1900 census she was described as a sewing girl. She graduated from St. Mary's Academy, a Catholic boarding school in Monroe, Michigan, in June 1905, when she was nearly twenty-five, shortly before entering the convent of the Benedictine Sisters in Covington, Kentucky, that summer. She took the religious name Mary Domitilla. From 1905 until 1920 Sister M. Domitilla taught in St. Walburg Academy in Covington and in Villa Madonna Academy, a boarding school for girls run by the Benedictine Sisters in nearby Crescent Springs, Kentucky, across the Ohio River from Cincinnati.

Sister Domitilla obtained her bachelor's degree in 1920 from St. Xavier College (Xavier University since 1930), a Jesuit college for men in Cincinnati, Ohio, by attending special extension classes scheduled for late afternoons, Saturdays, and summers, arranged for the sisters from neighboring congregations. In 1922–23 Sister Domitilla was at the Catholic Sisters College of the Catholic University of America in Washington, D.C., and received her master's degree in 1923.

Sr. Domitilla was the first dean when Villa Madonna College was founded in 1921 in Covington. She was the senior officer who directed the academic work of the new college and was the mathematics instructor from 1921 until 1929, when it graduated its first class of five students. In 1929 Sister Domitilla returned to Catholic University, where she completed the work for her PhD in 1932 with a dissertation in algebraic geometry and with minors in physics and education. Upon her return to Villa Madonna College she was professor and department head until 1943; she also taught physics 1934–37. Two years later the college became coeducational. In the late 1960s it relocated to Crestview Hills, Kentucky, and was renamed Thomas More College.

In 1943 Sister M. Domitilla was elected by her community to be the Prioress of St. Walburg Convent. She served two successive terms as administrator and religious leader. In this period she was a member of the board of trustees of Villa Madonna College. Under her direction the community entered hospital work by taking over the administration of a hospital in Kentucky and by staffing a hospital in Colorado. Sister Domitilla Thuener died at Villa Madonna in Covington, Kentucky, at age ninety-six in 1977.

TORRANCE, Esther (McCormick). August 12, 1909–January 3, 1978.
Barnard College (BA 1931), Cornell University (MA 1932), Brown University (PhD 1939). PhD diss.: Superposition on monotonic functions (directed by Jacob David Tamarkin).

Esther Ober McCormick was the elder of two surviving children of Ethel Mary (Ober) and Thomas Holmes McCormick. Her parents had both attended Hiram College in Ohio; her mother was there for one year, 1900–01, and her father graduated in 1902, shortly before their marriage. Their first daughter died in infancy. Esther McCormick was born in Fort Wayne, Indiana, where her father was a mathematics teacher at Fort Wayne High School. Her younger brother was born in New York City after the family moved there in the summer of 1913. Her father was a teacher of mathematics at the Commercial High School there and occasionally served as a substitute preacher. He continued to teach at the high school after the family moved to East Orange, New Jersey. Esther McCormick graduated from Eastern Grammar School in East Orange in 1923 and from East Orange High School in 1927.

Esther McCormick attended Barnard College, the coordinate college of Columbia University, where she majored in mathematics. She took several graduate courses at Columbia in addition to her undergraduate work. She joined the AMS in the summer of 1930 and attended two meetings of the society before receiving her bachelor's degree in February 1931. She immediately entered graduate school at Cornell University for the second term of the academic year 1930–31 and returned to Columbia for graduate work during the summer of 1931. She had a graduate scholarship at Cornell for 1931–32 and received her master's degree in February 1932. She stayed at Cornell through the end of that academic year before going to Brown University for the year 1932–33 with a scholarship and part-time work.

While a student at Cornell, McCormick met Charles Chapman Torrance, who had received a PhD in mathematics from Cornell in 1931 and was an instructor there from 1927 until 1931. C. C. Torrance (1902–1967) was born in Yonkers, New York, and had earned an ME in 1922 and an MA in 1927 from Cornell. He spent 1931–32 as an instructor at Stanford University and 1933–34 at the Institute for Advanced Study in Princeton. During 1933–34 Esther McCormick traveled in Europe before marrying Charles Torrance on June 19, 1934.

C. C. Torrance was a member of the faculty at Case School of Applied Science in Cleveland, Ohio, from 1934 until 1946 and was a contract mathematician at the Bureau of Ordnance of the Navy Department in Washington, D.C., 1944–45. Esther Torrance was an instructor at Oberlin College 1937–38. Both of the Torrances were in residence at the Institute for Advanced Study in Princeton the following year, and Esther Torrance finished the work for her PhD at Brown in 1939. Daughters were born in 1941 and 1942; both later earned doctorates, one in mathematics and the other in physics.

In 1946 C. C. Torrance moved to the Naval Postgraduate School in Annapolis, Maryland. When the school moved to Monterey, California, in December 1951, the Torrances moved to the West Coast, and C. C. Torrance continued as professor at the school. Esther Torrance taught in high school in Salinas, California, for a brief time in the late 1950s or early 1960s. Her husband died very soon after his retirement in 1967, and afterwards Esther Torrance moved to Fresno. She was an assistant professor at Fresno State College (now California State University, Fresno) from 1967 until she retired in 1972 after an unsuccessful bid for tenure. She moved back to Monterey after her retirement.

When the family lived in Maryland, Esther Torrance was active in a number of community organizations; these included the Anne Arundel Youth Commission, the Annapolis Planning Commission, and the Citizens Planning and Housing Association of Annapolis and Anne Arundel County. She was also a registered lobbyist in Congress for a housing coalition. After they moved to California, she would camp with her daughters for several weeks every summer in Yosemite National Park. In August 1977, a few months before her death, she climbed Mt. Whitney, the tallest mountain in the contiguous United States. After a series of strokes and heart attacks in December 1977, Esther Torrance died in early January 1978 at age sixty-eight in Monterey, California. Before her death she had been working for passage of the Equal Rights Amendment.

TORREY, Marian M. December 9, 1893–September 16, 1971.

Brown University (Women's College) (BA 1916), Brown University (MA 1917), Cornell University (PhD 1924). PhD diss.: Classification of monoidal involutions having a fixed tangent cone (directed by Virgil Snyder).

Marian Marsh Torrey was born in Malden, Massachusetts, the daughter of Anna Louise (Marsh) and Daniel Temple Torrey. Daniel T. Torrey, at one time a Congregational minister in Massachusetts, was by 1900 an insurance agent in Providence, Rhode Island, where Marian Torrey grew up. Four children were living in 1910 of seven who had been born by that time.

From 1909 through 1912 Marian Torrey attended the Northfield Seminary, a preparatory school for girls in Massachusetts, and then entered Women's College, Brown University, in Providence. She graduated as a mathematics major in 1916, remained at Brown, and received her master's degree the following year.

Torrey taught mathematics at the St. Johnsbury Academy in Vermont and then taught for two years at the Phebe Anna Thorne Model School at Bryn Mawr. While teaching at the Model School she also took two courses at Bryn Mawr College. She was an instructor at West Virginia University 1920–23. While on the faculty there, she attended two summer quarters at the University of Chicago. The second of these was in 1923 when she started a course with E. J. Wilczynski. She had planned to complete her studies at Chicago under his direction the following year, but that summer Wilczynski's health forced him to give up teaching. At the suggestion of R. G. D. Richardson at Brown, Torrey contacted Virgil Snyder at Cornell University to inquire about the possibility of finishing her work there. She enrolled at Cornell in the fall of 1923 and received her PhD the following June.

On February 13, 1924, while Torrey was finishing her doctoral work at Cornell, Richardson wrote a letter on her behalf to **Ruth Goulding Wood** at Smith College. In it, he said, "The department [at Brown] feels that we have never had, among the twenty or more girls whom we have sent out to teach in colleges, any stronger candidate. She comes from a cultured family, has a rare gift of getting along with people, and is very competent mathematically .... If Brown University would employ women, I would not hesitate to ask President Faunce to call her here at a good salary. She would do much better than many of the men whom we have at present on our staff" (R. G. D. Richardson Papers, Correspondence 1921–1925, Folder W, Brown University Archives).

After Torrey received her PhD in 1924, she was an instructor at the University of Illinois for a year before going to Goucher College as an instructor. She was promoted to assistant professor in 1927, to associate professor in 1932, and to professor in 1942. She spent a sabbatical year, 1931–32, studying at Columbia University. While at Goucher she served as chairman of the department 1943–57; assistant to the dean for academic affairs 1937–43; and head of a resident dormitory, Baldwin House, from 1942 until she retired in 1959. Torrey began attending meetings of the MAA before she received her doctorate and continued attending for at least thirty years. She was chair of the Maryland-District of Columbia-Virginia Section 1952–53.

Torrey's favorite activities included renovating old houses, bridge, bowling, and walking. She died in Towson, Maryland, at the age of seventy-seven in 1971. Goucher's annual banquet of the Mathematics Club and its prize for outstanding scholarship in mathematics are named in her honor.

TULLER, Annita. December 30, 1910–August 29, 1994.

Hunter College (BA 1929), Bryn Mawr College (MA 1930, PhD 1937). PhD diss.: The measure of transitive geodesics on certain three-dimensional manifolds (directed by Gustav Arnold Hedlund).

Annita Tuller was the daughter of Ida (Bick) and Morris Tuller, natives of Russia who were likely Jewish. She was born in 1910 in Brooklyn, New York, and a brother was born in 1916. Her father was a jeweler and her brother became a physician.

Annita Tuller received her primary and secondary education in the public schools of Brooklyn. She graduated from Erasmus Hall High School in 1925 at age fourteen and entered Hunter College that same year. She was a French major when she entered but changed her major to mathematics her first year; she graduated in 1929.

In the fall of 1929 Tuller entered Bryn Mawr College as a graduate scholar. After earning her master's degree at the end of the academic year 1929–30, she returned to Hunter, at age nineteen, as a substitute instructor in mathematics for a year. From 1931 to 1935 she was a mathematics and physics teacher at William Cullen Bryant High School in Long Island City, Queens, New York. Having been encouraged to pursue a doctorate by **Anna Pell Wheeler** during her first year at Bryn Mawr, she returned there for further graduate study; she was a resident fellow 1935–36 and a graduate scholar 1936–37. She completed her work for the PhD in 1937 with a dissertation in differential geometry directed by G. A. Hedlund, whom she found to be very encouraging.

Tuller became a tutor at Hunter College in the fall of 1937. On November 23, 1938, she married Morris Levine (1912–1983) of Brooklyn, who worked for, and later became circulation manager for, a newspaper publishing firm in New York. The Levines had two daughters, born in 1942 and in 1944. The younger daughter earned a PhD from Columbia University in language and modern literature.

Annita Tuller, who continued to use her maiden name professionally, remained on the faculty at Hunter College: as tutor 1937–39, instructor 1939–49, assistant professor 1949–61, and associate professor 1961–68. She was teaching at the Bronx campus of Hunter when it became an independent college, Herbert H. Lehman College (CUNY), in 1968. She then taught at Lehman as professor and retired as professor emeritus on February 1, 1971.

During her career at Hunter and at Lehman, Tuller taught a variety of undergraduate courses as well as graduate courses in geometry. She taught in an NSF summer institute in 1959 and an academic year institute in 1965–66. From 1958 to 1970, as a member of the MAA Speakers Bureau, she spoke in high schools in the New York metropolitan area. She was active in the Metropolitan New York Section of the History of Science Society, for which she was treasurer from 1959 for about a decade. Tuller wrote a widely used and admired text, *A Modern Introduction to Geometries*, which appeared in 1967.

When her children were in elementary school, Tuller was active in the PTA. Later she did volunteer work recording mathematics texts for the blind. She was a member of UNA-USA (United Nations Association of the United States of America), the US Committee for UNICEF, Common Cause, and AARP. Tuller was Jewish and was a registered Democrat.

Annita Tuller moved to a residence for senior citizens in Cupertino, California, after suffering a serious heart attack in early 1993. She died in 1994 at age eighty-three.

TURNER, Bird M. April 18, 1877–September 5, 1962.

WEST VIRGINIA UNIVERSITY (BA 1915, MA 1917), BRYN MAWR COLLEGE (PHD 1920). PhD diss.: Plane cubics with a given quadrangle of inflexions (directed by Charlotte Angas Scott).

Bird Margaret Turner was the second child and first of four daughters of Mary Jane (Douglas) and John Marion Turner. Her father was a farmer in the northwest panhandle of West Virginia from before her birth in Moundsville until his retirement.

Bird Turner graduated from the Moundsville high school in 1893 and taught in the Moundsville area for most of the next two decades. She taught at Wood Hill School, a county grade school, 1895–96; she taught fourth, seventh, and eighth grades in Moundsville 1896–1900; and she taught mathematics in the Moundsville high school 1900–13. Starting in 1900 she took summer courses toward her undergraduate degree: at West Virginia University in 1900, at Harvard University in 1907, at Bethany College in West Virginia in 1909 (spring and summer), and again at West Virginia University in the summers 1910 and 1912–14. She was a student assistant at West Virginia University 1913–15 and received her BA there in 1915, at age thirty-eight, with majors in mathematics and physics. She was a member of the English club and won the Chi Omega prize in economics. Turner was an instructor in the summer school at West Virginia University in 1915, returned to Moundsville as principal of the high school for the year 1915–16, and was instructor in the summer school at West Virginia again in the summer of 1916.

Turner began her graduate work as a scholar in mathematics at Bryn Mawr College in the academic year 1916–17. She was granted the (honorary) President M. Carey Thomas European fellowship for her first year's work. She received her master's degree from West Virginia University in 1917 and was assistant director of the Phebe Anna Thorne Model

School of Bryn Mawr College 1917–18; she lists the Model School as her affiliation for the January 1920 MAA meeting in New York City. Turner was a reader in mathematics at Bryn Mawr College 1918–19, and, as a resident fellow 1919–20, she finished her dissertation under the direction of Charlotte A. Scott. Turner was Charlotte Scott's sixth, and next-to-last, PhD student and was forty-three when she received her doctorate. Between the ages of forty-five and fifty-two she published three articles in the *American Journal of Mathematics* and two in the *Annals of Mathematics*.

After receiving her PhD, Turner was an instructor at the University of Illinois 1920–23. She then returned to West Virginia University and was assistant professor 1923–25, associate professor 1925–31, and professor 1931–47. In 1947, at age seventy, she retired as professor emeritus.

Four years after her retirement, Turner returned to Moundsville, where she was a member of the Moundsville Presbyterian Church and the Moundsville Women's Club. Bird Turner died in Moundsville, West Virginia, at age eighty-five in 1962.

TURNER, Mary (Haberzetle). June 4, 1912–November 16, 1983.

SAINT XAVIER COLLEGE FOR WOMEN (BS 1934), UNIVERSITY OF CHICAGO (MS 1936, PHD 1938).

PhD diss.: [Haberzetle, M.] Two new universal Waring theorems (directed by Leonard Eugene Dickson).

Mary Barbara Haberzetle was born in Chicago, Illinois, the first of two daughters of Helen (Martin) and William Thomas Haberzetle. Her mother, orphaned at an early age, worked to support herself when young and determined that her children would receive an excellent education. Her father was in business with his brother. The younger daughter received a master's degree and taught in the Chicago public schools for twenty-six years.

Mary Haberzetle received her primary and secondary education in parochial schools in Chicago; she attended St. Anthony Grammar School and Academy of our Lady (also called Longwood Academy), where she was the top scholar in her graduating class of 1930.

Haberzetle entered Saint Xavier College for Women (now Saint Xavier University) in Chicago, where she held a four-year scholarship, majored in mathematics, and graduated with honors in 1934. During her studies at Saint Xavier, she was especially encouraged to do graduate work and urged to apply for a fellowship at the University of Chicago by **Emily Pixley**, a mathematics professor at Saint Xavier and a 1933 Chicago PhD.

After graduating from Saint Xavier, Haberzetle worked as a social worker for the Cook County Welfare Department 1934–35. She began her graduate work at the University of Chicago in the autumn quarter of 1935 and received her master's degree the following year. She was a fellow her last two years of graduate study and received her doctorate in 1938.

Haberzetle taught at Mount Holyoke College in the first semester of 1938, replacing **Frances Baker** who was on leave, and at Queens College in New York City 1939–41. Between 1937 and 1941 she published four papers. Generalizations by others of the results in her 1941 paper appeared over the next twenty years.

In January 1940 Mary Haberzetle married M. Jonathan Turner (1915–1995), who had received a master's degree in mathematics from the University of Chicago in 1937. He was a mathematics instructor 1937–40 and received an MS in aeronautical engineering from New York University in 1941. He worked as an engineer for the United Aircraft Corporation in Stratford, Connecticut, from 1941 until 1949 when they moved to Seattle where he spent the rest of his career at the Boeing Company.

Three children were born while they were in Connecticut, sons in 1942 and 1944, and a daughter in 1947. All received PhD's from the Yale graduate school in the 1970s, their sons in physics and their daughter in physical chemistry. Their second son also received an MD from the University of Miami.

In 1960, when the children were in their teens and approaching college age, Mary Turner resumed her teaching by taking a position at Seattle University as an assistant professor; she was promoted to associate professor in 1963 and became a full professor in spring 1971 shortly before the onset of illness hastened her retirement in June that year.

Her sister wrote to one of the authors that Turner "had talent as an artist and enjoyed all the arts – painting, music, and weaving. She also enjoyed outdoor activities, hiking, mountain climbing, and snow-shoeing which she enjoyed during her retirement years." Mary Haberzetle Turner died in Seattle, Washington, in 1983 at age seventy-one.

VAN BENSCHOTEN, Anna L. August 12, 1866–September 18, 1927.
CORNELL UNIVERSITY (BS 1894, PHD 1908), UNIVERSITY OF CHICAGO (MS 1900). PhD diss.: The birational transformations of algebraic curves of genus four (directed by Virgil Snyder).

Anna Lavinia Van Benschoten was born in Elmira, New York, the daughter of Mary Jane (Pugsley) and Moses M. Van Benschoten. Her father served in the infantry during the Civil War; he died in 1872, when she was five. In 1870 Anna and her mother, a music teacher, were living with Anna's maternal grandfather in Binghamton, New York.

Anna Van Benschoten received her primary and secondary education in the public schools of Binghamton and graduated from Binghamton Central High School in 1886, the year after her mother died. It is unclear what Van Benschoten did after her high school graduation. She was listed as a boarder in an 1888 Binghamton, New York, directory. According to a 1928 letter written by Amy J. Douglass, who lived with Van Benschoten from the end of 1921 until her death in 1927, she traveled in Europe from June to December 1890.

Van Benschoten enrolled as an undergraduate at Cornell University in 1891. Most of the mathematics courses she took were in the general area of geometry, although she also took courses in algebra and a year-long seminar on mathematics pedagogy. After graduating from Cornell in 1894, she taught at Binghamton Central High School 1894–98. She studied during the summers of 1897 and 1898 and the following two academic years at the University of Chicago before receiving her master's degree in mathematics and astronomy in 1900. Van Benschoten spent 1900–01 traveling in Europe and taking courses in mathematics and astronomy in Göttingen. In particular, she was in Felix Klein's projective geometry lecture in winter 1900–01.

In 1901 Van Benschoten began her college teaching career as professor at Wells College, a women's college in Aurora, New York, assuming the position of professor that was vacated when **Annie MacKinnon (Fitch)** married. Van Benschoten was at Wells until 1920, except for leaves of absence. During most of her first decade there she was the only one giving instruction in mathematics. During most of her second decade she was joined by one instructor. Van Benschoten first took a leave in the academic year 1906–07 to resume her graduate studies at Cornell and received her doctorate in 1908. Beginning in 1915 Van Benschoten was often on leave and retired because of ill health after her final leave 1918–20. She spent part of the year 1918 at the University of Arizona teaching mathematics to "boys of the Student Army Corps" (Amy J. Douglass to Foster M. Coffin, February 27, 1928, Box 133, Deceased Alumni Files, #41-2-877, Division of Rare and Manuscript Collections, Cornell University Library) and moved to Whittier, California, in about 1921. According to Douglass, "She gave up continuous teaching because of impaired health due to Rheumatic Arthritis and during her six years of residence in Whittier she did private tutoring" and taught at Whittier College from January to April 1924. Her obituary indicates that she was at Whittier temporarily to replace a regular faculty member.

Douglass reported that in addition to her earlier travels in Europe, Van Benschoten made a third trip abroad to Norway, Sweden, Russia, and Germany, and that she also visited Alaska and Jamaica. Van Benschoten died in Whittier, California, in 1927 at age sixty-one; her remains were sent to Binghamton, New York, for interment.

VARNHORN, Mary C. July 8, 1914–November 9, 1988.

College of Notre Dame of Maryland (BA 1936), Catholic University of America (MA 1937, PhD 1939). PhD diss.: Some properties of quartic functions of one variable (directed by Edward Jerome Finan).

Mary Catherine Varnhorn was born in Baltimore, Maryland, the daughter of Eleanor (Levy) and John Henry Varnhorn. At the time of the 1920 census, she, her parents, and her brother were living in Baltimore with her mother's family. At that time her father was a commercial traveler. Later they owned their own home in Baltimore, and her father was described as a solicitor for a food company.

Mary Varnhorn did her undergraduate work at the College of Notre Dame of Maryland in Baltimore. After her graduation in 1936 she began her graduate studies at the Catholic University of America in Washington, D.C. She received her master's degree in 1937 and her PhD in 1939.

Varnhorn returned to the College of Notre Dame of Maryland as instructor for the year 1939–40. She then joined the faculty at Trinity College in Washington, D.C., in 1940. When Varnhorn arrived at Trinity College (now called Trinity Washington University), she joined two women religious. One left the following year and the other died suddenly in the fall of 1943, leaving the twenty-nine-year-old Mary Varnhorn alone in the department. At first, some courses were taught by Otto Ramler, professor at Catholic University, and then by various others who filled in. Varnhorn remained at Trinity for the rest of her career. She was instructor 1940–44, assistant professor 1944–45, associate professor 1945–58, and professor 1958–78. From 1945 until her retirement, she was chairman of the department, which usually consisted of at most three people. She was recognized especially for her commitment to teaching when she received, in 1965, the *Pro Ecclesia et Pontifice* Medal for twenty-five years of service at Trinity College.

Varnhorn maintained her residence in Baltimore throughout her life and commuted into Washington. After her formal retirement in 1978 she taught two courses at Trinity College as an adjunct in the fall of 1978–79, and she taught briefly at Loyola College in Baltimore and at the University of Baltimore. She was a member of the Duodecimal Society of America (since 1980 Dozenal Society of America), an organization for the conduct of research and education in the use of the base twelve number system. Varnhorn was at one time president of the Mother Seton Mission Crusade Unit, a group devoted to helping the needy. Mary Varnhorn died at age seventy-four in Baltimore in 1988. A memorial liturgy was celebrated at Trinity College.

VAUDREUIL, Sister Mary Felice. August 20, 1894–October 27, 1978.

DePaul University (BA 1921), Catholic University of America (PhD 1931). PhD diss.: Two correspondences determined by the tangents to a rational cuspidal quartic with a line of symmetry (directed by Aubrey Edward Landry).

Annette Vaudreuil was born in Chippewa Falls, Wisconsin, the eldest of five children of Lenora (Blair) and Ludger J. Vaudreuil. Her mother was born in Wisconsin; her father was born near Québec City in Canada and moved to Wisconsin as a boy. Later he was a lumber merchant in Chippewa Falls. Annette Vaudreuil's mother died in 1905, when Annette was ten and the youngest of the five children was two years old. Her father remarried in 1907. Five of six children born of this second marriage lived to maturity.

Annette Vaudreuil attended the Notre Dame grade school and high school in Chippewa Falls. She entered the Congregation of the School Sisters of Notre Dame after her graduation from McDonell Memorial High School in 1911, was received in 1913, and took first vows in 1915 and final vows in 1921.

As a novice Sister Mary Felice Vaudreuil was sent to teach in the high school in Escanaba, Michigan. After profession in 1915 she returned to Escanaba but, after one week of teaching, was recalled and sent to teach at the Academy of Our Lady in the Longwood area of Chicago. In October 1915 she enrolled at DePaul University in Chicago. By taking courses that met in the late afternoons, Saturdays, or in the summers, she was able to teach and complete the work for her bachelor's degree, which she received in August 1921. She then took graduate courses in the autumn-winter session at DePaul in French, Latin, and philosophy in 1921–22, and in philosophy, education, and mathematics in 1922–23. She also did graduate work at Creighton University in Nebraska, Marquette University in Wisconsin, and Loyola University of Chicago, presumably in the summers.

Sister Felice taught at the Academy of Our Lady until 1925, when she was transferred to St. Mary's College in Prairie du Chien, Wisconsin. In 1928 she was one of the first two faculty members at St. Mary's to be given a sabbatical leave for full-time study. She entered the Catholic University of America in Washington, D.C., and was in residence there from 1928 until she completed her work for the PhD in 1931 with a dissertation in algebraic geometry and minors in education and physics.

After receiving her doctorate, Sister Felice returned to Mount Mary College in Milwaukee, the successor to St. Mary's in Prairie du Chien. She taught mathematics, meteorology, and astronomy and was chairman of the mathematics department. She was lecturer for the graduate school at Loyola University in Chicago for three or four years in the mid-1930s and taught some graduate courses at Marquette University in Milwaukee. She was chairman of the Wisconsin Section of the MAA 1939–40 and 1945–46 and was secretary-treasurer 1953–60; she served on the national MAA Committee on High School Contests 1957–62. At the state level, she was a member of the Wisconsin Mathematics Council for which she was vice president 1951–52 and president 1952.

Among Sister Felice's many interests were astronomy and meteorology. In 1966 she was awarded a pin by the US government for twenty years of volunteer service to the US Weather Bureau as a weather observer. She was a member of the Milwaukee Astronomical Society and the National Geographic Society. She was described by a colleague at Mount Mary College as being very interested in music and the arts; she listened to the weekly broadcasts from the Metropolitan Opera while following along with the libretto. She was a gourmet cook, bird watcher, linguist, philosopher, seamstress, and bridge player.

Sister Felice retired from Mount Mary College to the motherhouse in Mequon, Wisconsin, in 1967. Hearing of the need for a mathematics teacher in a nearby college, she taught a course there for a few years. After her retirement she also studied to be a ham radio operator and was able to contact missionary sisters. In 1972 Sister Mary Felice Vaudreuil entered the Notre Dame Health Care Center in Elm Grove, Wisconsin, and died there six years later at age eighty-four.

VIVIAN, Roxana H. December 9, 1871–May 31, 1961.

Wellesley College (BA 1894), University of Pennsylvania (PhD 1901). PhD diss.: The poles of a right line with respect to a curve of order n.

Roxana Hayward Vivian was the daughter of Roxana (Nott), born in New Hampshire, and Robert Hayward Vivian, born in Nova Scotia, Canada, a tailor. She was born in Hyde Park (now part of Boston), Massachusetts, the eldest of three daughters.

Roxana Vivian attended public schools and graduated from Hyde Park High School in 1890. She then enrolled at Wellesley College, where she majored in Greek and mathematics before graduating in 1894. She held a secondary school teaching certificate and taught in the Stoughton, Massachusetts, public high school 1894–95. Vivian then taught Greek and mathematics 1896–98 at the Walnut Hill School, a private boarding school in Massachusetts, while also pursuing graduate work in those subjects at Wellesley.

Vivian was an alumnae fellow for women at the University of Pennsylvania 1898–1901 and received her PhD in 1901 with a minor in astronomy. She returned to Wellesley as instructor and was the first member of the mathematics department to possess a doctorate.

Roxana Vivian was granted a leave of absence from Wellesley for 1906–07 to teach in Turkey at the American College for Girls (also known as Constantinople Woman's College). She continued her leave for 1907–09 to remain as acting president of the school while the president was away raising funds for the college. In 1908, while still in Turkey, Vivian was promoted to associate professor at Wellesley. She returned to Wellesley in 1909, having weathered a precarious time in Turkey because of the Young Turks Revolution in 1908.

In 1913–15 Vivian had a partial leave of absence from Wellesley and reduced her teaching load to a one-hour course while she worked as financial secretary of the Women's Educational and Industrial Union (WEIU), a non-profit social and educational agency concerned with working women in Boston. In 1913–14 she also taught an extension course in statistics for Boston University, and in 1914 she coauthored a statistical article with the eminent botanist and biometrician J. Arthur Harris. Vivian continued her interest in statistics in social science and published articles and a book in that field.

In 1918 Vivian was promoted to professor and was appointed director of Wellesley's department of hygiene and physical education. She again had a leave from Wellesley's mathematics department and served as director until 1921, when she returned to the mathematics department. During 1925–26 she visited Cornell University and attended classes in several mathematical subjects. She left Wellesley in 1927.

Vivian spent the spring and summer of 1928 organizing and equipping the Eagle Wing School, a private college preparatory school for girls in Orleans, Massachusetts; her title there was principal. The school was unable to open because of financial circumstances. She spent the following year teaching mathematics in Oak Grove Seminary, a private school in Vassalboro, Maine. She then became professor at Hartwick College in Oneonta, New York, where she also served in the newly created position of dean of women 1929–30; she resigned her professorship and left the college at the end of the first semester 1930–31. Vivian was next instructor and dean of girls at Rye Public High School, also in New York, 1931–35 and held no professional position after that.

Vivian's reports to Wellesley over the years indicate that sometime after she left the college, she spent three or four winters in Chicago with her Wellesley freshman roommate and several winters in Stamford, Connecticut. In 1952, when she was eighty, she reported to Wellesley that she had spent two years in Greece and was then helping in an office of statistics at a Red Cross blood bank; she was also guiding visitors on Friday afternoons in Trinity Church, Boston. Vivian belonged to the Boston College Club and was an Episcopalian and a Republican. Roxana H. Vivian died at age eighty-nine in 1961 at the Sophia Snow House of the Roxbury Home for Aged Women in Boston.

WEEKS, Dorothy W. May 3, 1893–June 4, 1990.

Wellesley College (BA 1916), Massachusetts Institute of Technology (MS 1923, PhD 1930), Simmons College (MS 1925). PhD diss.: A study of the interference of polarized light by the method of coherency matrices (directed by Norbert Wiener).

Dorothy Walcott Weeks was born in Philadelphia, Pennsylvania, the second of three children of Mary (Walcott) and Edward Mitchell Weeks. Her father was an engraver who had studied at the Corcoran Art School in Washington, D.C., and at George Washington University, where he received an LLB. Her elder brother received his bachelor's and master's degrees from the Massachusetts Institute of Technology (MIT), and her younger sister graduated from Wellesley College. In 1900 the family moved from Cheltenham, Pennsylvania, to Washington, D.C., where Dorothy Weeks attended public schools and graduated from Western High School.

Weeks enrolled at Wellesley and graduated in 1916 with a mathematics major, having taken all the courses offered in physics, mathematics, and chemistry. She then moved home, taught at Fairmount Seminary, and took the tests required to become an examiner in the US Patent Office. In spring 1917, she was a statistical clerk with the Office of Farm Management in the Department of Agriculture. Later that year, after passing the exams, she became the third woman to be appointed a patent examiner. In order to qualify for a promotion, which she received, Weeks studied substantive patent law at George

Washington University in the evenings for a year. She also studied physics at Cornell University during the summers of 1918 and 1919. In 1920 she left the patent office to work at the National Bureau of Standards where she could attend classes given by physics faculty of the Johns Hopkins University. She returned to the patent office in June 1920 but joined the instructional staff of the physics laboratory at MIT that September as a laboratory assistant in the electrical laboratory. She was instructor 1922–24, received her MS in experimental physics in June 1923, and taught college preparatory courses in physics part time at the Buckingham School in Cambridge 1923–24. She left MIT in 1924 after a change in the administration made her position less attractive. While working at Filene's department store, she attended the Prince School of Retailing at Simmons College 1924–25 and received a master's degree. She became a supervisor for women, and then hiring supervisor, at the Jordan Marsh department store in Boston and remained there until September 1927. She then held a temporary position doing statistical work for a researcher at the Harvard Medical School.

In February 1928 Weeks returned to MIT for graduate work and studied theoretical physics in the mathematics department. In 1928–29 she was also a three-quarter-time instructor of physics at Wellesley and in 1929–30 held a Horton-Hallowell fellowship given by the Wellesley alumnae association. She was in Washington, D.C., the summers of 1929 and 1930; the first summer she taught mathematics at Western High School, and the second summer she worked as an expert technical advisor to the US Civil Service Commission. She received her PhD from MIT in 1930.

Weeks then went to Wilson College, a women's college in Chambersburg, Pennsylvania, to develop the physics department and serve as the professor and head. She spent most summers elsewhere: in 1931 she taught mathematics and chemistry at Central High School in Washington; in 1932 and part of 1934 she did research in Cambridge, England; and in 1935 she returned to MIT and began a research program in atomic spectroscopy. Through 1950, except for the war years, she returned to work in MIT's spectroscopy laboratory during her vacations and while on sabbatical.

During World War II, Weeks taught electricity and mechanics for an engineering science management war course at Wilson College and then was a technical aide in the liaison office of the US Office of Scientific Research and Development (OSRD) supervising the British reports section 1943–45. After the war, she returned to Wilson as professor of physics and head of the department and continued part time at the OSRD through the end of 1946. She was a Guggenheim fellow at MIT and at laboratories in England and northern Europe 1949–50 and was a consultant to the National Science Foundation 1953–56.

Weeks retired from Wilson in 1956 and moved to Wellesley to live with Louise McDowell, her former Wellesley physics teacher, mentor, and long-time friend. In Massachusetts Weeks became a physicist with the Ordnance Materials Research Office of the Army Materials Research Agency at the Watertown Arsenal. She soon became their technical representative to the Committee on Radioactive Shielding, and until 1962 she coordinated a project to develop shielding material for use against nuclear weapons. She left the arsenal in 1964 and was a part-time spectroscopist for the NASA-supported Solar Satellite Project at the Harvard College Observatory until her final retirement in 1976 at age eighty-three. She also was a physics lecturer at the Newton College of the Sacred Heart 1966–71.

Dorothy Weeks and Louise McDowell had traveled together over the years, and they purchased a summer cabin in the White Mountains of New Hampshire in 1946. McDowell required full-time care from late 1960 until her death in 1966. In 1961 Weeks bought McDowell's house in Wellesley and bequeathed it to the college. Weeks was active in the AAUW and received its achievement award in 1969. She was treasurer of the Washington Branch 1918–20, president of the Pennsylvania-Delaware Division 1938–40, and president of the Boston Branch 1963–67; she also was a member of awards and grants committees. She was active in the International Federation of University Women, and in the mid-1950s she was secretary, vice president, and then president of the Central Pennsylvania Section of the American Association of Physics Teachers. Weeks received honorary degrees from Regis College, MIT, and the Medical College of Philadelphia. The Wellesley alumnae association gave her its alumnae achievement award in 1983. Dorothy W. Weeks had been living in Wellesley before she died in Newton, Massachusetts, at age ninety-seven in 1990. Her papers are at MIT, and a memoir written after she retired is at the Center for History of Physics, American Institute of Physics.

WEISS, Marie J. September 21, 1903–August 19, 1952.

STANFORD UNIVERSITY (BA 1925, PHD 1928), RADCLIFFE COLLEGE (MA 1926). PhD diss.: Primitive groups which contain substitutions of prime order p and of degree $6p$ or $7p$ (directed by William Albert Manning).

Marie Johanna Weiss was born in Eugene, California, the youngest of three surviving children of Alice Hedwig (Buschke) and Frederick Weiss. Both parents were born in Germany and immigrated to the United States when they were about fifteen years old. At the turn of the century, they were living in Stanislaus County, California, where Frederick Weiss raised stock; by 1910 the family, now consisting of an eleven-year-old daughter, a nine-year-old son, and Marie, owned their own farm and were living in Stockton, California, where they remained throughout Marie's university years.

Marie J. Weiss attended public grammar and high schools in Stockton before going to Stanford University in the fall of 1921. In the summer of 1924, after her junior year, she served as an assistant in instruction in mathematics. She was elected to Phi Beta Kappa and graduated with great distinction in mathematics in 1925, one of eleven receiving BA's in mathematics at Stanford that year, eight of whom were women.

The following academic year Weiss studied at Radcliffe College and earned her master's degree. She then returned to Stanford as a graduate student and university fellow in mathematics during the two years 1926–28. She was again an assistant in instruction and taught theory of functions in the summer of 1927. Her dissertation in group theory is dated August 1927, and she received her PhD in June 1928. Weiss was appointed to a National Research Council fellowship in mathematics for 1928–29 and was reappointed for the following year; she spent both years studying at the University of Chicago.

In 1930 Marie J. Weiss joined the faculty of the mathematics department at H. Sophie Newcomb College, Tulane University, as assistant professor; the same year **Nola Anderson (Haynes)** joined the department as associate professor and chair. Weiss remained at Newcomb as assistant professor until 1936, except for 1934–35, when she was on leave to spend the year at Bryn Mawr as Emmy Noether scholar and resident scholar in mathematics.

Weiss returned to Newcomb College for the year 1935–36 but left in 1936 to take an assistant professorship at Vassar College. After two years at Vassar she returned to Newcomb as professor and chair when, in 1938, Nola Anderson left the Newcomb department to marry. Weiss remained in those positions during her remaining years at Newcomb. She turned her attention to preparing a highly successful text and to participating in many professional activities, in addition to administering the department and teaching. She was particularly active as a member of the MAA; she was a member of the board of trustees 1937–38, was associate editor of the *Monthly* 1940–46, and was governor-at-large 1950–52. She presented six papers at AMS meetings 1927–36 and was appointed to the AMS nominating committee in 1939. Weiss was also active at the national level in AAUP, serving on the council 1940–42 and as second vice president 1948–50.

Beginning in about February 1952 Marie Weiss became increasingly ill. While visiting her widowed mother and her brother in California in the summer of 1952, the illness worsened, and Marie Weiss died in San Francisco at age forty-eight of acute congestive heart failure. The Marie J. Weiss Memorial Scholarship Fund was established at Newcomb College in 1952.

WELLS, Mary Evelyn. August 20, 1881–October 7, 1965.
MOUNT HOLYOKE COLLEGE (BA 1904), UNIVERSITY OF CHICAGO (MS 1907, PHD 1915). PhD diss.: On inequalities of certain types in general linear integral equation theory (directed by Eliakim Hastings Moore).

Mary Evelyn Wells was born in LeRaysville, Pennsylvania, the second of three children of Delphine (Whitford) and William Henry Wells. Her father was listed as a farmer in Pennsylvania in the 1880 census. In 1900 the family was living in Naugatuck, Connecticut, where her father was a carpenter. In 1910 her mother was widowed, living in Naugatuck, and working at home as a dressmaker.

Mary Wells received most of her elementary education at home and in the public schools of Brushville and LeRaysville, Pennsylvania. She attended a little over a year of grammar school and four years of high school in Naugatuck and graduated from the academic course in the Naugatuck high school in 1900. That fall she entered Mount Holyoke College, from which she graduated four years later as a mathematics major. She stayed at Mount Holyoke as a department tutor in mathematics 1904–05.

In the fall of 1905, Wells entered the University of Chicago, where she studied for the next two years. She was a fellow the second year and received her master's degree in 1907. She became an instructor at Mount Holyoke College that fall and taught there for five years before returning to the University of Chicago as a graduate student in 1912. She stayed at Chicago two years with a fellowship from Mount Holyoke 1912–13 and a fellowship from Chicago 1913–14 during which she taught one quarter. Wells had also served as a reader in mathematics for the College Entrance Examination Board in New York City from 1908 to 1913.

During 1914–15 Mary Evelyn Wells was an acting associate professor at Oberlin College as a temporary replacement for **Mary Sinclair**, who was away for her first sabbatical. In 1915 Wells received her PhD and joined the faculty at Vassar College, where she remained except for leaves until her retirement in 1948. She was instructor 1915–20, assistant professor 1920–22, associate professor 1922–28, professor 1928–48, and emeritus professor after 1948. From 1936 to 1948 she was chairman of the department, and her teaching included a course in cryptography for the Navy during World War II.

Wells had a leave of absence from Vassar from 1926 to 1928. In 1926–27 she was an exchange professor and head of the department at Women's Christian College, University of Madras, India, and in 1927–28 she studied at the Istituto Fisico of the University of Rome, where Vito Volterra was professor. She had corresponded with him in 1925, before she left for India, and in 1928 she published a translation into English of Volterra's 1926 paper that describes his famous theory on population growth.

During 1936–37, Wells had a leave of absence from Vassar and returned to Women's Christian College, University of Madras. She served as a trustee of the University of Madras from 1930 to 1937 and again after 1948. She was also a director of St. Christopher's College in Madras 1928–37, of Women's Christian College in Madras 1927–37 and after 1948, and of Women's Christian College in Tokyo 1930–37.

In about 1940 Wells described herself as an Episcopalian whose favorite recreations were skating and climbing. For many years she maintained a home in Southport, Maine, where she spent some summers. In 1957 Wells was living in Naugatuck, where her widowed sister, a former teacher, made her home. In November 1958 she moved from Naugatuck to Wethersfield, Connecticut. It appears that she was living in Vermont by about 1959.

Mary Evelyn Wells died at age eighty-four in 1965 in Rutland, Vermont. The Mary E. Wells and Gertrude Smith Fund, honoring Wells and another long-time mathematics faculty member, was established at Vassar for students demonstrating excellence in mathematics.

WHEELER, Anna (Johnson) Pell. May 5, 1883–March 26, 1966.
University of South Dakota (BA 1903), University of Iowa (MA 1904), Radcliffe College (MA 1905), University of Chicago (PhD 1910). PhD diss.: [Pell, A. J.] I. Biorthogonal systems of functions, II. Applications of biorthogonal systems of functions to the theory of integral equations (directed by Eliakim Hastings Moore).

Anna Johnson was born in Calliope (now Hawarden), Iowa, the youngest of three surviving children of four born to Amelia (Friberg or Frieberg) and Andrew Gustav Johnson, both from Sweden. Her parents had each immigrated to the United States in 1872 and settled in Union Creek in Dakota Territory after their marriage. Her father first farmed but later was a furniture dealer and undertaker. In 1882 the family moved to Iowa, first to Calliope and then, in about 1891, to nearby Akron. Anna Johnson attended public school in Akron and graduated from high school there in 1899.

Anna Johnson entered the University of South Dakota in 1899 and graduated in 1903. While there Anna Johnson and her sister both roomed in the home of Emma and Alexander Pell, Anna's mathematics instructor. Pell recognized her mathematical talent, and with his encouragement and a scholarship Johnson went to the University of Iowa the following year and received her master's degree in 1904. She then went to Radcliffe College with a scholarship and received her second master's degree in 1905. Still on scholarship she stayed at Radcliffe another year.

In 1906 Anna Johnson went to the university in Göttingen on an Alice Freeman Palmer fellowship from Wellesley College. On July 9, 1907, she married Alexander Pell in Göttingen. Pell's first wife had died in 1904, and he and Johnson had remained in contact over the years. Pell was born Sergei Petrovich Degaev in 1857 in Moscow, and his life in Russia is depicted in *The Degaev Affair* by Richard Pipes. Toward the end of 1880 he joined the People's Will, a group responsible for the 1881 assassination of Tsar Alexander II. In December 1883 Degaev assassinated the head of the secret police and fled with his wife first to England and then to North America. In 1891 Degaev and his wife, using the names Alexander and Emma Pell, became US citizens, and Alexander Pell began his study of mathematics at Washington University in St. Louis. He was awarded his PhD from the Johns Hopkins University in 1897 and was appointed professor at South Dakota.

Alexander and Anna Pell returned to the United States in August 1907. She taught at South Dakota fall semester 1907 and returned alone to Göttingen in spring 1908. In August 1908 Alexander Pell resigned as dean of the College of Engineering at South Dakota and became assistant professor at the Armour Institute of Technology (now Illinois Institute of Technology) in Chicago. In December 1908, when she was almost ready to take her final examination with David Hilbert in Göttingen, Anna J. Pell returned to the United States without her degree. In January 1909 she enrolled at the University of Chicago.

Anna J. Pell received her PhD magna cum laude in 1910 from the University of Chicago. While she was officially a student of E. H. Moore, she wrote at the time that her dissertation, in the new field of functional analysis, had been written independently while in Göttingen. She remained in Chicago and taught a class at the university during fall 1910. In January 1911 Alexander Pell had a stroke, and Anna Pell substituted for him the remainder of the semester. His only teaching after the stroke was for a semester at Northwestern University in 1915–16. In fall 1911 Anna Pell became an instructor at Mount Holyoke College in Massachusetts. In 1914 she was promoted to associate professor. Four years later she moved to Bryn Mawr College in Pennsylvania as associate professor. Early in 1921 Alexander Pell died. That same year the third edition of *American Men of Science* appeared with Anna J. Pell starred. The following year the first of Anna Pell's eight PhD students earned her degree. Pell was appointed to the original board of trustees of the AMS in 1923 and was elected to serve on the council of the AMS for 1924–26. Also, in 1923 Pell became the first woman to deliver an invited address at a meeting of the AMS; four years later she became the first woman to deliver the AMS colloquium lectures.

Pell became head of the department at Bryn Mawr after Charlotte Scott's retirement in 1924 and was named Alumnae Professor of Mathematics the following year. In July 1925 she married Arthur Leslie Wheeler (1861–1932), a Connecticut-born, Yale-educated classicist who had been on the Bryn Mawr faculty since 1900 and who was widowed in 1915. Before their marriage, Arthur Wheeler had accepted a professorship at Princeton University. In 1927 Anna Pell Wheeler became a non-resident lecturer at Bryn Mawr and taught there part time while living in Princeton, New Jersey. She continued this arrangement for the next four years except for a year spent in Europe and directed three dissertations during that period. While still living in Princeton, she returned to her professorship in 1931. The following year Arthur Wheeler died and Anna Pell Wheeler moved back to Bryn Mawr, Pennsylvania. As head of the department at the college, Wheeler hired strong research faculty, including Emmy Noether, who joined the Bryn Mawr College faculty in 1933. Wheeler's last research publication appeared in 1935.

New Jersey College for Women and Mount Holyoke College awarded Wheeler honorary DSc degrees in 1932 and 1937, respectively. After her retirement in 1948 Anna Pell Wheeler continued to live in Bryn Mawr but spent summers in a cottage, named QED, that she and Arthur Wheeler had bought in the Adirondack Mountains of New York State. In 1953 she endowed a mathematics scholarship at the University of South Dakota honoring Alexander Pell. In 1960 anonymous gifts in honor of Anna Pell Wheeler and in memory of Charlotte Angas Scott established funds for prizes awarded annually by Bryn Mawr College in their names. Anna Pell Wheeler was eighty-two when she died in Bryn Mawr in 1966.

WHELAN, A. Marie. September 6, 1895–June 14, 1966.
GOUCHER COLLEGE (BA 1918), JOHNS HOPKINS UNIVERSITY (PHD 1923). PhD diss.: The theory of the binary octavic (directed by Frank Morley).

Anna Marie Whelan was born in Baltimore, Maryland, the sixth of eight children of Honora A. (Conroy) and James J. Whelan. Her parents had each emigrated from Ireland as children. In 1900 her father was a clerk; in 1920 he was a salesman at a clothiers.

After Marie Whelan graduated from Western High School in Baltimore in 1914, she entered Goucher College and graduated four years later. She began her graduate studies at the Johns Hopkins University in Baltimore in the fall of 1918 and, except for the year 1920–21 when she taught in Pullman, West Virginia, attended Johns Hopkins continuously until she received her doctorate in 1923. She attended the summer school in 1919 and held a university fellowship in mathematics 1919–20. A Johns Hopkins directory indicates that she also taught in the summer sessions of 1919 and 1920. Her subordinate subjects at Johns Hopkins were applied mathematics and geophysics.

After receiving her PhD in 1923, Marie Whelan, who sometimes used the name A. Marie Whelan professionally, taught for two years each at Olivet College in Michigan and Dominican College of San Rafael in California. She joined the faculty at Hunter College as an instructor in 1927 and was an assistant professor from 1932 until her retirement in 1965. During her first decade there she often was associated with the evening and extension sessions. In the 1930s she wrote several plays that were performed by mathematics clubs.

Anna Marie Whelan died in 1966 in Manhattan at age seventy. She was buried in New Cathedral Cemetery in Baltimore. Among her survivors was a sister, with whom she had lived in New York.

WHITE, Marion Ballantyne. March 28, 1871–January 30, 1958.
UNIVERSITY OF MICHIGAN (PHB 1893), UNIVERSITY OF WISCONSIN (MA 1906), UNIVERSITY OF CHICAGO (PHD 1910). PhD diss.: The dependence of the focal point on curvature in space problems of the calculus of variations (directed by Gilbert Ames Bliss).

Marion Ballantyne White was the third of five children of Jennie E. (McLaren) and Samuel Holmes White. Her mother was born in Glasgow, Scotland, and was a teacher living in Chicago before her marriage. Marion White's father, also a teacher, was born in New York and was a widower at the time of his marriage to Jennie McLaren. He held a bachelor's degree from the University of Michigan and a law degree from the Albany Law School. After finishing his education he was a principal of public schools in Chicago and then of the Peoria Normal School. He served in many important educational positions including the presidency of the National Education Association. Her siblings were three brothers and a sister. Marion White was born in Peoria, Illinois, and did her precollege work in the public schools there.

White was a student at Smith College 1888–89 but left after one year and taught in one of the Peoria public schools 1889–90. She enrolled in the University of Michigan in October 1890 and graduated three years later, the same year as her younger brother. The two years after her graduation, Marion White taught mathematics in the high school in Pueblo, Colorado. During 1895–99 and 1900–01 she taught mathematics in the Peoria high school. In the intervening year, 1899–1900, she traveled abroad with her mother and sister; the travels included Munich, where she studied in the winter.

Marion B. White was instructor of mathematics at the University of Illinois 1901–08. White earned her master's degree at the University of Wisconsin in 1906 by studying there during three summers. After leaving Illinois she did her doctoral work at the University of Chicago 1908–10 and was a fellow her last year there. While in Chicago, White was living with her widowed mother and her younger sister, who was a high school teacher. Her dissertation in the calculus of variations was the third directed by G. A. Bliss.

In 1910 White joined the faculty at the University of Kansas as assistant professor. Letters reproduced in the history of the mathematics department at the University of Kansas by G. Baley Price note that she was hired by John Wesley Young, possibly as an experiment, and was recommended for promotion to associate professor by the department in March 1914. That fall Marion White moved to Michigan State Normal College (now Eastern Michigan University) in Ypsilanti, where she had been appointed associate professor of mathematics and dean of women. She served as dean for two years and associate professor four years.

In 1918 White joined the faculty at Carleton College in Minnesota and remained there until her retirement in 1937. She was assistant professor 1918–30, dean of women 1922–24, and associate professor 1930–37. She was considered to be an especially successful teacher and was involved in college and community activities. She was active in the Carleton chapter of Phi Beta Kappa and the AAUW, and served terms in 1931–32 and 1936–37 on the executive committee of the Minnesota Section of the MAA. After her retirement White moved to Pasadena, California, where she spent the last two decades of her life. She died there in 1958 at age eighty-six.

WIDDER, Vera (Ames). December 12, 1909–May 18, 2004.

UNIVERSITY OF SASKATCHEWAN (BA 1931, MA 1932), BRYN MAWR COLLEGE (PHD 1938). PhD diss.: [Ames, V.] On systems of linear equations in Hilbert space with n parameters (directed by **Anna Pell Wheeler**).

Vera Adela Ames was born in Milestone, Saskatchewan, south of Regina, and was the daughter of Margaret Ophelia (Mooney) and Charles Edgar Ames, both originally from Ontario. Her mother became a seamstress; her father farmed near Milestone. There were five children; Vera was the second, and an only son was the third. The children attended Milestone School, the daily trips being made with horses, according to a local history. For grade twelve, which was not offered in the local school, Vera attended Regina College, at that time a Methodist high school and junior college associated with the University of Saskatchewan.

Vera Ames originally intended to take a teacher training program after finishing high school so that she could teach in a country school. However, her parents offered her the opportunity to continue her education at the university level. Ames attended the

University of Saskatchewan from 1928 until 1932, when she received her master's degree. In 1931, after completing the customary three-year program, she earned her BA with high honors in mathematics. That year she was also one of two awarded the University Gold Medal, given only to those "whose standing is considerably in advance of that required for High Honors or Great Distinction." She was an assistant in the mathematics department her last two years at Saskatchewan.

After receiving her master's degree, Ames was a resident fellow at Bryn Mawr College 1932–34. The following year she was an instructor at H. Sophie Newcomb College, substituting for **Marie Weiss**, who was studying with Emmy Noether at Bryn Mawr College that year. The next year Ames was a part-time teacher at the Baldwin School in Bryn Mawr, while continuing her graduate studies at the college. In 1936–37 she was a teacher at Miss Fine's School in Princeton, New Jersey. Ames returned as an instructor to the University of Saskatchewan in 1937 and remained there for two years. In 1938 she received her doctorate from Bryn Mawr with a dissertation directed by Anna Pell Wheeler on systems of linear equations in infinitely many unknowns.

Vera Ames and David Vernon Widder, who had met at Anna Pell Wheeler's summer home in the Adirondacks, were married on June 12, 1939. David V. Widder, who had taught at Bryn Mawr 1924–31, was born March 25, 1898, in Harrisburg, Pennsylvania. He received his BA in 1920, MA in 1923, and PhD in mathematics in 1924 from Harvard University and was on the faculty there at the time of their marriage. He remained on the faculty at Harvard until his retirement in 1968. He had a Fulbright fellowship in Italy 1955–56 and in Australia 1962–63.

The Widders had two children: a son born in about 1940, who earned a master's degree in engineering, and a daughter born in June 1951, who earned a PhD in biology. Vera Widder taught during and shortly after World War II and again after her first child was in school; she was an instructor at Cambridge Junior College 1942–44 and 1945–47, a lecturer at UCLA 1948–49, and a lecturer at Tufts College 1950–51. She also tutored as a volunteer as part of the Boston school integration program and in the Concord prison. She later recalled that soon after her marriage, G. D. Birkhoff, D. V. Widder's advisor, told her that "one career in a home is enough" (*AWM Newsletter* 12 (4): 10).

Both Widders were skilled pianists and bridge players. David Vernon Widder died on July 8, 1990. Vera Widder remained in their home in Arlington, Massachusetts, until 1998 when she moved to Sarasota, Florida, where she died at age ninety-four in 2004. Her estate plans included gifts to Bryn Mawr College and to the Unitarian Universalist Service Committee.

WILLIAMS, Emily (Coddington). October 21, 1873–August 8, 1952.

UNIVERSITY OF LONDON (BA 1896), COLUMBIA UNIVERSITY (MA 1898, PHD 1905), NEW YORK UNIVERSITY (LLB 1913). PhD diss.: [Coddington, E.] A brief account of the historical development of pseudospherical surfaces from 1827 to 1887.

Emily Matilda Coddington was born in New York City, the only child of Julia (Fellows) and Jefferson Coddington, both descended from early seventeenth-century colonial settlers. Her father was a graduate of Columbia College and an attorney who died in 1876 when Emily Coddington was a two-year-old. Four years later she and her mother were living with her maternal grandmother in Manhattan. The household also included six other family members and six servants.

Coddington studied at private schools in the New York City area. She matriculated at the University of London in June 1894, received her Intermediate Arts in 1895 and her BA in 1896. Also in 1896 she wrote what would be her master's thesis for a degree awarded two years later by Columbia University. She wrote both this thesis and her PhD dissertation on historical topics. There is no evidence that Coddington ever held a paid position, in mathematics or in any other field. However, she did maintain some contact with the mathematical community. She kept her membership in the AMS until her death and often attended meetings of the society in New York City until shortly before her marriage in 1917. She also attended the International Congress of Mathematicians in Rome in 1908 and in Cambridge, England, in 1912. During this period she also studied law. Coddington entered the Law School of New York University in 1909 and was admitted to the New York Bar in January 1912 before receiving her law degree in 1913.

On November 5, 1917, at the age of forty-four, Emily Coddington married William Henry Williams (1876–1943), an executive with an exporting business at the time of their marriage and chairman of the board of North Central Texas Oil Company at the time of his death. Both before and after her marriage Emily Coddington Williams lived in New York City, although she and her husband also had a summer home in Newport, Rhode Island. They had no children.

Emily Coddington Williams had wide-ranging interests. They included gardening, genealogy, and creative writing. Her writings included a one-act play, two novels, and a sketch of William Coddington, a colonial governor of Rhode Island.

During the course of her life Emily C. Williams made many trips to Europe. In April 1952 she made the last of these trips, on the Queen Mary. After having fractured her hip on board ship, Emily Williams died at the American Hospital in Paris in August 1952 at age seventy-eight. Her estate, which was estimated at $12,000,000, was divided among her two godchildren and charitable groups in New York City and Newport, Rhode Island.

WILLIAMS, Martha (Hathaway). January 25, 1914–July 22, 1989.

WELLESLEY COLLEGE (BA 1935), MASSACHUSETTS INSTITUTE OF TECHNOLOGY (MS 1936, PHD 1939). PhD diss.: [Plass, M. H.] Ruled surfaces in Euclidean four space (directed by Dirk Jan Struik).

Martha Eldora Hathaway was born in Montclair, New Jersey, the daughter of Sarah Emma (Vaughan) and Joseph Wood Hathaway. She had two younger brothers. Her father worked for many years for Western Electric Company in New York City. In 1918 he described his occupation as electrical engineer, in 1930 as accountant.

Martha Hathaway attended public schools in Montclair and graduated with honors in mathematics from the high school there in 1931. She then entered Wellesley College, where she was a Durant scholar and served as president of the Wellesley mathematics club her senior year, 1934–35. She graduated with honors in mathematics in 1935. On June 19, two days after her graduation, Martha Hathaway married Herbert Fitz Randolph Plass in Wellesley, Massachusetts. Herbert Plass was born in Newark, New Jersey, in 1912, and he earned bachelor's and master's degrees in physics from the Massachusetts Institute of Technology in 1934 and 1935, respectively, before entering medical school at Harvard University.

Martha H. Plass entered MIT in 1935 and finished the work for her master's degree a year later. She continued at MIT, where she minored in physics and where D. J. Struik supervised her doctoral dissertation. During the second semester 1937–38, she was an instructor at Wellesley and was the nominee of Wellesley for its institutional membership to the AMS. In the fall of 1938 she became an instructor at the University of Maryland. In 1939 Martha Plass received her PhD from MIT and Herbert R. Plass received his MD from Harvard; they also were divorced that year.

Martha Plass continued to teach at the University of Maryland as an instructor until 1940, when she married Jonathan Wilber Williams and moved to North Carolina. Jonathan W. Williams (1910–1970) earned his PhD in organic chemistry in 1935 from Northwestern University. He was on the faculty at the University of Maryland 1937–39 and at the University of North Carolina 1939–41. During World War II he worked at the US Naval Research Laboratory 1941–42 and with the National Defense Research Committee of the Office of Scientific Research and Development 1942–45.

Martha Williams was again an instructor at Maryland 1941–43. After the war, Jonathan Williams worked for the DuPont chemical company 1946–54 and then for the Haskell

Laboratory of Toxicology and Industrial Medicine, established by DuPont, 1954–60. In 1960 he took a position with the patent and licensing division of the US Department of the Interior.

Martha and Jonathan Williams had two children: a daughter born in about 1943 and a son born about five years later. The family lived in Wilmington, Delaware, by the early 1950s. Martha Williams was a member of the Canterbury Garden Club and the Delaware Daffodil Society and was president of the Delaware Federation of Garden Clubs 1962–65. She was president of the Wilmington branch of the AAUW 1961–62.

Martha Williams died in Wilmington in 1989 at age seventy-five. She was survived by her two brothers, her two children, and three grandsons. Memorial services were held in the First Unitarian Church in Sharpley, Delaware.

WILSON, Hazel (Schoonmaker). October 31, 1888–April 23, 1988.

WELLESLEY COLLEGE (BA 1911), RADCLIFFE COLLEGE (MA 1914), CORNELL UNIVERSITY (PHD 1927). PhD diss.: [Schoonmaker, H. E.] Non-monoidal involutions having a congruence of invariant conics (directed by Virgil Snyder).

Hazel Edith Schoonmaker was born in Philadelphia, Pennsylvania, the elder of two children of Christiana M. (Abele) and Arthur Twing Schoonmaker. Her father graduated from the Hahnemann Medical College of Philadelphia in 1894; shortly thereafter the family was living in Westfield, Massachusetts, where her father practiced medicine.

Hazel Schoonmaker graduated from high school in Westfield before she entered Wellesley College, from which she graduated in 1911 with a major in mathematics and a minor in German. She taught in high school in Massachusetts 1912–13 before entering Radcliffe College for graduate work. After receiving her master's degree in 1914, Schoonmaker taught in high school in New York for a year and was acting associate professor at Denison University in Granville, Ohio, the following year. She taught in high schools in New York 1916–17 and in Pennsylvania 1917–19. During 1920 she was professor at McKendree College in Lebanon, Illinois; in 1920–21 she was acting professor at Augustana College in Rock Island, Illinois; and in 1921–22 she was professor at Gulf Park College for Women, a two-year school in Long Beach, Mississippi. The summer of 1922 she studied at Columbia University and then spent three years as professor at Western College for Women in Oxford, Ohio.

Schoonmaker first attended Cornell University in the summer session of 1924; she began full-time study there in 1925 and stayed at Cornell until 1928. She was appointed a scholar for 1926–27 and received her PhD in 1927 with major geometry, first minor analysis, and second minor education. She remained at Cornell as resident doctor the following year, taking two mathematics courses and preparing typed notes for a course she had taken from Virgil Snyder.

After leaving Cornell, Schoonmaker was an assistant professor at the New Jersey College for Women, the public coordinate college for women associated with Rutgers University in New Brunswick. She then joined the faculty at Hartwick College in Oneonta, New York, where she remained for seven years until her marriage on July 14, 1938, to Levi Thomas Wilson (1885–1975). L. T. Wilson had received a PhD in mathematics from Harvard University in 1915, and in 1917 he began his long career at the US Naval Academy in Annapolis, Maryland.

From 1942 until 1951 Hazel Wilson taught and was chairman of the mathematics department at Annapolis High School; she studied at Johns Hopkins during the summer of 1945. In 1951 L. T. Wilson retired from the Naval Academy and became professor of mathematics and physics and head of the department of physics at Jacksonville State Teachers College (now Jacksonville State University) in Alabama. At the same time, Hazel Wilson retired from Annapolis High School and became associate professor of mathematics, also at Jacksonville State. In 1952 they both also became instructors at the University of Alabama, Gadsden Center. In 1956 L. T. Wilson went to Doane College in Crete, Nebraska, again as professor of mathematics and physics and head of the department. Hazel Wilson remained at Jacksonville and Gadsden Center until 1957, when she joined her husband at Doane as professor.

They both remained at Doane until 1959 when they moved to Jacksonville University in Florida, she as professor of mathematics and he as professor of physics and head of the department of physics and physical sciences. They both retired in 1964 as professors emeriti and later lived in St. Petersburg, Florida.

In 1974, nearly fifty years after receiving her PhD, Hazel S. Wilson responded to a request for a donation to Cornell by honoring her advisor with a significant contribution to the Virgil Snyder Book Endowment. Levi Wilson died in 1975. Hazel Wilson died in St. Petersburg, Florida, at age ninety-nine in 1988.

WOLF, Louise A. October 20, 1898–November 14, 1962.

UNIVERSITY OF WISCONSIN (BA 1931, MA 1933, PHD 1935). PhD diss.: Similarity of matrices in which the elements are real quaternions (directed by Mark Hoyt Ingraham).

Louise Adelaide Wolf was the elder of two daughters, both born in Milwaukee, Wisconsin, of Caroline (Kupperian) and John Theodore Wolf. Her mother was born in Germany and immigrated to the United States in 1892, and her father was born in Milwaukee; both were formally educated through elementary school. In the 1900 census, her father's occupation was listed as conductor on the street railroad; later he was a truck gardener. Her sister, **Margarete Wolf (Hopkins)**, who was thirteen years younger than Louise Wolf, also received a PhD in mathematics from Wisconsin in 1935.

Louise Wolf attended the 26th Avenue School and, for three and a half years, South Division High School, in Milwaukee. She attended Milwaukee-Downer College 1915–16 and took a variety of positions over the next dozen years. These included working in a dental office and in a public library in Milwaukee, teaching two years in a district school in Nevada, and working two years in Florida. She returned to college at the Milwaukee Extension Division of the University of Wisconsin in 1928, when her sister began college there. Having delayed her own college education, Louise Wolf was then able to help finance the education of her sister. They both moved to the main campus at Madison the following year to complete their undergraduate and graduate work. Louise Wolf was president of the junior mathematics club her senior year in Madison.

Louise Wolf taught a course as an assistant during the first semester of her senior year and most subsequent semesters while she was a student at Wisconsin. Typically she taught about three courses a year. She received her bachelor's degree in 1931 and finished the work for her master's degree in 1933. Louise Wolf and her sister, Margarete, finished the work for their doctorates in 1935, both as students of Mark H. Ingraham.

Louise Wolf immediately took a position with the University of Wisconsin's Extension Division. She was engaged in circuit teaching her first year and traveled regularly between Milwaukee, Fond du Lac, Sheboygan, and Manitowoc giving instruction. Starting in 1936, she taught at the Milwaukee Center, often referred to as the University of Wisconsin in Milwaukee. Around the beginning of War War II she was also a lecturer at the University of Wisconsin in Madison. In Milwaukee she was an instructor 1935–38, an assistant professor until 1951, and an associate professor until her retirement as associate professor emeritus in January 1962. In 1955 the Extension Division had merged with Wisconsin State College to form the University of Wisconsin-Milwaukee.

In 1938 Louise Wolf presented a talk that had been coauthored with her sister to an AMS meeting in New York City. In the 1940s Wolf became active in the Wisconsin Section of the MAA, initially serving on program committees. She was first elected secretary of the section in 1948 and served in that capacity for the next five years. She also served on committees of NCTM related to summer meetings in Wisconsin in 1940 and 1950. Louise A. Wolf died at age sixty-four in Milwaukee in 1962.

WOOD, Ruth G. January 29, 1875–May 5, 1939.
SMITH COLLEGE (BL 1898), YALE UNIVERSITY (PHD 1901). PhD diss.: Non-Euclidean displacements and symmetry transformations.

Ruth Goulding Wood was born in Central Falls, Rhode Island, the younger of two children of Kate Bassett (Pond) and Samuel Eugene Wood. Her father was a bookkeeper in the 1870s and manager of the yarn department of a manufacturing company in the late 1880s. Ruth's brother was born in 1873 and became a mechanical engineer.

Ruth Wood attended Pawtucket High School before entering Smith College in Northampton, Massachusetts. She took the literary course at Smith, with major subjects mathematics, German, and the sciences, and graduated in 1898. She then entered Yale University; she held a scholarship in 1899–1900 and a fellowship in 1900–01. She received her PhD in 1901.

Wood was an instructor at Mount Holyoke College 1901–02 before joining the two-person mathematics department at Smith College as instructor in 1902. She was at Smith her entire career except for the year 1908–09, when she was studying at the university in Göttingen. She was instructor 1902–09, associate professor 1909–14, and professor 1914–35. She became chairman of the department in 1922 and remained so until 1928. Her only sabbatical was in spring 1934 when she traveled in Egypt, Greece, and Turkey. In June 1935, at age sixty, she resigned from the Smith faculty and was named professor emeritus.

Wood often spoke to the mathematical club of Smith College about mathematics and her mathematically related experiences. In 1909 she spoke on her university experience at Göttingen. In 1918 she discussed her work the previous summer at the New York Life Insurance Company, and in fall 1928 she and **Susan Rambo** gave talks about their visit to the International Congress in Bologna. Wood had also attended international congresses in Cambridge, England, in 1912 and in Toronto in 1924. About a year after she retired, Wood reported in the *Smith Alumnae Quarterly* that the previous winter she had gone to California via the Panama Canal, visited Yosemite, and did not miss teaching. In February 1937 she sailed on a cruise to South America; she went down the west coast and crossed the Andes by auto and mountain lakes by boat.

In 1931 Wood moved to Florence, Massachusetts, a few miles from Northampton, to a house she had designed. The house included a greenhouse to accommodate her expertise as a gardener. She also had a summer home in Cummington, in the Berkshire foothills. Wood's mother, who had been living with her, died in May 1932.

Wood died at sixty-four in 1939 in Springfield, Massachusetts. Newspaper accounts indicate that upon the death of the last of three beneficiaries of a trust fund she had set up, the fund and accumulated interest were to go to the trustees of Smith College to ensure that at least one woman mathematics professor earned the highest salary of anyone on the faculty.

WORTHINGTON, Euphemia Richardson. December 22, 1881–August 30, 1969.
WELLESLEY COLLEGE (BA 1904), YALE UNIVERSITY (PHD 1908). PhD diss.: Some theorems on surfaces (directed by James Pelham Pierpont).

Euphemia Richardson Worthington was born in Troy, New York, the first of three daughters of Sophia Adelaide (Whidden), born in Nova Scotia, and John Worthington, of New York. Her father, a widower with three children by his previous marriage, was a coal dealer. Her sisters were born in 1883 and 1887; the older sister died as a young child.

Euphemia Worthington attended the Emma Willard School in Troy 1895–1900. She then entered Wellesley College and graduated in 1904. Worthington entered Yale University in 1905 and received her PhD in 1908.

The following year Worthington taught at the Emma Willard School. From 1909 to 1918 she was an instructor at Wellesley, her undergraduate alma mater. About a year after the United States entered World War I, Worthington left Wellesley to do war work.

She was assistant to the chief engineer of the Gallaudet Aircraft Corporation in East Greenwich, Rhode Island, from July 1918 to May 1919. She reported for the 1919 issue of the "Wellesley Class of 1904 Record Book": "This summer I have been working for an aircraft company near Providence and am finding my work, which is doing mathematical work for the chief engineer, very interesting and very far from monotonous, a decided change from teaching freshmen" (Wellesley College Archives). In May 1919 Worthington took, and passed, the first part of the associate examination of the Actuarial Society of America (ASA). Later in 1919 she was unemployed and living in Pasadena, California. She resumed her position at the Gallaudet Aircraft Corporation and worked there again from March 1920 to January 1922. In May 1920 she passed the second part of the ASA associate examination and apparently never attempted the remaining two parts.

In January 1922 Worthington was appointed instructor in mathematics at the Southern Branch of the University of California. This former state normal school conferred its first bachelor's degree in 1923 and changed its name to the University of California at Los Angeles in 1927. Worthington became assistant professor in 1925 and retired as assistant professor emeritus in 1949. She had leaves with two-thirds salary from March 25 to June 30 in 1929 and from January 1 to June 30 in 1932. She worked as a stress analyst in the engineering department of the Bell Aircraft Company in Buffalo, New York, in summer 1942. She was also designated an emeritus fellow by UCLA in 1954.

For some years after her retirement, Worthington spent summers in Carmel, California, and winters in Victoria, British Columbia, Canada, before eventually moving permanently to Victoria. Euphemia Worthington died in 1969 at age eighty-seven in Victoria.

WYANT, Kathryn. January 16, 1897–July 16, 1942.
UNIVERSITY OF MISSOURI (BS 1921, MA 1922, PHD 1929). PhD diss.: The ideals in the algebra of generalized quaternions over the field of rational numbers (directed by Gustav Eric Wahlin).

Emily Kathryn Wyant was born in Ipava, Illinois, the younger of two children of Alice Mary (Dillon) and Clarence Jacob Wyant. In 1900 the family was living in Dixon, in northern Illinois, and her father was a student. Kathryn Wyant attended a country grade school in Bader, Illinois, about twenty-five miles from Ipava. In 1910 the family was living in Bolivar, Missouri, where her father was a merchant. Kathryn graduated from the high school there in May 1914.

Kathryn Wyant entered the University of Missouri in June 1914 and did most of her undergraduate work in summer sessions there. She taught two winters in rural schools in Polk County, Missouri, and three in the high schools at Callao and Neosho, Missouri, before receiving her BS in education in June 1921. Wyant was an instructor in mathematics at the University of Missouri from 1921 until 1930. She completed her master's degree at Missouri in 1922 with a major in physics and a minor in mathematics. After having attended one summer session at the University of Chicago, she completed the work for her PhD at Missouri in 1929 with a major in mathematics and a minor in astronomy.

From 1930 to 1933 Wyant was professor of mathematics at Northeastern State Teachers College (now Northeastern State University) in Tahlequah, Oklahoma. From 1933 to 1934, Wyant did postgraduate work at the University of Chicago before joining, in 1934, the faculty at Athens College (now Athens State University) in Alabama. She served as chairman of the department of mathematics there until her early retirement in 1940 because of ill health.

Throughout her career Wyant was unusually active in professional and honorary societies. While she was at Missouri she was particularly involved with the mathematics honorary fraternity Pi Mu Epsilon, the graduate women's scientific fraternity Sigma Delta Epsilon, and the MAA. She was national vice president of Sigma Delta Epsilon in 1926 and president from June to December 1926. The following year she was chairman of the Missouri Section of the MAA. Also during this period Wyant became interested in some Diophantine problems and solutions that had appeared in *The Missouri Intelligencer* in

1823 and 1824, and, more than a decade after giving talks on those problems, her talk appeared as a pamphlet, "Some mathematical problems in Missouri history."

While at Northeastern State Teachers College, Wyant transformed the existing mathematics club into the first chapter of Kappa Mu Epsilon, a mathematics honor society for schools that place an emphasis on undergraduate education; she served as its first national president 1931–35 and its fourth national historian 1939–41. In 1936 she was a co-founder of the Tennessee Valley Mathematics Association for high school teachers, junior college teachers, and others interested in mathematics. She served as secretary-treasurer of the mathematics division of the Alabama Education Association 1936–37.

Wyant was a Methodist and was a member of the Order of the Eastern Star. In about 1940 she listed her hobbies as supervising mathematics fraternity work and driving her car and indicated that playing with her cocker spaniel was her favorite recreation. In 1940 Kathryn Wyant was living in Birmingham, Alabama; she was living outside of Athens, Alabama, at the time of her death at age forty-five in 1942. She is buried in Ipava, Illinois, the town of her birth.

YEATON, Marie M. (Johnson). March 1, 1898–March 19, 1978.

KNOX COLLEGE (BA 1920), UNIVERSITY OF IOWA (MS 1921), UNIVERSITY OF CHICAGO (PHD 1928). PhD diss.: [Johnson, M. M.] Tensors of the calculus of variations (directed by Gilbert Ames Bliss).

Marie Mathilda Johnson was born in Galesburg, Illinois, the younger of two children of Ellen (Olson) and Oscar William Johnson. Her father was proprietor of a grocery store. Marie Johnson attended grammar school and high school in Galesburg.

Johnson attended Knox College, also in Galesburg, and received her BA in 1920. She studied at the University of Iowa the following year and received her master's degree in 1921 with a major in mathematics and a minor in education. In 1921 Johnson became an instructor at Lake Forest College in Lake Forest, Illinois. During the summers of 1922 through 1926 she attended the University of Chicago. Her last year at Lake Forest, 1925–26, she was assistant professor and acting head of the department. After leaving Lake Forest she held a fellowship at the University of Chicago for a year and received her doctorate in 1928.

In 1927 Johnson accepted an instructorship at Pennsylvania State College but was released from that contract in order to go to Oberlin College in Ohio as an acting assistant professor replacing **Mary E. Sinclair** who was on leave. Johnson's position became permanent, and she was assistant professor at Oberlin from 1928 until 1943, when she was promoted to associate professor. She had a leave of absence during the academic year 1937–38 and spent the year in residence at the Institute for Advanced Study in Princeton.

On October 27, 1943, Marie Johnson married her Oberlin colleague, Chester Henry Yeaton (1886–1970), a widower who was professor of mathematics at the time of the marriage. Chester Yeaton had received his bachelor's degree at Bowdoin College in 1908, his master's degree at Harvard in 1909, and his PhD at Chicago in 1915. Marie Johnson Yeaton formally resigned from the Oberlin faculty effective March 1, 1944. She was acting associate professor the next winter term and again second semester 1951–52, after which Chester H. Yeaton retired. He died in Oberlin in 1970 at age eighty-three.

After her resignation from the faculty, Marie M. Yeaton continued her membership in the Oberlin Mathematics Club for which she had previously served as faculty advisor. She was also a member of the Oberlin Social Science Club; was president of the local chapter of Delta Kappa Gamma, an education honorary society; and was treasurer of the Oberlin branch of the AAUW. Marie Yeaton died in Oberlin, Ohio, at age eighty, after having been a resident of a nursing home for the previous five years.

YOUNG, Mabel M. July 18, 1872–March 4, 1963.
WELLESLEY COLLEGE (BA 1898), COLUMBIA UNIVERSITY (TEACHERS COLLEGE) (MA 1899), JOHNS HOPKINS UNIVERSITY (PHD 1914). PhD diss.: Dupin's cyclide as a self-dual surface (directed by Frank Morley).

Mabel Minerva Young was born in Worcester, Massachusetts, the daughter of Minerva (Tyler) and Willie C. Young. Her mother graduated from the Oread Institute, a private school and college for girls and women in Worcester. In 1880 the Youngs and their two daughters, seven-year-old Mabel and her two-year-old sister, were living in Worcester with Willie Young's parents. Willie C. Young was a machinist and bookkeeper and then president of the W. C. Young Manufacturing Company that made machinists' tools. He served as a Republican member of the Massachusetts state legislature in 1896 and 1897.

Mabel Young attended public schools in Worcester for her primary and secondary education. For the latter she completed a five-year course at Classical High School, from which she graduated as salutatorian. When she entered Wellesley College in 1894, she was unusually well prepared, especially in Greek, Latin, and mathematics.

During her four years at Wellesley, Young studied mainly English, mathematics, philosophy, and psychology before receiving her bachelor's degree in 1898. She spent the next year at Columbia University, where she did work in education and psychology. In 1899 she earned her MA and a master's diploma in education from Teachers College, Columbia, with a master's thesis, "The aims and methods of the teaching of mathematics in the secondary schools of Germany and the United States."

From 1899 to 1902, Young taught mainly English at the Northfield Seminary, a girls' preparatory school in Massachusetts. She also taught 1902–04. In 1904 she returned to Wellesley as an assistant in mathematics and was made an instructor in 1906. In 1904–08 she was enrolled in three hours a week of graduate courses in pure mathematics at Wellesley and did graduate work at Cornell University during the summer of 1908.

Young entered the Johns Hopkins University in 1909 and was in residence 1909–10 and again 1912–14, while she had leaves of absence from Wellesley 1909–11 and 1912–14. She was a fellow by courtesy at Johns Hopkins, with remission of tuition fees, 1913–14. In 1914 Young received her doctorate with a dissertation in algebraic geometry and with physics as a first subordinate subject.

The remainder of Young's career was spent at Wellesley. She remained an instructor until 1919; she was assistant professor 1919–24, associate professor 1924–30, professor 1930–41, and then emeritus professor. During the year 1930–31 she traveled for six months in the Mediterranean region while on a leave of absence for study and travel. She also attended the International Congress of Mathematicians in Oslo in the summer of 1936.

In 1932 **Helen Merrill** retired from Wellesley and the department chairmanship that she had held for sixteen years. The following year Young became department chairman and remained so until her retirement in 1941. She was the Lewis Atterbury Stimson professor for some years.

Mabel Young's interests included music and travel, and in about 1940 she described herself as Congregational and a Republican. She died in 1963 at age ninety in Wellesley.

Abbreviations

AAAS	American Association for the Advancement of Science
AARP	American Association of Retired Persons
AAUP	American Association of University Professors
AAUW	American Association of University Women
ACA	Association of Collegiate Alumnae
ACM	Association for Computing Machinery
AMP	Applied Mathematics Panel
AMS	American Mathematical Society
ASA	American Statistical Association
AWM	Association for Women in Mathematics
CUPM	Committee on the Undergraduate Program in Mathematics (MAA)
DAR	Daughters of the American Revolution
ICM	International Congress of Mathematicians
JFM	*Jahrbuch über die Fortschritte der Mathematik*
MAA	Mathematical Association of America
MR	*Mathematical Reviews*
NASA	National Aeronautics and Space Administration
NCTM	National Council of Teachers of Mathematics
NDRC	National Defense Research Committee
NSF	National Science Foundation
NRC	National Research Council
ONR	Office of Naval Research
SIAM	Society for Industrial and Applied Mathematics
SMSG	School Mathematics Study Group
WAVES	Women Accepted for Volunteer Emergency Service (US Naval Reserve)
WPA	Works Progress Administration (Work Projects Administration after 1939)

Archives and Manuscript Collections

The archives and manuscript collections that we used are arranged geographically below with those visited by one or both of the authors indicated by an asterisk (*). We thank all the archivists and librarians who aided us in our use of these materials, whether we visited in person or communicated by other means. We want to single out the following collections that were of particular importance in the gathering of information on the women in our study. One is the Helen Brewster Owens Papers at the Schlesinger Library, Radcliffe Institute, Harvard University, Cambridge, Massachusetts. The other, which was compiled by the authors and Uta C. Merzbach, is the Early Women Doctorates Collection (number 2006.3037) in the Mathematics Collections, Division of Information Technology and Communications, National Museum of American History, Smithsonian Institution, Washington, D.C. Names of other collections used are available on the website associated with this book. Included below are names of congregations of women religious and of colleges that have provided us with material from their archives, although there may not be a separate unit so designated. We have also been assisted by many people in public libraries, alumni offices, other university offices, mathematics departments, and secondary schools which are not included on the list below. In a number of cases, the name of the institution or the name of the repository has changed since we used the material. In these instances, we are using what we believe to be the current name.

Alabama

Athens. Athens State University, Archives

Montevallo. University of Montevallo, Carmichael Library, Archives and Special Collections

Arkansas

Fayetteville. University of Arkansas, University Libraries, Special Collections Division

California

*Berkeley. University of California, Berkeley, The Bancroft Library, University Archives

*Berkeley. University of California, Berkeley, The Bancroft Library, Regional Oral History Office

Los Angeles. Immaculate Heart Community

Los Angeles. University of California, Los Angeles, University Archives

Redlands. University of Redlands, University Archives

San Rafael. Dominican University of California

*Stanford. Stanford University, Special Collections, University Archives

Whittier. Whittier College, Wardman Library, College Archives

Colorado

*Colorado Springs. Colorado College, Tutt Library, Special Collections and Archives

Connecticut

*Middletown. Wesleyan University Library, Special Collections and Archives

*New Haven. Yale University Library, Sterling Memorial Library, Manuscripts and Archives

New London. Connecticut College, Connecticut College Archives

District of Columbia

*Washington. The Catholic University of America, American Catholic History Research Center and University Archives

Washington. George Washington University, Special Collections Research Center, University Archives

Washington. National Academy of Sciences, NAS Archives

*Washington. Smithsonian Institution, National Museum of American History, Division of Information Technology and Communications

*Washington. Trinity Washington University

Florida

DeLand. Stetson University, University Archives

Winter Park. Rollins College, Archives and Special Collections

Georgia

Athens. University of Georgia, Hargrett Rare Book and Manuscript Library, University Archives

Macon. Mercer University, Special Collections, Georgia Baptist History Depository, Tift College Archives

Rome. Shorter College, Shorter Museum and Archives

Idaho

Boise. Boise State University, Albertsons Library, Special Collections Department

Illinois

*Bloomington. Illinois Wesleyan University, Tate Archives and Special Collections, University Archives

*Chicago. Center for Research Libraries, College and University Course Catalogs

*Chicago. Loyola University Chicago, Cudahy Library, University Archives

*Chicago. University of Chicago, Special Collections Research Center, University of Chicago Archives

*Evanston. Northwestern University, University Archives

Galesburg. Knox College, Archives and Special Collections

Rockford. Rockford College, College Archives

*Urbana. University of Illinois at Urbana-Champaign, University of Illinois Archives

*Wheaton. Wheaton College, College Archives

Indiana

*Bloomington. Indiana University, Herman B. Wells Library, University Archives

*Bloomington. Indiana University, Lilly Library, Manuscripts
Notre Dame. University of Notre Dame, Archives of the University of Notre Dame

Iowa

Ames. Iowa State University, Iowa State University Library, Special Collections Department, University Archives
Dubuque. Clarke College, Archives Office
Iowa City. University of Iowa, University Libraries, Department of Special Collections, University Archives

Kansas

Concordia. Sisters of St. Joseph Archives
Leavenworth. Sisters of Charity of Leavenworth, Archives

Kentucky

Villa Hills. Benedictine Sisters of St. Walburg Monastery
St. Catharine. Dominican Sisters, St. Catharine of Sienna Archives
Berea. Berea College, Hutchins Library, Special Collections and Archives, Berea College Archives
Georgetown. Georgetown College, Ensor Learning Resource Center, Special Collections and Archives
Lexington. Transylvania University, Special Collections, Archives
Lexington. University of Kentucky, Margaret I. King Library, Special Collections and Digital Programs, University Archives
Louisville. Spalding University, The Library, University Archives
Nazareth. Sisters of Charity of Nazareth, Archives
St. Catharine. Dominican Sisters, St. Catharine of Sienna Archives

Louisiana

*New Orleans. Tulane University, Special Collections, University Archives
New Orleans. Xavier University of Louisiana, Library, Xavier Archives

Maryland

Baltimore. College of Notre Dame of Maryland, Loyola-Notre Dame Library, Archives
Baltimore. Goucher College, Special Collections and Archives, Goucher College Archives
*Baltimore. The Johns Hopkins University, Sheridan Libraries, Milton S. Eisenhower Library, Ferdinand Hamburger University Archives
*Baltimore. The Johns Hopkins University, Sheridan Libraries, Special Collections
Bethesda. National Library of Medicine, History of Medicine Division, Modern Manuscripts Collection
*College Park. American Institute of Physics, Center for History of Physics, Niels Bohr Library and Archives
*College Park. University of Maryland, Special Collections
*College Park. University of Maryland, Hornbake Library, Archives and Manuscripts

Massachusetts

Brighton. Congregation of the Sisters of St. Joseph of Boston, Boston CSJ Archives
*Cambridge. Harvard University, Pusey Library, Harvard University Archives
*Cambridge. Harvard University, Radcliffe Institute, Schlesinger Library
Medford. Tufts University, University Archives and Manuscript Collections
*Northampton. Smith College, College Archives
*Northampton. Smith College, Sophia Smith Collection
*South Hadley. Mount Holyoke College, Archives and Special Collections
*Wellesley. Wellesley College, Wellesley College Archives

Michigan

*Ann Arbor. University of Michigan, Bentley Historical Library, University Archives
Detroit. Wayne State University, Walter P. Reuther Library, University Archives
Hillsdale. Hillsdale College, Mossey Library, Hillsdale College Archives

Minnesota

Minneapolis. University of Minnesota, University Archives
Moorhead. Concordia College, Concordia College Archives
*Northfield. Carleton College, Carleton College Archives
Rochester. Sisters of Saint Francis

Mississippi

Cleveland. Delta State University, Charles W. Capps, Jr. Archives and Museum, University Archives

Nebraska

Fremont. Midland Lutheran College, College Archives

New Hampshire

Hanover. Dartmouth College, Rauner Special Collections Library, College Archives

New York

*Albany. New York State Archives
Aurora. Wells College, Wells College Archives
Bronx. Fordham University, University Libraries, Archives and Special Collections
*Brooklyn. Brooklyn College of the City University of New York, Archives and Special Collections
*Ithaca. Cornell University Library, Division of Rare and Manuscript Collections
*New York. The City College of the City University of New York, Archives and Special Collections
*New York. Columbia University, University Archives
*New York. Columbia University, Rare Books and Manuscripts Library
*New York. Hunter College of the City University of New York, Hunter College Libraries, Archives and Special Collections
*New York. New York University, Bobst Library, University Archives
Oneonta. Hartwick College, Paul F. Cooper, Jr. Archives, Hartwick College Records

*Poughkeepsie. Vassar College, Vassar College Libraries, Archives and Special Collections

North Carolina

*Durham. Duke University; Rare Book, Manuscript, and Special Collections Library; University Archives

*Greensboro. Guilford College, Friends Historical Collection

*Greensboro. University of North Carolina at Greensboro, University Archives and Manuscripts

*Raleigh. Meredith College, College Archives

Ohio

Athens. Ohio University, Robert E. and Jean R. Mahn Center for Archives and Special Collections, University Archives

Chardon. Sisters of Notre Dame, Notre Dame Education Center, Province Archives

Cincinnati. Sisters of Mercy Provincialate

Cincinnati. Sisters of Notre Dame de Namur, Ohio Province Archives

*Columbus. Ohio State University, University Archives

Hiram. Hiram College, Archives and Special Collections

Kent. Kent Sate University, Department of Special Collections and Archives, University Archives

Oberlin. Oberlin College, Oberlin College Archives

*Oxford. Miami University, University Libraries Special Collections Department, Miami University Archives

*Oxford. Miami University, Western College Memorial Archives

Tiffin. Heidelberg College, College Archives

Oklahoma

Norman. University of Oklahoma, Western History Collections, University Archives

Oregon

Forest Grove. Pacific University, Archives Department

Pennsylvania

*Bryn Mawr. Bryn Mawr College, Library, Special Collections, College Archives

Chambersburg. Wilson College, The Hankey Center, C. Elizabeth Boyd '33 Archives

Greensburg. Seton Hill University, University Archives

Indiana. Indiana University of Pennsylvania, Special Collections and University Archives

New Wilmington. Westminster College, College Archives

*Philadelphia. University of Pennsylvania, University Archives and Records Center

Pittsburgh. University of Pittsburgh, Archives Service Center, University Archives

University Park. Pennsylvania State University, Paterno Library, Special Collections Library, University Archives

Rhode Island

Kingston. University of Rhode Island, University Libraries, Special Collections, University Archives

*Providence. Brown University, John Hay Library, University Archives

South Carolina

Rock Hill. Winthrop University, Dacus Library, University Archives

Texas

*Austin. University of Texas at Austin, Center for American History, Archives of American Mathematics

*Austin. University of Texas at Austin, Center for American History, University of Texas Archives

San Antonio. Congregation of Divine Province, Archives

San Antonio. Incarnate Word Generalate, Archives

Utah

Salt Lake City. University of Utah, J. Willard Marriott Library, Special Collections Department, University Archives

Vermont

Burlington. University of Vermont, Special Collections, University Archives

Virginia

Charlottesville. University of Virginia, Special Collections Library

Farmville. Longwood University, Special Collections

Lynchburg. Lynchburg College Archives

Radford. Radford University, McConnell Library, Special Collections

Sweet Briar. Sweet Briar College, College Library, Archives

Washington

Seattle. University of Washington, Special Collections, University Archives

West Virginia

Morgantown. West Virginia University, West Virginia and Regional History Collection

Wisconsin

*Madison. University of Wisconsin–Madison, Steenbock Memorial Library, Archives

Madison. Wisconsin Historical Society, Library and Archives

Milwaukee. Marquette University, Special Collections and University Archives

Milwaukee. Mount Mary College, Archives

Canada

Nova Scotia. Halifax. Sisters of Charity of St. Vincent de Paul, Mount Saint Vincent

*Ontario. Toronto. University of Toronto, Thomas Fisher Rare Book Library, University Archives

*Saskatchewan. Saskatoon. University of Saskatchewan, University Archives

England

London. Imperial College, Corporate Records and College Archives

London. University of London, Senate House Library, History Collections

Germany

*Göttingen. Georg-August-Universität Göttingen, Universitätsarchiv Göttingen, Niedersächische Staats- und Universitätsbibliothek

Scotland

*Edinburgh. University of Edinburgh, Special Collections, Edinburgh University Archives

Selected Bibliography

Albisetti, James C. *Schooling German Girls and Women: Secondary and Higher Education in the Nineteenth Century.* Princeton, NJ: Princeton University Press, 1988.

American Mathematical Society. *Semicentennial Addresses of the American Mathematical Society.* Providence, RI: American Mathematical Society, 1938. Reprint, New York: Arno Press, 1980, and Providence, RI: American Mathematical Society, 1988.

Archibald, Raymond Clare. *A Semicentennial History of the American Mathematical Society 1888–1938.* Providence, RI: American Mathematical Society, 1938. Reprint, New York: Arno Press, 1980, and Providence, RI: American Mathematical Society, 1988.

Barnes, Mabel S., Judy Green, Jeanne LaDuke, Vivienne Malone-Mayes, and Olga Taussky-Todd. "Centennial Reflections on Women in American Mathematics." *AWM Newsletter* 18, no. 6 (1988): 4–12.

Bell, E. T. "Fifty Years of Algebra in America, 1888–1938." In *Semicentennial Addresses of the American Mathematical Society,* 1–34.

Bernstein, Dorothy L., M. Gweneth Humphreys, Anne F. O'Neill, and Mina Rees. "Women Mathematicians before 1950." *AWM Newsletter* 9 no. 4 (1979): 9–18.

Bowler, Sr. Mary Mariella. "A History of Catholic Colleges for Women in the United States of America." PhD dissertation, The Catholic University of America, 1933.

Bremner, Christina Sinclair. *Education of Girls and Women in Great Britain.* London: Swan Sonnenschein, 1897.

Bridges, Flora. "Coeducation in Swiss Universities." *Popular Science Monthly* 38 (1890–91): 524–30.

Case, Bettye Anne, and Anne M. Leggett. *Complexities: Women in Mathematics.* Princeton, NJ: Princeton University Press, 2005.

Clinton, Catherine. "Women & Harvard: The First 350 Years." *Harvard Magazine* 89 (Sept.-Oct. 1986): 123–28.

Cochell, Gary G. "The Early History of the Cornell Mathematics Department: A Case Study in the Emergence of the American Mathematical Research Community." *Historia Mathematica* 25 (1998): 133–53.

Cockey, Beale W. "Mathematics at Goucher: 1888–1979." Goucher College, 1979.

Cohen, Lucy M. "Early Efforts to Admit Sisters and Lay Women to The Catholic University of America: An Introduction." In *Pioneering Women at the Catholic University of America: Papers Presented at a Centennial Symposium, November 11, 1988,* edited by E. Catherine Dunn and Dorothy A. Mohler, 1–18. Hyattsville, MD: International Graphics (Distributed by the Catholic University of America Press, Washington, DC), 1990.

Dolan, Eleanor F., and Margaret P. Davis. "Antinepotism Rules in American Colleges and Universities: Their Effect on the Faculty Employment of Women." *Educational Record* 41 (1960): 285–95.

Duren, Peter, ed. *A Century of Mathematics in America,* 3 pts. Providence, RI: American Mathematical Society, 1988–89.

Duren, W. L., Jr. "Graduate Student at Chicago in the Twenties." *American Mathematical Monthly* 43 (1976): 243–48.

Eells, Walter Crosby. "Earned Doctorates for Women in the Nineteenth Century." *AAUP Bulletin* 42 (1956): 644–51.

Eisenmann, Linda. "Educating the Female Citizen in a Post-war World: Competing Ideologies for American Women, 1945–1965." *Educational Review* 54 (2002): 133–41.

Eschbach, Elizabeth Seymour. *The Higher Education of Women in England and America, 1865–1920.* New York: Garland Publishing, 1993.

Fenster, Della. "Role Modeling in Mathematics: The Case of Leonard Eugene Dickson (1874–1954)." *Historia Mathematica* 24 (1997): 7–24.

Fenster, Della Dumbaugh, and Karen Hunger Parshall. "Women in the American Mathematical Research Community: 1891–1906." In *The History of Modern Mathematics. Volume III: Images, Ideas, and Communities*, edited by David E. Rowe and Eberhard Knobloch, 229–61. Boston, MA: Academic Press, 1994.

Franklin, Christine Ladd. "The Education of Woman in the Southern States." In *Woman's Work in America*, edited by Annie Nathan Meyer, 89–106, 434–40. New York: H. Holt and Co., 1891. Reprint, New York: Arno Press, 1972.

Gleason, Philip. *Contending with Modernity: Catholic Higher Education in the Twentieth Century.* New York: Oxford University Press, 1995.

Green, Judy, and Jeanne LaDuke. "Women in American Mathematics: A Century of Contributions." In *A Century of Mathematics in America*, Part 2, edited by Peter Duren, with the assistance of Richard A. Askey and Uta C. Merzbach, 379–98.

Green, Judy, and Jeanne LaDuke. "Women in the American Mathematical Community: The Pre-1940 Ph.D.'s." *Mathematical Intelligencer* 9, no. 1 (1987): 11–23.

Green, Judy, and Jeanne LaDuke. "Contributors to American Mathematics: An Overview and Selection." In *Women of Science: Righting the Record*, edited by G. Kass-Simon and Patricia Farnes, 117–46. Bloomington: Indiana University Press, 1990.

Grinstein, Louise S. "Some 'Forgotten' Women of Mathematics: A Who Was Who." *Philosophia Mathematica* 13/14 (1976/77): 73–78.

Grinstein, Louise S., and Paul J. Campbell, eds. *Women of Mathematics: A Biobibliographic Sourcebook.* Westport, CT: Greenwood Press, 1987.

Haggerty, M. M. "Occupational Destination of Ph.D. Recipients." *Educational Record* 9 (1928): 209–18.

Harper, Charles A. *A Century of Public Teacher Education: The Story of the State Teachers Colleges as They Evolved from the Normal Schools.* Washington, DC: Hugh Birch–Horace Mann Fund for the American Association of Teachers Colleges, 1939. Reprint, Westport, CT: Greenwood Press, 1970.

Henrion, Claudia. *Women in Mathematics: The Addition of Difference.* Bloomington: Indiana University Press, 1997.

Hewitt, Gloria C. "The Status of Women in Mathematics." In *Expanding the Role of Women in the Sciences*, edited by Anne M. Briscoe and Sheila M. Pfafflin. *Annals of the New York Academy of Sciences* 323 (1979): 100–109.

Hutchinson, Emilie J. *Women and the Ph.D.: Facts from the Experiences of 1,025 Women Who Have Taken the Degree of Doctor of Philosophy Since 1877.* Greensboro, NC: North Carolina College for Women, 1929. Bulletin no. 2, Institute of Women's Professional Relations.

Johanson, Christine. *Women's Struggle for Higher Education in Russia, 1855–1900.* Kingston, ON: McGill-Queen's University Press, 1987.

Jones, Burton W., and Wolfgang J. Thron. "A History of the Mathematics Departments of the University of Colorado." Boulder, CO: 1979.

Kandel, I. L. "The Ph.D. Degree: The Need for Redefining Standards and Purposes." *The Journal of Higher Education* 10, no. 5 (May 1939): 233–36, 290.

Kenschaft, Patricia C. "The Students of Charlotte Angas Scott." *Mathematics in College* (1982): 16–20.

Keyser, C. J. "Mathematical Productivity in the United States." *Educational Review* 24 (1902): 346–57.

King, Amy. "Women Ph.D.'s in Mathematics in USA and Canada: 1886–1973." *Philosophia Mathematica* 13/14 (1976/77): 79–129.

Lewis, Albert C. "The Building of the University of Texas Mathematics Faculty, 1883–1938." In *A Century of Mathematics in America*, Part 3, edited by Peter Duren, with the assistance of Richard A. Askey, Uta C. Merzbach, and Harold M. Edwards, 205–39.

Maddison, Isabel. *Handbook of British, Continental, and Canadian Universities.* New York: Macmillan, 1896. Second ed., New York: Macmillan, 1899, subtitled *With Special Mention of the Courses Open to Women.*

Mazòn, Patricia Michelle. "Academic citizenship and the admission of women to German universities, 1865–1914." PhD dissertation, Stanford University, 1995.

Merzbach, Uta C. "Mathematics at Bryn Mawr: The First Fifty Years." Paper presented at the annual meeting of the History of Science Society, Chicago, December 1984.

Moore, Calvin C. *Mathematics at Berkeley: A History.* Wellesley, MA: A. K. Peters, 2007.

Murray, Margaret A. M. *Women Becoming Mathematicians: Creating a Professional Identity in Post-World War II America.* Cambridge, MA: MIT Press, 2000.

National Research Council. *Doctorate Production in United States Universities, 1920–1962: With Baccalaureate Origins of Doctorates in Sciences, Arts, and Professions.* Compiled by Lindsey R. Harmon and Herbert Soldz. Washington, DC: National Academy of Sciences, 1963.

National Research Council. *A Century of Doctorates: Data Analyses of Growth and Change.* Washington, DC: National Academy of Sciences, 1978.

National Research Council. *Climbing the Academic Ladder: Doctoral Women Scientists in Academe.* Washington, DC: National Academy of Sciences, 1979.

Newcomer, Mabel. *A Century of Higher Education for American Women.* New York: Harper, 1959. Reprint, Washington, DC: Zenger Publishing Co., 1975.

Oates, Mary J. "The Development of Catholic Colleges for Women, 1895–1960." *U.S. Catholic Historian* 7 (1988): 413–28.

Parshall, Karen Hunger. "Eliakim Hastings Moore and the Founding of a Mathematical Community in America, 1892–1902." *Annals of Science* 41 (1984): 313–33.

Parshall, Karen Hunger, and David E. Rowe. *The Emergence of the American Mathematical Research Community, 1876–1900: J. J. Sylvester, Felix Klein, and E. H. Moore*, volume 8 of *History of Mathematics*. Providence, RI: American Mathematical Society, 1994.

Perrone, Fernanda. "Women Academics in England, 1870–1930." *History of Universities* 12 (1993): 339–67.

Pollard, Lucille Addison. "Women on College and University Faculties: A Historical Survey and a Study of Their Present Academic Status." EdD dissertation, University of Georgia, 1965. Reprint, New York: Arno Press, 1977.

Poulson, Susan L., and Loretta P. Higgins. "Gender, Coeducation, and the Transformation of Catholic Identity in American Higher Education." *Catholic Historical Review* 89 (2003): 489–510.

Price, G. Baley. *History of the Department of Mathematics of the University of Kansas, 1866–1970.* Lawrence: Kansas University Endowment Association, University of Kansas, 1976.

Reingold, Nathan. "Refugee Mathematicians in the United States of America, 1933–1941: Reception and Reaction." *Annals of Science* 38 (1981): 313–38.

"Report of Committee on Endowment of Fellowship." *Publications of the Association of Collegiate Alumnae* Series II. no. 7 (1888).

Rice, Adrian. "Mathematics in the Metropolis: A Survey of Victorian London." *Historia Mathematica* 23 (1996): 376–417.

Richardson, R. G. D. "The Ph.D. Degree and Mathematical Research." *American Mathematical Monthly* 43 (1936): 199–215. Reprinted in *A Century of Mathematics in America*, Part 2, edited by Peter Duren, with the assistance of Richard A. Askey and Uta C. Merzbach, 361–78.

Robinson, Gilbert de B. *The Mathematics Department in the University of Toronto 1827–1978.* Toronto: Department of Mathematics, University of Toronto, 1979.

Robinson, Mabel Louise. "The Curriculum of the Woman's College." *Bureau of Education Bulletin* 1918, no. 6.

Rossiter, Margaret W. *Women Scientists in America: Struggles and Strategies to 1940.* Baltimore, MD: Johns Hopkins University Press, 1982.

Rossiter, Margaret W. *Women Scientists in America: Before Affirmative Action 1940–1972.* Baltimore, MD: Johns Hopkins University Press, 1995.

Simpson, Renate. *How the Ph.D. Came to Britain: A Century of Struggle for Postgraduate Education*, volume 54 of *Research into Higher Education Monographs.* Guildford, Surrey: Society for Research into Higher Education, 1983.

Singer, Sandra L. *Adventures Abroad: North American Women at German-Speaking Universities, 1868–1915.* Westport, CT: Praeger Publishers, 2003.

Solomon, Barbara Miller. *In the Company of Educated Women: A History of Women and Higher Education in America.* New Haven, CT: Yale University Press, 1985.

Stanford, Edna Cleo. "The History of the Mathematics Department at the University of Illinois." MA thesis, University of Illinois, 1940.

Talbot, Marion, and Lois Kimball Mathews Rosenberry. *The History of the American Association of University Women, 1881–1931.* Boston, MA: Houghton Mifflin, 1931.

Tarwater, J. Dalton, ed. *The Bicentennial Tribute to American Mathematics: 1776–1976.* Mathematical Association of America, 1977.

Taylor, James Monroe. *Before Vassar Opened: A Contribution to the History of the Higher Education of Women in America.* Boston, MA: Houghton Mifflin Co, 1914. The first two chapters of this book were originally published as "College Education for Girls in America before Vassar Opened." *Educational Review* 44: (1912) 217–33, 325–47. Reprint, Freeport, NY: Books for Libraries Press, 1972. Reprint, Farmingdale, NY: Dabor Social Science Publications, 1978.

Tobies, Renate. "Einführung: Einflußfaktoren auf die Karriere von Frauen in Mathematik und Naturwissenschaften." In *Aller Männerkultur zum Trotz: Frauen in Mathematik und Naturwissenschaften*, edited by Renate Tobies, 17–67.

Tobies, Renate. "Mathematikerinnen und ihre Doktorväter." In *Aller Männerkultur zum Trotz: Frauen in Mathematik und Naturwissenschaften*, edited by Renate Tobies, 131–58.

Tobies, Renate. "Zum Beginn des mathematischen Frauenstudiums in Preussen." *NTM: Schriftenreihe für Geschichte der Naturwissenschaften, Technik und Medizin.* (1991/92): 151–72.

Tobies, Renate, ed. *Aller Männerkultur zum Trotz: Frauen in Mathematik und Naturwissenschaften.* Frankfurt, Germany: Campus Verlag, 1997.

Tolley, Kimberley. *The Science Education of American Girls: A Historical Perspective.* New York: RoutledgeFalmer, 2003.

Tryon, Ruth Wilson. *Investment in Creative Scholarship: A History of the Fellowship Program of the American Association of University Women, 1890–1956.* Washington, DC: American Association of University Women, 1957.

Whitman, Betsey S. "Women in the American Mathematical Society before 1900." *AWM Newsletter* Part 1. 13, no. 4 (July-Aug. 1983): 10–14. Part 2. 13, no. 5 (Sept.-Oct. 1983): 7–9. Part 3. 13, no. 6 (Nov.-Dec. 1983): 9–12.

Woody, Thomas. *A History of Women's Education in the United States,* 2 vols. New York: Science Press, 1929. Reprint, New York: Octagon Books, 1966.

Zimmern, Alice. "Women in European Universities." *Forum* 19 (1895): 187–99.

Index to the Essay

Titles in This Series

34 **Judy Green and Jeanne LaDuke,** Pioneering women in American mathematics: The pre-1940 PhD's, 2009

33 **Eckart Menzler-Trott,** Logic's lost genius: The life of Gerhard Gentzen, 2007

32 **Karen Hunger Parshall and Jeremy J. Gray, Editors,** Episodes in the history of modern algebra (1800–1950), 2007

31 **Judith R. Goodstein, Editor,** The Volterra chronicles: The life and times of an extraordinary mathematician 1860–1940, 2007

30 **Michael Rosen, Editor,** Exposition by Emil Artin: A selection, 2006

29 **J. L. Berggren and R. S. D. Thomas, Editors,** Euclid's *Phaenomena*: A translation and study of a Hellenistic treatise in spherical astronomy, 2006

28 **Simon Altmann and Eduardo L. Ortiz, Editors,** Mathematics and social utopias in France: Olinde Rodrigues and his times, 2005

27 **Miklós Rédei, Editor,** John von Neumann: Selected letters, 2005

26 **B. N. Delone,** The St. Petersburg school of number theory, 2005

25 **J. M. Plotkin, Editor,** Hausdorff on ordered sets, 2005

24 **Hans Niels Jahnke, Editor,** A history of analysis, 2003

23 **Karen Hunger Parshall and Adrain C. Rice, Editors,** Mathematics unbound: The evolution of an international mathematical research community, 1800–1945, 2002

22 **Bruce C. Berndt and Robert A. Rankin, Editors,** Ramanujan: Essays and surveys, 2001

21 **Armand Borel,** Essays in the history of Lie groups and algebraic groups, 2001

20 Kolmogorov in perspective, 2000

19 **Hermann Grassmann,** Extension theory, 2000

18 **Joe Albree, David C. Arney, and V. Frederick Rickey,** A station favorable to the pursuits of science: Primary materials in the history of mathematics at the United States Military Academy, 2000

17 **Jacques Hadamard**Non-Euclidean geometry in the theory of automorphic functions, 1999

16 **P. G. L. Dirichlet (with Supplements by R. Dedekind),** Lectures on number theory, 1999

15 **Charles W. Curtis,** Pioneers of representation theory: Frobenius, Burnside, Schur, and Brauer, 1999

14 **Vladimir Maz'ya and Tatyana Shaposhnikova,** Jacques Hadamard, a universal mathematician, 1998

13 **Lars Gårding,** Mathematics and mathematicians: Mathematics in Sweden before 1950, 1998

12 **Walter Rudin,** The way I remember it, 1997

11 **June Barrow-Green,** Poincaré and the three body problem, 1997

10 **John Stillwell,** Sources of hyperbolic geometry, 1996

9 **Bruce C. Berndt and Robert A. Rankin,** Ramanujan: Letters and commentary, 1995

8 **Karen Hunger Parshall and David E. Rowe,** The emergence of the American mathematical research community, 1876–1900: J. J. Sylvester, Felix Klein, and E. H. Moore, 1994

7 **Henk J. M. Bos,** Lectures in the history of mathematics, 1993

For a complete list of titles in this series, visit the AMS Bookstore at **www.ams.org/bookstore/**.